Stochastic Mechanics
Random Media
Signal Processing and Image Synthesis
Mathematical Economics and Finance
Stochastic Optimization
Stochastic Control
Stochastic Models in Life Sciences

Applications of Mathematics

Stochastic Modelling and Applied Probability

35

Springer
New York
Berlin
Heidelberg
Hong Kong
London
Milan
Paris
Tokyo

Applications of Mathematics

1 Fleming/Rishel, **Deterministic and Stochastic Optimal Control** (1975)
2 Marchuk, **Methods of Numerical Mathematics,** Second Ed. (1982)
3 Balakrishnan, **Applied Functional Analysis,** Second Ed. (1981)
4 Borovkov, **Stochastic Processes in Queueing Theory** (1976)
5 Liptser/Shiryayev, **Statistics of Random Processes I: General Theory,** Second Ed. (1977)
6 Liptser/Shiryayev, **Statistics of Random Processes II: Applications,** Second Ed. (1978)
7 Vorob'ev, **Game Theory: Lectures for Economists and Systems Scientists** (1977)
8 Shiryayev, **Optimal Stopping Rules** (1978)
9 Ibragimov/Rozanov, **Gaussian Random Processes** (1978)
10 Wonham, **Linear Multivariable Control: A Geometric Approach,** Third Ed. (1985)
11 Hida, **Brownian Motion** (1980)
12 Hestenes, **Conjugate Direction Methods in Optimization** (1980)
13 Kallianpur, **Stochastic Filtering Theory** (1980)
14 Krylov, **Controlled Diffusion Processes** (1980)
15 Prabhu, **Stochastic Storage Processes: Queues, Insurance Risk, Dams, and Data Communication,** Second Ed. (1998)
16 Ibragimov/Has'minskii, **Statistical Estimation: Asymptotic Theory** (1981)
17 Cesari, **Optimization: Theory and Applications** (1982)
18 Elliott, **Stochastic Calculus and Applications** (1982)
19 Marchuk/Shaidourov, **Difference Methods and Their Extrapolations** (1983)
20 Hijab, **Stabilization of Control Systems** (1986)
21 Protter, **Stochastic Integration and Differential Equations** (1990)
22 Benveniste/Métivier/Priouret, **Adaptive Algorithms and Stochastic Approximations** (1990)
23 Kloeden/Platen, **Numerical Solution of Stochastic Differential Equations** (1992)
24 Kushner/Dupuis, **Numerical Methods for Stochastic Control Problems in Continuous Time,** Second Ed. (2001)
25 Fleming/Soner, **Controlled Markov Processes and Viscosity Solutions** (1993)
26 Baccelli/Brémaud, **Elements of Queueing Theory** (1994)
27 Winkler, **Image Analysis, Random Fields, and Dynamic Monte Carlo Methods: An Introduction to Mathematical Aspects** (1994)
28 Kalpazidou, **Cycle Representations of Markov Processes** (1995)
29 Elliott/Aggoun/Moore, **Hidden Markov Models: Estimation and Control** (1995)
30 Hernández-Lerma/Lasserre, **Discrete-Time Markov Control Processes: Basic Optimality Criteria** (1996)
31 Devroye/Györfi/Lugosi, **A Probabilistic Theory of Pattern Recognition** (1996)
32 Maitra/Sudderth, **Discrete Gambling and Stochastic Games** (1996)
33 Embrechts/Klüppelberg/Mikosch, **Modelling Extremal Events** (1997)
34 Duflo, **Random Iterative Models** (1997)

(continued after index)

Harold J. Kushner G. George Yin

Stochastic Approximation and Recursive Algorithms and Applications

Second Edition

With 31 Figures

Springer

Harold J. Kushner
Division of Applied Mathematics
Brown University
Providence, RI 02912, USA
Harold_Kushner@Brown.edu

G. George Yin
Department of Mathematics
Wayne State University
Detroit, MI 48202, USA
gyin@math.wayne.edu

Cover illustration: Cover pattern by courtesy of Rick Durrett, Cornell University, Ithaca, New York.

Mathematics Subject Classification (2000): 62L20, 93E10, 93E25, 93E35, 65C05, 93-02, 90C15

Library of Congress Cataloging-in-Publication Data
Kushner, Harold J. (Harold Joseph), 1933–
Stochastic approximation and recursive algorithms and applications / Harold J. Kushner, G. George Yin.
p. cm. — (Applications of mathematics ; 35)
Rev. ed. of: Stochastic approximation algorithms and applications, c1997.
ISBN 0-387-00894-2 (acid-free paper)
1. Stochastic approximation. 2. Recursive stochastic algorithms. 3. Recursive algorithms. I. Kushner, Harold J. (Harold Joseph), 1933–. Stochastic approximation algorithms and applications. II. Yin, George, 1954– III. Title. IV. Series.
QA274.2.K88 2003
519.2—dc21 2003045459

ISBN 0-387-00894-2 Printed on acid-free paper.

Printed in the United States of America.

9 8 7 6 5 4 3 2 1 SPIN 10922088

Typesetting: Pages created by the authors in LaTeX 2.09 using Springer's svsing.sty macro.

www.springer-ny.com

Springer-Verlag New York Berlin Heidelberg
A member of BertelsmannSpringer Science+Business Media GmbH

To Our Parents,

Harriet and Hyman Kushner

and

Wanzhen Zhu and Yixin Yin

Preface and Introduction

The basic stochastic approximation algorithms introduced by Robbins and Monro and by Kiefer and Wolfowitz in the early 1950s have been the subject of an enormous literature, both theoretical and applied. This is due to the large number of applications and the interesting theoretical issues in the analysis of "dynamically defined" stochastic processes. The basic paradigm is a stochastic difference equation such as $\theta_{n+1} = \theta_n + \epsilon_n Y_n$, where θ_n takes its values in some Euclidean space, Y_n is a random variable, and the "step size" $\epsilon_n > 0$ is small and might go to zero as $n \to \infty$. In its simplest form, θ is a parameter of a system, and the random vector Y_n is a function of "noise-corrupted" observations taken on the system when the parameter is set to θ_n. One recursively adjusts the parameter so that some goal is met asymptotically. This book is concerned with the qualitative and asymptotic properties of such recursive algorithms in the diverse forms in which they arise in applications. There are analogous continuous time algorithms, but the conditions and proofs are generally very close to those for the discrete time case.

The original work was motivated by the problem of finding a root of a continuous function $\bar{g}(\theta)$, where the function is not known but the experimenter is able to take "noisy" measurements at any desired value of θ. Recursive methods for root finding are common in classical numerical analysis, and it is reasonable to expect that appropriate stochastic analogs would also perform well.

In one classical example, θ is the level of dosage of a drug, and the function $\bar{g}(\theta)$, assumed to be increasing with θ, is the probability of success at dosage level θ. The level at which $\bar{g}(\theta)$ takes a given value v is sought.

The probability of success is known only by experiment at whatever values of θ are selected by the experimenter, with the experimental outcome being either success or failure. Thus, the problem cannot be solved analytically. One possible approach is to take a sufficient number of observations at some fixed value of θ, so that a good estimate of the function value is available, and then to move on. Since most such observations will be taken at parameter values that are not close to the optimum, much effort might be wasted in comparison with the stochastic approximation algorithm $\theta_{n+1} = \theta_n + \epsilon_n[v - \text{observation at } \theta_n]$, where the parameter value moves (on the average) in the correct direction after each observation. In another example, we wish to minimize a real-valued continuously differentiable function $f(\cdot)$ of θ. Here, θ_n is the nth estimate of the minimum, and Y_n is a noisy estimate of the negative of the derivative of $f(\cdot)$ at θ_n, perhaps obtained by a Monte Carlo procedure. The algorithms are frequently constrained in that the iterates θ_n are projected back to some set H if they ever leave it. The mathematical paradigms have posed substantial challenges in the asymptotic analysis of recursively defined stochastic processes.

A major insight of Robbins and Monro was that, if the step sizes in the parameter updates are allowed to go to zero in an appropriate way as $n \to \infty$, then there is an implicit averaging that eliminates the effects of the noise in the long run. An excellent survey of developments up to about the mid 1960s can be found in the book by Wasan [250]. More recent material can be found in [16, 48, 57, 67, 135, 225]. The book [192] deals with many of the issues involved in stochastic optimization in general.

In recent years, algorithms of the stochastic approximation type have found applications in new and diverse areas, and new techniques have been developed for proofs of convergence and rate of convergence. The actual and potential applications in signal processing and communications have exploded. Indeed, whether or not they are called stochastic approximations, such algorithms occur frequently in practical systems for the purposes of noise or interference cancellation, the optimization of "post processing" or "equalization" filters in time varying communication channels, adaptive antenna systems, adaptive power control in wireless communications, and many related applications. In these applications, the step size is often a small constant $\epsilon_n = \epsilon$, or it might be random. The underlying processes are often nonstationary and the optimal value of θ can change with time. Then one keeps ϵ_n strictly away from zero in order to allow "tracking." Such tracking applications lead to new problems in the asymptotic analysis (e.g., when ϵ_n are adjusted adaptively); one wishes to estimate the tracking errors and their dependence on the structure of the algorithm.

New challenges have arisen in applications to adaptive control. There has been a resurgence of interest in general "learning" algorithms, motivated by the training problem in artificial neural networks [7, 51, 97], the on-line learning of optimal strategies in very high-dimensional Markov decision processes [113, 174, 221, 252] with unknown transition probabilities, in learning automata [155], recursive games [11], convergence in sequential decision problems in economics [175], and related areas. The actual recursive forms of the algorithms in many such applications are of the stochastic approximation type. Owing to the types of simulation methods used, the "noise" might be "pseudorandom" [184], rather than random.

Methods such as infinitesimal perturbation analysis [101] for the estimation of the pathwise derivatives of complex discrete event systems enlarge the possibilities for the recursive on-line optimization of many systems that arise in communications or manufacturing. The appropriate algorithms are often of the stochastic approximation type and the criterion to be minimized is often the average cost per unit time over the infinite time interval.

Iterate and observation averaging methods [6, 149, 216, 195, 267, 268, 273], which yield nearly optimal algorithms under broad conditions, have been developed. The iterate averaging effectively adds an additional time scale to the algorithm. Decentralized or asynchronous algorithms introduce new difficulties for analysis. Consider, for example, a problem where computation is split among several processors, operating and transmitting data to one another asynchronously. Such algorithms are only beginning to come into prominence, due to both the developments of decentralized processing and applications where each of several locations might control or adjust "local variables," but where the criterion of concern is global.

Despite their successes, the classical methods are not adequate for many of the algorithms that arise in such applications. Some of the reasons concern the greater flexibility desired for the step sizes, more complicated dependence properties of the noise and iterate processes, the types of constraints that might occur, ergodic cost functions, possibly additional time scales, nonstationarity and issues of tracking for time-varying systems, data-flow problems in the decentralized algorithm, iterate-averaging algorithms, desired stronger rate of convergence results, and so forth.

Much modern analysis of the algorithms uses the so-called ODE (ordinary differential equation) method introduced by Ljung [164] and extensively developed by Kushner and coworkers [123, 135, 142] to cover quite general noise processes and constraints by the use of weak ergodic or averaging conditions. The main idea is to show that, asymptotically, the noise effects average out so that the asymptotic behavior is determined effectively by that of a "mean" ODE. The usefulness of the technique stems from the fact that the ODE is obtained by a "local analysis," where the dynamical term of the ODE at parameter value θ is obtained by averaging the Y_n as though the parameter were fixed at θ. Constraints, complicated state dependent noise processes, discontinuities, and many other difficulties can be handled. Depending on the application, the ODE might be replaced by a constrained (projected) ODE or a differential inclusion. Owing to its versatility and naturalness, the ODE method has become a fundamental technique in the current toolbox, and its full power will be apparent from the results in this book.

The first three chapters describe applications and serve to motivate the algorithmic forms, assumptions, and theorems to follow. Chapter 1 provides the general motivation underlying stochastic approximation and describes various classical examples. Modifications of the algorithms due to robustness concerns, improvements based on iterate or observation averaging methods, variance reduction, and other modeling issues are also introduced. A Lagrangian algorithm for constrained optimization with noise corrupted observations on both the value function and the constraints is outlined. Chapter 2 contains more advanced examples, each of which is typical of a large class of current interest: animal adaptation models, parametric optimization of Markov chain control problems, the so-called Q-learning, artificial neural networks, and learning in repeated games. The concept of state-dependent noise, which plays a large role in applications, is introduced. The optimization of discrete event systems is introduced by the application of infinitesimal perturbation analysis to the optimization of the performance of a queue with an ergodic cost criterion. The mathematical and modeling issues raised in this example are typical of many of the optimization problems in discrete event systems or where ergodic cost criteria are involved. Chapter 3 describes some applications arising in adaptive control, signal processing, and communication theory, areas that are major users of stochastic approximation algorithms. An algorithm for tracking time varying parameters is described, as well as applications to problems arising in wireless communications with randomly time varying channels. Some of the mathematical results that will be needed in the book are collected in Chapter 4.

The book also develops "stability" and combined "stability–ODE" methods for unconstrained problems. Nevertheless, a large part of the work concerns constrained algorithms, because constraints are generally present either explicitly or implicitly. For example, in the queue optimization problem of Chapter 2, the parameter to be selected controls the service rate. What is to be done if the service rate at some iteration is considerably larger than any possible practical value? Either there is a problem with the model or the chosen step sizes, or some bizarre random numbers appeared. Furthermore, in practice the "physics" of models at large parameter values are often poorly known or inconvenient to model, so that whatever "convenient mathematical assumptions" are made, they might be meaningless at large state values. No matter what the cause is, one would normally alter the unconstrained algorithm if the parameter θ took on excessive values. The simplest alteration is truncation. Of course, in addition to truncation, a practical algorithm would have other safeguards to ensure robustness against "bad" noise or inappropriate step sizes, etc. It has been somewhat traditional to allow the iterates to be unbounded and to use stability methods to prove that they do, in fact, converge. This approach still has its place and is dealt with here. Indeed, one might even alter the dynamics by introducing "soft" constraints, which have the desired stabilizing effect.

However, allowing unbounded iterates seems to be of greater mathematical than practical interest. Owing to the interest in the constrained algorithm, the "constrained ODE" is also discussed in Chapter 4. The chapter contains a brief discussion of stochastic stability and the perturbed stochastic Liapunov function, which play an essential role in the asymptotic analysis.

The first convergence results appear in Chapter 5, which deals with the classical case where the Y_n can be written as the sum of a conditional mean $g_n(\theta_n)$ and a noise term, which is a "martingale difference." The basic techniques of the ODE method are introduced, both with and without constraints. It is shown that, under reasonable conditions on the noise, there will be convergence with probability one to a "stationary point" or "limit trajectory" of the mean ODE for step-size sequences that decrease at least as fast as $\alpha_n / \log n$, where $\alpha_n \to 0$. If the limit trajectory of the ODE is not concentrated at a single point, then the asymptotic path of the stochastic approximation is concentrated on a limit or invariant set of the ODE that is also "chain recurrent" [9, 89]. Equality constrained problems are included in the basic setup.

Much of the analysis is based on interpolated processes. The iterates $\{\theta_n\}$ are interpolated into a continuous time process with interpolation intervals $\{\epsilon_n\}$. The asymptotics (large n) of the iterate sequence are also the asymptotics (large t) of this interpolated sequence. It is the paths of the interpolated process that are approximated by the paths of the ODE.

If there are no constraints, then a stability method is used to show that the iterate sequence is recurrent. From this point on, the proofs are a special case of those for the constrained problem. As an illustration of the methods, convergence is proved for an animal learning example (where the step sizes are random, depending on the actual history) and a pattern classification problem. In the minimization of convex functions, the subdifferential replaces the derivative, and the ODE becomes a differential inclusion, but the convergence proofs carry over.

Chapter 6 treats probability one convergence with correlated noise sequences. The development is based on the general "compactness methods" of [135]. The assumptions on the noise sequence are intuitively reasonable and are implied by (but weaker than) strong laws of large numbers. In some cases, they are both necessary and sufficient for convergence. The way the conditions are formulated allows us to use simple and classical compactness methods to derive the mean ODE and to show that its asymptotics characterize that of the algorithm. Stability methods for the unconstrained problem and the generalization of the ODE to a differential inclusion are discussed. The methods of large deviations theory provide an alternative approach to proving convergence under weak conditions, and some simple results are presented.

In Chapters 7 and 8, we work with another type of convergence, called *weak convergence*, since it is based on the theory of weak convergence of a sequence of probability measures and is weaker than convergence with probability one. It is actually much easier to use in that convergence can be proved under weaker and more easily verifiable conditions and generally with substantially less effort. The approach yields virtually the same information on the asymptotic behavior. The weak convergence methods have considerable theoretical and modeling advantages when dealing with complex problems involving correlated noise, state dependent noise, decentralized or asynchronous algorithms, and discontinuities in the algorithm. It will be seen that the conditions are often close to minimal. Only a very elementary part of the theory of weak convergence of probability measures will be needed; this is covered in the second part of Chapter 7. The techniques introduced are of considerable importance beyond the needs of the book, since they are a foundation of the theory of approximation of random processes and limit theorems for sequences of random processes.

When one considers how stochastic approximation algorithms are used in applications, the fact of ultimate convergence with probability one can be misleading. Algorithms do not continue on to infinity, particularly when $\epsilon_n \to 0$. There is always a stopping rule that tells us when to stop the algorithm and to accept some function of the recent iterates as the "final value." The stopping rule can take many forms, but whichever it takes, all that we know about the "final value" at the stopping time is information of a distributional type. There is no difference in the conclusions provided by the probability one and the weak convergence methods. In applications that are of concern over long time intervals, the actual physical model might "drift." Indeed, it is often the case that the step size is not allowed to go to zero, and then there is no general alternative to the weak convergence methods at this time.

The ODE approach to the limit theorems obtains the ODE by appropriately averaging the dynamics, and then by showing that some subset of the limit set of the ODE is just the set of asymptotic points of the $\{\theta_n\}$. The ODE is easier to characterize, and requires weaker conditions and simpler proofs when weak convergence methods are used. Furthermore, it can be shown that $\{\theta_n\}$ spends "nearly all" of its time in an arbitrarily small neighborhood of the limit point or set. The use of weak convergence methods can lead to better probability one proofs in that, once we know that $\{\theta_n\}$ spends "nearly all" of its time (asymptotically) in some small neighborhood of the limit point, then a *local analysis* can be used to get convergence with probability one. For example, the methods of Chapters 5 and 6 can be applied locally, or the local large deviations methods of [63] can be used. Even when we can only prove weak convergence, if θ_n is close to a stable limit point at iterate n, then under broad conditions the mean escape time (indeed, if it ever does escape) from a small neighborhood of that limit point is at least of the order of e^{c/ϵ_n} for some $c > 0$.

Section 7.2 is motivational in nature, aiming to relate some of the ideas of weak convergence to probability one convergence and convergence in distribution. It should be read only "lightly." The general theory is covered in Chapter 8 for a broad variety of algorithms, using what might be called "weak local ergodic theorems." The essential conditions concern the rates of decrease of the conditional expectation of the future noise given the past noise, as the time difference increases. Chapter 9 illustrates the relative convenience and power of the methods of Chapter 8 by providing proofs of convergence for some of the examples in Chapters 2 and 3.

Chapter 10 concerns the rate of convergence. Loosely speaking, a standard point of view is to show that a sequence of suitably normalized iterates, say of the form $(\theta_n - \bar{\theta})/\sqrt{\epsilon_n}$ or $n^{\beta}(\theta_n - \bar{\theta})$ for an appropriate $\beta > 0$, converges in distribution to a normally distributed random variable with mean zero and finite covariance matrix $\bar{V}$. We will do a little better and prove that the continuous time process obtained from suitably interpolated normalized iterates converges "weakly" to a stationary Gauss–Markov process, whose covariance matrix (at any time t) is $\bar{V}$. The methods use only the techniques of weak convergence theory that are outlined in Chapter 7.

The use of stochastic approximation for the minimization of functions of a very high-dimensional argument has been of increasing interest. Owing to the high dimension, the classical Kiefer–Wolfowitz procedures can be very time consuming to use. As a result, there is much current interest in the so-called random-directions methods, where at each step n one chooses a direction d_n at random, obtains a noisy estimate $\widehat{Y}_n$ of the derivative in direction d_n, and moves an increment $-\epsilon_n \widehat{Y}_n$. Although such methods have been of interest and used in various ways for a long time [135], convincing arguments concerning their value and the appropriate choices of the direction vectors and scaling were lacking. The paper [226] proposed a different way of getting the directions and attracted needed attention to this problem. The proof of convergence of the random-directions methods that have been suggested to date are exactly the same as that for the classical Kiefer–Wolfowitz procedure (as in Chapter 5). The comparison of the rates of convergence under the different ways of choosing the random directions is given at the end of Chapter 10, and shows that the older and newer methods have essentially the same properties, when the norms of the direction vectors d_n are the same. It is seen that the random-directions methods can be quite advantageous, but care needs to be exercised in their use.

The performance of the stochastic approximation algorithms depends heavily on the choice of the step size sequence ϵ_n, and the lack of a general approach to getting good sequences has been a handicap in applications. In [195], Polyak and Juditsky showed that, if the coefficients ϵ_n go to zero "slower" than $O(1/n)$, then the averaged sequence $\sum_{i=1}^{n} \theta_i/n$ converges to its limit at an optimal rate. This implies that the use of relatively large step sizes, while letting the "off-line" averaging take care of the increased noise effects, will yield a substantial overall improvement. These results have since been corroborated by numerous simulations and extended mathematically. In Chapter 11, it is first shown that the averaging improves the asymptotic properties whenever there is a "classical" rate of convergence theorem of the type derived in Chapter 10, including the constant $\epsilon_n = \epsilon$ case. This will give the minimal window over which the averaging will yield an improvement. The maximum window of averaging is then obtained by a direct computation of the asymptotic covariance of the averaged process. Intuitive insight is provided by relating the behavior of the original and the averaged process to that of a three-time-scale discrete-time algorithm where it is seen that the key property is the separation of the time scales.

Chapter 12 concerns decentralized and asynchronous algorithms, where the work is split between several processors, each of which has control over a different set of parameters. The processors work at different speeds, and there can be delays in passing information to each other. Owing to the asynchronous property, the analysis must be in "real" rather than "iterate" time. This complicates the notation, but all of the results of the previous chapters can be carried over. Typical applications are decentralized optimization of queueing networks and Q-learning.

Some topics are not covered. As noted, the algorithm in continuous time differs little from that in discrete time. The basic ideas can be extended to infinite-dimensional problems [17, 19, 66, 87, 144, 185, 201, 214, 219, 246, 247, 248, 277]. The function minimization problem where there are many local minima has attracted some attention [81, 130, 258], but little is known at this time concerning effective methods. Some effort [31] has been devoted to showing that suitable conditions on the noise guarantee that there cannot be convergence to an unstable or marginally stable point of the ODE. Such results are needed and do increase confidence in the algorithms. The conditions can be hard to verify, particularly in high-dimensional problems, and the results do not guarantee that the iterates would not actually spend a lot of time near such bad points, particularly when the step sizes are small and there is poor initial behavior. Additionally, one tries to design the procedure and use variance reduction methods to reduce the effects of the noise.

Penalty-multiplier and Lagrangian methods (other than the discussion in Chapter 1) for constrained problems are omitted and are discussed in [135]. They involve only minor variations on what is done here, but they are omitted for lack of space. We concentrate on algorithms defined on r-dimensional Euclidean space, except as modified by inequality or equality constraints. The treatment of the equality constrained problem shows that the theory also covers processes defined on smooth manifolds.

We express our deep gratitude to Paul Dupuis and Felisa Vazquèz-Abad, for their careful reading and critical remarks on various parts of the manuscript of the first edition. Sid Yakowitz also provided critical remarks for the first edition; his passing away is a great loss. The long-term support and encouragement of the National Science Foundation and the Army Research Office are also gratefully acknowledged.

Comment on the second edition. This second edition is a thorough revision, although the main features and the structure of the book remain unchanged. The book contains many additional results and more detailed discussion; for example, there is a fuller discussion of the asymptotic behavior of the algorithms, Markov and non-Markov state-dependent-noise, and two-time-scale problems. Additional material on applications, in particular, in communications and adaptive control, has been added. Proofs are simplified where possible.

Notation and numbering. Chapters are divided into sections, and sections into subsections. Within a chapter, (1.2) (resp., (A2.1)) denotes Equation 2 of Section 1 (resp., Assumption 2 of Section 1). Section 1 (Subsection 1.2, resp.) always means the first section (resp., the second subsection of the first section) in the chapter in which the statement is used. To refer to equations (resp., assumptions) in other chapters, we use, e.g., (1.2.3) (resp., (A1.2.3)) to denote the third equation (resp., the third assumption) in Section 2 of Chapter 1. When not in Chapter 1, Section 1.2 (resp., Subsection 1.2.3) means Section 2 (resp., Subsection 3 of Section 2) of Chapter 1.

Throughout the book, $|\cdot|$ denotes either a Euclidean norm or a norm on the appropriate function spaces, which will be clear from the context. A point x in a Euclidean space is a column vector, and the ith component of x is denoted by x_i. However, the ith component of θ is denoted by θ^i, since subscripts on θ are used to denote the value at a time n. The symbol $'$ denotes transpose. Moreover, both A' and $(A)'$ will be used interchangeably, e.g., both $g_n^{\epsilon,\prime}(\theta)$ and $(g_n^\epsilon(\theta))'$ denote the transpose of $g_n^\epsilon(\theta)$. Subscripts θ and x denote either a gradient or a derivative, depending on whether the variable is vector or real-valued. For convenience, we list many of the symbols at the end of the book.

Providence, Rhode Island, USA — Harold J. Kushner
Detroit, Michigan, USA — G. George Yin

Contents

1
Introduction: Applications and Issues

1.0 Outline of Chapter

This is the first of three chapters describing many concrete applications of stochastic approximation. The emphasis is on the problem description. Proofs of convergence and the derivation of the rate of convergence will be given in subsequent chapters for many of the examples. Since the initial work of Robbins and Monro in 1951, there has been a steady increase in the investigations of applications in many diverse areas, and this has accelerated in recent years, with new applications arising in queueing networks, wireless communications, manufacturing systems, in learning problems, repeated games, and neural nets, among others. We present only selected samples of these applications to illustrate the great breadth. The basic stochastic approximation algorithm is nothing but a stochastic difference equation with a small step size, and the basic questions for analysis concern its qualitative behavior over a long time interval, such as convergence and rate of convergence. The wide range of applications leads to a wide variety of such equations and associated stochastic processes.

One of the problems that led to Robbins and Monro's original work in stochastic approximation concerns the sequential estimation of the location of the root of a function when the function is unknown and only noise-corrupted observations at arbitrary values of the argument can be made, and the "corrections" at each step are small. One takes an observation at the current estimator of the root, then uses that observation to make a small correction in the estimate, then takes an observation at the new value of the estimator, and so forth. The fact that the step sizes are small is important for the convergence, because it guarantees an "averaging of the noise." The basic idea is discussed at the beginning of Section 1. The examples of the Robbins–Monro algorithm in Section 1 are all more or less classical, and have been the subject of much attention. In one form or another they include root finding, getting an optimal pattern classifier (a best least squares fit problem), optimizing the set point of a chemical processor, and a parametric minimization problem for a stochastic dynamical system via a recursive monte carlo method. They are described only in a general way, but they serve to lay out some of the basic ideas.

Recursive algorithms (e.g., the Newton–Raphson method) are widely used for the recursive computation of the minimum of a smooth known function. If the form of the function is unknown but noise-corrupted observations can be taken at parameter values selected by the experimenter, then one can use an analogous recursive procedure in which the derivatives are estimated via finite differences using the noisy measurements, and the step sizes are small. This method, called the *Kiefer–Wolfowitz procedure*, and its "random directions" variant are discussed in Section 2.

The practical use of the basic algorithm raises many difficult questions, and dealing with these leads to challenging variations of the basic format. A key problem in effective applications concerns the "amount of noise" in the observations, which leads to variations that incorporate variance reduction methods. With the use of these methods, the algorithm becomes more effective, but also more complex. It is desirable to have robust algorithms, which are not overly sensitive to unusually large noise values. Many problems have constraints in the sense that the vector-valued iterates must be confined to some given bounded set. The question of whether averaging the iterate sequence will yield an improved estimate arises. Such issues are discussed in Section 3, and the techniques developed for dealing with the wide variety of random processes that occur enrich the subject considerably. A Lagrangian method for the constrained minimization of a convex function, where only noise corrupted observations on the function and constraints are available, is discussed in Section 4. This algorithm is typical of a class of multiplier-penalty function methods.

Owing to the small step size for large time, the behavior of the algorithm can be approximated by a "mean flow." This is the solution to an ordinary differential equation (ODE), henceforth referred to as the mean ODE, whose right-hand side is just the mean value of the driving term. The limit points of this ODE turn out to be the limit points of the stochastic approximation process.

1.1 The Robbins–Monro Algorithm

1.1.1 Introduction

The original work in recursive stochastic algorithms was by Robbins and Monro, who developed and analyzed a recursive procedure for finding the root of a real-valued function $\bar{g}(\cdot)$ of a real variable θ. The function is not known, but noise-corrupted observations could be taken at values of θ selected by the experimenter.

If $\bar{g}(\cdot)$ were known and continuously differentiable, then the problem would be a classical one in numerical analysis and Newton's procedure can be used. Newton's procedure generates a sequence of estimators θ_n of the root $\bar{\theta}$, defined recursively by

$$\theta_{n+1} = \theta_n - \left[\bar{g}_\theta(\theta_n)\right]^{-1} \bar{g}(\theta_n), \tag{1.1}$$

where $\bar{g}_\theta(\cdot)$ denotes the derivative of $\bar{g}(\cdot)$ with respect to θ. Suppose that $\bar{g}(\theta) < 0$ for $\theta > \bar{\theta}$, and $\bar{g}(\theta) > 0$ for $\theta < \bar{\theta}$, and that $\bar{g}_\theta(\theta)$ is strictly negative and is bounded in a neighborhood of $\bar{\theta}$. Then θ_n converges to $\bar{\theta}$ if θ_0 is in a small enough neighborhood of $\bar{\theta}$. An alternative and simpler, but less efficient, procedure is to fix $\epsilon > 0$ sufficiently small and use the algorithm

$$\theta_{n+1} = \theta_n + \epsilon \bar{g}(\theta_n). \tag{1.2}$$

Algorithm (1.2) does not require differentiability and is guaranteed to converge if $\theta_0 - \bar{\theta}$ is sufficiently small. Of course, there are faster procedures in this simple case, where the values of $\bar{g}(\theta)$ and its derivatives can be computed.

Now suppose that the values of $\bar{g}(\theta)$ are not known, but "noise-corrupted" observations can be taken at selected values of θ. Due to the observation noise, the Newton procedure (1.1) cannot be used. One could consider using procedure (1.2), but with $\bar{g}(\theta_n)$ replaced by a good estimate of its value obtained by averaging many observations. It was recognized by Robbins and Monro [207] that taking an excessive number of observations at each of the θ_n and using the averages in place of $\bar{g}(\theta_n)$ in (1.2) were inefficient, since θ_n is only an intermediary in the calculation and the value $\bar{g}(\theta_n)$ is only of interest in so far as it leads us in the right direction. They proposed the algorithm

$$\theta_{n+1} = \theta_n + \epsilon_n Y_n, \tag{1.3}$$

where ϵ_n is an appropriate sequence satisfying

$$\epsilon_n > 0, \quad \epsilon_n \to 0, \quad \sum_n \epsilon_n = \infty, \tag{1.4}$$

and Y_n is a "noisy" estimate of the value of $\bar{g}(\theta_n)$. The condition $\sum_n \epsilon_n^2 < \infty$ was used, but it will be seen in Chapter 5 that it can be weakened considerably. The decreasing step sizes imply that the rate of change of θ_n slows down as n goes to infinity. The choice of the sequence $\{\epsilon_n\}$ is a large remaining issue that is central to the effectiveness of the algorithm (1.3). More will be said about this later, particularly in Chapter 11 where "nearly optimal" algorithms are discussed. The idea is that the decreasing step sizes would provide an implicit averaging of the observations. This insight and the associated convergence proof led to an enormous literature on general recursive stochastic algorithms and to a large number of actual and potential applications.

The form of the recursive linear least squares estimator of the mean value of a random variable provides another motivation for the form of (1.3) and helps to explain how the decreasing step size actually causes an averaging of the observations. Let $\{\xi_n\}$ be a sequence of real-valued, mutually independent, and identically distributed, random variables with finite variance and unknown mean value $\bar{\theta}$. Given observations ξ_i, $1 \le i \le n$, the linear least squares estimate of $\bar{\theta}$ is $\theta_n = \sum_{i=1}^{n} \xi_i / n$. This can be written in the recursive form

$$\theta_{n+1} = \theta_n + \epsilon_n \left[\xi_{n+1} - \theta_n\right], \tag{1.5}$$

where $\theta_0 = 0$ and $\epsilon_n = 1/(n+1)$. Thus the use of decreasing step sizes $\epsilon_n = 1/(n+1)$ yields an estimator that is equivalent to that obtained by a direct averaging of the observations, and (1.5) is a special case of (1.3). Additionally, the "recursive filter" form of (1.5) gives more insight into the estimation process, because it shows that the estimator changes only by $\epsilon_n(\xi_{n+1} - \theta_n)$, which can be described as the product of the coefficient of "reliability" ϵ_n and the "estimation error" $\xi_{n+1} - \theta_n$. The intuition behind the averaging holds even in more complicated algorithms under broad conditions.

In applications, one generally prefers recursive algorithms, owing to their relative computational simplicity. After each new observation, one need not recompute the estimator from all the data collected to date. Each successive estimate is obtained as a simple function of the last estimate and the current observation. Recursive estimators are widely used in applications in communications and control theory. More will be said about them in the next two chapters. Indeed, recursive stochastic algorithms had been used in the control and communications area for tracking purposes even before the work of Robbins and Monro. Forms similar to (1.3) and (1.5) were used for smoothing radar returns and in related applications in continuous time, with ϵ_n being held constant at a value ϵ, which had the interpretation of the inverse of a time constant. There was no general asymptotic theory apart from computing the stationary mean square values for stable linear algorithms, however.

In the general Robbins–Monro procedure, Y_n and θ_n take values in $I\!R^r$-Euclidean r-space, where Y_n is a "noise-corrupted" observation of a vector-valued function $\bar{g}(\cdot)$, whose root we are seeking. The "error" $Y_n - \bar{g}(\theta_n)$

might be a complicated function of θ_n or even of past values $\theta_i, i \le n$. In many applications, one observes values of the form $Y_n = g(\theta_n, \xi_n) + \delta M_n$, where $\{\xi_n\}$ is some correlated stochastic process, δM_n has the property that $E[\delta M_n | Y_i, \delta M_i, i < n] = 0$ and where (loosely speaking) Y_n is an "estimator" of $\bar{g}(\theta)$ in that $\bar{g}(\theta) = Eg(\theta, \xi_n) = EY_n$ or perhaps $\bar{g}(\theta) = \lim_m (1/m) \sum_{i=0}^{m-1} Eg(\theta, \xi_i)$. Many of the possible forms that have been analyzed will be seen in the examples in this book. The basic mathematical questions concern the asymptotic properties of the θ_n sequence, its dependence on the algorithm structure and noise processes, and methods to improve the performance.

Remark on the notation. θ is used as a parameter that is a point in $I\!R^r$ and θ_n are random variables in the stochastic approximation sequence. We use $\theta_{n,i}$ to denote the ith component of θ_n when it is vector-valued. To avoid confusion, we use θ^i to denote the ith component of θ. This is the only exception to the rule that the component index appears in the subscript.

1.1.2 Finding the Zeros of an Unknown Function

Example 1. For each real-valued parameter θ, let $G(\cdot, \theta)$ be an unknown distribution function of a real-valued random variable, and define $m(\theta) = \int yG(dy, \theta)$, the mean value under θ. Given a desired level $\bar{m}$, the problem is to find a (perhaps nonunique) value $\bar{\theta}$ such that $m(\bar{\theta}) = \bar{m}$. Since $G(\cdot, \theta)$ is unknown, some sampling and nonparametric method is called for, and a useful one can be based on the Robbins–Monro recursive procedure. For this example, suppose that $m(\cdot)$ is nondecreasing, and that there is a unique root of $m(\theta) = \bar{m}$.

Let θ_n denote the nth estimator of $\bar{\theta}$ (based on observations at times $0, 1, \ldots, n-1$) and let Y_n denote the observation taken at time n at parameter value θ_n. We define θ_n recursively by

$$\theta_{n+1} = \theta_n + \epsilon_n \left[\bar{m} - Y_n\right], \tag{1.6}$$

which can be written as

$$\theta_{n+1} = \theta_n + \epsilon_n \left[\bar{m} - m(\theta_n)\right] + \epsilon_n \left[m(\theta_n) - Y_n\right]. \tag{1.7}$$

Suppose that

$$E\left[Y_n - m(\theta_n) | Y_i, \theta_i, i < n, \theta_n\right] = 0 \tag{1.8}$$

with probability one, and that the "noise" terms $Y_n - m(\theta_n) \equiv \delta M_n$ have bounded variances $\sigma^2(\theta_n)$. Property (1.8) implies that the δM_n terms are what are known as *martingale differences*, that is, $E[\delta M_n | \delta M_i, i < n] = 0$ with probability one for all n; see Section 4.1 for extensions of the definition and further discussion. The martingale difference noise sequence is perhaps the easiest type of noise to deal with, and probability one convergence results are given in Chapter 5. The martingale difference property often arises as follows. Suppose that the successive observations are independent in the sense that the distribution of Y_n conditioned on $\{\theta_0, Y_i, i < n\}$ depends only on θ_n, the value of the parameter used to get Y_n. The "noise terms" δM_n are not mutually independent, since θ_n depends on the observations $\{Y_i, i < n\}$. However, they do have the martingale difference property, which is enough to get good convergence results.

Under reasonable conditions, it is relatively easy to show that the noise terms in (1.7) "average to zero" and have no effect on the asymptotic behavior of the algorithm. To see intuitively why this might be so, first note that since $\epsilon_n \to 0$ here, for large n the values of the θ_n change slowly. For small $\Delta > 0$ define m_n^Δ by

$$\sum_{i=n}^{n+m_n^\Delta-1} \epsilon_i \approx \Delta.$$

Then

$$\theta_{n+m_n^\Delta} - \theta_n \approx \Delta[\bar{m} - m(\theta_n)] + \text{“error,”} \tag{1.9a}$$

where the error is

$$\sum_{i=n}^{n+m_n^\Delta-1} \epsilon_i \delta M_i.$$

Equation (1.8) implies that $\{\delta M_n\}$ is a sequence of zero mean orthogonal random variables. That is, $E\delta M_i \delta M_j = 0$ for $i \neq j$. Thus, the variance of the error is

$$E\left[\sum_{i=n}^{n+m_n^\Delta-1} \epsilon_i \delta M_i\right]^2 = \sum_{i=n}^{n+m_n^\Delta-1} E\epsilon_i^2 \delta M_i^2 = \sum_{i=n}^{n+m_n^\Delta-1} O(\epsilon_i^2) = O(\Delta)\epsilon_n.$$

This bound and (1.9a) imply that over iterate intervals $[n, n+m_n^\Delta)$ for small Δ and large n, the mean change in the value of the parameter is much more important than the "noise." Then, at least formally, the difference equation (1.9a) suggests that the asymptotic behavior of the algorithm can be approximated by the asymptotic behavior of the solution to the ODE

$$\dot{\theta} = \bar{g}(\theta) = \bar{m} - m(\theta). \tag{1.9b}$$

The connections between the asymptotic behavior of the algorithm and that of the mean ODE (1.9b) will be formalized starting in Chapter 5, where such ODEs will be shown to play a crucial role in the convergence theory. Under broad conditions (see Chapter 5), if $\bar{\theta}$ is an asymptotically stable point of (1.9b), then $\theta_n \to \bar{\theta}$ with probability one.

Note that $m_n^\Delta \to \infty$ as $n \to \infty$, since $\epsilon_n \to 0$. In the sequel, the expression that the noise effects "locally average to zero" is used loosely to mean that the noise effects over the iterate interval $[n, n + m_n^\Delta]$ go to zero as $n \to \infty$, and then $\Delta \to 0$. This is intended as a heuristic explanation of the precise conditions used in the convergence theorems.

The essential fact in the analysis and the intuition used here is the "time scale separation" between the sequences $\{\theta_i, i \geq n\}$ and $\{Y_i - m(\theta_i), i \geq n\}$ for large n. The ODE plays an even more important role when the noise sequence is strongly correlated. The exploitation of the connection between the properties of the ODE and the asymptotic properties of the stochastic approximation was initiated by Ljung [164, 165] and developed extensively by Kushner and coworkers [123, 126, 127, 135, 142]; see also [16, 57].

Example 2. Chemical batch processing. Another specific application of the root finding algorithm in Example 1 is provided by the following problem. A particular chemical process is used to produce a product in a batch mode. Each batch requires T units of time, and the entire procedure is repeated many times. The rate of pumping cooling water (called θ) is controlled in order to adjust the temperature set point to the desired mean value $\bar{m}$. The mean value $m(\theta)$ is an unknown function of the pumping rate and is assumed to be monotonically decreasing. The process dynamics are not known well and the measured sample mean temperature varies from batch to batch due to randomness in the mixture and other factors. The Robbins–Monro procedure can be applied to iteratively estimate the root $\bar{\theta}$ of the equation $m(\bar{\theta}) = \bar{m}$ while the process is in operation. It can also be used to track changes in the root as the statistics of the operating data change (then ϵ_n will usually be small but not go to zero). Let $\{\theta_n\}$ denote the pumping rate and Y_n the sample mean temperature observed in the nth batch, $n = 0, 1, \ldots$ Then the algorithm is $\theta_{n+1} = \theta_n + \epsilon_n[Y_n - \bar{m}]$.

Suppose that the sample mean temperatures from run to run are mutually independent, conditioned on the parameter values; in other words,

$$P\{Y_n \leq y | \theta_i, Y_i, i < n, \theta_n\} = P\{Y_n \leq y | \theta_n\}.$$

Then the noise $[Y_n - m(\theta_n)] \equiv \delta M_n$ is a martingale difference since

$$E[\delta M_n | Y_i, \theta_i, i < n, \theta_n] = 0 = E[\delta M_n | \delta M_i, i < n].$$

There might be a more complicated noise structure. For example, suppose that part of the raw material being used in the process varies in a dependent way from batch to batch. If the noise "nearly averages out" over a number of successive batches, then one expects that $\{\theta_n\}$ would still converge to $\bar{\theta}$. The correlated noise case requires a more complex proof (Chapter 6) than that needed for the martingale difference noise case (Chapter 5).

In general, one might have control over more than one parameter, and several set points might be of interest. For example, suppose that there is a control over the coolant pumping rate as well as over the level of a catalyst that is added, and one wishes to set these such that the mean temperature is $\bar{m}_1$ and the yield of the desired product is $\bar{m}_2$. Let θ_n and $\bar{m}$ denote the *vector-valued* parameter used on the nth batch ($n = 0, 1, \ldots$) and the desired vector-valued set point, resp., and let Y_n denote the *vector* observation (sample temperature and yield) at the nth batch. Let $m(\theta)$ be the (mean temperature, mean yield) under θ. Then the analogous vector algorithm can be used, and the asymptotic behavior will be characterized by the asymptotic solution of the mean ODE $\dot{\theta} = m(\theta) - \bar{m}$. If $m(\theta) - \bar{m}$ has a unique root $\bar{\theta}$ which is a globally asymptotically stable point of the mean ODE, then θ_n will converge to $\bar{\theta}$.

1.1.3 Best Linear Least Squares Fit

Example 3. This example is a canonical form of recursive least squares fitting. Various recursive approximations to least-squares type algorithms are of wide use in control and communication theory. A more general class and additional applications will be discussed in greater detail in Chapter 3. This section provides an introduction to the general ideas.

Two classes of patterns are of interest, either pattern A or pattern $\bar{A}$ (pattern "not A"). A sequence of patterns is drawn at random from a given distribution on $(A, \bar{A})$. Let $y_n = 1$ if the pattern drawn on trial n ($n = 0, 1, \ldots$) is A, and let $y_n = -1$ otherwise. The patterns might be samples of a letter or a number. The patterns themselves are not observed but are known only through noise-corrupted observations of particular characteristics. In the special case where pattern A corresponds to the letter "A," each observable sample might be a letter (either "A" or another letter), each written by a different person or by the same person at a different time (thus the various samples of the same letter will vary), and a scanning procedure and computer algorithm are used to decide whether the letter is indeed "A." Typically, the scanned sample will be processed to extract "features," such as the number of separate segments, loops, corners, etc. Thus, at times $n = 0, 1, \ldots$, one can suppose that a random "feature" vector $\widetilde{\phi}_n$ is observed whose distribution function depends on whether the pattern is A or $\bar{A}$. For illustrative purposes, we suppose that the members of the sequence of observations are mutually independent. In particular,

$$P\left\{\widetilde{\phi}_n \in \cdot \mid \widetilde{\phi}_i, y_i, i < n, y_n\right\} = P\left\{\widetilde{\phi}_n \in \cdot \mid y_n\right\}.$$

The independence assumption is often restrictive. It can be weakened considerably and will not be used in subsequent chapters.

A linear pattern classifier has the following form. The output v_n equals $\widetilde{\phi}_n'\widetilde{\theta} + \widetilde{\theta}_0$, where the $\theta = (\widetilde{\theta}_0, \widetilde{\theta})$ is a parameter, and $\widetilde{\theta}_0$ is real valued. If $v_n \geq 0$, then the hypothesis that the pattern is A is accepted, otherwise it is rejected. Define $\phi_n = (1, \tilde{\phi}_n)$. The quality of the decision depends on the value of θ, which will be chosen to minimize some decision error. Many error criteria are possible; here we wish to select θ such that

$$E\left[y_n - \theta'\phi_n\right]^2 = E\left[y_n - v_n\right]^2 \tag{1.10}$$

is minimized. This criterion yields a relatively simple algorithm and serves as a surrogate for the probability that the decision will be in error. Suppose that there are a matrix Q (positive definite) and a vector S such that $Q = E\phi_n\phi_n'$ and $S = Ey_n\phi_n$ for all n. Then the optimal value of θ is $\bar{\theta} = Q^{-1}S$.

The probability distribution of (y_n, ϕ_n) is not known, but we suppose that a large set of sample values $\{y_n, \phi_n\}$ is available. This "training" sample will be used to get an estimate of the optimal value of θ. It is of interest to know what happens to the estimates as the sample size grows to infinity. Let $\widehat{\theta}_n$ minimize the mean square sample error[1]

$$\frac{1}{n}\sum_{i=0}^{n-1}\left[y_i - \theta'\phi_i\right]^2.$$

Suppose that for large enough n the matrix $\Phi_n = \sum_{i=0}^{n-1}\phi_i\phi_i'$ is invertible with probability one. Then, given n samples, the minimizing θ is

$$\widehat{\theta}_n = \Phi_n^{-1}\sum_{i=0}^{n-1} y_i\phi_i. \tag{1.11}$$

If the matrix Φ_n is poorly conditioned, then one might use the alternative $\Phi_n + \delta n A$ where $\delta > 0$ is small and A is positive definite and symmetric.

Equation (1.11) can be put into a recursive form by expanding

$$\widehat{\theta}_{n+1} = \Phi_{n+1}^{-1}\left[\sum_{i=0}^{n-1} y_i\phi_i + y_n\phi_n\right],$$

to get [169, pp. 18–20]

$$\widehat{\theta}_{n+1} = \widehat{\theta}_n + \Phi_{n+1}^{-1}\phi_n\left[y_n - \phi_n'\widehat{\theta}_n\right]. \tag{1.12}$$

The dimension of ϕ_n is generally large, and the computation of the inverse can be onerous. This can be avoided by solving for (or approximating) Φ_n^{-1} recursively. The matrix inversion lemma [169, pp. 18–20]

$$\left[A + BCD\right]^{-1} = A^{-1} - A^{-1}B\left[DA^{-1}B + C^{-1}\right]^{-1}DA^{-1} \tag{1.13}$$

---

[1] In Chapter 3, we also use a discounted mean square error criterion, which allows greater weight to be put on the more recent samples.

can be used to compute the matrix inverse recursively, yielding

$$\Phi_{n+1}^{-1} = \Phi_n^{-1} - \frac{\Phi_n^{-1}\phi_n\phi_n'\Phi_n^{-1}}{1+\phi_n'\Phi_n^{-1}\phi_n}, \tag{1.14}$$

$$\widehat{\theta}_{n+1} = \widehat{\theta}_n + \frac{\Phi_n^{-1}}{1+\phi_n'\Phi_n^{-1}\phi_n}\phi_n\left[y_n - \phi_n'\widehat{\theta}_n\right]. \tag{1.15}$$

Equations (1.14) and (1.15) are known as the *recursive least squares* formulas.

Taking a first-order (in Φ_n^{-1}) expansion in (1.12) and (1.14) yields a linearized least squares approximation θ_n:

$$\theta_{n+1} = \theta_n + \Phi_n^{-1}\phi_n\left[y_n - \phi_n'\theta_n\right], \tag{1.16}$$

$$\Phi_{n+1}^{-1} = \left[I - \Phi_n^{-1}\phi_n\phi_n'\right]\Phi_n^{-1}. \tag{1.17}$$

The convergence of the sequences in (1.14) and (1.15) is guaranteed by any conditions that assure the convergence of the least squares estimator, since they are equivalent. But the approximation in (1.16) and (1.17) is no longer a representation of the least squares estimator. To facilitate the proof of convergence to the least squares estimator, it is useful to first put (1.17) into a stochastic approximation form. Define $B_n = n\Phi_n^{-1}$. Then

$$B_{n+1} = B_n + \frac{1}{n}\left[-B_n\phi_n\phi_n' + I\right]B_n - \frac{B_n}{n}\phi_n\phi_n'\frac{B_n}{n}.$$

One first proves that B_n converges, and then uses this fact to get the convergence of the θ_n defined by (1.16), where we substitute $B_n/n = \Phi_n^{-1}$.

These recursive algorithms for estimating the optimal parameter value are convenient in that one computes the new estimate simply in terms of the old one and the new data. These algorithms are a form of the Robbins–Monro scheme with random matrix-valued ϵ_n. Many versions of the examples are in [166, 169].

It is sometimes convenient to approximate (1.16) by replacing the random matrix Φ_n^{-1} with a positive real number ϵ_n to yield the stochastic approximation form of linearized least squares:

$$\theta_{n+1} = \theta_n + \epsilon_n\phi_n\left[y_n - \phi_n'\theta_n\right]. \tag{1.18}$$

A concern in applications is that ϵ_n will decrease too fast and that, for practical purposes, the iterate will "get stuck" too far away from the solution. For this reason, and to allow "tracking" if the statistics change, ϵ_n might be held constant at some $\epsilon > 0$.

The mean ODE, which characterizes the asymptotic behavior of (1.18), is

$$\dot{\theta} = S - Q\theta = -\frac{1}{2}\frac{\partial}{\partial\theta}E\left[y_n - \phi_n'\theta\right]^2. \tag{1.19}$$

which is asymptotically stable about the optimal point $\bar{\theta}$. The right-hand equality is easily obtained by a direct computation of the derivative. The ODE that characterizes the limit behavior of the algorithm (1.16) and (1.17) is similar and will be discussed in Chapters 3, 5, and 9.

Comments on the algorithms. The algorithm (1.18) converges more slowly than does ((1.12), (1.14)) since the Φ_n^{-1} is replaced by some rather arbitrary real number ϵ_n. This affects the direction of the step as well as the norm of the step size. For large n, $\Phi_n^{-1} \approx Q^{-1}/n$ by the law of large numbers. The relative speed of convergence of (1.18) and ((1.14), (1.15)) is determined by the "eigenvalue spread" of Q. If the absolute values of the ratios of the eigenvalues are not too large, then algorithms of the form of (1.18) work well. This comment also holds for the algorithms of Section 3.1.

Note on time-varying systems and tracking. The discussion of the recursive least squares algorithm (1.14) and (1.15) is continued in Section 3.5, where a "discounted" or 'forgetting factor" form, which weight recent errors more heavily, is used to track time-varying systems. A second adaptive loop is added to optimize the discount factor, and this second loop has the stochastic approximation form.

Suppose that larger ϵ_n were used, say $\epsilon_n = 1/n^\gamma, \gamma \in (.5, 1)$, and that the Polyak averaging method, discussed in Subsection 3.3 and in Chapter 11, is used. Then under broad conditions, the rate of convergence to $\bar{\theta}$ of the average $\Theta_n = \sum_{i=0}^{n-1} \theta_i/n$ is nearly the same as what would be obtained with $\epsilon_n = Q^{-1}/n$. The recursive algorithms ((1.16), (1.17)) or (1.18) might be preferred to the use of (1.11) since the computational requirements are less, and the good asymptotic behavior of ((1.16), (1.17)) or of (1.18) (at least under Polyak averaging for the latter). See Chapters 9 and 11 and [55, 151]. When the sequence $\{y_n\}$ is not stationary, then the optimal value of ϵ varies with time, and can be "tracked" by an adaptive algorithm; see Chapter 3 for more detail.

The key to the value of the stochastic approximation algorithm is the representation of the right side of (1.19) as the negative gradient of the cost function. This emphasizes that, whatever the origin of the stochastic approximation algorithm, it can be interpreted as a "stochastic" gradient descent algorithm. For example, (1.18) can be interpreted as a "noisy" gradient procedure. We do not know the value of the gradient of (1.10) with respect to θ, but the gradient of the sample $-[y_n - \phi_n'\theta]^2/2$ is just $\phi_n[y_n - \phi_n'\theta]$, the dynamical term in (1.18), when $\theta = \theta_n$. The mean value of the term $\phi_n[y_n - \phi_n'\theta]$ is just the negative of the gradient of (1.10) with respect to θ. Hence the driving observation in (1.18) is just a "noise-corrupted" value of the desired gradient at $\theta = \theta_n$. This general idea will be explored more fully in the next chapter. In the engineering literature, (1.18) is often viewed as a "decorrelation" algorithm, because the mean value of the right side of (1.18) being zero means that the error $y_n - \phi_n' \theta_n$ is uncorrelated with the observation ϕ_n. As intuitively helpful as this decorrelation idea might be, the interpretation in terms of gradients is more germane.

Now suppose that Q is not invertible. Then the components of ϕ_n are linearly dependent, which might not be known when the algorithm is used. The correct ODE is still (1.19), but now the right side is zero on a linear manifold. The sequence $\{\theta_n\}$ might converge to a fixed (perhaps random) point in the linear manifold, or it might just converge to the linear manifold and keep wandering, depending largely on the speed with which ϵ_n goes to zero. In any case, the mean square error will converge to its minimum value.

1.1.4 Minimization by Recursive Monte Carlo

Example 4. Example 3 is actually a function minimization problem, where the function is defined by (1.10). The θ-derivatives of the mean value (1.10) are not known; however, one could observe values of the θ-derivative of samples $[y_n - \theta' \phi_n]^2/2$ at the desired values of θ and use these in the iterative algorithm in lieu of the exact derivatives. We now give another example of that type, which arises in the parametric optimization of dynamical systems, and where the Robbins–Monro procedure is applicable for the sequential monte carlo minimization via the use of noise-corrupted observations of the derivatives.

Let θ be an $\mathbb{R}^r$-valued parameter of a dynamical system in $\mathbb{R}^k$ whose evolution can be described by the equation

$$\bar{X}^{m+1}(\theta) = b(\bar{X}^m(\theta), \theta, \chi^m), \quad m = 0, 1, \ldots, \quad \bar{X}^0(\theta) = X_0, \tag{1.20}$$

where χ^m are random variables, the function $b(\cdot)$ is known and continuously differentiable in (x, θ), and the components of $\bar{X}^m(\theta)$ are $\bar{X}_i^m(\theta), i \leq k$. Given a real-valued function $F(\cdot)$, the objective is to minimize $f(\theta) = EF(\bar{X}^N(\theta), \theta)$ over θ, for a given value of N. Thus the system is of interest over a finite horizon $[0, N]$. Equation (1.20) might represent the combined dynamics of a tracking and intercept problem, where θ parameterizes the tracker controller, and the objective is to maximize the probability of getting within "striking distance" before the terminal time N.

Define $\bar{\chi} = \{\chi^m, m = 0, \ldots, N-1\}$ and suppose that the distribution of $\bar{\chi}$ is known. The function $F(\cdot)$ is assumed to be known and continuously differentiable in (x, θ), so that sample (noisy) values of the system state, the cost, and their pathwise θ-derivatives, can be simulated. Often the problem is too complicated for the values of $EF(\bar{X}^N(\theta), \theta)$ to be explicitly evaluated. If a "deterministic" minimization procedure were used, one would need good estimates of $EF(\bar{X}^N(\theta), \theta)$ (and perhaps of the θ-derivatives as well) at selected values of θ. This would require a great deal of simulation at values of θ that may be far from the optimal point.

A recursive monte carlo method is often a viable alternative. It will require simulations of the system on the time interval $[0, N]$ under various selected parameter values. Define $\bar{U}_j^m(\theta) = \partial \bar{X}^m(\theta)/\partial\theta^j$, with components $\bar{U}_{j,i}^m(\theta) = \partial \bar{X}_i^m(\theta)/\partial\theta^j, j \leq r$, where we recall that θ^j is the jth component of the vector θ. Then $\bar{U}_i^0(\theta) = 0$ for all i, and for $m \geq 0$,

$$\bar{U}_i^{m+1}(\theta) = b_x'(\bar{X}^m(\theta), \theta, \chi^m)\bar{U}_i^m(\theta) + b_{\theta^i}(\bar{X}^m(\theta), \theta, \chi^m). \tag{1.21}$$

Supposing that the operations of expectation and differentiation can be interchanged, define $\bar{g}(\theta) = (\bar{g}_i(\theta), i \leq r)$ by

$$\begin{aligned} \frac{\partial E[F(\bar{X}^N(\theta), \theta)]}{\partial\theta^i} &= E\frac{\partial F(\bar{X}^N(\theta), \theta)}{\partial\theta^i} \\ &= E\left[F_x'(\bar{X}^N(\theta), \theta)\bar{U}_i^N(\theta) + F_{\theta^i}(\bar{X}^N(\theta), \theta)\right] \\ &\equiv -\bar{g}_i(\theta). \end{aligned} \tag{1.22}$$

Let $\theta_n = (\theta_{n,1}, \ldots, \theta_{n,r})$ be the nth estimator of the minimizing value of θ, with θ_0 given.

The system used on the nth simulation $(n = 0, 1, \ldots)$ is defined by

$$X_n^{m+1} = b(X_n^m, \theta_n, \chi_n^m), \quad X_n^0 = X_0, \quad m = 0, 1, \ldots, N-1,$$

where the random sequence $\chi_n = \{\chi_n^m, m < N\}$ has the same distribution as $\bar{\chi}$, for each n. Define $U_{n,i}^m = \partial X_n^m/\partial\theta_{n,i}$, where as usual $\theta_{n,i}$ is the ith component of θ_n; it satisfies an equation like (1.21). Define

$$Y_{n,i} = -F_x'(X_n^N, \theta_n)U_{n,i}^N - F_{\theta^i}(X_n^N, \theta_n) \tag{1.23}$$

and $Y_n = (Y_{n,i}, i = 1, \ldots, r)$. Then a Robbins–Monro recursive monte carlo procedure for this problem can be written as

$$\theta_{n+1} = \theta_n + \epsilon_n Y_n = \theta_n + \epsilon_n \bar{g}(\theta_n) + \epsilon_n\left[Y_n - \bar{g}(\theta_n)\right]. \tag{1.24}$$

If the $\{\chi_n\}$ are mutually independent, then the noise terms $[Y_n - \bar{g}(\theta_n)]$ are ($I\!R^r$-valued) martingale differences. However, considerations of variance reduction (see Subsection 3.1) might dictate the use of correlated $\{\chi_n\}$, provided that the noise terms still "locally average to zero." The mean ODE characterizing the asymptotic behavior is $\dot{\theta} = \bar{g}(\theta)$.

If the actual observations are taken on a physical system rather than obtained from a simulation that is completely known, then one might not know the exact form of the dynamical equations governing the system. If a form is assumed from basic physical considerations or simply estimated via observations, then the calculated pathwise derivatives will not generally be the true pathwise derivatives. Although the optimization procedure might still work well and approximation theorems can indeed be proved, care must be exercised.

1.2 The Kiefer–Wolfowitz Procedure

1.2.1 The Basic Procedure

Examples 3 and 4 in Section 1, and the neural net, various "learning" problems, and the queueing optimization example in Chapter 2 are all concerned with the minimization of a function of unknown form. In all cases, noisy estimates of the derivatives are available and could be used as the basis of the recursive algorithm. In fact, in Examples 3 and 4 and in the neural net example of Chapter 2, one can explicitly differentiate the sample error functions at the current parameter values and use these derivatives as "noisy" estimates of the derivatives of the (mean) performance of interest at those parameter values. In the queueing example of Chapter 2, pathwise derivatives are also available, but for a slightly different function from the one we wish to minimize. However, these pathwise derivatives can still be used to get the desired convergence results. When such pathwise differentiation is not possible, a finite difference form of the gradient estimate is a possible alternative; see [271] for the suggested recursive algorithms for stock liquidation.

We wish to minimize the function $EF(\theta,\chi) = f(\theta)$ over the $I\!R^r$-valued parameter θ, where $f(\cdot)$ is continuously differentiable and χ is a random vector. The forms of $F(\cdot)$ and $f(\cdot)$ are not completely known. Consider the following finite difference form of stochastic approximation. Let $c_n \to 0$ be a finite difference interval and let e_i be the standard unit vector in the ith coordinate direction. Let θ_n denote the nth estimate of the minimum. Suppose that for each i, n, and random vectors $\chi_{n,i}^+, \chi_{n,i}^-$, we can observe the finite difference estimate

$$Y_{n,i} = -\frac{[F(\theta_n + c_n e_i, \chi_{n,i}^+) - F(\theta_n - c_n e_i, \chi_{n,i}^-)]}{2c_n}. \tag{2.1}$$

Define $Y_n = (Y_{n,1}, \ldots, Y_{n,r})$, and update θ_n by

$$\theta_{n+1} = \theta_n + \epsilon_n Y_n. \tag{2.2}$$

The form (2.2), with Y_n defined by (2.1) as an estimator of a finite difference, is known as the *Kiefer–Wolfowitz* algorithm [110, 250] because Kiefer and Wolfowitz were the first to formulate it and prove its convergence.

Define

$$\begin{aligned}\psi_{n,i} = &\left[f(\theta_n + c_n e_i) - F(\theta_n + c_n e_i, \chi_{n,i}^+)\right] \\ &- \left[f(\theta_n - c_n e_i) - F(\theta_n - c_n e_i, \chi_{n,i}^-)\right],\end{aligned}$$

and write

$$\frac{[f(\theta_n + c_n e_i) - f(\theta_n - c_n e_i)]}{2c_n} \equiv \gamma_{n,i} = f_{\theta^i}(\theta_n) - \beta_{n,i}, \tag{2.3}$$

where $-\beta_{n,i}$ is the bias in the finite difference estimate of $f_\theta(\theta_n)$, via central differences. Setting $\psi_n = (\psi_{n,1}, \ldots, \psi_{n,r})$ and $\beta_n = (\beta_{n,1}, \ldots, \beta_{n,r})$, (2.2) can be rewritten as

$$\theta_{n+1} = \theta_n - \epsilon_n f_\theta(\theta_n) + \epsilon_n \frac{\psi_n}{2c_n} + \epsilon_n \beta_n. \tag{2.4}$$

Clearly, for convergence to a local minimum to occur, one needs that $\beta_n \to 0$; the bias is normally proportional to the finite difference interval $c_n \to 0$. Additionally, one needs that the noise terms $\epsilon_n \psi_n/(2c_n)$ "average locally" to zero. Then the ODE that characterizes the asymptotic behavior is

$$\dot{\theta} = -f_\theta(\theta). \tag{2.5}$$

The fact that the effective noise $\psi_n/(2c_n)$ is of the order $1/c_n$ makes the Kiefer–Wolfowitz procedure less desirable than the Robbins–Monro procedure and puts a premium on getting good estimates of the derivatives via some variance reduction method or even by using an approximation to the original problem. Frequently, c_n is not allowed to go to zero, and one accepts a small bias to get smaller noise effects.

Variance reduction. Special choices of the driving noise can also help when the optimization is done via a simulation that the experimenter can control. This will be seen in what follows. Keep in mind that the driving noise is an essential part of the system, even if it is under the control of the experimenter in a simulation, since we wish to minimize an average value. It is not necessarily an additive disturbance. For example, if we wish to minimize the probability that a queueing network contains more than N customers at a certain time, by controlling a parameter of some service time distribution, then the "noise" is the set of interarrival and service times. It is a basic part of the system.

Suppose that $F(\cdot, \chi)$ is continuously differentiable for each value of χ. Write

$$\begin{aligned} &\frac{F(\theta_n - e_i c_n, \chi_{n,i}^-) - F(\theta_n + e_i c_n, \chi_{n,i}^+)}{2c_n} + f_{\theta^i}(\theta_n) \\ &= \frac{1}{2c_n}\left[F(\theta_n, \chi_{n,i}^-) - F(\theta_n, \chi_{n,i}^+)\right] \\ &\quad - \left[\frac{1}{2}\left(F_{\theta^i}(\theta_n, \chi_{n,i}^+) + F_{\theta^i}(\theta_n, \chi_{n,i}^-)\right) - f_{\theta^i}(\theta_n)\right] + \widetilde{\beta}_{n,i} \\ &\equiv \widetilde{\psi}_{n,i} + \widetilde{\beta}_{n,i}, \end{aligned} \tag{2.6}$$

where $\widetilde{\beta}_{n,i}$ is the bias in the finite difference estimate. Then the algorithm can be written as

$$\theta_{n+1} = \theta_n - \epsilon_n f_\theta(\theta_n) + \epsilon_n \widetilde{\psi}_n + \epsilon_n \widetilde{\beta}_n. \tag{2.7}$$

Equation (2.6) shows that it is preferable to use $\chi_{n,i}^+ = \chi_{n,i}^-$ if possible, since it eliminates the dominant $1/c_n$ factor in the effective noise. That is, $\widetilde{\psi}_{n,i}$

is the first term on the third line of (2.6) and is not inversely proportional to c_n.

The use of $\chi_{n,i}^+ = \chi_{n,i}^-$ can also be advantageous, even without differentiability. Fixing $\theta_n = \theta$, letting $EF(\theta \pm c_n e_i, \chi_{n,i}^\pm) = f(\theta \pm c_n e_i)$, and $\widetilde{F}(\theta, \chi) = F(\theta, \chi) - f(\theta)$, the variance of the effective noise is

$$E\left[\widetilde{F}(\theta + c_n e_i, \chi_{n,i}^+)\right]^2 + E\left[\widetilde{F}(\theta - c_n e_i, \chi_{n,i}^-)\right]^2 \\ -2E\left[\widetilde{F}(\theta + c_n e_i, \chi_{n,i}^+)\right]\left[\widetilde{F}(\theta - c_n e_i, \chi_{n,i}^-)\right],$$

divided by $4c_n^2$, which suggests that the larger the correlation between $\chi_{n,i}^\pm$, the smaller the noise variance will be when c_n is small.

If $\{(\chi_n^+, \chi_n^-), n = 0, 1, \ldots\}$ is a sequence of independent random variables, then the ψ_n and $\widetilde{\psi}_n$ are martingale differences for each n. Note that the noises ψ_n and $\widetilde{\psi}_n$ can be complicated functions of θ_n. In the martingale difference noise case, this θ_n-dependence can often be ignored in the proofs of convergence, but it must be taken into account if the $\chi_n^\pm$ are correlated in n.

Iterating on a subset of components at a time. Each iteration of (2.2) requires $2r$ observations. This can be reduced to $r + 1$ if one-sided differences are used. Since the one-sided case converges slightly more slowly, the apparent savings might be misleading. An alternative is to update only one component of θ at a time. In particular, it might be worthwhile to concentrate on the particular components that are expected to be the most important, provided that one continues to devote adequate resources to the remaining components. The choice of component can be quite arbitrary, provided that one returns to each component frequently enough. In all cases, the difference interval can depend on the coordinate direction.

If we wish to iterate on one component of θ at a time, then the following form of the algorithm can be used:

$$\theta_{nr+i+1} = \theta_{nr+i} + \epsilon_n e_{i+1} \widetilde{Y}_{nr+i}, \tag{2.8}$$

where

$$\widetilde{Y}_{nr+i} = \frac{F(\theta_{nr+i} - c_n e_{i+1}, \chi_{nr+i}^-) - F(\theta_{nr+i} + c_n e_{i+1}, \chi_{nr+i}^+)}{2c_n}.$$

The iteration in (2.8) proceeds as follows. For each $n = 0, 1, \ldots$, compute $\theta_{nr+i+1}, i = 0, \ldots, r-1$, from (2.8). Then increase n by one and continue. The mean value of $\widetilde{Y}_n$ is periodic in n, but the convergence theorems of Chapters 5 to 8 cover quite general cases of n-dependent mean values.

Comments. The iterate averaging method of Subsection 3.3 can be used to alleviate the difficulty of selecting good step sizes ϵ_n. As will be seen in

Chapter 11, the averaging method of Subsection 3.3 has no effect on the bias but can reduce the effects of the noise. In many applications, one has much freedom to choose the form of the algorithm. Wherever possible, try to estimate the derivative without the use of finite differences. The use of "common random numbers" $\chi_n^+ = \chi_n^-$ or other variance reduction methods can also be considered. In simulations, the use of minimal discrepancy sequences [184] in lieu of "random noise" can be useful and is covered by the convergence theorems. Small biases in the estimation of the derivative might be preferable to the asymptotically large noise effects due to the $1/c_n$ term. Hence, an appropriately small but fixed value of c_n should be considered. If the procedure is based on a simulation, then it is advisable to start with a simpler model and a larger difference interval to get a rough estimate of the location of the minimum point and a feeling for the general qualitative behavior and the best values of ϵ_n, either with or without iterate averaging.

1.2.2 Random Directions

Random directions. One step of the classical KW procedure uses either $2r$ or $r + 1$ observations, depending on whether two-sided or one-sided differences are used. Due to considerations of finite difference bias and rates of convergence, the symmetric two-sided difference is usually chosen. If a "sequential form" such as (2.8) is used, where one component of θ is updated at a time, then $2r$ steps are required to get a "full" derivative estimate. Whenever possible, one tries to estimate the derivative directly without recourse to finite differences, as, for example, in Example 4 and Section 2.5. When this cannot be done and the dimension r is large, the classical Kiefer–Wolfowitz method might not be practical. One enticing alternative is to update only one direction at each iteration using a finite difference estimate and to select that direction randomly at each step. Then each step requires only two observations.

In one form or another such methods have been in experimental or practical use since the earliest work in stochastic approximation. Proofs of convergence and the rate of convergence were given in [135], for the case where the direction was selected at random on the surface of the unit sphere, with the conclusion that there was little advantage over the classical method. The work of Spall [212, 213, 226, 227, 228, 229], where the random directions were chosen in a different way, showed advantages for such high dimensional problems and encouraged a reconsideration of the random directions method. The particular method used in [226] selected the directions at random on the vertices of the unit cube with the origin as the center. It will be seen in Chapter 10 that whatever advantages there are to this approach are due mainly to the fact that the direction vector has norm $\sqrt{r}$ instead of unity. Thus selection at random on the surface of a sphere with radius $\sqrt{r}$

will work equally as well. The proofs for the random directions methods discussed to date are essentially the same as for the usual Kiefer–Wolfowitz method and the "random directions" proof in [135] can be used. This will be seen in Chapters 5 and 10. In Chapter 10, when dealing with the rate of convergence, there will be a more extensive discussion of the method. It will be seen that the idea can be very useful but must be used with awareness of possible undesirable "side effects," particularly for "short" runs.

Let $\{d_n\}$ denote a sequence of random "direction" vectors in $I\!R^r$. It is not required that $\{d_n\}$ be mutually independent and satisfy $Ed_nd_n' = I$, where I is the identity matrix in $I\!R^r$, although this seems to be the currently preferred choice. In general, the values d_nd_n' must average "locally" to the identity matrix, but one might wish to use a variance reduction scheme that requires correlation among successive values. Let the difference intervals be $0 < c_n \to 0$. Then the algorithm is

$$\theta_{n+1} = \theta_n - \epsilon_n d_n \frac{[Y_n^+ - Y_n^-]}{2c_n}, \tag{2.9}$$

where $Y_n^\pm$ are observations taken at parameter values $\theta_n \pm c_n d_n$. The method is equally applicable when the difference interval is a constant, and this is often the choice since it reduces the noise effects and yields a more robust algorithm, even at the expense of a small bias.

Suppose that, for some suitable function $F(\cdot)$ and "driving random variables" $\chi_n^\pm$, the observations can be written in the form

$$Y_{n,i}^\pm = F(\theta_n \pm e_i c_n, \chi_n^\pm) = f(\theta_n \pm c_n d_n) + \psi_n^\pm, \tag{2.10}$$

where $\psi_n^\pm$ denotes the effective observation "noise." Supposing that $f(\cdot)$ is continuously differentiable, write (2.9) as

$$\theta_{n+1} = \theta_n - \epsilon_n d_n d_n' f_\theta(\theta_n) + \epsilon_n \beta_n + \epsilon_n \frac{d_n(\psi_n^- - \psi_n^+)}{2c_n}, \tag{2.11}$$

where β_n is the bias in the symmetric finite difference estimator of the derivative of $f(\cdot)$ at θ_n in the direction d_n with difference interval $c_n d_n$ used. Note that d_n and ψ_n cannot generally be assumed to be mutually independent, except perhaps in an asymptotic sense, since the observation noises $\psi_n^\pm$ depend on the parameter values at which the observations are taken.

Centering d_nd_n' about the identity matrix yields

$$\theta_{n+1} = \theta_n - \epsilon_n \left[f_\theta(\theta_n) - \beta_n\right] + \epsilon_n \frac{d_n(\psi_n^- - \psi_n^+)}{2c_n} + \epsilon_n \psi_n^d, \tag{2.12}$$

where

$$\psi_n^d = \left[I - d_n d_n'\right] f_\theta(\theta_n)$$

is the "random direction noise." The mean ODE characterizing the asymptotic behavior is the same as that for the Kiefer–Wolfowitz method, namely, the gradient descent form

$$\dot{\theta} = -f_\theta(\theta). \tag{2.13}$$

Comment on variance reduction. Recall the discussion in connection with (2.6) concerning the use of common driving random variables. If $\chi_n^+ = \chi_n^-$ then the term in (2.12) that is proportional to $1/c_n$ is replaced by $\epsilon_n d_n \psi_n$, where ψ_n is not proportional to $1/c_n$, and we have a form of the Robbins-Monro method.

1.3 Extensions of the Algorithms: Variance Reduction, Robustness, Iterate Averaging, Constraints, and Convex Optimization

In this section, we discuss some modifications of the algorithms that are motivated by practical considerations.

1.3.1 A Variance Reduction Method

Example 1 of Section 1 was a motivational problem in the original work of Robbins and Monro that led to [207], where θ represents an administered level of a drug in an experiment and $G(\cdot, \theta)$ is the unknown distribution function of the response under drug level θ. One wishes to find a level $\theta = \bar{\theta}$ that guarantees a mean response of $\bar{m}$. $G(\cdot, \theta)$ is the distribution function over the entire population. But, in practice, the subjects to whom the drug is administered might have other characteristics that allow one to be more specific about the distribution function of their response. Such information can be used to reduce the variance of the observation noise and improve the convergence properties. The method to be discussed is a special case of what is known in statistics as *stratified sampling* [204].

Before proceeding with the general idea, let us consider a degenerate example of a similar problem. Suppose that we wish to estimate the mean value of a particular characteristic of a population, say the weight. This can be done by random sampling; simply pick individuals at random and average their sample weights. Let us suppose a special situation, where the population is divided into two groups of equal size, with all individuals in each group having the same weight. Suppose, in addition, that the experimenter is allowed to select the group from which an individual sample is drawn. Then to get the average, one need only select a single individual from each group. Let us generalize this situation slightly. Suppose that each individual in the population is characterized by a pair (X, W), where X

takes two values A and B, and W is the weight. We are allowed to choose the group (A or B) from which any sample is drawn. If X is correlated with W, then by careful selection of the group membership of each successive sample, we can obtain an estimate of the mean weight with a smaller variance than that given by purely random sampling.

Now, return to the original stochastic approximation problem. Suppose that the subjects are divided into two disjoint groups that we denote for convenience simply by light (L) and heavy (H). Let the prior distribution that a subject is in class L or H be the known probabilities p_L and $p_H = 1 - p_L$, and let the associated but unknown response distribution functions be $G_L(\cdot, \theta), G_H(\cdot, \theta)$ with unknown mean values $m_L(\theta), m_H(\theta)$, resp., which are nondecreasing in θ. In Example 1 of Section 1, subjects are drawn at random from the general large population at each test and $G(\cdot, \theta) = p_L G_L(\cdot, \theta) + p_H G_H(\cdot, \theta)$, but there is a better way to select them.

To illustrate a variance reduction method, consider the special case where $p_L = 0.5$. Let $m(\cdot)$ be continuous and for each integer k let $\epsilon_n/\epsilon_{n+k} \to 1$ as $n \to \infty$. Since we have control over the class from which the subject is to be drawn, we can select them in any reasonable way, provided that the averages work out. Thus, let us draw every $(2n)$th subject at random from L and every $(2n+1)$st subject at random from H. Then, for $\theta_n = \theta$, the respective mean values of the first bracketed term on the right of (1.7) are $\bar{m} - m_L(\theta)$ and $\bar{m} - m_H(\theta)$, according to whether n is even or odd. For n even,

$$\theta_{n+2} = \theta_n + \epsilon_n \left[\bar{m} - m_L(\theta_n)\right] + \epsilon_{n+1} \left[\bar{m} - m_H(\theta_{n+1})\right] + \text{noise terms.}$$

The mean ODE that determines the asymptotic behavior is still (1.9b), the mean over the two possibilities. This is because ϵ_n becomes arbitrarily small as $n \to \infty$ that in turn implies that the rate of change of θ_n goes to zero as $n \to \infty$ and $\epsilon_n/\epsilon_{n+1} \to 1$, which implies that successive observations have essentially the same weight.

Let $\sigma_L^2(\theta)$ (resp., $\sigma_H^2(\theta)$) denote the variances of the response under L (resp., H), under parameter value θ. Then, for large n and $\theta_n \approx \theta$, the average of the variances of the Y_n and Y_{n+1} for the "alternating" procedure is approximately

$$p_L \sigma_L^2(\theta) + p_H \sigma_H^2(\theta).$$

Let E_H^θ, E_L^θ denote the expectation operators under the distributions of the two sub-populations. Then, for $\theta_n \approx \theta$, the average variance of each response under the original procedure, where the subjects are selected at random from the total population, is approximately

$$E_L^\theta \left[Y_n - p_L m_L(\theta) - p_H m_H(\theta)\right]^2 p_L + E_H^\theta \left[Y_n - p_L m_L(\theta) - p_H m_H(\theta)\right]^2 p_H,$$

which equals

$$p_L \sigma_L^2(\theta) + p_H \sigma_H^2(\theta) + p_L \left[m_L(\theta) - m(\theta)\right]^2 + p_H \left[m_H(\theta) - m(\theta)\right]^2.$$

Thus the variance for the "alternating" procedure is smaller than that of the original procedure, provided that $m_L(\theta) \neq m_H(\theta)$ (otherwise it is equal).

The "alternating" choice of subpopulation was made to illustrate an important point in applications of the Robbins–Monro procedure and indeed of all applications of stochastic approximation. The quality of the behavior of the algorithm (rate of convergence and variation about the mean flow to be dealt with in Chapter 10) depends very heavily on the "noise level," and any effort to reduce the noise level will improve the performance. In this case, the value of $E[Y_n|\theta_n = \theta]$ depends on whether n is odd or even, but it is the "local average" of the mean values that yields the mean ODE, which, in turn, determines the limit points.

This scheme can be readily extended to any value of p_H. Consider the case $p_H = 2/7$. There are several possibilities for the variance reduction algorithm. For example, one can work in groups of seven, with any permutation of $HHLLLLL$ used, and the permutation can vary with time. Alternatively, work in groups of four, where the first three are any permutation of HLL and the fourth is selected at random, with L being selected with probability $6/7$. If one form of the algorithm is well defined and convergent, all the suggested forms will be. The various alternatives can be alternated among each other, etc. Again, the convergence proofs show that it is only the "local averages" that determine the limit points.

1.3.2 Constraints

In practical applications, the allowed values of θ are invariably confined to some compact set either explicitly or implicitly. If the components of the parameter θ are physical quantities, then they would normally be subject to upper and lower bounds. These might be "flexible," but there are usually values beyond which one cannot go, due to reasons of safety, economy, behavior of the system, or other practical concerns. Even if the physics or the economics themselves do not demand *a priori* bounds on the parameters, one would be suspicious of parameter values that were very large relative to what one expects.

The simplest constraint, and the one most commonly used, truncates the iterates if they get too big. Suppose that there are finite $a_i < b_i$ such that if $\theta_{n,i}$ ever tries to get above b_i (resp., below a_i) it is returned to b_i (resp., a_i). Continuing, let $q(\theta)$ denote a measure of some penalty associated with operating the system under parameter value θ. It might be desired to minimize the total average cost subject to $q(\theta) \leq c_0$, a maximum allowable value. For this example, define the *constraint set* $H = \{\theta : a_i \leq \theta^i \leq b_i, q(\theta) - c_0 \leq 0\}$. Define $\Pi_H(\theta)$ to be the closest point in H to θ. Thus if $\theta \in H$, $\Pi_H(\theta) = \theta$. A convenient constrained or *projected* algorithm has the form

$$\theta_{n+1} = \Pi_H(\theta_n + \epsilon_n Y_n). \tag{3.1}$$

This and more general constrained algorithms will be dealt with starting in Chapter 5. The works [48, 50] use an alternative to projection by simply returning the state to a random point in a compact set if it strays too far.

1.3.3 Averaging of the Iterates: "Polyak Averaging"

When it can be proved that θ_n converges to the desired value $\bar{\theta}$, one might be tempted to suppose that the sample average

$$\Theta_n = \frac{1}{N_n} \sum_{i=n-N_n+1}^{n} \theta_i \tag{3.2}$$

is a "better" estimate of $\bar{\theta}$ than θ_n is, where $\{N_n\}$ is some sequence going to infinity and $n - N_n \geq 0$. This supposition might or might not be true, depending primarily on the values selected for ϵ_n. This issue is discussed at length in Chapter 11. It has been known for a long time that if $\epsilon_n = O(1/n)$, then the asymptotic behavior of Θ_n is no better than θ_n and can be worse in the sense of rate of convergence. It was shown by Polyak and Juditsky [194, 195] that if ϵ_n goes to zero slower than $O(1/n)$, then (3.2) is preferable. Proofs under weaker conditions are in [149, 151, 265, 267]. In [149, 151] and in Chapter 11, it is shown that the advantage of (3.2) is generic to stochastic approximation. An interpretation in terms of multiscale stochastic approximations is given, and this provides insight to why the method works.

When (3.2) is used, the step sizes ϵ_n are selected to be larger than they would ordinarily be so that there is more "jumping around" of the iterate sequence. The averaging (3.2) compensates for this "jumping around," and (under broad conditions) the rate of convergence of Θ_n is what one would get for the θ_n itself if the "optimal" matrix-valued sequence ϵ_n were used. This alleviates considerably the serious problem of selection of good values for the ϵ_n. A full discussion is in Chapter 11.

1.3.4 Averaging the Observations

The method of the previous section is concerned with the advantages of averaging the iterates. The basic stochastic approximation procedure has the original form (3.1) or its unprojected form. The averaged iterate (3.2) has better asymptotic performance if the step size goes to zero slowly enough. An alternative is to average the iterates as well as observations. Drop the constraints and consider the form

$$\theta_{n+1} = \bar{\theta}_n + n\epsilon_n \bar{Y}_n,$$

where

$$\bar{Y}_n = \frac{1}{n}\sum_{i=1}^{n} Y_i \text{ and } \bar{\theta}_n = \frac{1}{n}\sum_{i=1}^{n} \theta_i.$$

It can be shown that under appropriate conditions, the algorithm is asymptotically optimal [6, 216, 217, 273, 268]. This will be illustrated by the following simple example with $Y_n = A\theta_n + \xi_n$, where A is a Hurwitz matrix and $\{\xi_n\}$ is a sequence of mutually independent random variables having mean zero and bounded covariance Σ_0. Define $\bar{\xi}_n = \sum_{i=1}^n \xi_i/n$.

We can write

$$\bar{\theta}_{n+1} = \bar{\theta}_n + \frac{n}{n+1}\epsilon_n \bar{Y}_n = \left[I + \frac{n\epsilon_n}{n+1}A\right]\bar{\theta}_n + \frac{n\epsilon_n}{n+1}\bar{\xi}_n.$$

Define

$$K(n,i) = \begin{cases} \prod_{j=i+1}^{n}\left[I + \frac{j\epsilon_j}{j+1}A\right], & \text{if } i < n, \\ I, & \text{if } i = n. \end{cases}$$

Then

$$\begin{aligned} \bar{\theta}_{n+1} &= K(n,0)\bar{\theta}_1 + \sum_{i=1}^{n} \frac{i\epsilon_i}{i+1} K(n,i)\bar{\xi}_i \\ &= K(n,0)\bar{\theta}_1 + \sum_{i=1}^{n}\sum_{l=1}^{i} \frac{\epsilon_i}{i+1} K(n,i)\xi_l. \end{aligned}$$

If $\epsilon_n = 1/n^\gamma$ for $\gamma < 1$, then the first term on the last line goes to zero as $o(1/n)$ and n times the variance of the last term tends to $A^{-1}\Sigma_0(A^{-1})'$. This is its optimal asymptotic value (see Subsection 10.2.2 and Chapter 11), and is the same value that one gets with (3.2). Of course, we have neglected the nonlinear terms, but the development illustrates the great potential of averaging methods; see the references for more detail.

1.3.5 Robust Algorithms

In using the Robbins–Monro procedure for sequential monte carlo optimization, at the nth update, one draws the values of the random variables Y_n according to their distribution under parameter value θ_n. Consider Example 4 of Section 1 for specificity. If the values of the χ_n^m are unbounded, as they would be under a Gaussian distribution, then the values of Y_n might be unbounded. In this case, a single excessively large value of the observation, due to a rare occurrence of a large value of χ_n^m, might cause a large change in $\theta_{n+1} - \theta_n$. In applications, one might know very little about the system dynamics or even about the statistics of the driving noise at large parameter values. The modeling at large values is often determined by convenience or tradition. However, it is undesirable for single observations to have large effects on the iterate values, and a more robust procedure might be sought, analogous to the procedures of robust statistical analysis [102]. One such approach will now be described.

Let $\psi_i(\cdot), i \le r$, be bounded real-valued functions on the real line, and define $\psi(\theta) = (\psi_1(\theta^1), \ldots, \psi_r(\theta^r))$. Let $\psi_i(\cdot)$ be monotonically nondecreasing and satisfy $\psi_i(0) = 0, \psi_i(u) = -\psi_i(-u)$ and $\psi_i(u)/u \to 0$ as $u \to \infty$. One commonly used function is $\psi_i(u) = \min\{u, K_i\}$ for $u \ge 0$, where K_i is a given constant. The algorithm

$$\theta_{n+1} = \theta_n + \epsilon_n \psi(Y_n) \tag{3.3}$$

is suggested by procedures in robust statistics. It has the advantage that it reduces the effects of large values of the observation noise. Indeed, if an excessive sample value of a noise term appears in a simulation, one might ignore (rather than truncate) that sample by treating it as an outlier, a procedure similar to (3.3). If the stochastic approximation algorithm were used on an actual operating system and not in a simulation, and a large value of the observation appeared, one would generally seek to determine the physical reason behind the large value and not accept it without question. Thus, algorithms of the type (3.3) are actually incorporated into practice in one way or another and, as a practical matter without loss of much generality, one could suppose that Y_n are bounded, although we will not generally do so. Note that in (3.3), the observation is truncated, while the iterate is truncated in (3.1). Both forms of truncation can be used together. For a particular class of algorithms, Polyak and Tsypkin [196, 197] evaluated the optimal choice of $\psi(\cdot)$ from a minimax perspective.

1.3.6 Nonexistence of the Derivative at Some θ

In order to illustrate the procedure for nondifferentiable functions, return to Example 4 of Section 1, and suppose that the derivative of $F(\cdot)$ with respect to some component of θ does not exist at some point.[2] First, consider the particular case $F(x) = \sum_i |x_i|$. It does not explicitly depend on θ. Then $F_{x_i}(x) = \text{sign}(x_i)$ for $x_i \neq 0$, and it is undefined at $x_i = 0$. One needs some rule for selecting the value used in lieu of the derivative where the derivative does not exist. If $X_{n,i}^N = 0$ in the stochastic approximation algorithm at iteration n, then use any value in the interval $[-1, 1]$ for the derivative. In fact, this would be the set of possibilities if a finite difference method were used.

We concentrate on the correct form of the mean ODE. To get the correct right-hand side of the ODE at θ, one needs to account for the possible behavior of the Y_n when the iterate sequence varies in a small neighborhood of θ. Let $N_\delta(u)$ be a δ neighborhood of u. Fix θ and suppose that, for each

[2] Another example where the derivative does not exist at some points is illustrated in Figure 3.6.1.

i,

$$\lim_{\delta_1 \to 0} \lim_{\delta \to 0} \sup_{\widehat{\theta} \in N_{\delta_1}(\theta)} P\left\{\bar{X}_{n,i}^N(\widehat{\theta}) \in N_\delta(0)\right\} = 0. \tag{3.4}$$

Then the choice made when $\bar{X}_{n,i}^N(\theta_n) = 0$ does not affect the form of the ODE at θ.

If (3.4) fails, then a differential inclusion replaces the mean ODE $\dot{\theta} = \bar{g}(\theta)$ as follows. Recall the definition of $\bar{U}^N(\theta)$ above (1.21), and that $F(\cdot)$ does not depend explicitly on θ in this example. Let $G(\theta)$ denote the set of all possible values of the vector

$$\{-EF'_{x_i}(\bar{X}_i^N(\theta))\bar{U}_i^N(\theta), i \leq r\}$$

over the choices made for the derivatives $F_{x_i}(\cdot)$ when $\bar{X}_i^N(\theta) = 0$, and let co denote the convex hull. Define the convex upper semicontinuous set

$$\bar{G}(\theta) = \bigcap_{\delta > 0} \text{co} \left[\bigcup_{\widehat{\theta} \in N_\delta(\theta)} G(\widehat{\theta}) \right]. \tag{3.5}$$

If there is only one choice at θ and the mean values are continuous at θ, then $\bar{G}(\theta)$ contains only one point, namely, $\bar{g}(\theta)$. More generally, the mean ODE is replaced by the *differential inclusion*

$$\dot{\theta} \in \bar{G}(\theta). \tag{3.6}$$

Consider the following example for $r = 1$. For $\theta \neq 0$, let $G(\theta)$ contain only a single point $\bar{g}(\theta)$, where $\bar{g}(\cdot)$ is decreasing, continuous on $(-\infty, 0)$ and on $(0, \infty)$, with $\bar{g}(0^-) > 0$ and $\bar{g}(0^+) < 0$. Let $G(0) = [\bar{g}(0^+), \bar{g}(0^-)]$. Then all solutions of (3.6) tend to zero, despite the fact that the right side of (3.6) is multivalued at $\theta = 0$.

1.3.7 Convex Optimization and Subgradients.

Let $f(\cdot)$ be a real-valued convex function on $I\!R^r$. A vector γ is a *subgradient* of $f(\cdot)$ at θ if $f(\theta + x) - f(\theta) \geq \gamma' x$ for all $x \in I\!R^r$. Let $SG(\theta)$ denote the set of subgradients of $f(\cdot)$ at θ, which is also called the set of subdifferentials at θ. This set is closed and convex and satisfies

$$SG(\theta) = \bigcap_{\delta > 0} \text{co} \left[\bigcup_{\widehat{\theta} \in N_\delta(\theta)} SG(\widehat{\theta}) \right]. \tag{3.7}$$

Hence it is upper semicontinuous in θ. Suppose that we wish to minimize $f(\cdot)$ via recursive monte carlo. Let θ_n denote the sequence of estimators of the minimizer, and suppose that at each θ_n we can observe $Y_n = \gamma_n + \psi_n$, where ψ_n is the observation noise and $\gamma_n \in -SG(\theta_n)$. Then the Robbins–Monro algorithm for minimizing $f(\cdot)$ is

$$\theta_{n+1} = \theta_n + \epsilon_n Y_n = \theta_n + \epsilon_n \left[\gamma_n + \psi_n\right].$$

If $f(\cdot)$ is continuously differentiable at θ, then the derivative $f_\theta(\theta)$ is the unique member of $SG(\theta)$. Otherwise, $f_\theta(\theta)$ is a subgradient and is chosen in some way from $SG(\theta)$.

Suppose, for specificity (the condition will be weakened in Chapters 6 and 8), that $E[\psi_n|\theta_i, Y_i, i < n, \theta_n] = 0$ and ψ_n have uniformly bounded variances. Then the noise is a martingale difference and the equation that characterizes the limit points is the following form of (3.6):

$$\dot{\theta} \in -SG(\theta). \tag{3.8}$$

Convex Optimization: The Kiefer-Wolfowitz form. Now suppose that only finite difference estimators can be used, and the Kiefer-Wolfowitz procedure of Section 2 must be used. Suppose that the function $f(\cdot)$ to be minimized is continuous and convex but not everywhere differentiable. Recall the definition of $\gamma_{n,i}$ in (2.3). Then the form (2.4) is replaced by

$$\theta_{n+1} = \theta_n - \epsilon_n \gamma_n + \epsilon_n \frac{\psi_n}{2c_n}. \tag{3.9}$$

By the properties of convex functions, γ_n is a subgradient of the convex function of a real variable defined by $f(\theta_n + e_i \cdot)$ at some point in the interval $[-c_n, c_n]$. Thus, γ_n is an "approximation" to a subgradient of $f(\cdot)$ at $\theta = \theta_n$. More precisely, if $\theta_n \to \theta$ and $c_n \to 0$, then distance$(\gamma_n, SG(\theta)) \to 0$, where $SG(\theta)$ is the set defined in (3.7). The mean ODE characterizing the asymptotic behavior of the algorithm is replaced by the differential inclusion (3.8).

1.4 A Lagrangian Algorithm for Constrained Function Minimization

Suppose that $f(\cdot)$ is a strictly convex, continuously differentiable, and real-valued function, and $q_i(\cdot), i \le K$ are convex, continuously differentiable, and real-valued, all defined on $I\!R^r$. The objective is to minimize $f(\cdot)$ subject to $q_i(x) \le 0, i \le K$. None of the functions are known, but we can get noisy estimates of the functions and their derivatives. Suppose that there are constants $a_i < b_i, i \le r$, such that the constrained minimum $\bar{x}$ is interior to the box $\Pi[a_i, b_i]$. For $0 \le \lambda^i$, define the *Lagrangian* $L(x, \lambda) = f(x) + \sum_i \lambda^i q_i(x)$. Suppose that the gradient vectors $\{q_i(\bar{x}) : i$ such that $q_i(\bar{x}) =$

$0\}$ are linearly independent. A necessary and sufficiently condition for the existence of a unique minimum is that $q_i(\bar{x}) \leq 0$ and there are $\bar{\lambda}^i$ such that $L_x(\bar{x}, \bar{\lambda}) = 0$, with $\bar{\lambda}^i = 0$ if $q_i(\bar{x}) < 0$. This is a form of the *Kuhn-Tucker condition* [45, 278].

A pair $(\widetilde{x}, \widetilde{\lambda})$ with $0 \leq \widetilde{\lambda}$ is a *saddle point* of $L(\cdot)$ if for all x and all λ with nonnegative components $L(\widetilde{x}, \lambda) \leq L(\widetilde{x}, \widetilde{\lambda}) \leq L(x, \widetilde{\lambda})$. A necessary and sufficient condition for $\bar{x}$ to be a constrained minimum is that there is a $\bar{\lambda}$ with nonnegative components such that $(\bar{x}, \bar{\lambda})$ is a saddle point [278].

Consider a recursive algorithm for finding a saddle point. Let x_n, λ_n denote the nth estimates and use the following form of the Robbins-Monro procedure. Let G denote the box $\Pi[a_i, b_i]$. Suppose that there are known constants $A_i, i \leq K$, such that $\bar{\lambda}^i \leq A_i$. Then use

$$\begin{aligned} x_{n+1} &= \Pi_G \left[x_n + \epsilon_n \left(L_x(x_n, \lambda_n) + \xi_n^x \right) \right], \\ \lambda_{n+1,i} &= \lambda_{n,i} + \epsilon_n \left[q_i(x_n) + \xi_{n,i}^\lambda \right]_0^{A_i}, \quad i \leq K. \end{aligned} \tag{4.1}$$

Here ξ_n^x is the noise in the estimate of $L(x_n, \lambda_n)$ and $\xi_{n,i}^\lambda$ is the noise in the estimate of the value of $q_i(x_n)$. Using the methods in the subsequent chapters, the algorithm can be shown to converge to a saddle point under broad conditions. The original form is in [141] and convergence was also shown in [135]. Some numerical data is in [139]. If finite differences must be used to estimate the gradients, then the noise is of the form in the Kiefer-Wolfowitz procedure.

The so-called multiplier-penalty function methods of nonlinear programming [21] can also be adapted. Results for some such algorithms are in [135, 137, 140, 141].

2

Applications to Learning, Repeated Games, State Dependent Noise, and Queue Optimization

2.0 Outline of Chapter

This chapter deals with more specific classes of examples, which are of increasing importance in current applications in many areas of technology and operations research. They are described in somewhat more detail than those of Chapter 1. Each example should be taken as one illustration of a class of problems in a rapidly expanding literature. They demonstrate some of the great variety of ways that recursive stochastic algorithms arise. Section 1 deals with a problem in learning theory; in particular, the learning of an optimal hunting strategy by an animal, based on the history of successes and failures in repeated attempts to feed itself efficiently, and is typical of many "adaptive" models in biology. Section 2 concerns the "learning" or "training" of a neural network. In the training phase, a random series of inputs is presented to the network, and there is a desirable response to each input. The problem is to adjust the weights in the network to minimize the average distance between the actual and desired responses. This is done by a training procedure, where the weights are adjusted after each (input, output) pair is observed. Loosely speaking, the increments in the weights are proportional to stochastic estimates of the derivative of the error with respect to the weights.

Many problems in applications are of interest over a long time period. One observes a process driven by correlated noise and wishes to recursively adjust a parameter of the process so that some quantity is optimized. Suppose that for each parameter value, there is an associated stationary probability, and one wishes to minimize a criterion that depends on these stationary probabilities. This is to be done recursively. One observes the process over a (perhaps not very long) period of time at the current parameter value, appropriately adjusts the parameter value based on the measurements, and then continues to observe the same system over the next period of time, etc. For such problems, the current measurements are affected by the "state memory in the system," as well as by the external driving forces appearing during the current observational period. This involves the concept of Markov state-dependent noise. A versatile model for such a process is introduced in Section 3, and it is illustrated by a simple routing example.

Section 4 concerns recursive procedures for optimizing control systems or approximating their value functions. First, we consider the so-called Q-learning, where we have a controlled Markov chain model whose transition probabilities are not known and one wishes to learn the optimal strategy in the course of the system's operation. Then we consider the problem of approximating a value function by a linear combination of some basis functions. The process might be only partially observable and its law is unknown. It need not be Markov. In the next example, we are given a Markov control problem whose law is known, and whose control is parameterized. The problem is to find the optimal parameter value by working with a single long sample path. Although there are serious practical issues in the implementation of such algorithms, they are being actively investigated and are of interest since the law of the process is often not completely known or analytical computation is much too hard. All the noise types appear, including martingale difference, correlated, and Markov state-dependent noise.

The theme of optimization of an average cost over a long time interval is continued in Section 5, which concerns the optimization of the stationary performance of a queue with respect to a parameter of the service time distribution. The IPA (infinitesimal perturbation analysis) method is used to get the sample derivatives. Notice that the effective "memory" of the system at the start of each new iteration needs to be taken into account in the convergence analysis. Analogous models and issues arise in many problems in the optimization of queueing type networks and manufacturing systems.

The example in Section 6, called "passive stochastic approximation," is a type of nonlinear regression algorithm. One is given a sequence of "noisy" observations at parameter values that are determined *exogenously* (not subject to the control of the experimenter), and one seeks the parameter value at which the mean value of the observation is zero. The stochastic approximation algorithm is a useful nonparametric alternative to the classical method of fitting a nonlinear function (say in the mean square sense) and then locating the roots of that function. An interesting application of this idea to the problem of determining the concentration of a product in a chemical reactor subject to randomly varying inputs is in [274, 275]. Other interesting examples can be found in [16, 225, 250] and in Chapter 3.

Section 7 concerns learning in two-player repeated games. Each player has a choice of actions, knows its own payoffs, but not those of the other player. The player tries to learn its optimal strategy via repeated games and observations of the opponents strategies. Such problems have arisen in the economics literature. The results illustrate new ways in which complex recursive stochastic algorithms arise, as well as some of the phenomena to be expected.

2.1 An Animal Learning Model

This example concerns the purported learning behavior of an animal (say, a lizard) as it tries to maximize its reward per unit time in its food gathering activities. There are many variations of the problem that are covered by the basic results of convergence of stochastic approximation algorithms, and they illustrate the value of the general theory for the analysis of recursive algorithms arising in diverse fields. The basic model is taken from [177, 218]. The lizard has a fixed home location and can hunt in an arbitrary finite region. In the references, the region is restricted to a circle, but this is not needed.

Insects appear at random moments. Their weights are random and will be considered surrogates for the food value they contain. If an insect appears when the lizard is prepared to hunt (i.e., the lizard is at its home location), then the lizard makes a decision whether or not to pursue, with the (implicit) goal of maximizing its long-run return per unit time. We suppose that when the lizard is active (either pursuing or returning from a pursuit), no insect will land within the hunting range. It is not claimed that the model reflects conscious processes, but it is of interest to observe whether the learning behavior is consistent with some sort of optimization via a "return for effort" principle.

Next, we define the basic sequence of intervals between decision times for the lizard. At time zero, the lizard is at its home base, and the process starts. Let τ_1 denote the time elapsed until the appearance of the first insect. If the lizard pursues, let r_1 denote the time required to pursue and return to the home base and τ_2 the time elapsed after return until the next insect appears. If the lizard decides not to pursue, then τ_2 denotes the time elapsed until the appearance of the next insect. In general, let τ_n denote the time interval until the next appearance of an insect from either the time of appearance of the $(n-1)$st, if it was not pursued, or from the time of return from pursuing the $(n-1)$st, if it was pursued. If the lizard pursues, let r_n denote the time to pursue and return to the home base ready to hunt again. Let w_n denote the weight of the nth insect, and let J_n denote the indicator function of that insect being caught if pursued. Let I_n denote the indicator function of the event that the lizard decides to pursue at the nth opportunity. Let the time of completion of action on the $(n-1)$st decision be T_n. The pair (w_n, r_n) is known to the lizard when the insect appears. Define

$$W_n = \sum_{i=1}^{n-1} w_i I_i J_i,$$

and let $\theta_n = T_n/W_n$, which is the inverse of the sample consumption rate per unit time.

Suppose that the random variables (w_n, τ_n, r_n, J_n) for $n \geq 1$ are mutually independent in n and identically distributed, with bounded second moments, and $w_n > 0, \tau_n > 0$ with probability one. The four components can be correlated with each other. [In fact, the independence (in n) assumption can be weakened considerably without affecting the convergence of the learning algorithm.] Let there be a continuous function $p(\cdot)$ such that $E[J_n|r_n, w_n] = p(r_n, w_n) > 0$ with probability one, which is known to the lizard. The learning algorithm to be discussed performs as follows. Let $\bar{\theta} = \liminf_n ET_n/W_n$, where the infimum is over all (non-anticipative) strategies. The algorithm produces a sequence of estimates θ_n such that (Section 5.7) $\theta_n \to \bar{\theta}$ with probability one. The threshold $\bar{\theta}$ gives the optimal long-term decision rule: Pursue only if $r_n \leq \bar{\theta} w_n p(r_n, w_n)$. Thus, the optimal policy is to pursue only if the ratio of the required pursuit time to the expected gain is no greater than the threshold, and this policy is approximated asymptotically by the learning procedure.

The model can be generalized further by supposing that the estimates of $p(w_n, r_n)$ and of r_n are subject to "error." One introduces additional random variables to account for the errors.

Let $n \geq 1$, and let θ_1 be an arbitrary real number. If the lizard does not pursue at the nth opportunity, then

$$\theta_{n+1} = \frac{T_n + \tau_n}{W_{n+1}} = \frac{T_n + \tau_n}{W_n}. \tag{1.1a}$$

If the insect is pursued and captured, then

$$\theta_{n+1} = \frac{T_n + \tau_n + r_n}{W_{n+1}} = \frac{T_n + \tau_n + r_n}{W_n + w_n}. \tag{1.1b}$$

If the insect is pursued but not captured, then

$$\theta_{n+1} = \frac{T_n + \tau_n + r_n}{W_n}. \tag{1.1c}$$

Thus, if the insect is pursued,

$$\theta_{n+1} = \frac{T_n + \tau_n + r_n}{W_n + w_n} J_n + \frac{T_n + \tau_n + r_n}{W_n}(1 - J_n). \tag{1.1d}$$

The lizard compares the conditional expectations, given (r_n, w_n, θ_n) of the right sides in (1.1a) and (1.1d) and chooses the action that gives the minimum.

Define $\epsilon_n = 1/W_n$. Then ϵ_n decreases only after a successful pursuit. Either (1.2a) or (1.2b) will hold, according to the choice of nonpursuit or pursuit:

$$\theta_{n+1} = \theta_n + \epsilon_n \tau_n, \tag{1.2a}$$

$$\theta_{n+1} = \theta_n + \epsilon_n(\tau_n + r_n) - \epsilon_n \theta_n w_n J_n + O(\epsilon_n^2) J_n. \tag{1.2b}$$

Under the conditions to be imposed, $\epsilon_n \to 0$ with probability one and the term $O(\epsilon_n^2)$ in (1.2b) can be shown to be asymptotically unimportant relative to the other terms. It does not contribute to the limit, and for notational simplicity, it will be ignored henceforth. Thus, if the insect is pursued, neglecting the "small" $O(\epsilon_n^2)$ term, we have

$$\theta_{n+1} = \theta_n + \epsilon_n(\tau_n + r_n) - \epsilon_n \theta_n w_n p(r_n, w_n) + \epsilon_n \theta_n w_n (p(r_n, w_n) - J_n).$$

Finally, with $K_n = I_{\{r_n - \theta_n w_n p(r_n, w_n) < 0\}}$,

$$\begin{aligned} \theta_{n+1} = \theta_n & + \epsilon_n \tau_n + \epsilon_n \min\{r_n - \theta_n w_n p(r_n, w_n), 0\} \\ & + \epsilon_n \theta_n w_n \left(p(r_n, w_n) - J_n\right) K_n. \end{aligned} \tag{1.3}$$

The problem is interesting partly because the step sizes decrease randomly.

For fixed nonrandom θ, define the mean value

$$\bar{g}(\theta) = E\tau_n + E\min\{r_n - \theta w_n p(r_n, w_n), 0\},$$

and define the noise term

$$\xi_n = \tau_n + \min\{r_n - \theta_n w_n p(r_n, w_n), 0\} - \bar{g}(\theta_n) + \theta_n w_n \left(p(r_n, w_n) - J_n\right) K_n.$$

For nondegenerate cases, $\bar{g}(\cdot)$ is Lipschitz continuous, positive for $\theta = 0$, approximately proportional to $-\theta$ for large θ, and there is a unique value $\theta = \bar{\theta}$ at which $\bar{g}(\theta) = 0$. Rewrite the algorithm as

$$\theta_{n+1} = \theta_n + \epsilon_n \bar{g}(\theta_n) + \epsilon_n \xi_n. \tag{1.4}$$

The mean ODE that determines the asymptotic behavior is $\dot{\theta} = \bar{g}(\theta)$.

The value of ϵ_n and the final form of the algorithm are simply consequences of the representation used for the iteration; in particular, the definition $\epsilon_n = 1/W_n$ This representation allows us to put the iteration into the familiar form of stochastic approximation, so that a well-known theory can be applied.

2.2 A Neural Network

In the decision problem of Example 3 in Subsection 1.1.3, a sequence of patterns appeared, with the nth pattern denoted by y_n. The patterns themselves were not observable, but at each time n, one could observe random ϕ_n that is correlated with the y_n. We sought an affine decision rule (an affine function of the observables) that is best in the mean square sense. The statistics of the pairs (y_n, ϕ_n) were not known. However, during a training period many samples of the pairs (y_n, ϕ_n) were available, and a recursive linear least squares algorithm was used to sequentially get the optimal weights for the affine decision function. Thus, during the training period, we used a sequence of inputs $\{\phi_n\}$ and chose θ so that the outputs $v_n = \theta'\phi_n$ matches the sequence of correct decisions y_n as closely as possible in the mean square sense. Neural networks serve a similar purpose, but the output v_n can be a fairly general nonlinear function of the input [8, 97, 193, 205, 253]. The issue of large dimensionality is discussed in [171] via the random directions Kiefer–Wolfowitz procedure (see Chapters 5 and 10). A related "Kohonen" algorithm is treated in [73].

In this model, there are two layers of "neurons," the first layer having K neurons and the second (output layer) having only one neuron. The input at time n to each of the K neurons in the first layer is a linear combination of the observable random variables at that time. The output of each of these neurons is some nonlinear function of the input. The input to the neuron in the output layer is a linear combination of the outputs of the first layer, and the network output is some nonlinear function of the input to that output neuron. For a suitable choice of the nonlinear function (say the sigmoid), any continuous vector-valued map between input and output can be approximated by appropriate choices of the number of neurons and the weights in the linear combinations [97, 253]. In practice, the number of neurons needed for a good approximation can be very large, and much insight into the actual problem of concern might be required to get an effective network of reasonable size. Some more critical discussion from a statistical perspective is in [51, 205].

The training of the network can be described in terms of its input, output, and the desired and actual relationships between them. The definitions are illustrated in Figure 2.1.

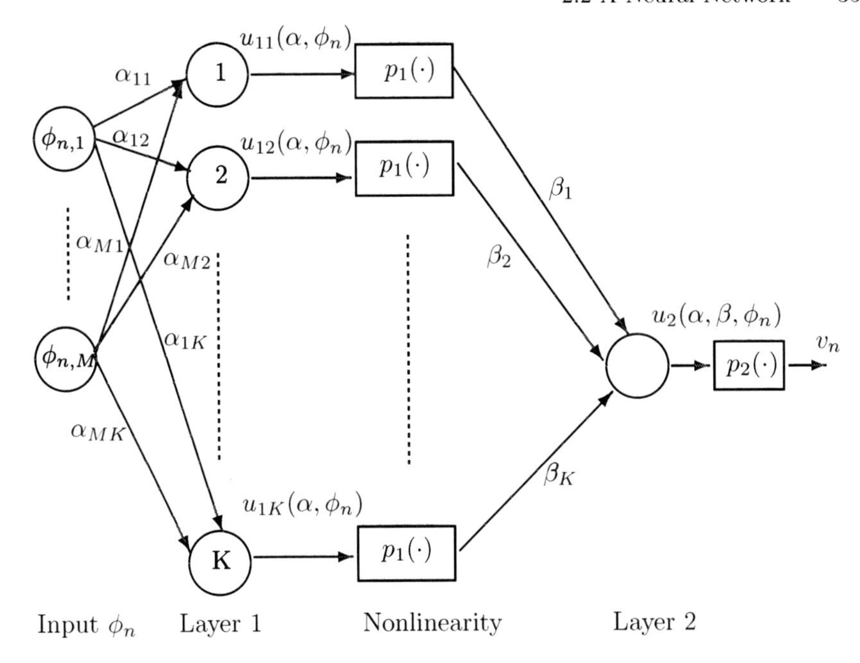

Figure 2.1. A two-layer neural network.

First we describe the network with fixed weights. There is a sequence of M-dimensional input vectors, the one at time n being $\phi_n = (\phi_{n,1}, \ldots, \phi_{n,M})$, with associated actual network outputs v_n (that can be observed) and desired outputs y_n (that are not generally observable in applications, at least after the training period). If the network is well designed, then v_n is a "good approximation" to y_n. There are weights α_{ij} such that the input at time n to "neuron" j in the first layer is

$$u_{1j}(\alpha, \phi_n) = \sum_{i=1}^{M} \alpha_{ij}\phi_{n,i}, \quad j = 1, \ldots, K.$$

The output of this neuron is $p_1(u_{1j}(\alpha, \phi_n))$, where $p_1(\cdot)$ is a real-valued antisymmetric nondecreasing and continuously differentiable function of a real variable, which is positive for positive arguments and is often chosen to be the sigmoid function. Denote the derivative by $\dot{p}_1(\cdot)$. There are weights β_i such that the input to the single second layer or "output" neuron is the linear combination of the outputs of the first layer

$$u_2(\alpha, \beta, \phi_n) = \sum_{j=1}^{K} \beta_j p_1(u_{1j}(\alpha, \phi_n)).$$

The output of the second layer neuron is $u(\alpha, \beta, \phi_n) = p_2(u_2(\alpha, \beta, \phi_n))$, where $p_2(\cdot)$ has the properties of $p_1(\cdot)$. Denote the derivative of $p_2(\cdot)$ by

$\dot{p}_2(\cdot)$. One wishes to "train" the network to minimize the error between the desired and the actual outputs.

Suppose that there is a "training" period in which the set of pairs (input, desired output) $\{(\phi_n, y_n), n = 1, \ldots\}$, as well as the actual outputs $\{v_n\}$ are available. For simplicity, let us suppose that the input-output pairs (y_n, ϕ_n) are mutually independent and that its distribution does not depend on n. The actual situations that can be handled are far more general and can be seen from the convergence proofs in the following chapters. Let $\theta = (\alpha, \beta)$ denote the r-vector of weights that is to be chosen, and let θ_n denote the value used for the nth training session. Define $v_n = u(\theta_n, \phi_n)$. The weights are to be recursively adjusted to minimize the mean square error $E[y - u(\theta, \phi)]^2$. Of course, the results are not confined to this error function. Define the sample mean square error $e(\theta, \phi, y) = (1/2)[y - u(\theta, \phi)]^2$ and the sample error $e_n = [y_n - v_n]$.

The basic adaptive algorithm has the form

$$\theta_{n+1,i} = \theta_{n,i} - \epsilon_n \frac{\partial e(\theta_n, \phi_n, y_n)}{\partial \theta^i} = \theta_{n,i} + \epsilon_n e_n \frac{\partial u(\theta_n, \phi_n)}{\partial \theta^i}, \quad i = 1, \ldots, r.$$

Using the formula

$$v = u(\alpha, \beta, \phi) = p_2(u_2(\alpha, \beta, \phi)) = p_2\left(\sum_{j=1}^{K} \beta_j p_1(u_{1j}(\alpha, \phi))\right),$$

the derivatives are evaluated from the formulas for repeated differentiation

$$\begin{aligned}\frac{\partial u(\theta, \phi)}{\partial \beta_j} &= \dot{p}_2(u_2(\alpha, \beta, \phi)) p_1(u_{1j}(\alpha, \phi)),\\ \frac{\partial u(\theta, \phi)}{\partial \alpha_{ij}} &= \dot{p}_2(u_2(\alpha, \beta, \phi)) \beta_j \dot{p}_1(u_{1j}(\alpha, \phi)) \phi^i,\end{aligned}$$

where ϕ^i denotes the ith component of the canonical input vector ϕ.

Under appropriate conditions, the ODE that characterizes the asymptotic behavior is

$$\dot{\theta}^i = \overline{g}^i(\theta) = -\frac{E\partial e(\theta, \phi, y)}{\partial \theta^i}, \quad i = 1, \ldots, r.$$

Being a gradient procedure, θ_n converges to a stationary point of this ODE. In applications there are usually many local minima. If it appears that the iteration is "hung up" at a local minimum, then several training runs might be needed, each with a different initial condition. The time required for training can be very long if there are many weights to be adjusted.

The neurons in the first layer can be connected to one another, and feedback can be incorporated from the output back to the first layer. An interesting application of this "interconnected" form to a problem in nonlinear filtering for processes defined by stochastic difference equations is in [170]. Interesting stochastic approximation algorithms motivated by the training of neural networks are in [7, 13].

2.3 State-Dependent Noise: A Motivational Example

Most of the noise processes that have appeared in the stochastic approximation algorithms up to this point in the book have been martingale differences, where $E[Y_n|\theta_0, Y_i, i < n]$ depended on θ_n but not otherwise random. More complicated noise processes are common occurrences. For example, suppose that the sequence of patterns in Example 3 of Section 1.1 is correlated or that the training data in the neural network example of Section 2 is correlated, but that the distributions do not depend on the stochastic approximation algorithm. Such noise processes are called *exogenous*. Example 4 of Section 1.1 provides a similar example, if the $\{\chi^m\}$ in (1.1.20) are correlated, but determined purely by effects "external" to the algorithm. In many important applications, the evolution of the effective noise process depends more intimately on the iterate (also to be called the *state*), and there is a reasonably long-term "memory" in this dependence. The following adaptive "routing" example taken from [136] illustrates the point in a simple way. The adaptive queueing problem discussed in Section 5 and the other references listed there show that such models are of considerable interest. Indeed, they are current canonical forms used in the optimization of queueing and manufacturing systems where the cost criterion is an average performance over the infinite time interval. The time-varying parameter tracking problem in [150] (see also Chapter 3) is another example of a complicated model in stochastic approximation that can be treated with this type of noise model. The example and the specific assumptions given in this section are for motivational purposes only.

To proceed, let us consider the following example. Suppose that calls arrive at a switch randomly, but at discrete instants $n = 1, 2, \ldots$ No more than a single call can arrive at a time and

$$P\{\text{arrival at time } n \mid \text{data to but not including time } n\} = \mu > 0.$$

The assumptions concerning single calls and discrete time make the formulation simpler, but the analogous models in continuous time are treated in essentially the same manner. To have a clear sequencing of the events, suppose that the calls are completed "just before" the discrete instants, so that if a call is completed at time n^-, then that circuit is available for use by a new call that arrives at time n. There are two possible routes for each call. The ith route has N_i lines and can handle N_i calls simultaneously. The sets of call lengths and interarrival times are mutually independent, and $\lambda_i > 0$ is the probability that a call is completed at the $(n + 1)$st instant, given that it is in the system at time n and handled by route i, and the rest of the past data. The system is illustrated in Figure 3.1.

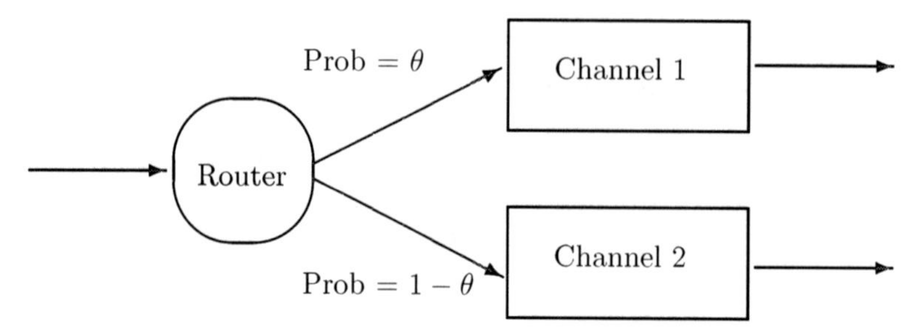

Figure 3.1. The routing system.

The routing law is "random," and it is updated by a stochastic approximation procedure with constant step size ϵ. Let $\xi_n^\epsilon = (\xi_{n,1}^\epsilon, \xi_{n,2}^\epsilon)$ denote the occupancies of the two routes at time n. If a call arrives at time $n+1$ then it is sent first to route 1 with probability θ_n^ϵ and to route 2 with probability $1 - \theta_n^\epsilon$. If all lines of the selected route are occupied at that time, the call is redirected to the other route. If the alternative route is also full, the call is lost from the system. Let $J_{n,i}^\epsilon$ be the indicator function of the event that a call arrives at time $n+1$, is sent to route i, and is accepted there. Our updating rule for θ_n^ϵ is

$$\theta_{n+1}^\epsilon = \Pi_{[a,b]} \left[\theta_n^\epsilon + \epsilon_n Y_n^\epsilon\right] = \Pi_{[a,b]} \left[\theta_n^\epsilon + \epsilon \left((1 - \theta_n^\epsilon) J_{n,1}^\epsilon - \theta_n^\epsilon J_{n,2}^\epsilon\right)\right], \quad (3.1)$$

where $0 < a < b < 1$ are truncation levels and $\Pi_{[a,b]}$ denotes the truncation. One could also use decreasing step sizes $\epsilon_n \to 0$, but in such problems one normally wishes to allow tracking of the optimal value of θ when the statistics change, in which case we cannot have $\epsilon_n \to 0$.

An examination of the right-hand side $\bar{g}(\theta)$ of the mean ODE in (3.4) shows that $\bar{g}(\theta) = 0$ is equivalent to (under stationarity) the statement that there is equal probability that each route is full at the time of a call's arrival. Thus, the adaptive algorithm (3.3) given below serves to equate the probabilities of being full in the long run. This might be called a "fairness to the user" criterion. Many other design goals can be realized with appropriate forms of the algorithm.

The occupancies ξ_n^ϵ (and the random acceptances and routing choices) determine the effective noise in the system, and the evolution of $\{\xi_n^\epsilon\}$ depends on $\{\theta_n^\epsilon\}$ in a complicated way and with significant memory. The dependence is of Markovian type in that

$$P\left\{\xi_{n+1}^\epsilon = \tilde{\xi} \mid \xi_i^\epsilon, \theta_i^\epsilon, i \le n\right\} = P\left\{\xi_{n+1}^\epsilon = \tilde{\xi} \mid \theta_n^\epsilon, \xi_n^\epsilon\right\}. \quad (3.2)$$

Define $v_i = (1 - \lambda_i)^{N_i}$. If a call is assigned to route 1 at time $n+1$ and is not accepted there then (a) the call arrives at $(n+1)$; (b) it is assigned to route 1; (c) route 1 is full at time n and there are no departures on the interval $[n, n+1)$. Thus, we have the conditional expectations

$$\begin{aligned} E\left[J_{n,1}^\epsilon \mid \theta_i^\epsilon, \xi_i^\epsilon, i \le n\right] &= \mu\theta_n^\epsilon \left[1 - v_1 I_{\{\xi_{n,1}^\epsilon = N_1\}}\right] \equiv \gamma_1(\theta_n^\epsilon, \xi_n^\epsilon), \\ E\left[J_{n,2}^\epsilon \mid \theta_i^\epsilon, \xi_i^\epsilon, i \le n\right] &= \mu\left(1 - \theta_n^\epsilon\right)\left[1 - v_2 I_{\{\xi_{n,2}^\epsilon = N_2\}}\right] \equiv \gamma_2(\theta_n^\epsilon, \xi_n^\epsilon). \end{aligned}$$

Define

$$\begin{aligned} g(\theta, \xi) &= (1-\theta)\gamma_1(\theta, \xi) - \theta\gamma_2(\theta, \xi), \\ \delta M_n &= \left[(1-\theta_n^\epsilon)J_{n,1}^\epsilon - \theta_n^\epsilon J_{n,2}^\epsilon\right] - g(\theta_n^\epsilon, \xi_n^\epsilon). \end{aligned}$$

Thus, for the Y_n defined in (3.1),

$$E\left[Y_n^\epsilon | \theta_i^\epsilon, \xi_i^\epsilon, i \le n\right] = g(\theta_n^\epsilon, \xi_n^\epsilon).$$

Rewrite the algorithm as

$$\theta_{n+1}^\epsilon = \Pi_{[a,b]}\left[\theta_n^\epsilon + \epsilon\left(g(\theta_n^\epsilon, \xi_n^\epsilon) + \delta M_n\right)\right]. \tag{3.3}$$

The δM_n are martingale differences, but the conditional mean $g(\theta_n^\epsilon, \xi_n^\epsilon)$ is not simple, since the time evolution of the sequence $\{\xi_n^\epsilon\}$ depends heavily on the time evolution of the iterate sequence $\{\theta_n^\epsilon\}$. Nevertheless, as will be seen in Chapter 8, effective methods are available to show that the effects of the noise disappear in the limit as $\epsilon \to 0$ and $n \to \infty$.

To study the asymptotic properties of θ_n^ϵ, it is useful to introduce the "fixed-θ" process as follows. Let $\{\xi_n(\theta)\}$ denote the Markov chain for the occupancy levels that would result if the parameter θ_n^ϵ were held constant at the value θ. The $\{\xi_n(\theta)\}$ is an ergodic Markov chain; we use E^θ to denote the expectations of functionals with respect to its unique stationary measure.

If ϵ is small, then θ_n^ϵ varies slowly and, loosely speaking, the "local" evolution of the $g(\theta_n^\epsilon, \xi_n^\epsilon)$ can be treated as if the θ_n^ϵ were essentially constant. That is, for small ϵ and Δ and $\theta_n^\epsilon = \theta$, a law of large numbers suggests that we have approximately

$$\epsilon \sum_{i=n}^{n+\Delta/\epsilon} g(\theta_i^\epsilon, \xi_i^\epsilon) \approx \Delta E^\theta g(\theta, \xi_n(\theta)).$$

Then the mean ODE that characterizes the limit behavior for small ϵ and large time is

$$\dot{\theta} = \bar{g}(\theta), \text{ for } \theta \in [a, b], \tag{3.4}$$

where

$$\bar{g}(\theta) = E^\theta g(\theta, \xi_n(\theta))$$

is a continuous function of θ. This will be established in Chapter 8. If the solution of (3.4) tries to exit the interval $[a, b]$, then it is stopped on the boundary.

Comment. In this example, the ultimate goal of the algorithm is equating the stationary probabilities that the routes will be full on arrival of a call. Yet the iteration proceeds in real time using the history of the rejections for the updating, and each update uses a different value of the parameter θ. So, the observations are not unbiased estimates of the "mean error function" $\bar{g}(\theta)$ at any value of θ. The process $\{\xi_n\}$ is analogous to a sufficient statistic in that it encapsulates the effects of the past on the future performance of the algorithm. The heuristic discussion shows that because θ_n varies slowly for small ϵ, the stationary averages at the current parameter values are actually being estimated increasingly well as ϵ decreases and time increases. This idea will be formalized in Chapters 8 and 9, in which the Markov property (3.2) will play an important role. An increasing number of applications of stochastic approximation concern long-term averages, and use observations on a single sample path taken at instants of time going to infinity, each observation being taken at a new parameter value. In these cases, the "effective memory" process ξ_n and their fixed-θ versions $\xi_n(\theta)$ will play a crucial role. The examples in the next two sections illustrate various applications of the state-dependent noise model to the optimization of discrete event systems.

2.4 Learning Optimal Controls and Value Functions

In this section, three classes of learning algorithms for controlled processes will be discussed. They concern problems of estimating a cost function or an optimal control when analytical methods cannot be used effectively. They are all based on simulation. Such methods are called for if the transition probabilities are not known, but the processes can be simulated or physical samples observed under chosen controls. Even if the transition probabilities are known, the problem of analytic computation might be too hard, and one might seek an approximation method that exploits the possibilities of simulation. The intention in each case is to introduce the subject by focusing on one basic approach. The entire area is under active study, and much remains to be done to get broadly useful algorithms. There is a growing literature on these topics and a recent survey and additional references are in [208].

The problem of Subsection 4.1 concerns finding an optimal feedback control and value function for a Markov chain by the method of Q-learning. It involves martingale difference noise. If the number of (state, control) pairs is large, as it usually is, then this procedure will converge slowly.

The problem of Subsection 4.2 concerns the approximation of a value function for a fixed control, and involves correlated "exogenous" noise. Such approximation methods for the value function can be used as the basis of recursive algorithms which approximate an optimal control by following the logic of the approximation in policy space procedure [22] or some other gradient-type procedure to get a sequence of controls which (hopefully) approximate the optimal one. Consider the following sequential approach. For a given control, get a good approximation to the value function via the method of Subsection 4.2. Then, assuming that this approximation is the true value function, apply the approximation in policy space method to get an improved control, then approximate the value function under the new control by a method such as that in Subsection 4.2, etc. Such methods go by the generic name of *actor-critic* [208]. Obviously, one cannot actually compute the approximation of Subsection 4.2 of the value function for each control. References [112, 113] concern a two-time-scale stochastic approximation approach to this problem. Here, the control is parameterized and the iteration on the control occurs relatively infrequently, which allows the algorithm for computing the approximation to the value function to nearly converge between control updates. The method essentially approximates the cost function for the current parameter value via a procedure like that in Subsection 4.2, and then uses that approximation to get the next value by a scheme such as approximation in policy space. One expects such two-time-scale methods to be slow, unless a lot of attention is given to the details concerning the time between control updates and the nature of the update itself. The paper [221] is concerned with various efficient adaptations of the actor-critic method and successfully illustrates the ideas on several "balancing" type problems.

The problem of Subsection 4.3 concerns the optimization of a stationary cost for a controlled Markov chain, and involves Markov state-dependent noise of the type introduced in Section 3. The paper [174] concerns getting the optimal parameter value for a parameterized control for the average cost per unit time problem. It uses the behavior of the process between returns to some fixed state to estimate the derivatives of the stationary cost with respect to the parameter. The control parameter is updated at the start of each such cycle. If the state space is large, then the return times and the variances of the estimators will generally be large as well. The paper [172] applies the approximation methods to a problem in communications and presents data showing the convergence. The book [232] contains an interesting historical perspective on the origin of the various methods.

2.4.1 Q-Learning

This example concerns the adaptive optimization of a control system governed by a Markov chain, a problem that is of increasing interest in robotic learning and artificial intelligence. Let $p(i, j|d)$ be transition probabilities for a controlled finite-state Markov chain, where d denotes the control variable and it takes values in a finite set $U(i)$ for each i. Let u_n denote the control action taken at time n and ψ_n the state of the chain at time n. Let

$\mathcal{F}_n$ denote the minimal σ-algebra generated by $\{\psi_j, u_j, j < n, \psi_n\}$. The control action taken at each time n might simply be a function of the state ψ_n at that time. More generally, it can be a random function of the past history. In the latter case, it is assumed to be *admissible* in the sense that it is chosen according to a probability law depending on the history up to the present. That is, it is selected by specifying the conditional probabilities

$$P\{u_n = d | \mathcal{F}_n\}.$$

We use π to denote an admissible control policy.

Given initial state i, admissible control policy π, and discount factor $0 < \beta < 1$, the cost criterion is

$$W(i, \pi) = E_i^\pi \sum_{n=0}^{\infty} \beta^n c_n,$$

where c_n is the real-valued cost realized at time n and E_i^π denotes the expectation under initial state i and policy π. It is supposed that the current cost depends only on the current state and control action in that for all c

$$P\{c_n = c | \psi_j, u_j, j < n, \psi_n = i, u_n = d\} = P\{c_n = c | \psi_n = i, u_n = d\},$$

where the right side does not depend on n. Henceforth, for notational convenience, $c_{n,id}$ is used to denote the random cost realized at time n when (state, control)$=(i, d)$. Conditioned on the current state being i and the current control action being d, we suppose that the current cost $c_{n,id}$ has a finite mean $\bar{c}_{id}$ and a uniformly bounded variance.

Then there is an optimal control that depends only on the current state. Define $V(i) = \inf_\pi W(i, \pi)$. Then the Bellman (dynamic programming) equation for the minimal cost is

$$V(i) = \min_{d \in U(i)} \left[\bar{c}_{id} + \beta \sum_j p(i, j|d) V(j) \right]. \tag{4.1}$$

The minimizing controls in (4.1) are an optimal policy. In other words, when in state i, the optimal control is the (or any, if nonunique) minimizer in (4.1) (see [18, 198]).

The so-called Q-learning algorithm, to be described next, was motivated by the problem of adaptive optimization of the Markov chain control problem, when the transition probabilities are not known, but the system can be simulated or observed under any choice of control actions. The idea originated in [251], and proofs of convergence are in [234, 252]; the former is quite general; see also [5]. The analogous algorithm for the average cost per unit time problem is discussed in [1, 173].

The algorithm involves recursively estimating the so-called Q-functions $\bar{Q}_{id}$, where $\bar{Q}_{id}$ is the cost given that we start at state i, action $d \in U(i)$

is taken, and then the optimal policy is used henceforth. This is to be distinguished from the definition of $V(i)$, which is the value given that we start at state i and use the optimal policy at all times. One must estimate the quantity $\bar{Q}_{id}$ for each state and control action, but this is easier than estimating the transition probabilities for each i, j, d. Since $\min_{d \in U(i)} \bar{Q}_{id} = V(i)$, an application of the principle of optimality of dynamic programming yields the relationship

$$\bar{Q}_{id} = \bar{c}_{id} + \beta \sum_j p(i, j|d) \min_{v \in U(j)} \bar{Q}_{jv}. \tag{4.2}$$

A stochastic approximation algorithm is to be used to recursively estimate the Q-values. The dimension of the problem is thus the total number of (state, action) pairs. The algorithm can use either a single simulated process, an actual physical control process, or multiple simultaneous processes. For simplicity, suppose that only a single control process is simulated. At each step, one observes the value i of the current state of the Markov chain and then selects a control action d. Then the value of the state at the next time is observed. At that time, the Q-value for only that particular (state, action) pair (i, d) is updated. Thus the algorithm is asynchronous.

Suppose that the pair (i, d) occurred at time n, and the next state ψ_{n+1} is observed. Let $Q_n = \{Q_{n,xd}; x, d : d \in U(x)\}$ be the current estimator of $\bar{Q}$. Then we update as

$$Q_{n+1,id} = Q_{n,id} + \tilde{\epsilon}_{n,id} \left[c_{n,id} + \beta \min_{v \in U(\psi_{n+1})} Q_{n,\psi_{n+1}v} - Q_{n,id} \right], \tag{4.3}$$

where $\tilde{\epsilon}_{n,id}$ is the step size and is chosen as follows. Let ϵ_n be a sequence of positive real numbers such that $\sum \epsilon_n = \infty$, and let w_{id} be positive real numbers. Define $\epsilon_{n,id} = w_{id}\epsilon_n$. The step size used at the current iteration depends on the number of times that the particular current (state, action) pair has been seen before. If the current (state, action) pair is (i, d), and this pair has been seen $(k-1)$ times previously in the simulation, then the step size used in the current updating of the estimate of $\bar{Q}_{id}$ is $\tilde{\epsilon}_{n,id} = \epsilon_{k,id}$. More formally, let $k(i, d, n)$ denote the number of times the pair (i, d) has been seen before or at time n, and define $\tilde{\epsilon}_{n,id} = \epsilon_{k(i,d,n),id}$. If (i, d) is the (state, action) pair seen at time n, then for $(j, v) \neq (i, d)$, there is no update at n and we have $Q_{n+1,jv} = Q_{n,jv}$. In Chapter 12, where the proof of convergence will be given, the values of the estimators of the $\bar{Q}_{id}$ will be truncated at $\pm B$ for some suitably large B. As will be seen, this simplifies the proofs and causes no loss of generality. In fact $|\bar{Q}_{id}| \leq \max_{i,d} |\bar{c}_{i,d}|/(1-\beta)$. Constant step sizes can also be used, and this is done in the convergence proof in Chapter 12. The proof for the decreasing step size case is virtually the same.

For a vector $Q = \{Q_{id}; i, d : d \in U(i)\}$, define

$$T_{id}(Q) = E\left[c_{n,id} + \beta \min_{v \in U(\psi_{n+1})} Q_{\psi_{n+1}v} \middle| \mathcal{F}_n, u_n = d, \psi_n = i\right].$$

By the Markov property,

$$T_{id}(Q) = \bar{c}_{i,d} + \beta \sum_j p(i,j|d) \min_{v \in U(j)} Q_{jv}.$$

With the definition of the "noise"

$$\delta M_n = \left[c_{n,id} + \beta \min_{v \in U(\psi_{n+1})} Q_{n,\psi_{n+1}v}\right] - T_{id}(Q_n),$$

we can write

$$Q_{n+1,id} = Q_{n,id} + \widetilde{\epsilon}_{n,id}\left[T_{id}(Q_n) - Q_{n,id} + \delta M_n\right]. \tag{4.4}$$

Note that $E[\delta M_n|\mathcal{F}_n, u_n = d] = 0$, and that the conditional variance $E[\delta M_n^2|\mathcal{F}_n, u_n = d]$ is bounded uniformly in n, ω.

The map $T(\cdot)$ is a Lipschitz continuous contraction with respect to the sup norm and $\bar{Q}$ is the unique fixed point [18, 190]. Suppose that each (state, action) pair is visited infinitely often with probability one. Then one can show that Q_n converges to $\bar{Q}$ with probability one. Let $N_{id}(n, n+m)$ be the number of times that the pair (i,d) is returned to in the time interval $[n, n+m]$. Suppose that for any pairs (i,d) and (j,v), the ratios

$$N_{id}(n, n+m)/N_{jv}(n, n+m)$$

are bounded away from both zero and infinity as n and m go to infinity. Then, it is shown in Chapter 12 that for the constrained algorithm, there is a diagonal matrix $D(t)$ whose diagonal components $D_{id}(t)$ are bounded and strictly positive such that the mean ODE that characterizes the limit points is

$$\dot{Q}_{id} = D_{id}(t)\left(T_{id}(Q) - Q_{id}\right) + z_{id}, \tag{4.5}$$

where the $z_{id}(t)$ term is zero if $-B < Q_{id}(t) < B$; otherwise it serves only to keep $Q_{id}(\cdot)$ from leaving the interval $[-B, B]$ if the dynamics try to force it out. If B is large enough, then the limit point is $\bar{Q}$.

2.4.2 *Approximating a Value Function*

In this subsection, we will illustrate a class of algorithms that is of current interest for recursively approximating a discounted cost function for a stochastic process, as a function of the initial condition. If the process is controlled, then the control is fixed. The class of approximating algorithms is known as $TD(\lambda)$, $0 < \lambda \le 1$ [235] (TD standing for temporal difference).

We will concentrate on the class $TD(1)$, and discuss some of the basic ideas. The classes $\lambda < 1$ will only be commented on briefly.

We will start with a Markov model. This Markov assumption can be weakened considerably, which is an advantage of the approach. Let $\{X_n\}$ denote a Markov chain with values in some compact topological space, a time-invariant transition function, and a unique invariant measure $\mu(\cdot)$. Let E_μ denote the expectation with respect to the measure of the stationary process. For a bounded and continuous real-valued function $k(\cdot)$, the cost is

$$W(x) = E_x \sum_{n=0}^{\infty} \beta^n k(X_n, X_{n+1}), \quad 0 < \beta < 1.$$

We wish to approximate $W(x)$ by a finite sum $\widetilde{W}(x,w) = \sum_{i=1}^{p} w_i \phi_i(x) = w'\phi(x)$, where the real-valued functions $\phi_i(\cdot)$ are bounded and continuous and the optimal weight w is to be determined by an adaptive algorithm. The approximation is in the sense of

$$\min_w e(w), \quad \text{where} \quad e(w) = E_\mu \left[W(X_0) - w'\phi(X_0)\right]^2 /2, \tag{4.6}$$

where X_0 denotes the canonical "stationary random variable." It is always supposed that $E_\mu \phi(X_0)\phi'(X_0)$ is positive definite. A simulation based recursive algorithm of the gradient descent type will be used to get the optimal value of w. Whenever possible, one constructs such adaptive algorithms so that they are of the gradient descent type. Such simulation-based adaptive methods are motivated by the difficulty of computing cost functions for large or continuous state spaces.

Following the usual method [235], define the sequence

$$C_{n+1} = \beta\lambda C_n + \phi(X_{n+1}), \quad 0 \le \lambda \le 1, \tag{4.7}$$

and the difference

$$d_n = k(X_n, X_{n+1}) + \beta\widetilde{W}(X_{n+1}, w_n) - \widetilde{W}(X_n, w_n). \tag{4.8}$$

Until further notice, set $\lambda = 1$. The difference d_n is a "noisy" estimate of an approximation error in that

$$W(X_n) - E_{X_n}\beta W(X_{n+1}) = E_{X_n} k(X_n, X_{n+1}).$$

The adaptive algorithm is (which defines Y_n)

$$w_{n+1} = w_n + \epsilon_n d_n C_n = w_n + \epsilon_n Y_n, \tag{4.9}$$

where $\epsilon_n > 0, \sum_n \epsilon_n = \infty$. We will now heuristically justify the gradient descent assertion and motivate the convergence by deriving the mean ODE.

For large n, (4.9) can be well approximated by

$$w_{n+1} = w_n + \epsilon_n \left[k(X_n, X_{n+1}) + \beta\widetilde{W}(X_{n+1}, w_n) - \widetilde{W}(X_n, w_n) \right] \times \sum_{l=-\infty}^{n} [\lambda\beta]^{n-l}\phi(X_l). \tag{4.10}$$

The mean ODE is determined by the stationary mean value of the terms in (4.10), with w_n held fixed. Suppose that the iterates are bounded. Then, by Theorem 8.2.1 or 8.2.2 and the uniqueness of the stationary measure, the mean ODE is

$$\begin{aligned} \dot{w} &= \bar{g}(w) \\ &= E_\mu \left[k(X_n, X_{n+1}) + \beta\widetilde{W}(X_{n+1}, w) - \widetilde{W}(X_n, w) \right] \sum_{l=-\infty}^{n} [\lambda\beta]^{n-l}\phi(X_l). \end{aligned} \tag{4.11}$$

For the probability one convergence case, use Theorem 6.1.1.

To see why this limit form is correct, first note that w_n varies slowly for large n, Supposing that w_n is fixed at w for many iterates, the law of large numbers for Markov chains implies that the sample average of a large number of the terms Y_n in (4.10) is approximated by its stationary mean. Now, fixing $w_n = w$ and computing the stationary expectation of the terms in the coefficient of ϵ_n on right side of (4.10) yields

$$\begin{aligned} E_\mu \left[\beta\widetilde{W}(X_{n+1}, w) - \widetilde{W}(X_n, w) \right] \sum_{l=-\infty}^{n} [\lambda\beta]^{n-l}\phi(X_l) \\ = -\left[w' E\phi(X_n) \right] \phi(X_n). \end{aligned} \tag{4.12}$$

Since $k(\cdot)$ is real-valued,

$$E_\mu k(X_n, X_{n+1}) \sum_{l=-\infty}^{n} [\lambda\beta]^{n-l}\phi(X_l) = \sum_{l=0}^{\infty} [\lambda\beta]^l E_\mu \phi(X_0) E_{X_0} k(X_l, X_{l+1}). \tag{4.13}$$

Thus, for $\lambda = 1$, we see that the stationary mean of Y_n, with w_n fixed at w, is

$$\text{gradient}_w e(w) = -E_\mu \left[E_{X_0} \sum_{n=0}^{\infty} \beta^n k(X_n, X_{n+1}) - w'\phi(X_0) \right] \phi(X_0). \tag{4.14}$$

Hence the mean ODE is

$$\dot{w} = -\text{gradient}_w e(w) = \bar{g}(w), \tag{4.15}$$

and the algorithm is of the gradient descent type. If there is a concern that the iterates might not be bounded, then use the constrained algorithm

$$w_{n+1} = \Pi_H \left[w_n + \epsilon_n d_n C_n \right],$$

where H is a constraint set, perhaps a hyperrectangle. The mean ODE is just the projected form of (4.11). If H is a hyperrectangle, and the optimal point is inside H, then the solution to the ODE converges to it; otherwise, it converges to the closest point in H.

If the algorithm is not constrained, then the gradient descent characterization can be used to prove stability of the algorithm by using a perturbed Liapunov function, based on the Liapunov function $e(w)$ for (4.15). This is covered by Theorem 6.7.3.

An alternative formulation: less memory. In lieu of (4.6), consider the problem

$$\min e_0(w), \quad \text{where } e_0(w) = E_\mu \left[E_{X_0} k(X_0, X_1) - w'\phi(X_0)\right]^2 /2. \tag{4.16}$$

Here the cost to be approximated is simply $E_{X_0} k(X_0, X_1)$, which takes only one step of the future into account. A convergent recursive algorithm for computing the minimizer is

$$w_{n+1} = w_n + \epsilon_n \left[k(X_n, X_{n+1}) - w_n'\phi(X_n)\right] \phi(X_n), \tag{4.17}$$

which is also of the gradient descent form, since (with w_n fixed at w) the stationary mean of the right side of (4.17) is just the gradient of $e_0(w)$.

One can interpolate between the criteria (4.6) and (4.16), by constructing minimization problems where the cost takes more and more of the future into account in a discounted way. This is a motivation for the algorithm *TD*$(\lambda), \lambda < 1$; see [232, 235].

Comments and extensions. The evaluations of the mean values in (4.12) and (4.13) involve a lot of cancellations, using the fact that the stationary means of various different random variables that appeared are of opposite sign. Although many of the stationary means cancel, the associated random variables do not. This suggests a high noise level and slow convergence. The algorithm (4.9) could be constrained: For example, we could restrict $w_{n,i}$ to a finite interval $[a_i, b_i]$ or subject to other constraints. The time invariance of the transition function could be replaced by periodicity.

For a particularly useful extension, the chain needs to be only partly observable. For example, let $X_n = (\widehat{X}_n, \widetilde{X}_n)$, where only $\widehat{X}_n$ together with the costs $k(X_n, X_{n+1})$ are observable. Then, with $\phi(\widehat{X}_n)$ used in (4.7) in lieu of $\phi(X_n)$, the algorithm (4.9) finds the minimizer of $E_\mu \left[W(X_0) - w'\phi(\widehat{X})\right]^2$. The process need not be Markov, provided that some stationarity and mixing conditions hold. Then μ is replaced by the steady state probability and the E_x in the definition of $W(x)$ is replaced by conditioning on the data to the initial time. Such robustness is very useful, since "practical" processes often have "hidden states" or are not Markov. This is also covered by either Theorem 8.2.1 or 8.2.2 for the weak convergence case and Theorem 6.1.1 for the probability one convergence case.

2.4.3 Parametric Optimization of a Markov Chain Control Problem

Problem formulation. In this subsection, we are concerned with a controlled Markov chain and average cost per unit time problem, where the control is parameterized. We seek the optimal parameter value and do not use value function approximations of the type of Subsection 4.2. The approach is based on an estimate of the derivative of the invariant measure with respect to the parameter. This is an example of an optimization problem over an infinite time interval. The procedure attempts to approximate the gradient of the stationary cost with respect to the parameter. It is not a priori obvious how to estimate this gradient when the times between updates of the parameter must be bounded. We will exploit a useful representation of this derivative. The problem of minimizing a stationary cost is continued in the next section, for a queueing model, where an analog of the pathwise derivative of Example 4 of Subsection 1.1.4 will be used.

The process is a finite-state Markov chain $\{X_n\}$ with time-invariant and known transition probabilities $p(x, y|\theta)$ that depend continuously and differentiably on a parameter θ that takes values in some compact set. We will suppose that θ is real-valued and confined to some interval $[a, b]$. The general vector-valued case is treated simply by using the same type of estimate for each component of θ. Write $p_\theta(x, y|\theta)$ for the θ-derivative. Let the chain be ergodic for each θ, and denote the unique invariant measure by $\mu(\theta)$. Let $E_{\mu(\theta)}$ denote expectation under the stationary probability and E_x^θ the expectation, given parameter value θ and initial condition x. The objective is to obtain

$$\min_\theta e(\theta), \quad e(\theta) = E_{\mu(\theta)} k(X_0, X_1, \theta) = \sum_{x,y} \mu(x, \theta) p(x, y|\theta) k(x, y, \theta), \tag{4.18}$$

where $k(x, y, \cdot)$ is continuously differentiable in θ for each value of x and y, and $\mu(x, \theta)$ is the stationary probability of the point x. Our aim is the illustration of a useful general approach to optimization of an average cost per unit time criterion. The best approach is not clear at this time. But any algorithm must have some way of estimating the derivative of the stationary cost.

The invariant measures are hard to compute, so the minimization is to be done via simulation. Suppose that we try to minimize (4.18) by formally differentiating with respect to θ, and use the formal derivative as the basis of a gradient procedure. The gradient of $e(\cdot)$ at θ is $T_1(\theta) + T_2(\theta) + T_3(\theta)$, where (the subscript θ denotes derivative with respect to θ)

$$T_1(\theta) = \sum_{x,y} \mu(x, \theta) p(x, y|\theta) k_\theta(x, y, \theta),$$

$$T_2(\theta) = \sum_{x,y} \mu(x,\theta) p_\theta(x,y|\theta) k(x,y,\theta),$$

$$T_3(\theta) = \sum_{x,y} \mu_\theta(x,\theta) p(x,y|\theta) k(x,y,\theta).$$

For a gradient descent procedure the mean ODE would be

$$\dot{\theta} = -[T_1(\theta) + T_2(\theta) + T_3(\theta)]. \tag{4.19}$$

Write the stochastic algorithm as $\theta_{n+1} = \theta_n + \epsilon_n Y_n$, where Y_n is to be determined. Then, if θ_n is fixed at an arbitrary value θ, we need at least that the long term average of the Y_n is the right side of (4.19). The actual stochastic algorithm will be constructed with this in mind. If θ_n is held fixed at θ, then $T_1(\theta)$ is easy to estimate since, by the law of large numbers for a Markov chain, it is just the asymptotic average of the samples $k_\theta(X_n, X_{n+1}, \theta)$.

If $k(x,y,\theta)$ does not depend on the second variable y, then $T_2(\theta)$ does not appear. A naive approach to approximating $T_2(\theta)$ for the algorithm would be to simulate $p_\theta(X_n, X_{n+1}|\theta_n)k(X_n, X_{n+1}, \theta_n)$. If $\theta_n \approx \theta$, then by the law of large numbers for Markov chains, the sample average of many such terms will be close to the stationary expectation

$$\sum_{x,y} \mu(x,\theta) p(x,y|\theta) p_\theta(x,y|\theta) k(x,y,\theta).$$

But this is not $T_2(\theta)$ owing to the presence of the term $p(x,y|\theta)$. Thus, the correct form to simulate is $L(X_n, X_{n+1}, \theta_n)k(X_n, X_{n+1}, \theta_n)$, where $L(x,y,\theta) = p_\theta(x,y|\theta)/p(x,y|\theta)$ (which are assumed to be bounded). The stationary average of $L(X_n, X_{n+1}, \theta))k(X_n, X_{n+1}, \theta)$ is $T_2(\theta)$, as desired.

The term $T_3(\theta)$ cannot normally be computed explicitly and even its existence needs to be established for more general chains. A useful method of approximation by simulation is not at all obvious. The existence, characterization, and approximation, of this derivative were the subjects of [136, 148, 242, 262]. The method to be employed will be based on the results of [136], which covers the particular case of interest here. For a real-valued function $q(\cdot)$ of the chain and θ, define the stationary mean $\lambda_q(\theta) = \sum_x \mu(x,\theta) q(x,\theta)$. Then a useful representation of the derivative of $\lambda_q(\cdot)$ with respect to θ is: (see [136, Theorem 3])

$$\frac{d}{d\theta}\lambda_q(\theta) = \sum_{n=0}^{\infty} \mu'(\theta) P_\theta(\theta) P^n(\theta) \left[Q(\theta) - \lambda_q(\theta)\mathbf{e}\right], \tag{4.20}$$

where $\mathbf{e}$ is the vector whose components are all unity and $Q(\theta)$ denotes the vector $\{q(x,\theta)\}$. Note that the component of (4.20) that involves $\lambda_q(\theta)$ is zero. The sum (4.20) converges since $P^n(\theta) \to 0$ at a geometric rate.

In order to use the representation (4.20) as a basis for estimating $T_3(\theta)$, we need to simulate a sequence whose asymptotic expectation is (4.20).

Define $q(x,\theta) = E_x^\theta k(x, X_1, \theta)$. Before stating the algorithm, let us note that for $n \geq l$,

$$\begin{aligned} & E_x^\theta q(X_{n+1},\theta) L(X_l, X_{l+1}, \theta) \\ & \quad = \sum_{x_l} p^{(l)}(x, x_l|\theta) \sum_{x_{l+1}} p(x_l, x_{l+1}|\theta) L(x_l, x_{l+1}, \theta) \\ & \quad \times \sum_w p^{(n-l)}(x_{l+1}, w|\theta) q(w,\theta), \end{aligned} \tag{4.21}$$

where $p^{(n)}(\cdot)$ denotes the n-step transition probability. Averaging (4.21) with respect to the invariant measure yields

$$\sum_{x,y,w} \mu(x,\theta) p_\theta(x,y|\theta) p^{(n-l)}(y,w|\theta) q(w,\theta), \tag{4.22}$$

which is the $(n-l)$th summand of (4.20).

The above comments suggest one way of approximating (4.20) and a possible algorithm. Let $0 < \beta < 1$, but be close to unity. Consider the algorithm for approximating the optimal value of θ:

$$\begin{aligned} \theta_{n+1} &= \theta_n - \epsilon_n k_\theta(X_n, X_{n+1}, \theta_n) \\ & \qquad -\epsilon_n L(X_n, X_{n+1}, \theta_n) k(X_n, X_{n+1}, \theta_n) \\ & \qquad -\epsilon_n \left(k(X_n, X_{n+1}, \theta_n) - \lambda_n\right) C_n, \end{aligned} \tag{4.23}$$

$$\begin{aligned} \lambda_{n+1} &= \lambda_n + \epsilon_n' \left[k(X_n, X_{n+1}, \theta_n) - \lambda_n\right], \\ C_{n+1} &= \beta C_n + L(X_n, X_{n+1}, \theta_n), \end{aligned}$$

where we use $\epsilon_n' = q\epsilon_n$ for some $q > 0$. The update is done when X_{n+1} is observed.

Comments on the algorithm (4.23). The variable λ_n is intended to be an estimate of the optimal stationary cost. For large n, θ_n varies slowly. Suppose that it is fixed at a value θ. Then λ_n will converge to the stationary value $E_{\mu(\theta)} k(X_0, X_1, \theta)$. This suggests that the λ_n in the first line of (4.23) would not affect the mean values, asymptotically, since $E_{\mu(\theta)} L(X_0, X_1, \theta) = 0$. However, the "centering" role of λ_n in (4.23) is important in that it serves to reduce the variance of the estimates. The first coefficient of ϵ_n in the first equation in (4.23) is used to estimate $T_1(\theta)$, the second to estimate $T_2(\theta)$, and the third to estimate $T_3(\theta)$. Using (4.21) and the "discounted" definition of C_n, assuming that $\theta_n \approx \theta$ and $\lambda_n \approx \lambda_q(\theta)$, the stationary average of the third term is a discounted form of (4.20), namely,

$$\sum_{n=0}^{\infty} \beta^n \mu'(\theta) P_\theta(\theta) P^n(\theta) \left[Q(\theta) - \lambda_q(\theta)\mathbf{e}\right].$$

Thus for β close to unity, we have an approximation to (4.20) and $T_3(\theta)$. The noise in the algorithm (4.23) is state-dependent in the sense introduced in Section 3. For the constrained form of the algorithm any of the methods in the state-dependent-noise Section 8.4 can be used to justify the mean ODE and the convergence to a point close to the optimum for β close to unity.

Notes: Extension to diffusions or discretized diffusions. Representations analogous to (4.20) exist for Markov processes on more general state spaces [242], and this can be used as the basis of a recursive algorithm, just as (4.20) was. An alternative approach, based on Girsanov transformation techniques, is in [262]. It uses estimates computed over finite intervals $[n\Delta, n\Delta+\Delta)$. Reference [262] contains an extensive discussion of derivative estimators for jump-diffusions, where the drift function is subject to control. The method also works for reflected jump diffusions if the reflection direction and jumps are not controlled. In such "continuous state space" problems, one must usually approximate in the simulations, since the actual continuous-time trajectory cannot usually be simulated. The reference and [148] discusses discrete-time and Markov chain approximations and show that the derivative estimators will converge as the discretization parameter goes to zero.

2.5 Optimization of a GI/G/1 Queue

We now consider another approach to the optimization of a stationary cost function. The example concerns the optimization of the performance of a single server queue. The times of customer arrivals is a renewal process with bounded mean. For fixed θ, let $X_i(\theta)$ denote the time that the ith customer spends in the system, and let $K(\theta)$ be a known bounded real-valued function with a continuous and bounded gradient. The service time distribution is controlled by a real-valued parameter θ, which is chosen to minimize the sum of the average waiting time per customer and a cost associated with the use of θ. This is the cost function

$$L(\theta) = \lim_N \frac{1}{N} \sum_{i=1}^{N} EX_i(\theta) + K(\theta) \equiv \widehat{L}(\theta) + K(\theta). \tag{5.1}$$

The aim is to get the minimizing θ in the finite interval $[a, b]$. Generally, the values of $\widehat{L}(\theta)$ are very hard to compute. A viable alternative to the direct optimization of the unavailable (5.1) uses stochastic approximation. One would observe the queue over a long time period, and use the observational data (which are the times of arrival, departure, and service for each customer, up to the present). The data would be used to estimate the derivative of the cost function with respect to θ at the current value of

θ, yielding a stochastic approximation analog of the deterministic gradient descent procedure. We use "estimates of the derivative" in a loose sense, since each so-called estimate will be (perhaps strongly) biased. However, their cumulative effect yields the correct result.

This problem has attracted much attention (see [52, 78, 145, 160, 161], among others) and is typical of many current efforts to apply stochastic approximation to queueing and manufacturing systems and networks.

2.5.1 *Derivative Estimation and Infinitesimal Perturbation Analysis: A Brief Review*

The parameter θ is assumed to be real-valued throughout this section. A basic issue is that we can only update the estimators at finite times, although the cost function is of the "ergodic" type. Before proceeding further, let us review Example 4 of Section 1.1, where we were able to compute the "pathwise derivative" $\bar{U}^N(\theta)$. The derivative is said to be *pathwise*, because for each ω, the sample value $\bar{U}^N(\theta,\omega)$ is the derivative of the sample value $\bar{X}^N(\theta,\omega)$ with respect to θ. Let $\delta\theta$ be a "small" perturbation of θ. To get the formula (1.1.21) for the derivative at $\theta_n = \theta$, one *implicitly* solves for $\bar{X}^N(\theta)$ and $\bar{X}^N(\theta+\delta\theta)$, using the *same* vector of driving noise $\bar{\chi}$, computes the ratio $[\bar{X}^N(\theta+\delta\theta)-\bar{X}^N(\theta)]/\delta\theta$, and then lets $\delta\theta$ go to zero. By the rules of calculus, this procedure is what yields the final explicit formula (1.1.21). For the method to work, it is essential that the driving forces $\bar{\chi}$ be the same in the computations of $\bar{X}^N(\theta)$ and the $\bar{X}^N(\theta+\delta\theta)$ for all small $\delta\theta$. If explicit pathwise differentiation were not possible, then one might have to resort to a finite difference estimator, which suffers from the fact that its variance is inversely proportional to the square of the finite difference interval. As an aside, recall that under appropriate smoothness assumptions, finite differences with "common" random variables used for the simulations for the parameters θ_n and $\theta_n+\delta\theta$ are often almost as good as the pathwise derivative, as shown in the discussion connected with (1.2.6).

There is an analogy to this pathwise derivative approach that yields a very useful method for getting good estimates of derivatives of the desired functional of the queue. The approach for the queueing problem is not obvious, because the optimization is to be done via observations on one sample path (i.e., done "on line" using data from the physical process), and there are no "natural driving forces" analogous to the $\bar{\chi}$ that would allow the computation of the path for a perturbed value of θ. It turns out that one can construct suitable "driving forces" from the observations and then get a direct analog of what was done in Example 4 of Section 1.1 by a clever choice of the probability space. This is the subject of the important field *infinitesimal perturbation analysis*, initiated by Y.-C. Ho. The subject of IPA has rapidly grown, and there is a large amount of literature: see, for example, [84, 99, 101] and the references therein.

A few introductory comments on IPA will be made in the next paragraph, and then the results needed for the queueing problem at hand will be cited. The reader is referred to the literature for further reading. See also [145, 262, 148] for other results concerning pathwise derivatives that are useful in stochastic approximation.

Introduction to IPA: Computation of a pathwise derivative. Let θ be fixed until further notice. In this application, one observes a sample path of the physical process over a long time interval and seeks to construct an estimate of the derivative $L_\theta(\theta)$ from the sample data. Suppose, as a pure thought experiment, that one could observe simultaneously paths of the process corresponding to parameter values θ and $\theta + \delta\theta$, for all arbitrarily small $\delta\theta$, where the sequences $\{X_j(\theta), X_j(\theta + \delta\theta), j < \infty\}$ are defined on the same sample space. Suppose also that the path derivative exists for almost all realizations. In other words, the limit

$$\lim_{\delta\theta \to 0} \frac{X_j(\theta + \delta\theta) - X_j(\theta)}{\delta\theta} = D_j$$

exists with probability one for each j. Suppose also that

$$\frac{\partial E X_j(\theta)}{\partial \theta} = E \frac{\partial X_j(\theta)}{\partial \theta} = E D_j. \tag{5.2}$$

Then D_j is an unbiased estimator of $\partial E X_j(\theta)/\partial\theta$.

The idea is to estimate $\partial X_j(\theta)/\partial\theta$ at the given value of θ from a *single* sample path, with the parameter value θ used, analogously to what was done in Example 4 of Section 1.1. There are two substantial parts of the work. The first is to get the derivative of the sample path, and the second is to verify the validity of the interchange of differentiation and expectation in (5.2). The reward of success is that one has a useful estimator, a pathwise derivative, which is based only on the observed data and does not involve finite differences. In applications, each of these steps might involve a great deal of work. When these pathwise derivatives can be calculated, it is frequently found that their variance is smaller (often much smaller) than what one would obtain under alternative approaches. From the point of view of stochastic approximation, the smaller the variance of the estimator, the better the procedure. In view of these remarks, it is clear that IPA and related methods are important for applications to stochastic approximation.

Since we are only concerned with expectations, the precise nature of the probability space that is used to compute the pathwise derivative is unimportant to the final result. Any convenient space can be used, provided that the limit formulas depend only on the observations. We will now give a simple example to illustrate this, where we compute the sample θ-derivative of a random variable T with parameterized distribution function $F(\cdot|\theta)$.

For illustrative purposes, suppose that the distribution functions corresponding to parameter values θ and (for small $\delta\theta$) $\theta + \delta\theta$ are strictly monotone increasing. To compute a pathwise derivative, the same probability space must be used for all parameter values and this can be done in the following way. By the definition of a distribution function, the random variable χ defined by $\chi = F(T|\theta)$ has a uniform distribution on the interval $[0,1]$, and under our assumption on strict monotonicity, the values of χ and T can be computed uniquely one from the other and we can write $T = F^{-1}(\chi|\theta)$.

Now, with χ defined by $\chi = F(T|\theta)$, we can define the random variable $T(\theta+\delta\theta) = F^{-1}(\chi|\theta+\delta\theta)$, which has the distribution $F(\cdot|\theta+\delta\theta)$. The random variable χ serves as the "driving force," common to all values of $\delta\theta$, and it is constructed directly from the observed value of T. Thus, if the inverse function $F^{-1}(\cdot|\theta)$ is smooth enough in θ, we can get the representation of the pathwise derivative

$$\lim_{\delta\theta\to 0} \frac{F^{-1}(\chi|\theta+\delta\theta) - F^{-1}(\chi|\theta)}{\delta\theta} \equiv D.$$

Consider the simplest example, where $F(T|\theta) = 1 - e^{-T/\theta}$, the exponential distribution with mean θ. Then

$$F^{-1}(\chi|\theta) = -\theta \log(1-\chi),$$

and the derivative is simply $-\log(1-\chi)$.

In the problem under consideration, the derivative of the distribution of the sojourn time is not easily obtainable, but the values of the sojourn time of the ith customer depend on the service times of the customers that were in the queue and on the residual service time for the customer in service (if any) at the moment when that ith customer arrived. Let $\{T_n\}$ denote the sequence of (assumed mutually independent) service times, each of which has the distribution $F(\cdot|\theta)$. Define $\chi_n = F(T_n|\theta)$. Then χ_n are mutually independent and uniformly distributed on the unit interval, and they are independent of the interarrival intervals and the initial condition. The pathwise derivative $\partial T_n(\theta)/\partial\theta$ can be used to get suitable estimators of the θ-derivative of the mean customer sojourn times. The derivation is involved; we simply copy the estimators for our problem from the references, because we are more concerned with their use than their derivation.

2.5.2 The Derivative Estimate for the Queueing Problem

Some of the conditions that we will state will be used when the convergence is proved in Chapter 9. First, the sequence of service times will be defined in terms of "driving noises" that do not depend on θ. Then the formula for the pathwise derivative at parameter value θ will be stated and adapted for use in the stochastic approximation procedure. Let the parameterized service time distribution $F(\cdot|\theta)$ be weakly continuous in θ: That is for each bounded and continuous real-valued function $f(\cdot)$, $\int f(x)F(dx|\theta)$ is continuous in θ. Define the inverse function $F^{-1}(\cdot|\theta)$ by

$$F^{-1}(\chi|\theta) = \inf\{\zeta : F(\zeta|\theta) \geq \chi\},\ \chi \in [0,1],$$

and, for each value of χ, let the inverse have a θ-derivative that is continuous in θ (uniformly in χ). We denote the θ-derivative by $F_\theta^{-1}(\chi|\theta)$. Let $\{\zeta_i(\theta), i < \infty\}$ denote the sequence of service times, and define $\chi_n(\theta) = F(\zeta_n(\theta)|\theta)$ and the derivative $Z_n(\theta) = F_\theta^{-1}(\chi_n(\theta)|\theta)$. Define the cost for the first m customers, initialized at an arbitrary initial queue occupancy x_0 as

$$L_m(\theta, x_0) = \widehat{L}_m(\theta, x_0) + K(\theta) = \frac{1}{m}\sum_{i=1}^{m} EX_i(\theta) + K(\theta). \tag{5.3}$$

Suppose that the supremum over $\theta \in [a,b]$ of the mean service times is less than the mean interarrival time. Then the busy periods have finite mean length for each $\theta \in [a,b]$. Suppose that:

the distribution function of the interarrival times is continuous. (5.4)

Consider the estimator

$$\widehat{Z}_m(\theta) = \frac{1}{m}\sum_{i=1}^{m}\sum_{j=v_i(\theta)}^{i} Z_j(\theta), \tag{5.5}$$

where $v_i(\theta)$ is the index of the first arrival in the busy period in which customer i arrives. The *index* of a customer is defined to be n if that customer is the nth to arrive from time zero on. If the initial occupancy of the queue is zero, then (5.5) is the pathwise θ-derivative of

$$\frac{1}{m}\sum_{i=1}^{m} X_i(\theta).$$

Under broad conditions, it is an unbiased estimator of the θ-derivative of the mean value $\widehat{L}_m(\theta, x_0)$, and for each initial condition

$$\widehat{Z}_m(\theta) \to \widehat{L}_\theta(\theta) \tag{5.6}$$

with probability one and in mean as $m \to \infty$. The proof of this fact is one of the major issues in IPA. Proofs under various conditions and further references are in [84, 161, 231].

Henceforth, in this section, to simplify notation and not to worry about the possibly separate indices for arrivals and departures, we suppose that the queue starts empty. The conditions and results are the same in general.

The reader should keep in mind that we assume that the service time distribution is known for each θ of interest, and this will rarely be the case.

One needs to know that the derivative estimators are robust enough so that they either approximate the derivatives of the true distributions without serious error or their use will still lead to an improved system. Limited simulations have shown that one can often use simple approximations and still improve the system, but more work needs to be done on this issue. The references [148, 262] use a system model that is a parameterized stochastic differential equation and show that the pathwise derivatives of a "numerical" approximation converge to the pathwise derivatives of the original process as the approximation parameter goes to its limit.

The stochastic approximation algorithm. When we use stochastic approximation to minimize the cost (5.1) via the use of the IPA estimator, we update the parameter after the departure of each successive group of N customers. Other choices of updating times can be used, for example, update at the end of successive busy periods or at the end of the first busy period following the departure of the next N customers, or a random mixture of all of these methods, and so forth. Every reasonable choice of updating times has the same limit properties [145]. We use the "customer group number" instead of real time as the index of the stochastic approximation. For fixed θ, define the estimator on the nth interval (this contains the departures $[nN+1, \ldots, nN+N]$)

$$\widehat{Y}_n(\theta) = -\frac{1}{N} \sum_{i=nN+1}^{nN+N} \sum_{j=v_i(\theta)}^{i} Z_j(\theta). \tag{5.7}$$

Recall that $v_{nN}(\theta)$ is the index of the first arrival in the busy period in which arrival (equivalently, departure) nN occurs. Then, by (5.6)

$$\lim_n \frac{1}{nN} \sum_{i=1}^{nN} \sum_{j=v_i(\theta)}^{i} Z_j(\theta) = \widehat{L}_\theta(\theta), \tag{5.8}$$

with probability one and in expectation for each initial condition.

The step size in the algorithm will be a small constant ϵ. Hence, the θ_n^ϵ depend on ϵ, and so do the service times and queue lengths. In the physical system, θ_0^ϵ is used up to the time of departure of the Nth customer, then θ_1^ϵ is used up to the time of departure of the $2N$th customer, and so forth. For the actual physical system with the time varying parameter, let ζ_i^ϵ denote the actual service time of the ith customer and Z_i^ϵ the derivative of the inverse function at ζ_i^ϵ. Let v_i^ϵ be the index of the first arrival in the busy period in which customer i arrives. To update θ_n^ϵ, we use the estimator

$$\widehat{Y}_n^\epsilon = -\frac{1}{N} \sum_{i=nN+1}^{nN+N} \sum_{j=v_i^\epsilon}^{i} Z_j^\epsilon. \tag{5.9}$$

A reasonable stochastic approximation algorithm for updating θ is

$$\theta^\epsilon_{n+1} = \Pi_{[a,b]}\left[\theta^\epsilon_n + \epsilon\widehat{Y}^\epsilon_n - \epsilon K_\theta(\theta^\epsilon_n)\right]. \tag{5.10}$$

One could use a decreasing sequence $\epsilon_n \to 0$ in place of ϵ, but in such applications one generally uses a small and constant value, which would allow tracking of slow changes in the parameter.

The statistical structure of $\{\widehat{Y}^\epsilon_n\}$ is complicated. We know something about the unbiasedness and consistency properties for large N when the parameter does not change, but $\widehat{Y}^\epsilon_n$ is not an unbiased estimator of the derivative of $-\widehat{L}(\cdot)$ at $\theta = \theta_n$. It might be close to an unbiased estimator for large N, but N should not and need not be too large in a practical algorithm. Letting N be too large simply slows down the rate at which the new data is absorbed into the overall averaging, and the convergence holds for any N. It is not *a priori* clear that the algorithm will converge to the correct limit. However, the general methods to be developed in Chapter 8 will allow us to prove (in Chapter 9) the convergence under reasonably weak conditions. Again, the key is the slow rate of change of θ_n when ϵ is small, which allows us to effectively combine many successive $\widehat{Y}^\epsilon_n$ with essentially the same parameter value. Indeed, if θ_n varies slowly, as it would when ϵ is small, then (5.8) suggests that for large δ/ϵ and small ϵ and δ, the sum of $(\epsilon/\delta)\widehat{Y}^\epsilon_i$ for $i \in [n, n+\delta/\epsilon)$ would be a good estimator of the derivative at $\theta = \theta^\epsilon_n$. This will be formalized in Chapters 8 and 9. An important variation is the decentralized algorithm (Chapter 12), where there are several processors, each taking its own measurements, updating its own subset of parameters at intervals that are not necessarily synchronized between the processors, and passing information on the updates to other processors from time to time [145, 154, 239].

Other examples using IPA. Reference [241] contains a detailed discussion of stochastic approximation with various forms of IPA estimators, together with simulation data that both justifies the approach and illustrates the qualitative properties of the convergence.

There is a large literature on IPA-type estimators, sometimes in conjunction with a stochastic approximation-type procedure. From the point of view of stochastic approximation, the structure of many of these problems is similar, and the same proof (say those of Chapters 8 and 9) can be used. Typical applications are single queues, queueing networks, and various production scheduling problems. We only list a few references to give some flavor of the developments. In most cases, the results of this book simplify and extend the stochastic approximation parts of the development and allow the use of more flexible algorithms. Additionally, one can treat the multiclass problem [186], admission control [243], flow control in a closed network [244], routing in an open network [239], and routing in a network [100], as well as balancing the noise and bias and allocating total computational budget in optimization of steady state simulation models with likelihood ratio, IPA, or finite difference estimators [162].

Reference [46] concerns the optimization of the performance of an unreliable machine. The machine has a capacity constraint and is used to produce several types of parts, each with a given demand rate. The control problem is the decision concerning what to produce at any given time. The decision rule is determined by thresholds in the space of available inventory, and these thresholds are to be optimized. An appropriate cost function is given, and IPA-type estimators for the desired pathwise derivatives are constructed. A related problem, where there are two unreliable machines working in tandem, is treated in [260, 272]. The formulation of the stochastic approximation part is dealt with by the more efficient methods of this book in [145]. Optimization of an inventory system is discussed in [86]. Reference [245] concerns the optimization of a high-speed communications network. Messages may appear at any node and are to be routed through the network to the appropriate destination. Traffic is managed by a "token" system. With this method, each message moving through the system must be accompanied by a token. When the message arrives at the destination, the token is sent back to some input node to be reused. The control parameters are the probabilities p_{ij} that a token arriving with a message at destination i will be sent to arrival node j. If an arriving message finds a token waiting at the arrival node, then the message is routed to the appropriate destination. Otherwise, the message is queued at the arrival node pending the appearance of a free token there. A common performance criterion is the average time for the transmissions. One wants to choose p_{ij} to minimize this quantity. The optimizing probabilities satisfy the Kuhn–Tucker condition for a certain constrained optimization problem. An algorithm of the stochastic approximation type is constructed in [245] such that the stationary points of the mean ODE are just the Kuhn–Tucker points. Reference [4] obtains IPA estimators of first and second derivatives with respect to service time parameters for a closed queueing network.

2.6 Passive Stochastic Approximation

Let θ and Y_n be $\mathbb{R}^r$-valued. In a basic form of the Robbins–Monro algorithm, one observes $Y_n = g(\theta, \xi_n)$ at values of the parameter θ selected by the experimenter and seeks the value $\bar{\theta}$ such that $\bar{g}(\bar{\theta}) = Eg(\bar{\theta}, \xi_n) = 0$. In some applications, the values of θ_n are externally generated and cannot be chosen by the experimenter. Suppose that $\{\theta_n\}$ is a random sequence, not selected by the experimenter, who still wishes to find the root (assumed unique for simplicity) of the function $\bar{g}(\cdot)$ using measurements $Y_n = \bar{g}(\theta_n) + \xi_n$ where ξ_n is a "noise" sequence.

In [95], Härdle and Nixdorf suggested an approach to the problem and termed it *passive stochastic approximation.* The approach can be traced back to an early work of Révész [202]. Problems with real-valued θ are treated in [95], and multivariate cases are handled in [181]. The main idea is to combine the stochastic approximation methods with nonparametric estimation procedures. Now let θ and Y_n be $\mathbb{R}^r$-valued and $g(\cdot)$ an $\mathbb{R}^r$-valued function on $\mathbb{R}^r$. Let $K(\cdot)$ be a real-valued kernel function on $\mathbb{R}^r$; it is non-negative, symmetric about the origin, and it takes its maximum value at $\theta = 0$ and decreases monotonically to zero as any component of the argument increases to infinity. The root of the equation $\bar{g}(\theta) = 0$ is approximated by the sequence $\{\varpi_n\}$ defined by

$$\varpi_{n+1} = \varpi_n + \frac{\epsilon_n}{h_n^r} K\left(\frac{\theta_n - \varpi_n}{h_n}\right) Y_n, \tag{6.1}$$

where h_n represents the "window" width. The kernel function $K(\cdot)$ plays a crucial role. If ϖ_n and θ_n are far apart, $K((\theta_n - \varpi_n)/h_n)$ will be very small, and the measurement Y_n has little effect on the iteration.

For robustness purposes, one may use the constant step size and constant window width (see [275]) algorithm

$$\varpi_{n+1}^\epsilon = \varpi_n^\epsilon + \frac{\epsilon}{\delta^r} K\left(\frac{\theta_n^\epsilon - \varpi_n^\epsilon}{\delta}\right) Y_n, \quad \text{for } n \geq 0. \tag{6.2}$$

The rate of convergence of ϖ_n^ϵ to $\bar{\theta}$ is slower than that for the classical Robbins–Monro method and depends on the smoothness of $\bar{g}(\cdot)$ in θ. As the number of continuous θ-derivatives increases to infinity, the rate approaches that of the Robbins–Monro process [275]. A similar result holds when the constant step size is replaced by decreasing step sizes.

An alternative approach is to fit a nonlinear curve to the data and then use the fitted function to estimate the root. Nevertheless, this approach can be quite costly in its data requirements.

See [95, 202] for applications to nonlinear regression problems. An application to chemical process control is in [275], and [274] contains the results of numerical experiments on a binary distillation column model.

2.7 Learning in Repeated Stochastic Games

The theory of learning optimal strategies in repeated stochastic games leads to recursive stochastic algorithms whose mean ODEs can have quite interesting behavior. Many such algorithms have a special structure (e.g., they often satisfy the Kamke condition, which is defined above Theorem 4.4.1) and implies properties of the ODE that can be exploited to either prove convergence or to get insight into the asymptotic behavior. We will discuss only one relatively simple class of such problems. Many recursive algorithms that are based on either explicit or implicit competitive behavior have properties similar to those of repeated stochastic games; see, for example, the proportional-fair sharing recursive algorithm in Chapter 3, which is used to allocate resources in time-varying wireless systems. The reference [38] concerns learning in another type of (implicit) game that arises in decentralized access control in communications. It has the characteristics of a game owing to the way that the various users affect each other's performance. Such problems have arisen in the economics literature [79]. There is a large literature concerning other uses of recursive algorithms to study "learning" phenomena in economics; for example, see [69, 175].

A game model. Consider a cooperative game with two players, each having two possible actions. Each player knows only its own payoff matrix at each time but keeps a record of the actions used by the other player. The hope is that each player will learn its optimal strategy, if there is one. Let $\theta_{n,i}$ denote the fraction of the first n plays in which player i used its first action, called a_{i1}. Then θ_n can be written recursively as

$$\theta_{n+1,i} = \theta_{n,i} + \frac{1}{n+1}\left[I_{\{a_{i1}\text{used at time } n+1\}} - \theta_{n,i}\right]. \tag{7.1}$$

At each time, player i first observes the payoff that it would receive under each of its actions, supposes that the other player chooses its action at random according to the historical record to date, and then chooses the action that maximizes its conditional average return. This scenario has been called *fictitious play* [11, 79]. Although the game is called cooperative, the players do not coordinate their actions. The following format and assumptions are taken from [11].

The actual payoffs to each player depend on the strategies of both players Let $\{v_{kl}; k, l = 1, 2\}$ be given real numbers. Suppose that player 1 uses a_{1k} and player 2 uses action a_{2l} at time n. Then the deterministic part of the payoff to each player is v_{kl}. A small random perturbation is added to this deterministic payoff. For small $\delta > 0$, the actual payoff to player 1 is $u^{\delta}_{n,1,kl} = v_{kl} + \delta\eta_{n,1k}$ and that to player 2 is $u^{\delta}_{n,2,kl} = v_{kl} + \delta\eta_{n,2l}$, where $(\eta_{n,ik}; i = 1, 2, k = 1, 2), n = 1, 2 \ldots,$ are sequences of random variables which are identically distributed, have zero mean, and are independent in n. Thus, the part v_{kl} of the payoff is common to both players, but the random part is not: It plays the role of the effect of "private information." One of the interesting issues concerns the effect of such small "noisy" perturbations on the asymptotic behavior. Each player knows its own possible payoffs, but not those of the other player.[1]

Define the mean payoff to each player under each of its possible actions, given that the other player acts at random according to its historical record:

$$\bar{u}^{\delta}_{n+1,1,k}(\theta_{n,2}) = u^{\delta}_{n+1,1,k1}\theta_{n,2} + u^{\delta}_{n+1,1,k2}(1 - \theta_{n,2}).$$

[1] In [11], The difference $\widetilde{\eta}_{i,n} = \eta_{n,i1} - \eta_{n,i2}$ is assumed to have a density that is continuous and positive on the entire real line, with the density at value η going to zero fast enough so that it is as $o(1/|\widetilde{\eta}|)$.

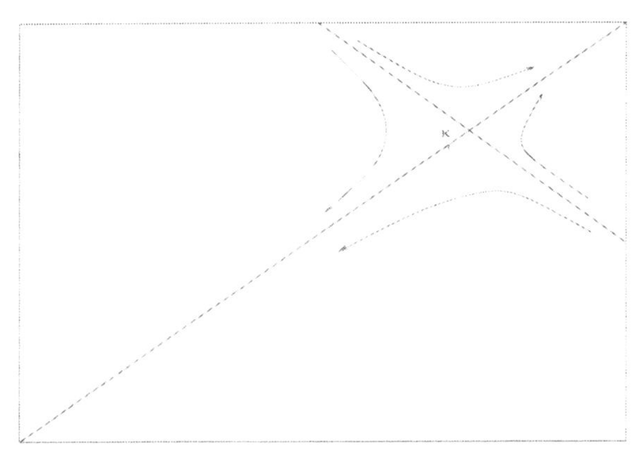

Figure 7.1. Phase plot for the perturbed game.

$$\bar{u}^{\delta}_{n+1,2,l}(\theta_{n,1}) = u^{\delta}_{n+1,2,1l}\theta_{n,1} + u^{\delta}_{n+1,2,2l}(1-\theta_{n,1}).$$

Define the conditional probability that a_{11} is better for player 1 than is a_{12}:

$$\bar{h}^{\delta}_1(\theta_{n,2}) = P\left\{\bar{u}^{\delta}_{n+1,1,1}(\theta_{n,2}) \geq \bar{u}^{\delta}_{n+1,1,2}(\theta_{n,2})\right\}$$

and define $\bar{h}^{\delta}_2(\theta_{n,1})$ for player 2 analogously.

The mean ODE for (1.1) is

$$\dot{\theta}^1 = \bar{g}_i(\theta) = \bar{h}^{\delta}_1(\theta^2) - \theta^1, \quad \dot{\theta}^2 = \bar{g}_2(\theta) = \bar{h}^{\delta}_2(\theta^2) - \theta^2. \tag{7.2}$$

Following [11], we say that θ is a *Nash distribution equilibrium* if $\bar{h}^{\delta}_1(\theta^2) = \theta^1, \bar{h}^{\delta}_2(\theta^1) = \theta^2$.

Discussion. The paper [11] contains additional references, a detailed discussion of convergence and nonconvergence, as well as of many other issues that arise, interpretations of the results in terms of the "information patterns," and of the differences between the perturbed and unperturbed repeated games. As $\delta \to 0$, the Nash distribution equilibria for (7.2) converge to those for the system with $\delta = 0$. If there are more than two possible actions for each player or if there are more than two players, the behavior can be quite complicated, even chaotic. For example, there are three player, 2-action games whose sample average frequencies θ_n converge to a non-degenerate limit cycle.

Example. This example is taken from [11]. Let $v_{11} > v_{21}$, $v_{22} > v_{12}$, $v_{11} \geq v_{12}$. Then, as $\delta \to 0$ there are three Nash distribution equilibria; $\theta = (1,1), \theta = (0,0)$ and $\theta = (p^*, p^*)$, where $p^* = K/(1+K)$ for $K = (v_{22} - v_{12})/(v_{11} - v_{21})$. The points $\theta = (0,0)$ and $\theta = (1,1)$ are stable. The mixed equilibrium at (p^*, p^*) is unstable. Then θ_n converges (with probability one) to one of the pure strategies. [The noise in (7.1) destabilizes the mixed strategy, as can be shown by using Theorem 5.8.1.] Let $K > 1$. Then, as $\delta \to 0$, the domain of attraction of $(0, 0)$ is much larger than that of $(1, 1)$. This suggests that convergence is "more likely" to the point $(0, 0)$, the "risk dominating strategy." The phase plot for the ODE is given in Figure 7.1.

3

Applications in Signal Processing, Communications, and Adaptive Control

3.0 Outline of Chapter

Adaptive control and communications theory provide numerous examples of the practical and effective use of stochastic approximation-type algorithms, and there is a large amount of literature concerning them. Despite the identity in form, whether the algorithms are actually called *stochastic approximation* depends on the traditions of the field, but the most effective techniques of proofs are often based on those of stochastic approximation. Our intention is to give the flavor of some of the applications. Proofs of convergence for some cases will appear in Chapter 9. Perhaps the most basic problem is the recursive estimation of the coefficients (parameters) of a linear system, when the input sequence is known, but only noise corrupted observations of its output can be measured. The output might simply be a weighted sum of the inputs, a weighted sum of the inputs and recent outputs, or a weighted sum of the inputs and recent outputs plus a weighted sum of the recent elements of a "driving" white noise sequence. The forms of the algorithms are essentially those of the pattern classification or "minimum mean squares" estimation problem of Example 3 in Section 1.1, although the noise terms might be correlated and the signals generated by a stochastic system. Algorithms very similar to those used for the parameter identification problem appear in economics as models of learning or adaptive "rational expectations" estimators [32, 33, 69, 116, 175, 176].

In practice, the values of the parameters to be estimated often change with time, and the basic algorithm must be adjusted to allow "tracking." A constant step size ϵ is generally used, which leads to the nontrivial problem of the choice of ϵ. One very useful approach involves recursively estimating the best value of ϵ by a superimposed adaptive (stochastic approximation) procedure (more will be said about this in Section 2). This procedure is effective and leads to new theoretical issues. An application of the method to adaptive interference suppression in a wireless communications system is in [115].

When the parameters are not changing with time, the averaging method of Subsection 1.3.3 applied to a simple stochastic approximation algorithm gives a rate of convergence that is nearly equal to the optimal least squares value, and this is discussed in Section 3. Section 4 outlines several applications in communication theory, including the adaptive noise and interference cancellation algorithms and the channel equalization problem.

Sections 5 and 6 concern problems that arise in mobile communications where the capacity of the channels that connect the mobiles with the base station vary rapidly and randomly with time. Adaptive antenna arrays increase the capacity and reliability of wireless communications systems. The algorithms for adapting the weights are a "discounted" form of those in Subsection 1.1.3. The discounting weighs recent observations more than older observations, and helps track the optimal weights as the conditions change. The performance is sensitive to the discount factor. In Section 5, another adaptive loop, of the stochastic approximation type, is added to adapt the discount factor. The idea is similar to that of the parameter tracking problem of Section 2. Numerical data illustrate the value of the approach.

Section 6 deals with the problem where time is divided into small scheduling intervals or slots, and in each slot one of the users is assigned to transmit. The possible transmission rates vary randomly with the user and with time. The assignment must balance the possibly conflicting goals of efficient use of the medium and assuring fairness. The so-called proportional fair-sharing algorithm is one of the current approaches, and is a form of stochastic approximation. The mean ODE has some special properties that facilitate analysis.

3.1 Parameter Identification and Tracking

3.1.1 The Classical Model

The first problem to be described is the identification of the unknown parameters of a linear dynamical system with observable stochastic inputs [49, 55, 92, 165] and [166, 169, 225, 263, 264], but only noise-corrupted measurements of the outputs are available. The problem is close to the classical linear regression problem in statistics with correlated measurements and is a fundamental and canonical model in control and communication theory. It is a canonical model in that the algorithms used and the associated convergence theorems are identical to those arising in many other applications, such as adaptive noise cancellation, channel equalization, and adaptive antenna arrays. A few of the simplest forms will be discussed. The problem in Section 5 is closely related to those in this section.

Let $\{\psi_n\}$ be a sequence of real-valued random variables that are the inputs to a dynamical system, whose output at time n is $v_n = \sum_{i=0}^{k_1} \bar{\alpha}_i \psi_{n-i}$. The parameters $\bar{\alpha}_i$ are to be estimated, and the system order $k_1 + 1$ might or might not be known, but until further notice suppose that $k_1 = k$, a known integer. The output v_n is not observable, but the measured output $y_n = v_n + \nu_n$ and the inputs $\{\psi_i, i \leq n\}$ are available at time n. The ν_n is a "corrupting noise." It is convenient to put the problem into vector form. Define the vectors $\phi_n = (\psi_n, \ldots, \psi_{n-k})$ and $\bar{\theta} = (\bar{\alpha}_0, \ldots, \bar{\alpha}_k)$. Then the output can be written as $v_n = \bar{\theta}'\phi_n$. The observable output is $y_n = \bar{\theta}'\phi_n + \nu_n$. Let $\theta = (\alpha_0, \ldots, \alpha_k)$ be an estimate of the system parameter. If the sequence $\{\phi_n, y_n\}$ is stationary, then it makes sense to look for the value of θ that minimizes the mean square output prediction error $Ee^2(\theta) = E|y_n - \theta'\phi_n|^2$. The problem is then formally identical to that of Example 3 of Section 1.1, and one can use any of the algorithms (1.1.11), ((1.1.14), (1.1.15)) (recursive least squares), ((1.1.16),(1.1.17)) (linearized least squares) or (1.1.18) (classical stochastic approximation). It will be seen in Chapter 9 that the methods of Chapters 6 and 8 can be used to prove convergence and to characterize the limits of these algorithms.

If the "driving" process $\{\phi_n, y_n\}$ is not stationary, then we would like to assure convergence under conditions $\{\phi_n, y_n\}$ that are as weak as possible. An alternative error criterion is (assuming that the limit exists)

$$F(\theta) = \lim_{n,m} \frac{1}{m} E \sum_{i=n}^{n+m-1} e_i^2(\theta), \tag{1.1}$$

where the limit is as n and m go to infinity simultaneously. The limit exists if the driving process has asymptotically periodic second order statistics, or if there is a transient period but the process is asymptotically stationary, or under any condition that guarantees that the averages of the second order statistics over long time intervals converge. There is a recursive "gradient descent" algorithm for minimizing $F(\cdot)$, and it is of interest to know whether the iterates θ_n given by that algorithm will converge to $\bar{\theta}$. Indeed, the θ_n given by the algorithms to be discussed will converge to $\bar{\theta}$ under broad conditions even if the limit (1.1) does not exist; see, for example, Section 6.8 (where the mean ODE is replaced by a mean differential inclusion) and [166, 169, 225].

The algorithms (1.1.11) and its recursive variant ((1.1.14), (1.1.15)) still make sense even if the limit in (1.1) does not exist. Indeed, (1.1.11) still yields the minimizer of the finite sum

$$\frac{1}{n}\sum_{i=0}^{n-1} e_i^2(\theta), \tag{1.2}$$

and ((1.1.16), (1.1.17)) is still a linearized form of (1.1.11). We can motivate (1.1.18) as being a simpler form of ((1.1.16), (1.1.17)). It can also be viewed as a type of gradient descent procedure, where the step size is a positive number ϵ_n and the estimator of the gradient with respect to the parameter evaluated at θ_n is the "noisy value"

$$\text{grad}_\theta [y_n - \theta'\phi_n]^2/2 = -e_n(\theta)\phi_n$$

evaluated at $\theta = \theta_n$. It will turn out that it is a good enough estimator of the gradient to assure that the iterates will converge to the minimizing parameter value under reasonable conditions.

Suppose that the limits

$$\lim_{n,m} \frac{1}{m} \sum_{i=n}^{n+m-1} E_n \phi_i \phi_i' = Q, \tag{1.3a}$$

$$\lim_{n,m} \frac{1}{m} \sum_{i=n}^{n+m-1} E_n \phi_i \nu_i = S \tag{1.3b}$$

exist in the sense of convergence in probability, where E_n denotes the expectation given $\{\theta_0, \phi_i, \nu_i, i \le n\}$, and Q and S are a non random matrix and vector, respectively. Then, under mild additional conditions (see Chapters 6, 8, and 9), the mean ODE that characterizes the limit points is

$$\dot{\theta} = -\text{grad}_\theta F(\theta) = S - Q(\theta - \bar{\theta}). \tag{1.4}$$

This ODE has the desired gradient descent form. The "noisy" gradient descent interpretation of (1.1.18), as formalized in (1.1.19), is the key to the asymptotic properties. If the system input ϕ_n and the observation noise ν_n are correlated, then $S \neq 0$, and the estimators θ_n and their limits will be biased.

More generally, y_n and ϕ_n need not be of the assumed forms, where $\phi_n = (\psi_n, \ldots, \psi_{n-k})$ is the set of the most recent $(k+1)$ inputs to the system and $y_n = v_n + \nu_n$ is the system output plus corrupting noise. First suppose that $k_1 > k$. Then, due to the effects of the "unmodeled terms"

$$\sum_{i=k+1}^{k_1} \bar{\alpha}_i \psi_{n-i},$$

there will be an additional bias term. In general, y_n might be the output of a general linear or nonlinear system, and $\{\phi_n\}$ a sequence of vector-valued "regressors" that are related to the system inputs ψ_n but not necessarily defined by $\phi_n = (\psi_n, \ldots, \psi_{n-k})$; we might want to find the value of θ that provides the best linear fit in the sense of minimizing either $Ee_n^2(\theta)$ (stationary case) or the $F(\cdot)$ defined by (1.1) provided that the limit exists. Then, if y_n replaces ν_n in the conditions, we still get convergence to the minimizer, under some additional conditions, and the mean ODE characterizing the limit points is

$$\dot{\theta} = \bar{S} - Q\theta, \tag{1.5}$$

where

$$\bar{S} = \lim_{n,m} \frac{1}{m} \sum_{i=n}^{n+m-1} E_n \phi_i y_i. \tag{1.6}$$

Commonly, in applications of linear systems identification theory, the actual physical model will have order $k_1 = \infty$, but with the "tails" having small effect, and there might also be small nonlinearities. Then the choice of the proper value of the model order k can be a major problem [206]. One needs to consider the assumed model as being a convenient approximation. In that regard, one can at best get a good approximation to the original model, but not the exact model. This approximation perspective stems from the earliest days of recursive parameter estimation [120]. The sequences $\{y_n\}$ and $\{\nu_n\}$ (resp, $\{\theta_n\}$ and $\{\phi_n\}$) can be vector (resp., matrix) valued. The mean square error criterion is convenient due to its computational simplicity and association with "energy." It is used in the examples here since it is the most common error function in applications, and the criteria for convergence are relatively easy to state. However, it is by no means the only possibility.

If the mean values of $(\phi_n \phi_n', \phi_n \nu_n)$ depend on time or if the limits in (1.3) or (1.6) do not exist, then the ODE might be replaced by a differential inclusion, where Q and S are replaced by set-valued functions and the equality $=$ is replaced by the set inclusion $\in$. Under reasonable conditions, the limit points of the solutions to the differential inclusion will also be the limit points of the sequence $\{\theta_n\}$; see Chapters 6, 8, and 9.

Delays. Owing to physical considerations in the applications, the inputs that are used to compute an output might be observed subject to a delay $D > 0$. For example, suppose that the output observed at time n has the form $y_n = \sum_{i=-D}^{k_1} \bar{\alpha}_i \psi_{n-i} + \nu_n$, but that the only inputs known at n are $\psi_i, i \leq n$. [Alternatively, suppose that the output at time n has the form $\sum_{i=0}^{k_1} \bar{\alpha}_i \psi_{n-i}$, with the only inputs known then being $\psi_i, i \leq n - D$.] Using the definition $\phi_n = (\psi_n, \ldots, \psi_{n-k})$, any of the algorithms ((1.1.14), (1.1.15)), ((1.1.16), (1.1.17)), or (1.1.18), can still be used, but the convergence will be to an incorrect value. Suppose that an upper bound $\Delta \geq D$ is known. A common alternative is to delay the estimate by Δ and use the "two-sided" estimator defined by $\sum_{i=-\Delta}^{k} \alpha_i \psi_{n-i}$. The error $e_n(\theta)$

is then defined by

$$e_n(\theta) = y_n - \sum_{i=-\Delta}^{k} \alpha_i \psi_{n-i}.$$

Redefining $\phi_n = (\psi_{n+\Delta}, \ldots, \psi_{n-k})$ and $\theta = (\alpha_{-\Delta}, \ldots, \alpha_k)$, the "two-sided" algorithms take any of the forms (1.1.11), ((1.1.14), (1.1.15)), ((1.1.16), (1.1.17)), or (1.1.18).

3.1.2 ARMA and ARMAX Models

An ARMA model. (Autoregressive and moving average model [49, 165, 166, 169, 225].) Let the system output at time n take the form

$$y_n = \sum_{i=0}^{k_1} \bar{\alpha}_i \psi_{n-i} + \sum_{i=1}^{k_2} \bar{\beta}_i y_{n-i} + \nu_n. \tag{1.7}$$

This is a common way to represent the input-output relationship in linear systems theory. The values of $\{y_i, \psi_i, i \leq n\}$ are known at time n, and ν_n is the unknown corrupting noise. The orders k_1, k_2 are not known and assumed orders r_1, r_2 are used. Define $\phi_n = (\psi_n, \ldots, \psi_{n-r_1}, y_{n-1}, \ldots, y_{n-r_2})$. Then with parameter estimator $\theta = (\alpha_0, \ldots, \alpha_{r_1}, \beta_1, \ldots, \beta_{r_1})$, the estimator of y_n at time n is

$$\phi_n' \theta = \sum_{i=0}^{r_1} \alpha_i \psi_{n-i} + \sum_{i=1}^{r_2} \beta_i y_{n-i}. \tag{1.8}$$

Let θ_n denote the value of the parameter estimator at time n. Stability of the y-system is often assumed. By *stability*, we mean that the difference equation $x_n = \sum_{i=1}^{k_2} \bar{\beta}_i x_{n-i}$ is asymptotically stable. Equivalently, the roots of the polynomial $1 - \sum_{i=1}^{k_2} \bar{\beta}_i z^i$ lie strictly outside the unit circle in the complex plane. The algorithms are still of the form used in Subsection 1.1.

An ARMAX model. (Autoregressive and moving average model with additional unobservable white noise inputs [49, 165, 166, 169, 225].) Let us generalize the model (1.7) to the form

$$y_n = \sum_{i=0}^{r_1} \bar{\alpha}_i \psi_{n-i} + \sum_{i=1}^{r_2} \bar{\beta}_i y_{n-i} + \sum_{i=1}^{r_3} \bar{\gamma}_i w_{n-i} + w_n. \tag{1.9}$$

The parameter values are not known, but $\{y_i, \psi_i, i \leq n\}$ is known at time n. The sequence $\{w_n\}$ consists of mutually independent and zero mean random variables, which are independent of $\{\psi_n\}$ and whose values are not observable except indirectly via the output measurements $\{y_n\}$, and they must be estimated. This constitutes the major problem in the analysis. The orders r_i are assumed known.

Let $\widehat{w}_n$ denote the estimator of w_n, and define

$$\begin{aligned}\phi_n &= (\psi_n, \ldots, \psi_{n-r_1}, y_{n-1}, \ldots, y_{n-r_2}, \widehat{w}_{n-1}, \ldots, \widehat{w}_{n-r_3}),\\ \bar{\theta} &= (\bar{\alpha}_0, \ldots, \bar{\alpha}_{r_1}, \bar{\beta}_1, \ldots, \bar{\beta}_{r_2}, \bar{\gamma}_1, \ldots, \bar{\gamma}_{r_3}),\end{aligned}$$

and the generic parameter estimator

$$\theta = (\alpha_0, \ldots, \alpha_{r_1}, \beta_1, \ldots, \beta_{r_2}, \gamma_1, \ldots, \gamma_{r_3}).$$

The estimator of w_n (at parameter estimate θ) is

$$\widehat{w}_n = y_n - \sum_{i=0}^{r_1} \alpha_i \psi_{n-i} - \sum_{i=1}^{r_2} \beta_i y_{n-i} - \sum_{i=1}^{r_3} \gamma_i \widehat{w}_{n-i} = y_n - \phi_n' \theta. \qquad (1.10)$$

Letting the parameter estimator at time n be θ_n, define the estimator of y_n, given the prior measurements, by

$$\widehat{y}_n = \phi_n' \theta_n. \qquad (1.11)$$

By the mutual independence properties of the elements of $\{w_n\}$, w_n is independent of the past data $\{\phi_i, w_{i-1}, i \leq n\}$. Hence $\widehat{y}_n$ estimates the "predictable" part of y_n. The objective is the minimization of either $E|\widehat{w}_n|^2$ or

$$\lim_{n,m} \frac{1}{m} \sum_{i=n}^{n+m-1} E|\widehat{w}_i|^2, \qquad (1.12)$$

depending on the (stationarity or limit) assumptions that one makes. These error functions are appropriate since $\widehat{w}_n = y_n - \widehat{y}_n$. The algorithm (1.1.18) now takes the form

$$\theta_{n+1} = \theta_n + \epsilon_n \phi_n \widehat{w}_n. \qquad (1.13)$$

This form is made complicated by the fact that the $\widehat{w}_{n-i}$ components of ϕ_n depend on all past values of the state in a complicated manner. Indeed, this is another form of the state-dependent noise problem first introduced in Section 2.3, and the state dependence needs to be accounted for in the analysis. A related algorithm, for the problem of channel equalization, is discussed in Section 4, and convergence is shown in Section 9.4.

3.2 Tracking Time Varying Systems: An Adaptive Step Size Algorithm

3.2.1 The Algorithm

Recall the linear system model in Subsection 1.1 where $v_n = \sum_{i=0}^{k_1} \bar{\alpha}_i \psi_{n-i}$, and suppose that the parameters $\bar{\alpha}_i$ and/or the probability distributions of ψ_n or ν_n vary slowly with time. Alternatively, for the more general regression model mentioned in Subsection 1.1, suppose that the probability distribution of the (observation, regressor) = (y_n, ϕ_n) vary slowly with time. Then, the optimal value of the estimator θ will also change with time. Such variations are a common fact of life in applications, and one wishes to "track" the changing optimal values as well as possible. The importance and prevalence of such tracking problems have led to an enormous literature (see [16, 56, 90, 91, 92, 93, 224, 255], among many others). The problem occurs in communication theory in dealing with adaptive equalizers, for time-varying channels, adaptive noise cancellation or signal enhancement systems, adaptive quantizers and other applications, where the generated signals, corrupting noise, and channel characteristics change with time. We will describe a recent approach to this problem that has proved useful and that leads to a nonclassical stochastic approximation problem. Data which illustrates the value of the algorithm, will be given in the next subsection.

Let us work with the simplest time-varying parameter tracking problem, where the observation at time n is given by

$$y_n = \phi_n' \bar{\theta}_n + \nu_n, \tag{2.1}$$

and $\bar{\theta}_n$ is the value of the slowly time-varying physical parameter at n. The values of $y_i, \phi_i, i \leq n$, are assumed to be known at time n. The algorithms used for tracking $\bar{\theta}_n$ take many forms, depending on what one is willing to assume about the time variations and the amount of computation that is acceptable; e.g., standard stochastic approximation with a fixed step size ϵ, recursive or linearized least squares with "forgetting factors," or adaptations of Kalman–Bucy filters. More information is in the cited references. In this example, we will use the stochastic approximation algorithm (1.1.18) with a fixed step size:

$$\theta_{n+1}^{\epsilon} = \theta_n^{\epsilon} + \epsilon \phi_n [y_n - \phi_n' \theta_n^{\epsilon}], \tag{2.2a}$$

where θ_n^ϵ is the estimate of $\bar{\theta}_n$. The rate of variation of the time-varying parameters and/or the probability distributions is assumed to be "slow." The algorithm for tracking time variations in Section 5 is based on a recursive least squares form, which could be adapted for use on the current problem, although there does not seem to be an advantage in doing it.

For effective use of (2.2a), one needs to select the proper value of ϵ. The optimal value depends on the "rates of variation" of $\bar{\theta}_n$ and of the probability distribution of ν_n, y_n, ϕ_n, and a great deal of effort has been put into this problem (see [16, 93] among many others). The faster the distributions and/or $\bar{\theta}_n$ vary, the larger ϵ should be. The larger the effects of the observation noise ν_n, the smaller ϵ should be. Thus, the proper value depends on the details of the physical model. In practical applications, one does not always know the exact form of the model, and one generally would not know the relevant probability distributions. Indeed, the actual model for $\{y_n, \nu_n, \phi_n\}$ used in the analysis might have been selected for convenience, with the true model being at least partially unknown.

In a sense, one has two estimation problems to contend with. The first is the estimation of $\bar{\theta}_n$. The second is the estimation of the optimal value of ϵ. A very useful approach to this problem was suggested in [16, p. 160]. The idea is to iteratively estimate the correct step size by use of an adaptive algorithm that is analogous to the "gradient descent algorithm" (2.2a). This superimposed adaptive algorithm would use "noisy" estimates of the derivative of the mean square error with respect to the step size ϵ. Thus, there are two adaptive algorithms working together. Let ϵ_n denote the current best estimate of the optimal value of ϵ and replace the ϵ in (2.2a) by ϵ_n, yielding the actual algorithm

$$\theta_{n+1} = \theta_n + \epsilon_n \phi_n \left[y_n - \phi_n' \theta_n \right]. \tag{2.2b}$$

But ϵ_n would not go to zero unless the parameter to be tracked is a constant. This idea was nicely exploited in [36], where many good examples arising in sonar signal processing were given, along with supporting simulation data. These simulations and those in [150] support the value of the approach. Proofs were given in [150] using general convergence theory for stochastic approximation processes with state-dependent noise; these convergence results are stated in Chapter 8. (The terminology used here is slightly different from that in [150].) In applications, the optimal value of ϵ is not necessarily small, even for slow variations. This fact supports the importance of the adaptive approach to step-size selection. (Note that earlier efforts to estimate the best value of ϵ have assumed that it is small.) Simulations show that if there are occasional discontinuities in the optimal value of ϵ, then the adaptive algorithm will track the changing values. For the algorithm to track the parameter $\bar{\theta}_n$ well, $\bar{\theta}_n$ would have to vary slowly in comparison with the process $\{\phi_n, \nu_n\}$. However, whether or not it varies slowly, under suitable conditions the algorithm to be described will converge to the value of ϵ that gives the best fit in the least squares sense.

The adaptive algorithm. For convenience in motivating the algorithm, suppose that for each small enough ϵ, the process $\{y_n, \phi_n, \theta_n^\epsilon\}$ is stationary, although the conditions needed for the proof of convergence are weaker [150]. The step sizes ϵ and ϵ_n as used will always be kept in a given constraint interval $[\epsilon_-, \epsilon_+]$ where $0 < \epsilon_- < \epsilon_+$ are small enough. We introduce the basic algorithm via a formal discussion, which is the actual motivation in [16, 36]. For fixed $\epsilon_n = \epsilon$, define the error $e_n(\epsilon) = y_n - \phi_n' \theta_n^\epsilon$. The scheme suggested in [16, p. 160] and exploited in [36, 150] is to choose the ϵ that minimizes the stationary value of the mean square error

$$E[y_n - \phi_n' \theta_n^\epsilon]^2/2 = Ee_n^2(\epsilon)/2. \tag{2.3}$$

Loosely speaking, let V_n^ϵ denote the "derivative" $\partial \theta_n^\epsilon / \partial \epsilon$ for the stationary process. The stationary θ_n^ϵ-process is not a classical function of ϵ, although its distribution depends on ϵ. Nevertheless, it can be shown (see [150]) that the V_n^ϵ as used in what follows can be interpreted as the desired derivative, in that the convergence results hold and its limit has the interpretation of a mean square derivative. Reference [150] also discusses an effective finite difference form of the algorithm for adapting ϵ that does not need such derivative variables. [See Section 5 for an application of the finite difference form.] Formally differentiating (2.3) with respect to ϵ, assuming stationarity, and setting the derivative to zero yields

$$0 = -E[y_n - \phi_n' \theta_n^\epsilon] \phi_n' V_n^\epsilon = -E e_n(\epsilon) \phi_n' V_n^\epsilon. \tag{2.4}$$

Dropping the expectation E in (2.4), yields $-e_n(\epsilon)\phi_n' V_n^\epsilon$, the "stochastic gradient" of (2.3) with respect to ϵ at time n. Formally differentiating (2.2a) with respect to ϵ yields

$$V_{n+1}^\epsilon = V_n^\epsilon - \epsilon \phi_n \phi_n' V_n^\epsilon + \phi_n [y_n - \phi_n' \theta_n^\epsilon].$$

Let ϵ_n, θ_n denote the actual sequences of step sizes and the estimates of $\bar{\theta}_n$, resp., resulting from the true operating adaptive algorithm, which will now be defined. Let $e_n = y_n - \phi_n' \theta_n$. For $0 < \mu \ll \epsilon_-$, this suggests the algorithm (as used in [16, 36, 150])

$$\theta_{n+1} = \theta_n + \epsilon_n \phi_n e_n, \tag{2.5}$$

$$\epsilon_{n+1} = \Pi_{[\epsilon_-, \epsilon_+]} \left[\epsilon_n + \mu e_n \phi_n' V_n \right], \tag{2.6}$$

$$V_{n+1} = V_n - \epsilon_n \phi_n \phi_n' V_n + \phi_n [y_n - \phi_n' \theta_n], \quad V_0 = 0. \tag{2.7}$$

Note that the actual stochastic approximation algorithm is (2.6), where the step size is μ. The sequence $\{\phi_n, y_n, V_n\}$ is the driving noise process for (2.6). The dependence of V_n on $\{\epsilon_n\}$ and $\{\theta_n\}$ is quite complicated. It is not the classical type of noise in stochastic approximation, but it is what we call *state-dependent noise*, where the "state" process is $\{\epsilon_n\}$. It is worth noting that the performance in applications is a great deal less sensitive to the value of μ than the original algorithm (2.2a) is to the choice of ϵ.

A comment on the truncation levels $\epsilon_\pm$. While needed in the mathematical proofs of convergence, the lower level of truncation ϵ_- does not seem to play an important role in applications. Even if ϵ_- is set to zero (hence violating the assumption $0 < \mu \ll \epsilon_-$), the observed performance of the algorithm is not much affected. The choice of the upper level ϵ_+ is *very* important for good behavior. The optimal value of ϵ is often close to the point where the algorithm (2.2a) becomes unstable. For values of ϵ that are too close to the instability region, there can be erratic behavior of the iterates θ_n, and even of ϵ_n in the adaptive algorithm. For fast speed of tracking and good transient behavior, one often seeks the largest value of

ϵ that is consistent with good asymptotic variance of the errors in $\theta_n - \bar{\theta}_n$. But as ϵ gets close to the instability point, the sample paths of θ_n tend to exhibit wide excursions. For this reason, one should not let ϵ_+ be too large at first. If the step sizes ϵ_n hover around the upper limit ϵ_+ and the algorithm is well behaved, then increase the value of ϵ_+. See [115] for an application to a problem in signal processing.

3.2.2 Some Data

We give some typical data for the adaptive step-size algorithm. The symbol $N(0,1)$ denotes a normally distributed random variable with zero mean and unit variance. These random variables are assumed to be mutually independent. Assume (2.1), with five unknown parameters and $\phi_n = (\psi_n, \psi_{n-1}, \ldots, \psi_{n-4})$, where $\psi_{n+1} = .5\psi_n + .5N(0,1)$ and $\theta_n = (\theta_{n,1}, \ldots, \theta_{n,5})$. One of the following models for the time-varying parameters will be used:

$$\bar{\theta}_{n+1,i} = (1 - .001)\bar{\theta}_{n,i} + .1N(0,1), \tag{2.8}$$

$$\bar{\theta}_{n+1,i} = (1 - .001)\bar{\theta}_{n,i} + \sqrt{.1}N(0,1), \tag{2.9}$$

$$\bar{\theta}_{n+1,i} = (1 - .0003)\bar{\theta}_{n,i} + \sqrt{.03}N(0,1). \tag{2.10}$$

Also, $y_n = \phi_n' \bar{\theta}_n + .5N(0,1)$. The stationary variances are the same for (2.9) and (2.10), but the parameters are "more identifiable" if the time variations are given by (2.10) since the correlation decreases more slowly. Note that in both cases the stationary variance of each component of θ_n is 50, rather large. Thus the parameters wander over a very wide range. The stationary variance for (2.8) is 5. Tables 2.1 and 2.2 concern (2.8), Tables 2.3 and 2.4 use (2.9), and the rest of the tables are for (2.10). In all cases, the algorithm begins with $\epsilon_0 = .01$.

Table 2.1 gives the sample mean square error for the fixed-ϵ algorithm (2.2a), so that we can see what the best value of ϵ is. The simulations involved 40 runs each of length 25,000. The sample averages for ϵ around 0.2 were very consistent, varying only slightly among the runs. Those for very small ϵ varied somewhat more, because the $\bar{\theta}_n$ are more highly correlated for these cases. Nevertheless, there was still a reasonable consistency. As ϵ increased above .25, the variations among the runs became more erratic. For example, when $\epsilon = .275$, most of the sample averages were between .8 and 1.0. But, out of 40 runs, there were values 17.9, 5.1, 4.6, 3.2, and several around 2. These general observations hold for the various random number generators that were used. What seems to be appearing as ϵ grows are error distributions with "heavy tails," as we expect from the basic structure of the algorithm (2.2a). Owing to these heavy tails, the estimates of the steady state mean square error becomes harder to obtain as ϵ gets beyond 0.275. This variability suggests that caution must be used with the upper truncation level. A truncation level of $\epsilon_+ = 0.3$ gave excellent adaptation results for our cases. The use of "feedback and averaging" analogous to what is discussed in the next section also worked well and is discussed in [150].

Tables 2.2, 2.4, and 2.6 give the average of the last 500 values of the adapted ϵ, and the values were close to the average. The variation in these averages between the runs was negligible. The adapted values of ϵ were generally very close to the optimal value, despite the flatness of the performance curve near that point. Note that the optimum value of ϵ is certainly far from infinitesimal, justifying an adaptive approach. As μ decreases, the rate of adaptation decreases. Note that the three values of μ in Table 2.2 differ by a factor of 100, with comparable results. It was also noted in [36] that the performance is much more sensitive to the values of ϵ (for the nonadapted algorithm) than to the value of μ for the adapted algorithm. The results for the finite difference form of the adaptive algorithm were about the same as those tabulated.

Tables 2.3 and 2.5 (resp., Tables 2.4 and 2.6) repeat Table 2.1 (resp., Table 2.2) for systems (2.9) and (2.10), respectively.

We emphasized problems where the parameter changes rapidly so that a large value of ϵ is needed. The method also works well for slower variations. If the parameter $\bar{\theta}_n$ were held fixed (i.e., time invariant), then ϵ_n would converge to zero as desired.

Table 2.1. Mean Square Errors
Fixed ϵ and (2.8) used

ϵ	Error
0.050	2.149
0.200	0.765
0.225	0.757
0.250	0.810
0.300	3.317

Table 2.2. SA Runs
Adaptive Algorithm, (2.8) Used

μ	Estimate $\bar{\epsilon}$
0.0100	0.208
0.0010	0.214
0.0001	0.214

Table 2.3. Mean Square Errors
Fixed ϵ and (2.9) Used

ϵ	Error
0.050	20.35
0.100	11.01
0.200	6.23
0.225	5.69
0.250	5.27
0.275	5.42
0.300	9.00

Table 2.4. SA Runs
Adaptive Algorithm, (2.9) Used

μ	Estimate $\bar{\epsilon}$
0.0100	0.230
0.0010	0.227
0.0001	0.272

Table 2.5. Mean Square Errors	
Fixed ϵ and (2.10) Used	
ϵ	Error
0.050	6.84
0.175	2.14
0.225	1.94
0.250	2.04
0.300	3.19

Table 2.6. SA Runs	
Adaptive Algorithm, (2.10) Used	
μ	Estimate $\bar{\epsilon}$
0.010	0.23
0.001	0.26

3.3 Feedback and Averaging in the Identification Algorithm

Return to the linear system model discussed at the beginning of Subsection 1.1, the assumptions used there, and the stochastic approximation algorithm (1.1.18). The averaging method of Subsection 1.3.3 can be used to get nearly optimal performance. Suppose that

$$\frac{\epsilon_n}{\epsilon_{n+1}} = 1 + o(\epsilon_n). \tag{3.1}$$

Then $n\epsilon_n \to \infty$. The rate of convergence of $\Theta_n = \sum_{i=0}^{n-1} \theta_i/n$ to $\bar{\theta}$ will be nearly optimal in the following sense. Let $S = 0$. Recall the definition of the least squares estimator $\widehat{\theta}_n$ defined by (1.1.11). It can be shown [164] that $nE[\widehat{\theta}_n - \bar{\theta}][\widehat{\theta}_n - \bar{\theta}]'$ converges to some finite covariance matrix V. By the results of Chapter 11, the sequence $\sqrt{n}(\Theta_n - \bar{\theta})$ converges in distribution to a normal random variable with mean zero and covariance matrix $\bar{V}$, which is optimal because it is the normalized asymptotic covariance of the least squares estimator. The computational requirements for (1.1.18) with iterate averaging are much less than those for the least squares (or linearized least squares) algorithm for problems of the dimensions that occur in practice; see also [56, 106].

Rewriting algorithm (1.1.18) and using the recursion for the average Θ_n just defined yields

$$\theta_{n+1} = \theta_n + \epsilon_n \phi_n \left[y_n - \phi_n' \theta_n \right], \tag{3.2}$$

$$\Theta_{n+1} = \Theta_n + \left(1 - \frac{1}{n+1}\right) \Theta_n + \frac{\theta_n}{n+1}. \tag{3.3}$$

Note the "two time-scale" form: Under the condition (3.1), for large n the process $\{\theta_n, n \geq N\}$ changes much faster than does $\{\Theta_n, n \geq N\}$. Indeed, the process $\{\theta_n\}$ plays the role of a driving noise process in (3.3). In fact, one might call the pair (3.2) and (3.3) a *three time-scale* problem, if the (faster) time scale of (y_n, ϕ_n) is counted. A major problem in the analysis arises from the $1/(n+1)$-time-dependence in the iteration (3.3), due to the $1/(n+1)$ term. The analysis of such processes falls outside the classical methods of stochastic approximation. The general averaging algorithm will be dealt with in Chapter 11.

Averaging with feedback. Algorithm (3.2)–(3.3) is "off-line" in the sense that the "best" values Θ_n are not used in the primary algorithm (3.2). One can ask whether there is any way of introducing Θ_n into the primary algorithm with profit. It can be shown that using $\phi_n'\Theta_n$ in place of $\phi_n'\theta_n$ in (3.2) will eliminate the advantages of averaging.

Now, let us feed the average of the estimates back into the primary algorithm. For $k > 0$, redefine θ_n by

$$\theta_{n+1} = \theta_n + \epsilon_n \phi_n [y_n - \phi_n' \theta_n] + \epsilon_n k [\Theta_n - \theta_n]. \tag{3.4}$$

Indeed, a feedback factor greater than the order of ϵ_n reduces the advantages of averaging. The algorithm (3.4) is a compromise. The rate of convergence of the average Θ_n is slower than it would be without the feedback, but the rate of convergence of θ_n is faster than without the feedback. One can also use $\epsilon_n = \epsilon$, and the feedback will improve the performance if k and the window of averaging are chosen appropriately. An analysis of the rate of convergence of the averaging/feedback algorithm, together with numerical data that shows its potential value, can be found in [151, 152].

The averaging can be time weighted by introducing various "forgetting factors" for data from the distant past by using

$$\Theta_{n+1} = (1 - b_n)\Theta_n + b_n \theta_n, \tag{3.5}$$

where

$$0 < b_n \to 0, \ \sum_{n=1}^{\infty} b_n = \infty,$$

or where

$$\Theta_{n+1} = (1 - \alpha\epsilon)\Theta_n + \alpha\epsilon\theta_n, \tag{3.6}$$

where $\alpha > 0$ is small; see [151, 152, 180].

3.4 Applications in Communications Theory

In communications systems, the transmitted signal is usually received in a corrupted form. The corruption is due to both the "filtering" effects of the channel through which the signal was transmitted and the effects of noise. Typically, the different frequency components of the transmitted signal are delayed and attenuated differently. Then the signal that is received at time n depends on the transmitted signal over some nonzero interval of time. The noise might be background noise which is added to the received signal at the point of reception, for example, atmospheric noise that appears at an antenna. It might be circuit or receiver noise due to the motions of electrons in the system. Other types of interference include echoes, noise introduced in the channel, and "cross talk" between channels. One would like to compensate for these effects by appropriate filtering of the received signals. When the exact probabilistic structure of the noise and the properties of the transmission channel are not known, some sort of adaptive processing is called for. We will briefly discuss several common problems that have been resolved by adaptive processing, where stochastic approximation-type algorithms are used.

3.4.1 Adaptive Noise Cancellation and Disturbance Rejection

Let $\{s_n\}$ denote a signal process, $\{\mu_n\}$ the corrupting noise process, and $y_n = s_n + \mu_n$ the received signal at time n. One would like to reduce or eliminate the effects of the corrupting noise μ_n. Suppose that at time n one can also observe another signal ψ_n such that the processes $\{\mu_n, \psi_n\}$ and $\{s_n\}$ are mutually independent, but $\{\mu_n\}$ is correlated with $\{\psi_n\}$ and $E\mu_n = E\psi_n = 0$. Then there is the possibility of using these independence and correlation properties to reduce the effect of μ_n on y_n, that is, use $\{\psi_n\}$ to partially cancel the effects of $\{\mu_n\}$. For a given integer k, define $z_n = \sum_{i=0}^{k} \alpha_i \psi_{n-i} = \theta' \phi_n$, where $\theta = (\alpha_0, \ldots, \alpha_k)$ and $\phi_n = (\psi_n, \ldots, \psi_{n-k})$. With the error defined by $e_n(\theta) = y_n - z_n$, one would like to find the value of θ that minimizes the right-hand side of (1.1). The adaptive algorithm (2.2a), where θ_n^ϵ is the estimate of the best value of θ at iterate n, is commonly used for this purpose. Suppose that there is a vector $\bar{S}$ such that the limit

$$\lim_{n,m} \frac{1}{m} \sum_{i=n}^{n+m-1} E_n \phi_i y_i = \bar{S} \tag{4.1}$$

exists in the sense of convergence in probability, and that (1.3a) holds for a matrix $Q > 0$. Then as in Subsection 1.1, under mild additional conditions, the mean ODE that characterizes the asymptotic behavior of θ_n^ϵ is

$$\dot{\theta} = \bar{S} - Q\theta, \tag{4.2}$$

and the corresponding stationary point is the minimizing (optimal) value of the parameter.

Applications. Many applications can be found in [96, 255]. First, consider the case where the signal y_n is received by a directional antenna, and the sequence $\{\mu_n\}$ is noise originating either in the atmosphere or in space.

Let there be an auxiliary antenna that simply receives other samples ψ_n of background noise, hopefully correlated with the disturbance μ_n. Then the adaptive noise cancellation algorithm can be used as discussed.

The interference to be reduced or canceled need not be what one would call noise. An example is the familiar 60-cycle interference in electrical circuits. In this case, one would try to make measurements of the interference as closely as possible to its source. Since our filter is defined in discrete time and the original signal is defined in continuous time, one needs to sample the original signal at times $n\Delta, n = 0, 1 \ldots$, for some $\Delta > 0$. Alternatively, one could work with a continuous-time adaptive filter, with essentially the same mathematical development used. It is useful to interpret the final (adaptive) filter in terms of its transfer function. If k were large enough and the samples were taken close enough together in real time, then one would see a "notch" at 60 cycles; that is, in the frequency domain, the adaptive filter acts to reduce the energy in a neighborhood of 60 cycles. This interpretation is obviously very important in applications, because it gives an intuitive and well understood physical meaning to the adaptation, and it is dealt with extensively in [96, 255] and other electrical engineering-oriented books and articles. Many of the practical applications are in continuous rather than discrete time, but the conditions and convergence proofs are nearly the same in continuous time and can usually be found by drawing the natural analogs.

Echo cancellation provides another class of examples where algorithms of the type of (2.2a) are used. Take, for example, wire communications, where there are two communicating people, one at either end, but there is "impedance mismatch" somewhere in the circuit. For example, in a "wire circuit" two different types of wire might be connected together. The mismatch causes a reflection of the signal, so that what each person receives at any time is a weighted sum of the signal from the other person and a delayed recent signal of their own (both signals possibly being distorted by the channel). Consider the simplest case, where only the signal from one side, say the left, is reflected back. Then the auxiliary measurements ψ_n are samples of the signal sent from the left, perhaps delayed. These are then weighed by the adaptive filter and subtracted from the received signal. Since multiple echoes can occur as well as "re-echoes" from the other side, the problem can become quite complicated.

There are many applications in vibration and noise control. Consider the problem of vibration control in an aircraft. The vibration typically originates in the engines, and one would like to reduce the audible noise effects of that vibration in the passenger cabin. Place one or more microphones in a suitable location(s) to get a noise sequence $\{\psi_n\}$ that is correlated as highly as possible with that affecting the cabin. The (negative of the) output z_n of the adaptive filter is a new noise process, which is induced in the cabin via some vibrating mechanism. Thus, the total effect is the difference $y_n - z_n$. The adaptive mechanism adjusts the weights so that they converge to the optimum, which gives the energy minimizing linear combination of the noises. In this case, there is no signal process s_n. The subject of noise and vibration control is of great current interest. An interesting reference to recent applications is [80].

In the preceding discussion, it was assumed that the adaptive filter output z_n could be subtracted immediately from y_n, which is not always true, even approximately. In the cited airplane example, the signal z_n produced by the adaptive filter might be electronic, and it causes another mechanism, perhaps a loudspeaker, to create the actual mechanical vibrations. One must take the mechanical and electronic "transfer function" properties of the driven mechanism into account in order to get the final noise effects. This complicates the adaptation problem and leads to many challenging problems that are yet to be resolved; see e.g., [133].

One of the best known successful applications of noise cancellation methods is the application by Widrow [255] to the problem of getting a good measurement of the heartbeat of a fetus, which is quite faint in comparison with the power of the beat of the mother's heart. In this case, y_n is the sum of the sounds from the hearts of the fetus and the mother, taken at an appropriate location on the mother's body. The auxiliary measurement ψ_n denotes the sound of both hearts, but taken at a location on the mother's body where the effects of the fetus' heartbeat are much weaker. Thus, the sound of the mother's heart is the "noise," which has much more power than the signal, namely, the heartbeat of the fetus. Yet the adaptive noise cancellation system effectively eliminates the bulk of the noise.

3.4.2 Adaptive Equalizers

The message that a user desires to send over a telecommunication channel is usually coded to remove redundant information and perhaps allows for the correction of errors at the receiver. Let $\{\psi_n\}$ denote the sequence of signals actually sent. The channel will then distort this sequence and may also add noise. Let y_n denote the channel output at time n, and let ψ_n denote the signal transmitted at time n. A channel equalizer is a filter whose input is the channel output and whose output is an estimate of the transmitted signal, perhaps delayed. Owing to time variations in both the channel and the distributions of the transmitted signal process, equalizers that adapt to the changes are an important part of many communications systems. The adaptation algorithms are of the stochastic approximation type.

There are two basic ways by which the equalizer is "trained." In the first method, a "training sequence" $\{\psi_n\}$ of actual transmitted signals is available at the output (for a sufficiently long time) so that a comparison can be made between the received signal and the true transmitted signal. After training, the resulting equalizer is used without change until another training sequence becomes available, and the process is repeated. In the second method, known as "blind equalization," there are at most short and occasional training sequences, and in between the adaptive system uses a suitable function of the outputs of the equalizer as surrogates for the unknown inputs ψ_n.

A training sequence is available: A moving average model. Suppose that the output of the equalizer (the estimator of ψ_n) has the form

$$\widehat{\psi}_n = \sum_{i=-D}^{k} \beta_i y_{n-i}, \tag{4.3}$$

where $D \geq 0$. One wishes to choose $\theta = (\beta_{-D}, \ldots, \beta_k)$ such that $E|\psi_n - \widehat{\psi}_n|^2$ is minimized. Define $\phi_n = (y_{n+D}, \ldots, y_{n-k})$. Note that if θ is a constant, then the gradient of $|\psi_n - \widehat{\psi}_n|^2/2$ with respect to θ is $-\phi_n[\psi_n - \widehat{\psi}_n]$. Then, following the line of thought used to get the algorithms in Example 3 of Section 1.1 or in the previous examples of this chapter, we use the algorithm

$$\theta_{n+1}^\epsilon = \theta_n^\epsilon + \epsilon\phi_n \left[\psi_n - \widehat{\psi}_n^\epsilon\right], \quad \widehat{\psi}_n^\epsilon = \phi_n'\theta_n^\epsilon, \tag{4.4}$$

where a constant step size ϵ is used, as is common in practice. Here, $\theta_n^\epsilon = (\beta_{n,-D}^\epsilon, \ldots, \beta_{n,k}^\epsilon)$ is the estimator of the best value of θ at time n, and $\widehat{\psi}_n^\epsilon = \phi_n'\theta_n^\epsilon$ is the estimator of ψ_n given by the adaptive system. The problem is the same as that considered in Subsection 1.1; under appropriate averaging conditions the mean ODE characterizing the limit points is

$$\dot{\theta} = -\text{grad}_\theta E|\psi_n - \widehat{\psi}_n|^2/2.$$

3.4.3 An ARMA Model, with a Training Sequence

The form of the equalizer used in the preceding paragraph is simple to analyze and behaves well in practice if D and k are large enough. To reduce the order, a model with feedback can be used. Consider the more general model for the input-output relationship of the channel:

$$\psi_n = \sum_{i=1}^{r_1} \bar{\alpha}_i \psi_{n-i} + \sum_{i=0}^{r_2} \bar{\beta}_i y_{n-i}, \tag{4.5}$$

where the values of the parameters $\bar{\theta} = (\bar{\alpha}_1, \ldots, \bar{\alpha}_{r_1}, \bar{\beta}_0, \ldots, \bar{\beta}_{r_2})$ are unknown, but $\bar{\beta}_0 > 0$ and the orders are assumed known. The way that equation (4.5) is written might suggest that it defines ψ_n as a function of the variables on the right side, but it is actually intended to define y_n in terms of the other variables. Note that there is no additive noise in (4.5). Analogously to (4.5), the equalizer with a fixed parameter

$$\theta = (\alpha_1, \ldots, \alpha_{r_1}, \beta_0, \ldots, \beta_{r_2})$$

has the form

$$\widehat{\psi}_n = \sum_{i=1}^{r_1} \alpha_i \widehat{\psi}_{n-i} + \sum_{i=0}^{r_2} \beta_i y_{n-i}, \tag{4.6}$$

where $\widehat{\psi}_n$ is the estimator of ψ_n. The value of θ is to be chosen. Define

$$\phi_n^\epsilon = (\widehat{\psi}_{n-1}^\epsilon, \ldots, \widehat{\psi}_{n-r_1}^\epsilon, y_n, \ldots, y_{n-r_2}),$$

$$\phi_n = (\psi_{n-1}, \ldots, \psi_{n-r_1}, y_n, \ldots, y_{n-r_2}),$$

and let θ_n^ϵ be the nth estimate of $\bar{\theta}$. With the time-varying parameter, the output of the adaptive equalizer (the estimator of ψ_n) is

$$\widehat{\psi}_n^\epsilon = (\phi_n^\epsilon)' \theta_n^\epsilon. \tag{4.7}$$

Again we wish to minimize $E|\psi_n - \widehat{\psi}_n^\epsilon|^2$. With the given definitions, the algorithm is

$$\theta_{n+1}^\epsilon = \theta_n^\epsilon + \epsilon \phi_n^\epsilon \left[\psi_n - \widehat{\psi}_n^\epsilon \right]. \tag{4.8}$$

The form is quite complicated since both ϕ_n^ϵ and the error $\psi_n - \widehat{\psi}_n^\epsilon$ have a complex dependence on the history of the past iterate sequence. This is another example of state-dependent noise. If the model were Markov, then it would be of the type introduced in Sections 2.3 and 2.4.

A linear systems interpretation. Some insight into the algorithm (4.8) can be gained by putting it into a form where the role of the transfer function of the linear system (4.5) is apparent. Let q^{-1} denote the unit delay operator. That is, $q^{-1} y_n = y_{n-1}$. Define the transfer functions

$$A(q^{-1}) = 1 - \sum_{i=1}^{r_1} \bar{\alpha}_i q^{-i},$$

$$B(q^{-1}) = \sum_{i=0}^{r_2} \bar{\beta}_i q^{-i},$$

$$T(q^{-1}) = \frac{A(q^{-1})}{B(q^{-1})}.$$

Then $A(q^{-1})\psi_n = B(q^{-1}) y_n$, and $T(q^{-1})$ is the transfer function of the channel. Suppose that the roots of the polynomials $A(z)$ and $B(z)$ in the complex plane are strictly outside the unit circle, so that the systems $x_n - \sum_{i=1}^{r_1} \bar{\alpha}_i x_{n-i} = 0$ and $\sum_{i=0}^{r_2} \bar{\beta}_i x_{n-i} = 0$ are asymptotically stable. [The requirement of stability of the first system can be a serious restriction in practice.] Then, asymptotically, we can write

$$\psi_n = T^{-1}(q^{-1}) y_n. \tag{4.9}$$

Thus, the ideal equalizer has a transfer function that is just the inverse of that of the channel, which is hardly surprising.

Define $\widetilde{\theta}^\epsilon_n = \theta^\epsilon_n - \bar{\theta}$ and let us write the error $\psi_n - \widehat{\psi}^\epsilon_n$ in terms of the sequence $\{(\phi^\epsilon_n)'\widetilde{\theta}^\epsilon_n\}$. First note that we can write

$$(\phi^\epsilon_n)'\widetilde{\theta}^\epsilon_n = \sum_{i=1}^{r_1}(\alpha^\epsilon_{n,i} - \bar{\alpha}_i)\widehat{\psi}^\epsilon_{n-i} + \sum_{i=0}^{r_2}(\beta^\epsilon_{n,i} - \bar{\beta}_i)y_{n-i}, \tag{4.10}$$

where $\alpha^\epsilon_{n,i}$ and $\beta^\epsilon_{n,i}$ are the components of θ^ϵ_n. Equation (4.10) is sometimes referred to as the *equation error.*

We can now write

$$\psi_n - \widehat{\psi}^\epsilon_n = -(\phi^\epsilon_n)'\widetilde{\theta}^\epsilon_n + \sum_{i=1}^{r_1}\bar{\alpha}_i(\psi_{n-i} - \widehat{\psi}^\epsilon_{n-i}). \tag{4.11}$$

Symbolically,

$$\psi_n - \widehat{\psi}^\epsilon_n = -A^{-1}(q^{-1})[(\phi^\epsilon_n)'\widetilde{\theta}^\epsilon_n],$$

and using this we can rewrite (4.8) in terms of the $\widetilde{\theta}^\epsilon_n$ as

$$\widetilde{\theta}^\epsilon_{n+1} = \widetilde{\theta}^\epsilon_n - \epsilon\phi^\epsilon_n A^{-1}(q^{-1})[(\phi^\epsilon_n)'\widetilde{\theta}^\epsilon_n]. \tag{4.12}$$

The form (4.12) is an approximation since it ignores the initial condition and contains terms $(\phi^\epsilon_n)'\widetilde{\theta}^\epsilon_n$ for negative values of n. If the system has been in operation for a long time before the adaptive adjustment began or if n is large, then the approximation has negligible effect provided that $\sup_n E|\phi_n|^2 < \infty$ and the sequence $\{\theta_n\}$ is pathwise bounded. The latter bound will be shown to be the case in Chapter 9. The form of the algorithm (4.12) indicates the dependence of the limiting behavior of $\{\theta^\epsilon_n\}$ on the transfer function $A(q^{-1})$. Proof of the stability of the algorithm requires that $A(q^{-1})$ be "strictly positive real" in that the real part of $A(z)$ is strictly positive on the unit circle in the complex plane. This positive real condition and suitable conditions on the sequence $\{\psi_n\}$ guarantee that there is a positive definite (but not necessarily symmetric) matrix A such that the mean ODE is $\dot{\widetilde{\theta}} = -A\widetilde{\theta}$. Under somewhat weaker conditions on the sequence $\{\psi_n\}$, one gets $\dot{\widetilde{\theta}} \in -G\widetilde{\theta}$ in lieu of the ODE, where G is a compact convex set of positive definite matrices, and convergence will still occur (see Chapter 9). The strict positive real condition can be a restriction in practical applications. The fundamental role played by the strict positive real condition in applications to systems theory, particularly to the stability problem, was apparently first appreciated and systematically examined by Ljung [165], although it was used earlier in the formal analysis of various parameter identification algorithms. A simple illustration of its role is in [225, Section 6.2]. Conditions playing the same role (although not obviously) as the strict positive real condition had appeared earlier in related problems in statistics and Toeplitz forms (e.g., [88, Theorem a, p. 20]).

Blind equalization. If the value of ψ_n at some time n is not available to the adaptive system, then it is often estimated as follows. Suppose that ψ_n takes values in a finite set $\mathcal{D}$. For example, for a binary signal, one can use $\mathcal{D} = \{+1, -1\}$. The final decision concerning the value of the transmitted signal ψ_n is given by a $\mathcal{D}$-valued decision function $q(\cdot)$, where $q(\widehat{\psi}_n^\epsilon)$ is the decision. Then $q(\widehat{\psi}_n^\epsilon)$ replaces ψ_n in (4.4). The analysis of such algorithms is quite difficult, as can be seen from the development in [15, 16].

3.5 Adaptive Antennas and Mobile Communications

Problem formulation. Adaptive antenna arrays have been used for a long time with great success [54, 254]. The output of each antenna in the array is multiplied by a weight, then the outputs are added; see Figure 5.1. With appropriate choice of the weights, the effects of noise, and other interfering signals can be reduced and the directional properties controlled. The weights are adjusted adaptively, using measurements of the output, and perhaps some pilot or reference signal. The algorithm for adjusting the weights is typically some form of recursive least squares scheme or an approximation to such a scheme. In this section, we are concerned with the use of adaptive arrays in mobile communications [74, 236, 257]; in particular, optimizing reception at the base station of a single cell system with r antennas. The updates of the antenna weights are to be done in discrete time.

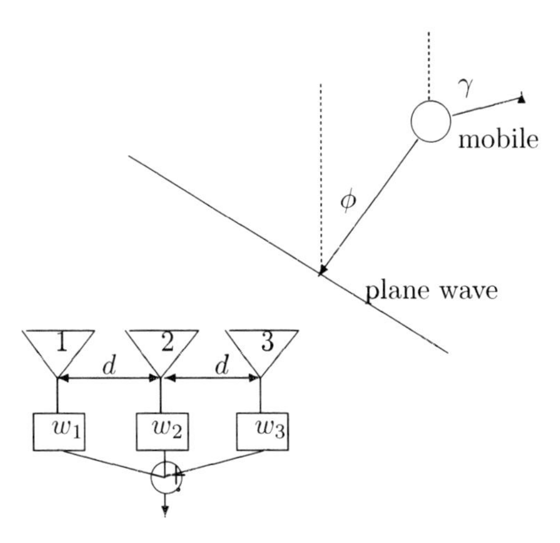

Figure 5.1. A Three Antenna Array.

The usual algorithms for adaptive arrays depend on parameters that are held fixed no matter what the operating situation, and the performance can strongly depend on the values of these parameters. We are concerned with the adaptive optimization of such parameters by the addition of another adaptive loop. The method can improve the operation considerably. It is common to work with complex-valued signals and weights, to conveniently account for phase shifts. To simplify the notation, we concatenate the real and complex parts of the antenna outputs and weights, so that, unless otherwise noted, we use real-valued variables.

Let $x_{n,i}, i \leq r$, denote the output of antenna i at measurement time n; it is the sum of the signals due to all of the mobiles, plus additive noise. The system is optimized for each mobile separately, and we select one. Let $w_{n,i}$ denote the weight assigned to antenna i at that time. Define the vectors $X_n = \{x_{n,i}, i \leq r\}$, $w_n = \{w_{n,i}, i \leq r\}$. Let $\{\bar{s}_n\}$ denote a real-valued and known pilot training sequence from the particular mobile that is being tracked. The algorithm to be presented also works well with partially blind adaptation. The weighted output is $w_n' X_n$. Let $s_{n,j}$ denote the signal from mobile j at the nth sampling time. Thus, $\bar{s}_n = s_{n,j}$ for the chosen mobile j.

The algorithm for adapting the antenna weights. For $\alpha \in (0, 1]$ and fixed weight vector w, define the discounted cost

$$J_n(\alpha, w) = \sum_{l=1}^{n} \alpha^{n-l} e_l^2(w), \quad e_n(w) = \bar{s}_n - w' X_n. \tag{5.1}$$

Typically, one uses $\alpha < 1$ to allow tracking of changing circumstances. The minimizing value of w in (5.1) is (compare with (1.1.11))

$$w_n(\alpha) = P_n \sum_{l=1}^{n} \alpha^{n-l} X_l \bar{s}_l, \quad P_n = \left[\sum_{l=1}^{n} \alpha^{n-l} X_l X_l' \right]^{-1}. \tag{5.2}$$

The recursive least squares algorithm for computing w can be written as (see [166])

$$\begin{aligned}
w_{n+1}(\alpha) &= w_n(\alpha) + L_n e_n(\alpha), \quad e_n(w_n(\alpha)) = \bar{s}_n - w_n'(\alpha) X_n, \\
L_{n+1} &= \frac{P_n X_{n+1}}{\alpha + X_{n+1}' P_n X_{n+1}}, \\
P_{n+1} &= \frac{1}{\alpha} \left[P_n - \frac{P_n X_{n+1} X_{n+1}' P_n}{\alpha + X_{n+1}' P_n X_{n+1}} \right].
\end{aligned} \tag{5.3}$$

P_n and L_n also depend on α. The recursive least squares algorithm ((1.1.14), (1.1.15)) is the special case, where $\alpha = 1$. The results of extensive simulations of this algorithm for mobile communications were reported in [257]. The quantity α is known as a *forgetting factor*, in that it allows more weight to be put on the more recent observations.

The desired performance in the sense of mean square detection error or bit error rate can be quite sensitive to the value of α, Consider the error criterion $Ee_n^2(w_n(\alpha))$ for large n. If the mobiles do not move and the variance of the additive noise constant, then the minimizing value of α will be unity. Suppose that $\{X_n, \bar{s}_n\}$ is "locally" stationary. Then, if the mobiles (particularly the one being tracked) are moving rapidly and the additive noise level is small, the optimal value of (in the sense of minimizing a mean square detection error) α will be relatively small. In practice, the optimal value might vary rapidly, perhaps changing significantly many times per second, as the operating conditions change. In what follows, an adaptive algorithm for finding or tracking the optimal discount factor will be developed. The procedure is similar to that for adaptively optimizing the step size in Section 2, except that it is based on a finite difference approximation. Keep in mind that that the purpose of the algorithm for adapting α is to improve the desired performance. It does this since it is based on a mean square detection error minimization. As noted in the comments on mean bit error rates below, the desired performance is improved.

The adaptive algorithm for α. The time-varying parameter tracking algorithm of Section 2 was based on a "derivative," or "mean-square derivative." In our case, the dependence of the algorithm (5.3) on α is quite complicated. To avoid dealing with the rather messy forms that would result from a differentiation of the right sides of (5.3) with respect to α, we simply work with a finite difference form. Typically, $Ee_n^2(w_n(\alpha))$ is strictly convex and continuously differentiable. The value increases sharply as α increases beyond its optimal value, and increases more slowly as α decreases below its optimal value. It is somewhat insensitive to α around the optimal value. The proof of convergence of the algorithm to be developed in discussed below Theorem 8.4.7.

Let $\delta > 0$ be a small difference interval and let α_n denote the value of α at the nth update. Run algorithm (5.3) for both $\alpha_n^+ = \alpha_n + \delta/2$ and $\alpha_n^- = \alpha_n - \delta/2$. The algorithm for adapting α is, for small $\epsilon > 0$,

$$\alpha_{n+1} = \alpha_n - \epsilon \frac{[e_n^+]^2 - [e_n^-]^2}{\delta}, \tag{5.4}$$

where

$$\begin{aligned}
e_n^\pm &= \bar{s}_n - X_n' w_n^\pm, \\
w_{n+1}^\pm &= w_n^\pm + L_n^\pm e_n^\pm, \\
L_{n+1}^\pm &= \frac{P_n^\pm X_{n+1}}{\alpha_n^\pm + X_{n+1}' P_n^\pm X_{n+1}}, \\
P_{n+1}^\pm &= \frac{1}{\alpha_n^\pm}\left[P_n^\pm - \frac{P_n^\pm X_{n+1} X_{n+1}' P_n^\pm}{\alpha_n^\pm + X_{n+1}' P_n^\pm X_{n+1}}\right].
\end{aligned} \tag{5.5}$$

The initial conditions w_0 and P_0 are obtained by a least-squares solution using a small initial block of data.

If α is fixed, then for small $\delta > 0$, the (stationary) expectation of the coefficient of ϵ in (5.4) should be close to the derivative of $Ee_n^2(w_n(\alpha))$ with respect to α. The intuitive idea behind the algorithm is that the value of α changes much more slowly than that of w, so that we are essentially in the stationary regime. In this case, we clearly have a stochastic algorithm driven by a process whose values are noisy estimates of a finite difference estimate of the negative of a derivative. The basic stochastic approximation algorithm is (5.4), since it is ϵ that is small. The quantities $(X_n, w_n^{\pm}, \bar{s}_n, L_n^{\pm}, P_n^{\pm})$ play the role of noise. Owing to the way that the evolution of this noise is tied to that of α_n we have another example of state-dependent noise.

A model for the signals from the mobiles. The behavior of the algorithm (5.4) and (5.5) will be illustrated by simulation data, a subset of those from [41]. We model a simple situation, which exhibits the essential features. The mobiles move in two dimensions and there are three evenly spaced antennas, with spacing $d > \lambda/2$, where λ is the carrier wavelength (in the simulation, the carrier is 800×10^6 Hz). There is one interfering mobile. The signal from neither mobile is scattered. [The behavior is similar for a larger number, and for scattering signals.] The effects of additional interferers with uniform scattering (Rayleigh fading) are modeled by adding complex-Gaussian (independent) noise to each antenna. In this example, the sets of signals from the different mobile are independent of each other and the members of each set are mutually independent. The known pilot signals $\{\bar{s}_n\}$, for the tracked or desired user, are binary-valued $(+1, -1)$.

The (complex-valued) antenna signature of the mobile corresponding to a plane wave arriving at an angle ϕ to the normal to the plane of the antennas (see Figure 5.1) is given by the vector (antenna 1 is the reference antenna)

$$c(\phi) = \left[1,\ \exp\left(-i\frac{2\pi}{\lambda}d\,\sin\phi\right),\ \exp\left(-i\frac{2\pi}{\lambda}2d\,\sin\phi\right)\right]. \tag{5.6}$$

The Doppler frequency of mobile j at time t is

$$\omega_j^d(t) = -\frac{2\pi}{\lambda}v_j(t)\cos(\phi_j(t) - \gamma_j(t)), \tag{5.7}$$

where $\gamma_j(t)$ is the angle of the travel of mobile j (see the figure), $v_j(t)$ its speed, and $\phi_j(t)$ the angle of arrival of its plane wave. The (complex-valued) component of the received signal at the antenna array at sampling time nh due to mobile j is given by

$$\bar{x}_j(nh) = \frac{1}{d_j^2(nh)}s_{n,j}\exp\left(i\left[\psi_j(0) + \int_0^{nh}\omega_j^d(s)ds\right]\right)c(\phi_j(nh)), \tag{5.8}$$

where $\psi_j(0)$ is the initial phase, $d_j^2(t)$ is the distance between mobile j and the array and $c_j(\phi)$ is (5.5) for mobile j. In the figures, the time between samples is $h = 4 \times 10^{-5}$.

In applications, the process X_n will rarely be stationary. For example, see the mobility model (5.8), where the dominant effect is that of the Doppler frequency. Some additional details on the averaging of signals arising from the model (5.8) can be found after Theorem 8.4.7.

Bit error rates. The algorithm tracks the optimal value of α. But the value of α is secondary to the actual performance in terms of bit error rates. For the example plotted, the bit error rate for the adaptive procedure reached a high of .005 during the high Doppler frequency section, and dropped to below .002 for the low and moderate Doppler frequency section. The error rates for the fixed-α algorithm depended on the choice of α. For $\alpha = .96$, a not unreasonable value, the error rate was about ten times as high. The best fixed value is $\alpha = 0.84$, for which the error rate is still about 20% higher. Of course, one rarely knows the optimal value, and in all the runs taken, the bit error rate for the adapted-α algorithm was always better than that for any fixed value.

Sample paths of α_n. It is interesting to see how the value of α that yields the best performance is tracked. Note that if the mobiles are stationary, where $\alpha = 1$ is optimal, then the values of α_n in the adaptive procedure remained close to unity. In the plot, the signal to interference ratio is approximately 5.3db, and the interference to noise ratio is approximately 1.3db, all measured at the antennas. The direction and velocity of each mobile evolved as a semi-Markov process, each moving independently of the others. They were constant for a short random interval, then there was sudden acceleration or deceleration in each coordinate, and so forth. Mobile 2 is always the desired one. Only the associated piecewise constant Doppler shifts are given, since that is the most important factor. Note the 10^4 factor in the vertical scale for the Doppler frequency. The path of α_n "tracks" the Doppler frequency of the desired user in all cases, and that there is an improvement (sometimes significant) in the performance, over that corresponding to the use of constant values of α. The performance is much less sensitive to the value of ϵ than to the value of α, which supports the conclusions in [150].

Except for the brief transient periods, the values of α are close to the optimum. Changing the value of ϵ up or down by a factor of four had little effect on the overall performance. When smaller ϵ is used, the paths of α are smoother, the transient period is longer, but the overall performance is very similar. Note that the behavior of the Doppler frequency of the interfering mobile has negligible effect. Except for the cases of very high Doppler frequencies, the performance is approximately the same if blind adaptation is used, with the pilot signal being used for initialization and then only intermittently.

When the additive noise is increased, the optimal value of α increases. In all cases, the adaptive algorithm outperformed the constant α forms,

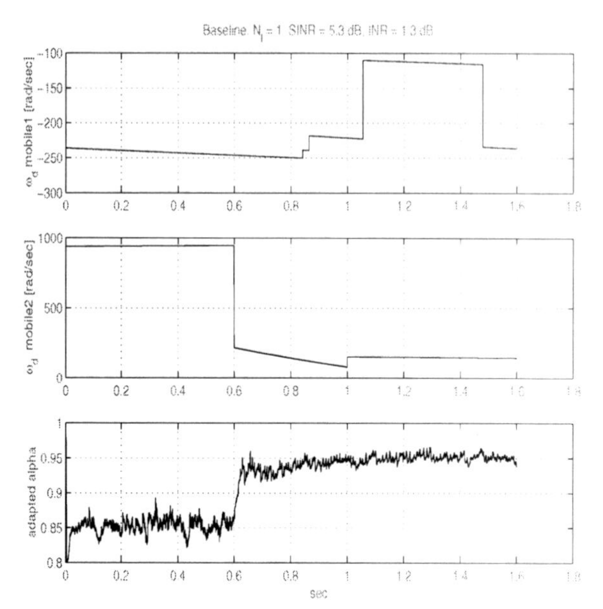

Figure 5.2. Tracking the optimal value of α.

sometimes significantly. If the variance of the additive noise is decreased, then the optimal values of α decrease, and the adapted values are still close to the optima. When the noise is smaller, other properties of the paths of the desired and interfering mobiles play a greater role, although the dominant influence is still the Doppler frequency of the desired mobile; see [41] for additional detail.

3.6 Mobile Communications with Time Varying Channels: Proportional Fair Sharing

We now consider another problem in wireless communications. There is a single base station transmitter, and r users competing to transmit data to their mobile destinations. Time is divided into small scheduling intervals (also called slots). In each interval one of the users is chosen to transmit. Due to movement of the mobiles, the characteristics of the connecting channels, hence the possible rates of transmission of the individual users in the slots, are randomly time varying. If user i is selected in interval n, then it transmits $r_{i,n}$ units of data, where $\{r_{i,n}, n < \infty\}$ is bounded and correlated in n. We will suppose that each user has an infinite backlog of data. This corresponds to the situation where r is the maximum number of allowed users, the system is full and each user has a large queue. More general queueing models can also be handled [146, 147].

Recent cellular systems [14] allow for the scheduling of the users, based on estimates of the current rates. The rates are determined by the use of pilot signals sent from the mobiles to the base station. One can get good estimates of the signal to noise ratio (SNR) for each user, and use this to determine the possible rates. The time slot of the system in [14] has duration 1.67 ms. Since the time between measurement and prediction of the rate is very short, fairly accurate rate predictions can be made. The great variation of rate as a function of SNR can be seen in Table 6.1. The gains that can be realized by exploiting the channel fluctuations can be seen from the table and from [2, 14].

SNR (in db)	-12.5	-9.5	-8.5	-6.5	-5.7	-4.0
Rate	0.0	38.4	76.8	102.6	153.6	204.8
SNR (in db)	-1.0	1.3	3.0	7.2	9.5	
Rate	307.2	614.4	921.6	1228.8	1843.2	

Table 6.1. Rate vs. SNR for 1% packet loss, taken from [14].

The selection of the user for transmission at any time is based on a balance between the current possible rates and "fairness." One cannot choose the user with the highest rate at each slot, since users with generally lower rates will be seriously penalized. The question then is how to share the slots fairly. Fair sharing will lower the total throughput over the maximum possible, but it will provide more acceptable levels to users with poorer SNRs. The algorithm to be considered (which is that used in the system discussed in [14]) performs this sharing by comparing the given rate for each user with its average throughput to date, and selecting the one with the maximum ratio. It is known as *proportional fair sharing* (PFS), and apparently was originally developed for the problem of allocating connections over multiple links on the Internet [107]. Let the end of time slot (equivalently, scheduling interval) n be called time n. At time n, it is assumed that the possible rates $\{r_{n,i+1}, i \leq r\}$ for the next time slot are known. Let $I_{n+1,i}$ denote the indicator function of the event that user i is chosen at time n (to transmit in slot $n+1$). One definition of the throughput for user i up to time n is the sample average

$$\theta_{n,i} = \frac{1}{n} \sum_{l=1}^{n} r_{l,i} I_{l,i}. \tag{6.1}$$

With the definition $\epsilon_n = 1/(n+1)$, (6.1) can be written in recursive form as

$$\theta_{n+1,i} = \theta_{n,i} + \epsilon_n \left[I_{n+1,i} r_{n+1,i} - \theta_{n,i} \right] = \theta_{n,i} + \epsilon_n Y_{n,i}. \tag{6.2}$$

An alternative definition of throughput discounts past values. For small $\epsilon > 0$, and discount factor $1 - \epsilon$, the discounted throughput is defined by

$$\theta^{\epsilon}_{n,i} = (1-\epsilon)^n \theta^{\epsilon}_{0,i} + \epsilon \sum_{l=1}^{n} (1-\epsilon)^{n-l} r_{l,i} I^{\epsilon}_{l,i}. \tag{6.3}$$

This can be written as

$$\theta^{\epsilon}_{n+1,i} = \theta^{\epsilon}_{n,i} + \epsilon \left[I^{\epsilon}_{n+1,i} r_{n+1,i} - \theta^{\epsilon}_{n,is} \right] = \theta^{\epsilon}_{n,i} + \epsilon Y^{\epsilon}_{n,i}, \tag{6.4}$$

where $I^{\epsilon}_{n+1,i}$ denotes the indicator function of the event that user i is chosen at time n with this algorithm. The value of ϵ is chosen to balance the needs of estimating throughput (requiring a small value of ϵ) with the ability to track changes in the channel characteristics (requiring a larger value of ϵ).

Let $d_i, i \leq r$, be positive numbers, which can be as small as we wish. For rule (6.2), the user chosen at time n is that determined by

$$\arg \max_{i \leq r} \{ r_{n+1,i}/(d_i + \theta_{n,i}) \}, \tag{6.5}$$

which defines proportional fair sharing. In the event of ties, for specificity, randomize among the possibilities. Recall that θ^i denotes the ith component of the nonrandom vector θ. Define the functions $\bar{h}_i(\cdot), i \leq r$, by the stationary expectation:

$$\bar{h}_i(\theta) = E r_i I_{\{r_i/(d_i+\theta^i) \geq r_j/(d_j+\theta^j), j \neq i\}} \tag{6.6}$$

Then the mean ODE is

$$\dot{\theta}^i = \bar{h}_i(\theta) - \theta^i, \quad i \leq r. \tag{6.7}$$

The ODE (6.7) is cooperative in the sense of Section 4.4. This is because, if the current throughput of any user $j \neq i$ is suddenly increased, the value of the average instantaneous rate $\bar{h}_i(\theta)$ for user i cannot be decreased. The monotonicity property of Theorem 4.4.1 can be used to show, under reasonable conditions, that the mean ODE has a unique and globally asymptotically stable limit point $\bar{\theta}$; see Chapter 9 and [146, 147]. The algorithm and convergence results are typical of a large family of related algorithms, with each corresponding to some concave utility function, which in turn is actually maximized by the associated algorithm.

Figure 6.1 plots the solution to the ODE for a two mobile problem with Rayleigh fading at rate 60Hz, and mean rates of 572 and 128 bits per slot, together with the paths for a simulation with $\epsilon = .0001$.

An example with discontinuous dynamics. The convergence proof requires that $\bar{h}(\cdot)$ be Lipschitz continuous. The following example concerns a system where there are two users, and the rates are fixed at r_1, r_2, not random. We can still do a weak converge analysis of $\theta^n(\cdot)$, although the mean ODE has discontinuous dynamics. The limit flow is shown in Figure 6.2. The vector field of the flow is unique for θ not on the line with slope r_2/r_1. On that line it takes values in the convex hull of the flow vectors at the neighboring points. There is still a unique limit $\bar{\theta} = (r_1/2, r_2/2)$ which is globally asymptotically stable. Numerical data show that, under "practical" conditions, the paths of $\theta^{\epsilon}(\cdot)$ follow the solution of the ODE very well.

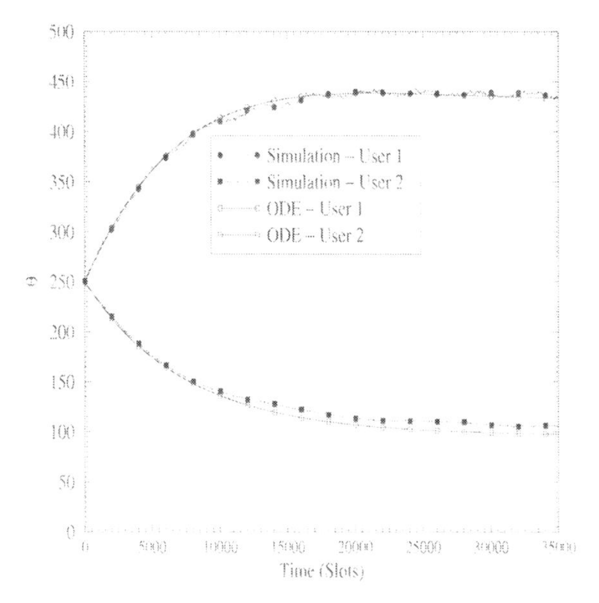

Figure 6.1. A sample path vs. solution to the ODE.

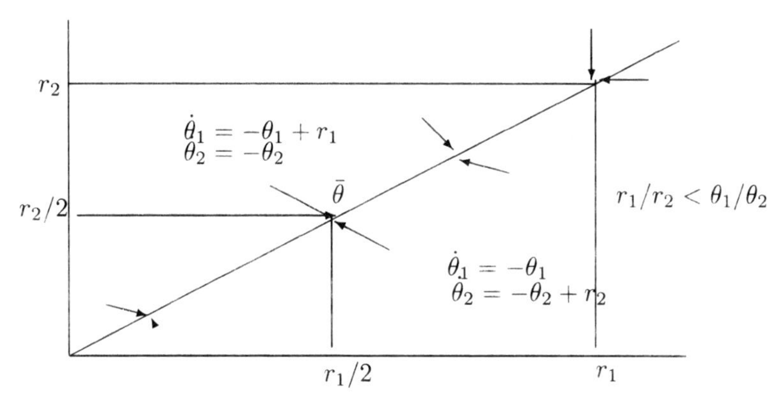

Figure 6.2. r_1, r_2 constant. Discontinuous dynamics.

An extension: Discretized or quantized rates. The basic algorithm is flexible and can be adjusted to accommodate special needs, with essentially the same proofs. Suppose, for example, that the variables $r_{n,i}$ are the theoretical rates in that there is in principle a transmission scheme that could realize them, perhaps by adjusting the symbol interval and coding in each scheduling interval. In applications, it might be possible to transmit at only one of a discrete set of rates. The algorithms and results are readily adjusted to accommodate this need. Continue to make the assignments using (6.5), based on the values $r_{n+1,i}, i \leq r$, but use the true transmitted rate, called $r^d_{n,i}$ in computing the throughput. Then (6.2) becomes

$$\theta_{n+1,i} = \theta_{n,i} + \epsilon_n \left(I_{n+1,i} r^d_{n+1,i} - \theta_{n,i} \right) = \theta_{n,i} + \epsilon_n Y_{n,i}, \tag{6.8}$$

The rule (6.5) maximizes a utility function. Define the utility function

$$U(\theta) = \sum_i \log(d_i + \theta^i). \tag{6.9}$$

Then, under (6.2), (6.5) maximizes $U(\theta_{n+1}) - U(\theta_n)$ to first order in the ϵ_n: By a first order Taylor expansion with remainder, we have

$$U(\theta_{n+1}) - U(\theta_n) = \epsilon_n \sum_i \frac{r_{n+1,i} I_{n+1,i} - \theta_{n,i}}{d_i + \theta_{n,i}} + O(\epsilon_n^2). \tag{6.10}$$

Since $\sum_i I_{n+1,i} \equiv 1$, maximizing the first order term requires that $I_{n+1,i}$ be chosen as in (6.5). The analogous result holds if θ_n^ϵ replaces θ_n. The rule (6.5) also maximizes $\lim_n U(\theta_n)$ [146, 147].

Other utilities. In view of the above comments, it can be said that the assignment algorithm (6.5) is based on the utility function (6.9), but any strictly concave utility function can be used instead yielding another assignment rule. Consider, for example,

$$U(\theta) = \sum_i c_i (\theta^i + d_i)^\gamma, \text{ where } 0 < \gamma < 1, \ d_i \geq 0. \tag{6.11}$$

Then

$$\dot{U}(\theta) = \gamma \sum_i \frac{c_i \dot{\theta}^i}{(\theta^i + d_i)^{1-\gamma}}. \tag{6.12}$$

The chosen user is

$$\arg\max_{i \leq r} \{ c_i r_{n+1,i} / (\theta_{n,i} + d_i)^{1-\gamma} \}. \tag{6.13}$$

The mean ODE is

$$\dot{\theta}^i = E r_{n,i} I_{\{c_i r_{n,i}/(\theta^i + d_i)^{1-\gamma} \geq c_j r_{j,n}/(\theta^j + d_j)^{1-\gamma}, j \neq i\}}. \tag{6.14}$$

There is still a unique globally asymptotically stable limit point. As γ increases, more attention is given to the users with the higher current rates.

Multiple transmitters. Up to now, there was only a single transmitter to be assigned. Similar results hold when there are multiple resources to be assigned. To illustrate the possibilities, consider the following special case, where there are two transmitting antennas, perhaps operating at different frequencies. The rates are $r_{n,ij}$ for user i in channel j at scheduling interval n.

One can allow many alternatives for use of the channels. The simplest is to assign by using (6.5) for each channel separately. Then, depending on the rates, both channels might be assigned to one user, or each might be assigned to a different user. Equivalently, one assigns so as to maximize the first order term in $U(\theta_{n+1}) - U(\theta_n)$. For the two user case, (6.2) is changed to

$$\theta_{n+1,1} = (1 - \epsilon_n)\theta_{n,1} + \epsilon_n r_{n+1,11} I_{\{r_{n+1,11}/r_{n+1,21} \geq (d_1+\theta_{n,1})/(d_2+\theta_{n,2})\}}$$
$$+\epsilon_n r_{n+1,12} I_{\{r_{n+1,12}/r_{n+1,22} \geq (d_1+\theta_{n,1})/(d_2+\theta_{n,2})\}},$$

with the analogous formula for the other user.

Next consider an alternative that allows the resources to be used more efficiently by admitting other options. We allow the above choices, based on assigning each channel independently. But we also allow the possibility that the two antennas are assigned to the same user, with (for example) space-time coding [233]. This simply adds another possible rate. The full algorithm increases the channel capacity over what space-time coding by itself could achieve. Let $r^c_{n,i}$ denote the rate using space-time coding when both channels are assigned to user i and let $I^c_{n,i}$ denote the indicator of this event. Let $I_{n+1,ij}$ denote the indicator function of the event that user i is assigned to channel j in interval n but space-time coding is not used. Then, up to the first order in ϵ_n, $U(\theta_{n+1}) - U(\theta_n)$ equals ϵ_n times

$$\frac{r_{n+1,11}I_{n+1,11}}{d_1+\theta_{n,1}} + \frac{r_{n+1,12}I_{n+1,12}}{d_1+\theta_{n,1}} + \frac{r_{n+1,21}I_{n+1,21}}{d_2+\theta_{n,2}} + \frac{r_{n+1,22}I_{n+1,22}}{d_2+\theta_{n,2}}$$
$$+\frac{r^c_{n+1,1}I^c_{n+1,1}}{d_1+\theta_{n,1}} + \frac{r^c_{n+1,2}I^c_{n+1,2}}{d_2+\theta_{n,2}} - \frac{\theta_{1,n}}{d_1+\theta_{1,n}} - \frac{\theta_{2,n}}{d_2+\theta_{2,n}}.$$

This yields a slightly more complicated form of the arg max rule of (6.5). The method of analysis is essentially the same and there is a unique globally asymptotically stable limit point in all cases. See [146, 147] for additional details.

4
Mathematical Background

4.0 Outline of Chapter

In this chapter we collect a number of mathematical ideas that will be frequently used in the sequel. The martingale process, discussed in Section 1, is one of the basic processes in stochastic analysis. It appears frequently as a "noise" term in the decomposition of the stochastic approximation algorithms, which will be used to facilitate the analysis. This was already apparent in many examples in Chapters 1 and 2, where the noise had the martingale difference property. Our method for analyzing the asymptotic properties of the stochastic approximation process involves showing that the asymptotic paths are those of the ordinary differential equation (ODE) determined by the "mean" dynamics of the algorithm. Some basic facts concerning ODEs and the Liapunov function method of proving their stability are discussed in Section 2. The limit points can be restricted to the set of chain recurrent points (Section 4). Section 3 concerns ODEs in which the dynamics are projected onto a compact constraint set. The solutions of such ODEs will appear as "limits" of the paths of constrained stochastic approximation algorithms. Many of the ODE's that arise have special properties that can be exploited in analyzing the asymptotic properties. One such monotonicity property is discussed in Section 4. The concept of chain recurrence, which is helpful in narrowing the set of possible limit points, is introduced. In Section 5, the stochastic analog of the Liapunov function method for proving stability is discussed. When the iterates in the stochastic approximation algorithms are unconstrained, such stability results will prove useful for showing that the paths are bounded, which is the first step in the proof of convergence. The concept of perturbed Liapunov function is introduced. More complex forms of such functions play an essential role in proving stability and convergence in Sections 5.5, 6.7 and in Chapter 10, particularly for correlated noise processes.

4.1 Martingales, Submartingales, and Inequalities

Let $(\Omega, \mathcal{F}, P)$ denote a probability space, where Ω is the sample space, $\mathcal{F}$ a σ-algebra of subsets of Ω, and P a probability measure on $(\Omega, \mathcal{F})$. The symbol ω denotes the canonical point in Ω. All subsequent random variables will be defined on this space. Let $\{M_n\}$ be a sequence of random variables that can be either real or vector-valued and that satisfies $E|M_n| < \infty$ for each n. If

$$E\left[M_{n+1}|M_i, i \le n\right] = M_n \text{ w.p.1 for all } n, \tag{1.1}$$

then $\{M_n\}$ is said to be a *martingale* or a martingale sequence. The difference $\delta M_n = M_{n+1} - M_n$ is called a *martingale difference.* By the definition of a martingale, if $E|M_n|^2 < \infty$ for each n, the martingale differences are uncorrelated in that for $m \neq n$,

$$E\left[M_{n+1} - M_n\right]\left[M_{m+1} - M_m\right]' = 0.$$

To provide a simple example, suppose that a gambler is playing a sequence of card games, and M_n is his "fortune" after the nth game. If $\{M_n\}$ is a martingale process, we say that the game is "fair" in that the expectation of $M_{n+1} - M_n$ conditioned on the past fortunes $\{M_i, i \le n\}$ is zero. In applications, the M_n themselves are often functions of other random variables. For example, we might have $M_n = f_n(\xi_i, i \le n)$ for some sequence of random variables $\{\xi_n\}$ and measurable functions $\{f_n(\cdot)\}$. In the gambling example, ξ_n might represent the actual sequence of cards within the nth game. Then we could say that the game is fair if $E\left[M_{n+1} - M_n|\xi_i, i \le n, M_0\right] = 0$. This suggests that for use in applications, it is convenient to define a martingale somewhat more generally, as follows. Let $\{\mathcal{F}_n\}$ be a sequence of sub-σ-algebras of $\mathcal{F}$ such that $\mathcal{F}_n \subset \mathcal{F}_{n+1}$, for all n. Suppose that M_n is measurable with respect to $\mathcal{F}_n$ [e.g., in the preceding example, $\mathcal{F}_n$ might be determined by $\{\xi_i, i \le n, M_0\}$]. Write the expectation conditioned on $\mathcal{F}_n$ as $E_{\mathcal{F}_n}$. If

$$E_{\mathcal{F}_n} M_{n+1} = M_n \text{ w.p.1 for all } n, \tag{1.2}$$

then we say that either $\{M_n, \mathcal{F}_n\}$ is a *martingale* or $\{M_n\}$ is an $\mathcal{F}_n$-martingale. If we simply say that $\{M_n\}$ is a martingale and do not specify $\mathcal{F}_n$, then we implicitly suppose (as we can always do) that it is just the

σ-algebra generated by $\{M_i, i \leq n\}$. Martingales are one of the fundamental types of processes in stochastic analysis, and there is a very large literature dealing with them; see, for example, [34, 183]. Note that if an $\mathcal{F}_n$-martingale is vector-valued, then each of the real-valued components is also an $\mathcal{F}_n$-martingale. Similarly, a finite collection of real-valued $\mathcal{F}_n$-martingales is a vector-valued $\mathcal{F}_n$-martingale.

Let M_n be real-valued, and replace (1.2) by

$$E_{\mathcal{F}_n} M_{n+1} \leq M_n \text{ w.p.1 for all } n. \tag{1.3}$$

Then we say either that $\{M_n, \mathcal{F}_n\}$ is a *supermartingale* or that $\{M_n\}$ is an $\mathcal{F}_n$-supermartingale. If the $\mathcal{F}_n$ are understood, then we might just say that $\{M_n\}$ is a supermartingale. If the inequality has the form

$$E_{\mathcal{F}_n} M_{n+1} \geq M_n \text{ w.p.1 for all } n,$$

then the process is called a *submartingale*. In the applications in this book, martingale processes occur when we decompose each member of a sequence of random variables into a part "depending on the past" and an "unpredictable" part. For example, let $\{Y_n\}$ be a sequence of random variables with $E|Y_n| < \infty$ for each n. Write

$$Y_n = (Y_n - E\left[Y_n | Y_i, i < n\right]) + E\left[Y_n | Y_i, i < n\right].$$

The first "unpredictable" part is a martingale difference, because the process defined by the sum

$$M_n = \sum_{j=0}^{n} (Y_j - E\left[Y_j | Y_i, i < j\right])$$

is a martingale. There are many useful inequalities and limit theorems associated with martingale-type processes that facilitate the analysis of their sample paths.

Martingale inequalities and a convergence theorem. Let $\{M_n, \mathcal{F}_n\}$ be a martingale, which is supposed to be real-valued with no loss in generality. The following key inequalities can be found in [34, Chapter 5], [77, Chapter 1] and [183, Chapter IV.5]. Let $q(\cdot)$ be a non-negative nondecreasing convex function. Then for any integers $n < N$ and $\lambda > 0$,

$$P_{\mathcal{F}_n} \left\{ \sup_{n \leq m \leq N} |M_m| \geq \lambda \right\} \leq \frac{E_{\mathcal{F}_n} q(M_N)}{q(\lambda)}, \tag{1.4}$$

which will be useful in getting bounds on the excursions of stochastic approximation processes. Commonly used forms of $q(\cdot)$ are $q(M) = |M|$, $q(M) = |M|^2$, and $q(M) = \exp(\alpha M)$ for some positive α. We will also need the inequality

$$E_{\mathcal{F}_n} \left[\sup_{n \leq m \leq N} |M_m|^2 \right] \leq 4 E_{\mathcal{F}_n} |M_N|^2. \tag{1.5}$$

If $\{M_n, \mathcal{F}_n\}$ is a non-negative supermartingale, then for integers $n < N$

$$P_{\mathcal{F}_n}\left\{\sup_{n\leq m\leq N} M_m \geq \lambda\right\} \leq \frac{M_n}{\lambda}. \tag{1.6}$$

Let y^- denote the negative part of the real number y, defined by $y^- = \max\{0, -y\}$. Let $\{M_n, \mathcal{F}_n\}$ be a real-valued submartingale with $\sup_n E|M_n| < \infty$. Then the martingale convergence theorem [34, Theorem 5.14] states: $\{M_n\}$ *converges with probability one, as* $n \to \infty$. A supermartingale $\{M_n\}$ converges with probability one if $\sup_n E[M_n]^- < \infty$.

Stopping times. Let $\mathcal{F}_n$ be a sequence of nondecreasing σ-algebras. A random variable τ with values in $[0, \infty]$ (the set of extended non-negative real numbers) is said to be an $\mathcal{F}_n$-*stopping time* (or simply a stopping time if the σ-algebras are evident) if $\{\tau \leq n\} \in \mathcal{F}_n$ for each n. Let $\mathcal{F}_n$ be the σ-algebra determined by a random sequence $\{\xi_i, i \leq n\}$. Then, if τ is an $\mathcal{F}_n$-stopping time, whether or not the event $\{\tau \leq n\}$ occurred can be "determined" by watching ξ_i up to and including time n. If a stopping time is not defined at some ω, we always set its value equal to infinity at that ω. Let $\{M_n, \mathcal{F}_n\}$ be a martingale (resp., a sub- or supermartingale) and let τ be a bounded (uniformly in ω) $\mathcal{F}_n$-stopping time. Define

$$\tau \wedge n = \min\{\tau, n\}.$$

Then $\{M_{\tau\wedge n}, \mathcal{F}_n\}$ is a martingale (resp., a sub- or supermartingale).

Continuous-time martingales. The definitions of martingale and sub- and supermartingale extend to continuous time. Let $M(t)$ be a random process satisfying $E|M(t)| < \infty$ for each $t \geq 0$, and let $\mathcal{F}_t$ be a nondecreasing sequence of σ-algebras such that $M(t)$ is $\mathcal{F}_t$-measurable. If $E_{\mathcal{F}_t}[M(t+s) - M(t)] = 0$ with probability one for each $t \geq 0$ and $s > 0$, then $\{M(t), \mathcal{F}_t\}$ is said to be a martingale. If the sequence of σ-algebras $\mathcal{F}_t$ is understood, then it might be omitted. In this book, essentially all we need to know about continuous parameter martingales are the following facts. Some additional material on the topic of Theorem 1.2 is in Section 7.2.4.

Theorem 1.1. *A continuous-time martingale whose sample paths are locally Lipschitz continuous with probability one on each bounded time interval is a constant with probability one.*

By *locally Lipschitz continuous with probability one*, we mean that for each $T > 0$ there is a random variable $K(T) < \infty$ with probability one such that for $t \leq t + s \leq T$,

$$|M(t+s) - M(t)| \leq K(T)s.$$

The result will be proved for real-valued $M(t)$ (with no loss of generality). First we assume that $E|M(t)|^2 < \infty$ for each t and that the Lipschitz constant is bounded by a constant K, and show that $|M(t)|^2$ is also a martingale. For $\Delta > 0$, we can write

$$\begin{aligned} &E_{\mathcal{F}_t} M^2(t+\Delta) - M^2(t) \\ &\quad = 2E_{\mathcal{F}_t} M(t)\left[M(t+\Delta) - M(t)\right] + E_{\mathcal{F}_t}\left[M(t+\Delta) - M(t)\right]^2. \end{aligned}$$

Then, using the martingale property on the first term on the second line above and the Lipschitz condition on the second term yields that the expression is bounded in absolute value by $K^2\Delta^2$. Now, adding the increments over successive intervals of width Δ and letting $\Delta \to 0$ show that $E_{\mathcal{F}_t} M^2(t+s) - M^2(t) = 0$ for any $s \geq 0$. Consequently, $|M(T+t) - M(T)|^2, t \geq 0$, is a martingale for each T. Since we have proved that the expectation of this last martingale is a constant and equals zero at $t = 0$, it is identically zero, which shows that $M(t)$ is a constant. If $E|M(t)|^2$ is not finite for all t, then a "stopping time" argument can be used to get the same result; that is, for positive N, work with $M(\tau_N \wedge \cdot)$ where $\tau_N = \min\{t : |M(t)| \geq N\}$ to show that $M(\tau_N \wedge \cdot)$ is a constant for each N. An analogous stopping time argument can be used if the Lipschitz constant is random.

Definition. Let $W(\cdot)$ be an $\mathbb{R}^r$-valued process with continuous sample paths such that $W(0) = 0$, $EW(t) = 0$, for any set of increasing real numbers $\{t_i\}$ the set $\{W(t_{i+1}) - W(t_i)\}$ is mutually independent and the distribution of $W(t+s) - W(t)$, $s > 0$, does not depend on t. Then $W(\cdot)$ is called a vector-valued *Wiener process* or *Brownian motion,* and there is a matrix Σ, called the covariance, such that $EW(t)W'(t) = \Sigma t$, and the increments are normally distributed [34].

The next theorem gives a convenient criterion for verifying that a process is a Wiener process. The criterion will be discussed further in Chapter 7.

Theorem 1.2. [68, Chapter 5, Theorem 2.12]. *Let $\{M(t), \mathcal{F}_t\}$ be a vector-valued martingale with continuous sample paths and let there be a matrix Σ such that for each t and $s \geq 0$,*

$$E_{\mathcal{F}_t}\left[M(t+s) - M(t)\right]\left[M(t+s) - M(t)\right]' = \Sigma s \text{ w.p.1.}$$

Then $M(\cdot)$ is a Wiener process with zero mean and covariance parameter Σ.

The Borel–Cantelli Lemma. Let A_n be events (i.e., sets in $\mathcal{F}$) and suppose that

$$\sum_n P\{A_n\} < \infty. \tag{1.7}$$

Then the *Borel-Cantelli Lemma* [34] states that for almost all ω only finitely many of the events A_n will occur.

Inequalities. Let X be a real-valued random variable. Then *Chebyshev's inequality* (see [34]) states that for any integer m and $\delta > 0$,

$$P\{|X| \geq \delta\} \leq \frac{E|X|^m}{\delta^m}. \tag{1.8}$$

For an integer k, let $X_i, i \leq k$, be real-valued random variables. *Hölder's inequality* states that for positive $p_i, i \leq k$, satisfying $\sum_i 1/p_i = 1$,

$$E|X_1 \cdots X_k| \leq E^{1/p_1}|X_1|^{p_1} \cdots E^{1/p_k}|X_k|^{p_k}. \tag{1.9}$$

The special case $k = 2, p_1 = p_2 = 2$ is called the *Schwarz inequality*. There is an analogous inequality for sums. Let $X_{i,n}$ be real-valued random variables and let $a_n \geq 0$ with $\sum_n a_n < \infty$. Then

$$\begin{aligned} E\left|\sum_n a_n X_{1,n} \cdots X_{k,n}\right| \\ \leq \left(E\sum_n a_n |X_{1,n}|^{p_1}\right)^{1/p_1} \cdots \left(E\sum_n a_n |X_{k,n}|^{p_k}\right)^{1/p_k}. \end{aligned} \tag{1.10}$$

Let $f(\cdot)$ be a convex function and $\mathcal{F}_0$ a σ-algebra, and suppose that $E|X| < \infty$ and $E|f(X)| < \infty$. Then *Jensen's inequality* is

$$\begin{aligned} &Ef(X) \geq f(EX) \quad \text{or with conditioning,} \\ &E_{\mathcal{F}_0} f(X) \geq f(E_{\mathcal{F}_0} X) \text{ w.p.1.} \end{aligned} \tag{1.11}$$

Burkholder's inequality and higher moment conditions for martingales. An extension of (1.4) based on Burkholder's inequality will be useful in Chapter 5 to prove a zero asymptotic growth property. Return to the notation above (1.2), and define the martingale $M_n = \sum_{i=1}^n \epsilon_i \xi_i$, where $\epsilon_n \geq 0$ is $\mathcal{F}_{n-1}$-measurable and $\sum_{n \leq N} \epsilon_n \leq 1$. Let $\sup_i E|\xi_i|^p < \infty$ for some even integer $p > 1$. By Burkholder's theorem [230, Theorem 6.3.10], there is a constant B (not depending on p) such that for each N,

$$E\left[\sup_{n \leq N} |M_n|^p\right]^{1/p} \leq \frac{Bp^{5/2}}{p-1} E\left[\left(\sum_{n=1}^N (M_n - M_{n-1})^2\right)^{p/2}\right]^{1/p}. \tag{1.12}$$

Define $m = p/2$, and let $s, \alpha_1, \ldots, \alpha_s$ be arbitrary positive integers such that $\sum_{i=1}^s \alpha_i = m$. Then, for some K depending on $\sup_n E|\xi_n|^p$,

$$KE\left[\left(\sum_{n=1}^N (M_n - M_{n-1})^2\right)^{p/2}\right] \leq KE\sum_P \sum_{i \leq N} \epsilon_i^{2\alpha_1} \sum_{i \leq N} \epsilon_i^{2\alpha_2} \cdots \sum_{i \leq N} \epsilon_i^{2\alpha_s}, \tag{1.13}$$

where $\sum_P$ is the sum over all such partitions of $[0, m]$. Consider a typical partition. Rewrite the inner sums in (1.13) as

$$E \sum_{i \leq N} \epsilon_i \epsilon_i^{2\alpha_1 - 1} \sum_{i \leq N} \epsilon_i \epsilon_i^{2\alpha_2 - 1} \cdots \sum_{i \leq N} \epsilon_i \epsilon_i^{2\alpha_s - 1},$$

Now, use Hölder's inequality to get the bound on (1.13):

$$E \left[\sum_{i \leq N} \epsilon_i \right]^{l_s} \left[\sum_{i \leq N} \epsilon_i \epsilon_i^{(2\alpha_1 - 1)q_1} \right]^{1/q_1} \cdots \left[\sum_{i \leq N} \epsilon_i \epsilon_i^{(2\alpha_s - 1)q_s} \right]^{1/q_s},$$

where $l_s = \sum_{i=1}^{s} 1/p_i$ and $1/p_i + 1/q_i = 1$. Choose the q_i such that $\sum_{i=1}^{s} 1/q_i = 1$ and the exponents are equal in that, for some $c > 0$, $(2\alpha_i - 1)q_i = c$ for all i. Then $l_s = s - 1$ and $\sum_{i=1}^{s} 1/q_i = 1 = \sum_{i=1}^{s} (2\alpha_i - 1)/c = (2m - s)/c$. Choosing s so that c is the smallest yields the bound on (1.12)

$$K_1 E \left[\sum_{i \leq N} \epsilon_i \right] \sum_{i \leq N} \epsilon_i \left[\epsilon_i\right]^m, \tag{1.14}$$

where K_1 depends on K, B and p. See [10, Proposition 4.2] for a similar calculation, and also [135, Example 6, Section 2.2] where the ξ_n are mutually independent.

4.2 Ordinary Differential Equations

4.2.1 *Limits of a Sequence of Continuous Functions*

Denote the Euclidean r-space by $\mathbb{R}^r$ with x being the canonical point. If $r = 1$, we use $\mathbb{R}$ instead of $\mathbb{R}^1$. For $b > a$, let $C^r[a, b]$ (resp., $C^r[0, \infty)$, $C^r(-\infty, \infty)$) denote the space of $\mathbb{R}^r$-valued continuous functions on the interval $[a, b]$ (resp., on $[0, \infty)$, $(-\infty, \infty)$). The metric used will be the sup norm if the interval is finite and the "local" sup norm if it is infinite: That is, a sequence $\{f_n(\cdot)\}$ in $C^r(-\infty, \infty)$ converges to zero if it converges to zero uniformly on each bounded time interval in the domain of definition.

Definition. Let $\{f_n(\cdot)\}$ denote a set of $\mathbb{R}^r$-valued functions on $(-\infty, \infty)$. The set is said to be *equicontinuous* in $C^r(-\infty, \infty)$ if $\{f_n(0)\}$ is bounded and for each T and $\epsilon > 0$, there is a $\delta > 0$ such that for all n

$$\sup_{0 \leq t - s \leq \delta,\ |t| \leq T} |f_n(t) - f_n(s)| \leq \epsilon. \tag{2.1}$$

There is an obviously analogous definition for the other time intervals. Note that (2.1) implies that each $f_n(\cdot)$ is continuous. The *Arzelà–Ascoli* Theorem ([58, p. 266], [209, p. 179]) states the following result.

Theorem 2.1. *Let $\{f_n(\cdot)\}$ be a sequence of functions in $C^r(-\infty,\infty)$, and let the sequence be equicontinuous. Then there is a subsequence that converges to some continuous limit, uniformly on each bounded interval.*

There is a simple extension of the concept of equicontinuity to a class of noncontinuous functions that will be useful. Suppose that for each n, $f_n(\cdot)$ is an $\mathbb{R}^r$-valued measurable function on $(-\infty,\infty)$ and $\{f_n(0)\}$ is bounded. Also suppose that for each T and $\epsilon > 0$, there is a $\delta > 0$ such that

$$\limsup_n \sup_{0\le t-s\le\delta,\ |t|\le T} |f_n(t) - f_n(s)| \le \epsilon. \tag{2.2}$$

Then we say that $\{f_n(\cdot)\}$ is *equicontinuous in the extended sense.* The functions $f_n(\cdot)$ might not be continuous, but we still have the following extension of Theorem 2.1, whose proofs are virtually the same as that of Theorem 2.1.

Theorem 2.2. *Let $\{f_n(\cdot)\}$ be defined on $(-\infty,\infty)$ and be equicontinuous in the extended sense. Then there is a subsequence that converges to some continuous limit, uniformly on each bounded interval.*

Example of equicontinuity in the extended sense. Let $\{f_n(\cdot)\}$ be equicontinuous in the original sense (2.1). Define $\bar{f}_n(\cdot)$ by $\bar{f}_n(t) = f_n(k/n)$ on the interval $[k/n,(k+1)/n)$. Then $\{\bar{f}_n(\cdot)\}$ is not continuous but $\{\bar{f}_n(\cdot)\}$ is equicontinuous in the extended sense.

All convergence theorems use some notion of compactness in one way or another. Equicontinuity is just such a notion, and the Arzelà–Ascoli Theorem played an essential role in obtaining relatively simple proofs of the convergence (with probability one) of stochastic approximation processes in [135]. The same basic idea, which uses a sequence of continuous-time interpolations of the stochastic approximation iterates θ_n with interpolation intervals ϵ_n, will play a fundamental role in this book. Define the interpolation $\theta^0(\cdot)$ of the stochastic approximation process θ_n as: $\theta^0(t) = \theta_0$ for $t \le 0$, and for $t \ge 0$,

$$\theta^0(t) = \theta_n \text{ on } [t_n, t_{n+1}), \quad \text{where } t_n = \sum_{i=0}^{n-1} \epsilon_i. \tag{2.3}$$

Define the sequence of *shifted processes* $\theta^n(\cdot) = \theta^0(t_n + \cdot)$. The tail behavior of the sequence $\{\theta_n\}$ is captured by the behavior of $\theta^n(\cdot)$ for large n.

In Chapters 5 and 6 it is shown, under reasonable conditions, that for almost all sample points ω, the set of paths $\{\theta^n(\cdot,\omega)\}$ is equicontinuous in the extended sense. The extended Arzelà–Ascoli Theorem (Theorem 2.2) can then be used to extract convergent subsequences whose limits satisfy the "mean" or "average" ODE (ordinary differential equation). Then the asymptotic properties of the ODE will tell us what we wish to know about the tail behavior of θ_n. This way of getting the ODE was introduced and used heavily in [135]; it is a very useful approach to the analysis of stochastic approximation algorithms. To illustrate the role of equicontinuity in getting useful limit and approximation results, we next apply the Arzelà–Ascoli Theorem to the problem of the existence of a solution to an ODE.

Example: Existence of the solution to an ODE. Given $X(0) \in \mathbb{R}$ and a continuous and bounded real-valued function $\bar{g}(\cdot)$ from $\mathbb{R}$ to $\mathbb{R}$, for $\Delta > 0$ define the sequence $\{X_n^\Delta\}$ by $X_0^\Delta = X(0)$ and

$$X_{n+1}^\Delta = X_n^\Delta + \Delta \bar{g}(X_n^\Delta), \quad n \geq 0.$$

Define the piecewise linear interpolation $X^\Delta(\cdot)$ by

$$X^\Delta(t) = \frac{(t - n\Delta)}{\Delta} X_{n+1}^\Delta + \frac{(n\Delta + \Delta - t)}{\Delta} X_n^\Delta, \quad \text{on } [n\Delta, n\Delta + \Delta). \tag{2.4}$$

Then we can write

$$X^\Delta(t) = X(0) + \int_0^t \bar{g}(X^\Delta(s))ds + \rho^\Delta(t),$$

where the interpolation error $\rho^\Delta(\cdot)$ goes to zero as $\Delta \to 0$. The sequence of functions $\{X^\Delta(\cdot)\}$ is equicontinuous. Hence, by the Arzelà–Ascoli Theorem, there is a convergent subsequence in the sense of uniform convergence on each bounded time interval, and it is easily seen that any limit $X(\cdot)$ must satisfy the ODE $\dot{X} = \bar{g}(X)$, with initial condition $X(0)$.

ODEs and stochastic approximation. Let us recall the discussion in Example 1 of Subsection 1.1.2. Our approach to the study of the asymptotic properties of the stochastic approximation sequence involves, either explicitly or implicitly, the asymptotic properties of an ODE that represents the "mean" dynamics of the algorithm. Thus, we need to say a little about the asymptotic behavior of ODEs. First, we reiterate the intuitive connection in a simple example. Write the stochastic approximation as $\theta_{n+1} = \theta_n + \epsilon_n Y_n$, and suppose that $\sup_n E|Y_n|^2 < \infty$. Suppose, for simplicity here, that there is a continuous function $\bar{g}(\cdot)$ such that $\bar{g}(\theta_n) = E[Y_n | Y_i, i < n, \theta_0]$. Then

$$\theta_{n+m+1} - \theta_n = \sum_{i=n}^{m} \epsilon_i \bar{g}(\theta_i) + \sum_{i=n}^{m} \epsilon_i \left[Y_i - \bar{g}(\theta_i)\right].$$

Since the variance of the second term (the "noise" term) is of the order of $\sum_{i=n}^m O(\epsilon_i^2)$, we might expect that as time increases, the effects of the noise will go to zero, and the iterate sequence will eventually follow the "mean trajectory" defined by $\bar{\theta}_{n+1} = \bar{\theta}_n + \epsilon_n \bar{g}(\bar{\theta}_n)$. Suppose that this is true. Then, if we start looking at the θ_n at large n when the decreasing ϵ_n

are small, the stochastic approximation algorithm behaves similarly to a finite difference equation with small step sizes. This finite difference equation is in turn approximated by the solution to the mean ODE $\dot{\theta} = \bar{g}(\theta)$. Additionally, when using stability methods for proving the boundedness of the trajectories of the stochastic approximation, the stochastic Liapunov function to be used is close to what is used to prove the stability of the mean ODE. These ideas will be formalized in the following chapters.

4.2.2 Stability of Ordinary Differential Equations

Definition: Stability. A set $A \subset H$ is said to be *locally stable in the sense of Liapunov* if for each $\delta > 0$ there is a $\delta_1 > 0$ such that all trajectories starting in $N_{\delta_1}(A)$ never leave $N_\delta(A)$. If the trajectories ultimately go to A, then it is said that A is *asymptotically stable in the sense of Liapunov.* If this holds for all initial conditions, then the asymptotic stability is said to be *global.*

Liapunov functions. Let $\bar{g}(\cdot) : \mathbb{R}^r \mapsto \mathbb{R}^r$ be a continuous function. The classical method of Liapunov stability [156] is a useful tool for determining the asymptotic properties of the solutions of the ODE $\dot{x} = \bar{g}(x)$.

Suppose that $V(\cdot)$ is a continuously differentiable and real-valued function of x such that $V(0) = 0$, $V(x) > 0$ for $x \neq 0$ and $V(x) \to \infty$ as $|x| \to \infty$. For $\lambda > 0$, define $Q_\lambda = \{x : V(x) \leq \lambda\}$. The time derivative of $V(x(\cdot))$ is given by

$$\dot{V}(x(t)) = V_x'(x(t))\bar{g}(x(t)) \equiv -k(x(t)). \tag{2.5}$$

For a given $\lambda > 0$, let $V(x(0)) \leq \lambda$, and suppose that $k(x) \geq 0$ for $x \in Q_\lambda$. The following conclusions are part of the theory of stability via the Liapunov function method:

- The inequality $\dot{V}(x(t)) = -k(x(t)) \leq 0$ for $x(t)$ in Q_λ implies that $V(x(\cdot))$ is nonincreasing along this trajectory and hence

$$x(t) \in Q_\lambda \text{ for all } t < \infty.$$

- Similarly, since $V(\cdot)$ is non-negative,

$$V(x(0)) \geq V(x(0)) - V(x(t)) = -\int_0^t \dot{V}(x(s))ds = \int_0^t k(x(s))ds.$$

  The last equation and the non-negativity of $k(\cdot)$ imply that $0 \leq \int_0^\infty k(x(s))ds < \infty$. Furthermore, since $k(\cdot)$ is continuous and the part of the path of the solution $x(\cdot)$ on $[0, \infty)$ that is in Q_λ is Lipschitz continuous, as $t \to \infty$ we have

$$x(t) \to \{x \in Q_\lambda : k(x) = 0\}.$$

If for each $\delta > 0$ there is an $\epsilon > 0$ such that $k(x) \geq \epsilon$ if $|x| \geq \delta$, then the convergence takes place even if $k(\cdot)$ is only measurable.

In this book, we will not generally need to know the Liapunov functions themselves, only that they exist and have appropriate properties. For existence of Liapunov functions under particular definitions of stability, see [119, 256].

Example and extension. Consider the two-dimensional system:

$$\dot{x} = Ax = \begin{bmatrix} x_2 \\ -x_1 - x_2 \end{bmatrix}. \tag{2.6}$$

If $V(x) = |x|^2 = x_1^2 + x_2^2$, then at $x(t) = x$, $\dot{V}(x) = V_x'(x)Ax = -2x_2^2$. By the Liapunov stability theory, we know that $x(t) \to \{x : x_2 = 0\}$. The given ODE is that for an undriven electrical circuit consisting of a resistor, a capacitor, and an inductor in a loop, where x_1 is the charge on the capacitor and x_2 is the current. The Liapunov function is proportional to the total energy. The Liapunov function argument says that the energy in the system decreases as long as the current through the resistor is nonzero, and that the current goes to zero. It does not say directly that the energy goes to zero. However, we know from the ODE for the circuit that the current cannot be zero unless the charge on the capacitor is zero. It is not hard to show by a direct analysis that $x_2(t) \to 0$ implies that $x_1(t) \to 0$ also. The proof of this follows from the fact that as long as the charge is not zero, the current cannot remain at zero. Thus the system cannot remain forever arbitrarily close to the set where $x_2 = 0$ unless $x_1(t)$ is eventually also arbitrarily close to zero.

The "double" limit problem arises since $k(x) = 0$ does not imply that $x = 0$. The analogous argument for more complicated problems would be harder, and a useful way to avoid it will now be discussed.

Definition. A set $\Lambda \in \mathbb{R}^r$ is an *invariant set* for the ODE $\dot{x} = \bar{g}(x)$ if for each $x_0 \in \Lambda$, there is a solution $x(t), -\infty < t < \infty$, that lies entirely in Λ and satisfies $x(0) = x_0$. The forward limit set or more simply the *limit set* for a given initial condition $x(0)$ is the set of limit points $\cap_{t \geq 0} \{x(s), s \geq t\}$ of the trajectory with initial condition $x(0)$.

If the trajectory is bounded and $\bar{g}(\cdot)$ depends only on x, then the limit set is a compact invariant set [89]. Recall that $x(t)$ does not necessarily converge to a unique point. For example, for the ODE $\ddot{u} + u = 0$ where u is real valued, the limit set is a circle.

Limit sets and the invariant set theorem. The Liapunov function method for proving stability and exhibiting the limit points works best when $k(x) = 0$ only at isolated points. When this is not the case, the following result, known as *LaSalle's Invariance Theorem* (see [156]), which improves on the assertions available directly from the Liapunov function analysis, often helps to characterize the limit set. In the next section, it will be seen that the set of possible limit points can be reduced even further.

Theorem 2.3. *For a given* $\lambda > 0$, *assume the conditions on the Liapunov function preceding* (2.5) *in the set* Q_λ, *and let* $V(x(0)) \leq \lambda$. *Then, as* $t \to \infty$, $x(t)$ *converges to the largest invariant set contained in the set* $\{x : k(x) = 0, V(x) \leq V(x(0))\}$.

Thus, in accordance with LaSalle's Invariance Theorem, we need to look for the largest bounded set B on which $k(x) = 0$ such that for each $x \in B$, there is an entire trajectory on the doubly infinite time interval $(-\infty, \infty)$ that lies all in B and goes through x. To apply this result to the example (2.6), note that there is no bounded trajectory of the ODE that satisfies $x_2(t) = 0$ for all $t \in (-\infty, \infty)$, unless $x(t) \equiv 0$.

Suppose that there is a continuously differentiable real-valued function $f(\cdot)$, bounded below with $f(x) \to \infty$ as $|x| \to \infty$ and such that $\dot{x} = \bar{g}(x) = -f_x(x)$. Then we can say more about the limit sets; namely, that set of stationary points $\{x : \bar{g}(x) = 0\}$ is a collection of disjoint compact sets and the limit trajectory must be contained in one of these sets. For the proof, use $f(\cdot)$ as a Liapunov function so that $k(x) = -|f_x(x)|^2$.

4.3 Projected ODE

In applications to stochastic approximation, it is often the case that the iterates are constrained to lie in some compact set H in the sense that if an iterate ever leaves H, it is immediately returned to the closest (or some other convenient) point in H. A common procedure simply truncates, as noted in the examples in Chapters 1 to 3. In other applications, there are physical constraints that the parameter θ_n must satisfy. Owing to the pervasive practical use of constraints, much of the development in this book will concern algorithms in which the state is constrained to a compact set in some way. The simplest constraint is just a truncation or projection of each component separately. This is condition (A3.1). Condition (A3.2) defines a more general constraint set, and (A3.3) defines a constraint set that is a submanifold of $\mathbb{R}^r$.

(A3.1) H is a hyperrectangle. In other words, there are real numbers $a_i < b_i, i = 1, \ldots, r$, such that $H = \{x : a_i \leq x_i \leq b_i\}$.

For $x \in H$ satisfying (A3.1), define the set $C(x)$ as follows. For $x \in H^0$, the interior of H, $C(x)$ contains only the zero element; for $x \in \partial H$, the boundary of H, let $C(x)$ be the infinite convex cone generated by the outer normals at x of the faces on which x lies.

Then the *projected* ODE (or ODE whose dynamics are projected onto

H) is defined to be

$$\dot{x} = \bar{g}(x) + z, \quad z(t) \in -C(x(t)), \tag{3.1}$$

where $z(\cdot)$ is the *projection* or *constraint term*, the minimum force needed to keep $x(\cdot)$ in H. Let us examine (3.1) more closely. If $x(t)$ is in H^0 on some time interval, then $z(\cdot)$ is zero on that interval. If $x(t)$ is on the interior of a face of H (i.e., $x_i(t)$ equals either a_i or b_i for a unique i) and $\bar{g}(x(t))$ points "out" of H, then $z(\cdot)$ points inward, orthogonal to the face. If $x(t)$ is on an edge or corner of H, with $\bar{g}(x(t))$ pointing "out" of H, then $z(t)$ points inward and takes values in the convex cone generated by the inward normals on the faces impinging on the edge or corner; that is, $z(t)$ takes values in $-C(x(t))$ in all cases. In general, $z(t)$ is the smallest value needed to keep $x(\cdot)$ from leaving H. For example, let $x_i(t) = b_i$, with $\bar{g}_i(x(t)) > 0$. Then, $z_i(t) = -\bar{g}_i(x(t))$.

The function $z(\cdot)$ is not unique in that it is defined only for almost all t. Apart from this it is determined by H and $\bar{g}(\cdot)$. If $\dot{x} = \bar{g}(x)$ has a unique solution for each $x(0)$, then so does (3.1).

Figure 3.1 illustrates the unconstrained and constrained flow lines. In applications, the actual constraint is often flexible, and one tries to use constraints that do not introduce unwanted limit points.

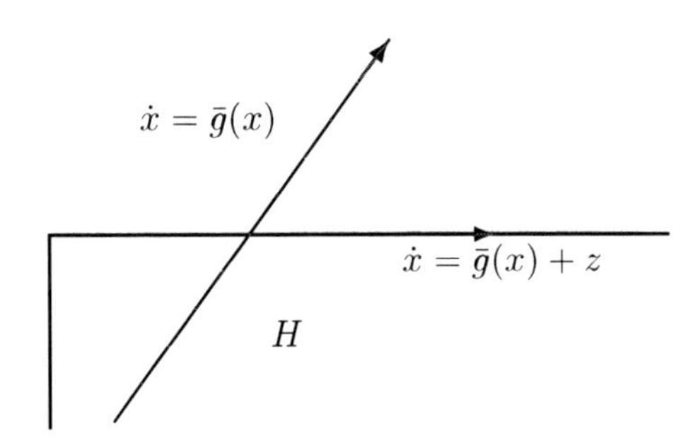

Figure 3.1. Constrained and unconstrained flow lines.

More general constraint sets. In the preceding discussion, we let the constraint set H be a hyperrectangle in order to simplify the discussion that introduced the projected ODE. In applications, one might have additional hard constraints on the state. The following two cases will be of particular importance, the first being motivated by nonlinear programming and is an extension of (A3.1).

(A3.2) Let $q_i(\cdot), i = 1, \cdots, p$, be continuously differentiable real-valued functions on $I\!R^r$, with gradients $q_{i,x}(\cdot)$. Without loss of generality, let

$q_{i,x}(x) \neq 0$ if $q_i(x) = 0$. Define $H = \{x : q_i(x) \leq 0, i = 1, \cdots, p\}$. Then H is connected, compact and nonempty.

A constraint $q_i(\cdot)$ is said to be *active* at x if $q_i(x) = 0$. Define $A(x)$, the set of (indices of the) *active constraints* at x, by $A(x) = \{i : q_i(x) = 0\}$. Define $C(x)$ to be the convex cone generated by the set of outward normals $\{y : y = q_{i,x}(x), i \in A(x)\}$. Suppose that for each x with nonempty $A(x)$, the set $\{q_{i,x}(x), i \in A(x)\}$ is linearly independent. If there are no active constraints at x, then $C(x)$ contains only the zero element.

(A3.3) H is an $\mathbb{R}^{r-1}$ dimensional connected compact surface with a continuously differentiable outer normal. In this case, define $C(x), x \in H$, to be the linear span of the outer normal at x.

If one cares to define the associated stochastic approximation algorithm, any of the usual types of constraints used in the theory of nonlinear programming can be handled, including (A3.2) applied to the manifold in (A3.3).

Note. If under (A3.2), there is only one active constraint (indexed by i) at t, and $\bar{g}(x(t))$ points out of H, then the right-hand side of (3.1) is just the projection of $\bar{g}(x(t))$ onto the tangent plane to the surface at $x(t)$. That is, onto the plane which is orthogonal to the gradient $q_{i,x}(x(t))$ at $x(t)$. Let $\bar{g}(x) = -f_x(x)$, for a continuously differentiable real-valued $f(\cdot)$. Then the constraints in (A3.1) or (A3.2) can give rise to spurious stationary points on the boundary, but this is the only type of singular point which can be introduced by the constraint. The function $f(\cdot)$ can still be used as a Liapunov function, since the derivative along the trajectory at a point x is $f_x'(x)[-f_x(x) + z] \leq 0$.

Note on the invariant set for the projected ODE. The result on invariant sets, Theorem 2.3, holds for the constrained process (3.1), but is less useful because the entire set H might be an invariant set if H is bounded. For this reason, when we work with the constrained algorithm, we simply use the limit points of (3.1) rather than the largest bounded invariant set. When H is compact, let L_H denote the set of limit points of (3.1), over all initial conditions in H:

$$L_H = \lim_{t \to \infty} \bigcup_{x \in H} \{x(s), s \geq t : x(0) = x\}.$$

Let S_H denote the set of points in H where $\bar{g}(x) + z = 0$ for some $z \in -C(x)$. These are called the *stationary points*. Interior to H, the stationarity condition is just $\bar{g}(x) = 0$.

Upper semicontinuity of $C(x)$. An "infinitesimal" change in x cannot increase the number of active constraints. Thus, loosely speaking, this property can be stated as $\lim_{y \to x} C(y) \subset C(x)$. More precisely, let $N_\delta(x)$ be a δ-neighborhood of x. Then

$$\bigcap_{\delta>0} \mathrm{co}\left[\bigcup_{y \in N_\delta(x)} C(y)\right] = C(x), \tag{3.2}$$

where co(A) denotes the closed convex hull of the set A. A set-valued function $C(\cdot)$ satisfying (3.2) is said to be *upper semi-continuous.*

In the analysis of the stochastic approximation algorithms, (3.1) appears in the integral form:

$$\begin{aligned} x(t+\tau) &= x(t) + \int_t^{t+\tau} \bar{g}(x(s))ds + Z(t+\tau) - Z(t), \\ x(t) &\in H, \quad \text{for all } t, \; Z(0) = 0, \end{aligned} \tag{3.3}$$

where $x(\cdot)$ is Lipschitz continuous and the following conditions hold: For $\tau > 0$,

$$Z(t+\tau) - Z(t) = 0 \text{ if } x(s) \in H^0 \text{ for almost all } s \in [t, t+\tau], \tag{3.4}$$

$$Z(t+\tau) - Z(t) \in -\mathrm{co}\left[\bigcup_{t \le s \le t+\tau} C(x(s))\right]. \tag{3.5}$$

The following theorem relating the forms (3.3) and (3.1) will be needed.

Theorem 3.1. *Assume one of the constraint set conditions* (A3.1), (A3.2), *or* (A3.3). *Let $\bar{g}(\cdot)$ be bounded on H and let* (3.3)–(3.5) *hold, where $x(\cdot)$ is Lipschitz continuous. Then $Z(\cdot)$ is absolutely continuous. There is a measurable function $z(\cdot)$ such that $z(t) \in -C(x(t))$ for almost all t and*

$$Z(t+\tau) - Z(t) = \int_t^{t+\tau} z(s)ds. \tag{3.6}$$

Differential inclusions. Recall the examples (1.3.6), (1.3.8), or the example of the figure of the proportional-fair sharing example in Chapter 3. In such cases, the ODE is replaced by a differential inclusion

$$\dot{x} \in G(x), \tag{3.7}$$

where the set $G(x)$ is convex for each x and it is upper semicontinuous in x. One might also have the constrained form $\dot{x} \in G(x) + z$. There are obvious analogs of the stability theorems to such models: see, for example, Section 5.4.

4.4 Cooperative Systems and Chain Recurrence

4.4.1 Cooperative Systems

Definitions. For vectors $x, y \in I\!R^r$, we write $x \geq y$ (resp., $x > y$) if $x_i \geq y_i$ for all i (resp., $x_i \geq y_i$ for all i and, in addition $x \neq y$). If $x_i > y_i$ for all i, then we write $x \gg y$. Let H, K be subsets of $I\!R^r$. Then write $H \geq K$ (resp., $H \gg K$) if $x \geq y$ (resp., $x \gg y$) for all $x \in H, y \in K$. Consider a dynamical system in $I\!R^r$: $\dot{x} = \bar{g}(x)$ where $\bar{g}(\cdot)$ is continuous. Where helpful for clarity, we might write $x(t|y)$ for the solution when the initial condition is y.

The function $\bar{g}(\cdot)$ is said to satisfy the *Kamke condition* [223] if for any x, y, and i, satisfying $x \leq y$ and $x_i = y_i$, we have $\bar{g}_i(x) \leq \bar{g}_i(y)$. Such systems are called *cooperative.* The mean ODEs for many stochastic approximation problems arising as models where there is some form of competition among participants lead to cooperative systems. The proof of convergence for such cases depends on a monotonicity property of the solution to the ODE. The importance of this concept was introduced in [98] and further discussion and applications are in [11, 12].

The Kamke condition implies the following monotonicity result. Its proof in [223, Proposition 1.1] assumes continuous differentiability of $\bar{g}(\cdot)$. But, it used only the Kamke condition, uniqueness of solutions, and the continuous dependence of the path on perturbations in the sense that $x(t|y + \delta y)$ converges to $x(t|y)$ on any bounded t-interval as $\delta y \to 0$.

Example. The Kamke condition is satisfied by the ODEs arising in the proportional-fair sharing example of Chapter 3. This example is concerned with the problem of allocating channel resources in mobile communications when the channel quality for each user varies randomly with time. The monotonicity property plays an important role in the analysis of the algorithm. The condition holds simply because if the throughputs for users $j \neq i$ are larger, then user i is more likely to get the current resource; see Sections 3.6 and 9.5 and [146] for more detail.

Theorem 4.1. [223, Proposition 1.1] *Let $\bar{g}(\cdot)$ be continuous and satisfy the Kamke condition. Let the solution to the ODE be unique for each initial condition, and continuous with respect to perturbations, as defined above. If $x(0) \leq y(0)$ (resp., $<, \ll$), then $x(t|x(0)) \leq x(t|y(0))$ (resp., $<, \ll$).*

4.4.2 Chain Recurrence

In Chapters 5 and 6, it will be shown that the stochastic approximation processes θ_n and $\theta^n(\cdot)$ converge with probability one to an invariant or limit set of the mean ODE. In the absence of other information, we may have to assume that the invariant set is the largest one. But, sometimes the largest invariant or limit sets contain points to which convergence clearly cannot occur. Consider the following example.

Example. Let x be real valued, with $\dot{x} = \bar{g}(x)$, where $\bar{g}(x) = x(1-x), H = [0,1]$. Then the entire interval $[0,1]$ is an invariant set for the ODE. It is clear that if any arbitrarily small neighborhood of $x = 1$ is entered infinitely often with probability $\mu > 0$, then (with probability one, relative to that set), the only limit point is $x = 1$. Furthermore, if each small neighborhood of $x = 0$ is exited infinitely often with probability $\mu > 0$, then it is obvious that the limit point will be $x = 1$ (with probability one, relative to that set). One does not need a sophisticated analysis to characterize the limit in such a case, and an analogous simple analysis can often be done in applications. But the idea of *chain recurrence* as introduced by Benaïm [9] can simplify the analysis in general, since it can be shown that the convergence must be to the subset of the invariant set that is chain recurrent, as defined below. In this example, the only chain recurrent points are $\{0,1\}$.

Definition: Chain recurrence. Let $x(t|y)$ denote the solution to either the ODE $\dot{x} = \bar{g}(x)$ or the constrained ODE $\dot{x} = \bar{g}(x) + z, z \in -C(x)$, at time t given that the initial condition is y. A point x is said to be *chain recurrent* [9, 10, 89] if for each $\delta > 0$ and $T > 0$ there is an integer k and points $u_i, T_i, 0 \le i \le k$, with $T_i \ge T$, such that

$$|x - u_0| \le \delta,\ |y_1 - u_1| \le \delta,\ \ldots,\ |y_k - u_k| \le \delta,\ |y_{k+1} - x| \le \delta, \tag{4.1}$$

where $y_i = x(T_{i-1}|u_{i-1})$ for $i = 1, \ldots, k+1$. We also say that two points x and $\bar{x}$ are *chain connected* if, with the above terminology,

$$|x - u_0| \le \delta,\ |y_1 - u_1| \le \delta,\ \ldots,\ |y_k - u_k| \le \delta,\ |y_{k+1} - \bar{x}| \le \delta \tag{4.2}$$

and with a similar perturbed path taking $\bar{x}$ to x. Not all points in the limit or invariant set are chain recurrent. See Figure 4.1 for an illustration of chain connectedness.

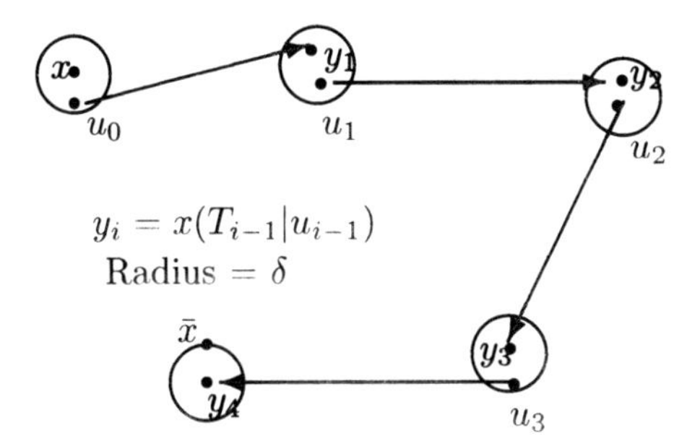

Figure 4.1. An example of chain connectedness.

Example. Figure 4.2 illustrates an example of chain recurrence. The flow lines are drawn. We suppose that $\bar{g}(\cdot)$ is Lipschitz continuous and $\bar{g}(x) = 0$ at the points $\{a, b, c, d, e\}$. Hence these points are stationary. Suppose that the point e is attracting for points interior to the rectangle. The square with corners $\{a, b, c, d\}$ is an invariant set. But the only chain recurrent points (in some neighborhood of the box) are $\{e\}$ and the lines connecting $\{a, b, c, d\}$. The limit points for the ODE (with initial conditions in some neighborhood of the box) are only $\{a, b, c, d, e\}$.

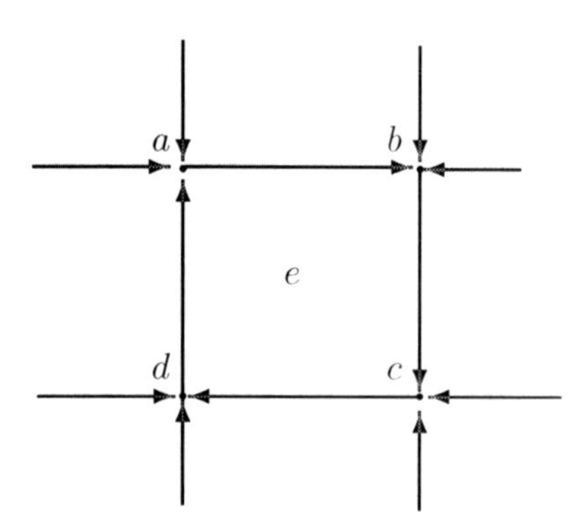

Figure 4.2. An example of chain recurrence vs. invariance.

4.5 Stochastic Stability and Perturbed Stochastic Liapunov Functions

The following theorem is a direct application of the martingale inequalities and the martingale convergence theorem of Section 1, and it is a natural analog of the Liapunov function theorem for ODEs. It is one of the original theorems in stochastic stability theory, first proved in [43, 109, 121, 122]. The basic idea will be adapted in various ways to prove convergence or to get bounds on the trajectories of the stochastic approximation sequences.

Theorem 5.1. *Let $\{X_n\}$ be a Markov chain on $I\!R^r$. Let $V(\cdot)$ be a real-valued and non-negative function on $I\!R^r$ and for a given $\lambda > 0$, define the set $Q_\lambda = \{x : V(x) \le \lambda\}$. Suppose that for all $x \in Q_\lambda$*

$$E\left[V(X_{n+1})|X_n = x\right] - V(x) \le -k(x) \tag{5.1}$$

for all n, where $k(x) \ge 0$ and is continuous on Q_λ. Then, with probability one, for each integer ν,

$$P\left\{\sup_{\nu \le m < \infty} V(X_m) \ge \lambda \middle| X_\nu\right\} I_{\{X_\nu \in Q_\lambda\}} \le \frac{V(X_\nu)}{\lambda}. \tag{5.2}$$

Equivalently, if $X_\nu \in Q_\lambda$, then the path on $[\nu, \infty)$ stays in Q_λ with a probability at least $1 - V(X_\nu)/\lambda$. Let Ω_λ denote the set of paths that stay in Q_λ from some time ν on. Then for almost all $\omega \in \Omega_\lambda$, $V(X_m)$ converges and $X_m \to \{x : k(x) = 0\}$. That is, for each $\omega \in \Omega_\lambda - N$ where N is a null set, the path converges to a subset of $\{x : k(x) = 0\}$, which is consistent with $V(x) =$constant.

Outline of proof. (See [3, 109, 122] for more detail.) The argument is a stochastic analog of what was done in the deterministic case. For notational simplicity, we suppose that $\nu = 0$ and $X_0 = x \in Q_\lambda$. Define $\tau_{Q_\lambda} = \min\{n : X_n \notin Q_\lambda\}$ (or infinity, if $X_n \in Q_\lambda$ for all n). Since we have not assumed that $k(x)$ is non-negative for $x \notin Q_\lambda$, it is convenient to work with the stopped process $\widetilde{X}_n$ defined by $\widetilde{X}_n = X_n$ for $n < \tau_{Q_\lambda}$, and $\widetilde{X}_n = X_{\tau_{Q_\lambda}}$ for $n \geq \tau_{Q_\lambda}$. Thus, from time τ_{Q_λ} on, the value is fixed at $X_{\tau_{Q_\lambda}}$, which is the value attained on first leaving Q_λ. Define $\widetilde{k}(x)$ to equal $k(x)$ in Q_λ and to equal zero for $x \notin Q_\lambda$. Now,

$$E\left[V(\widetilde{X}_{n+1})|\widetilde{X}_n\right] - V(\widetilde{X}_n) \leq -\widetilde{k}(\widetilde{X}_n) \quad \text{for all } n. \tag{5.3}$$

Thus, $\{V(\widetilde{X}_n)\}$ is a non-negative supermartingale and (5.2) is implied by (1.6). By the supermartingale convergence theorem, $\{V(\widetilde{X}_n)\}$ converges to some random variable $\widetilde{V} \geq 0$. Iterating (5.3) and using the non-negativity of $V(\cdot)$, for all x we have

$$V(x) \geq V(x) - E_x V(\widetilde{X}_n) \geq E_x \sum_{m=0}^{n-1} \widetilde{k}(\widetilde{X}_m), \tag{5.4}$$

where E_x denotes the expectation given that $X_0 = x$. In view of (5.3), $E_x \sum_{m=0}^{\infty} \widetilde{k}(\widetilde{X}_m) < \infty$. Thus, by the Borel–Cantelli Lemma, with probability one for any $\epsilon > 0$, $\widetilde{X}_n$ can spend only a finite amount of time more than a distance of ϵ from $\{x : \widetilde{k}(x) = 0\}$. □

Sometimes the right side of (5.1) is replaced by $-k(x) +$ "small term," and the "small term" can be used to guide the construction of a "perturbed" Liapunov function for which the conditional difference is nonpositive. The proof of Theorem 5.1 yields the following result.

Theorem 5.2. *Let $\{X_n\}$ be a Markov chain and $V(\cdot)$ a real-valued and non-negative function on $\mathbb{R}^r$. Let $\{\mathcal{F}_n\}$ be a sequence of σ-algebras, which is nondecreasing and where $\mathcal{F}_n$ measures at least $\{X_i, i \leq n\}$. Let δV_n be $\mathcal{F}_n$-measurable random variables such that $\delta V_n \to 0$ with probability one and $E|\delta V_n| < \infty$ for each n. Define $V_n(x) = V(x) + \delta V_n$. Suppose that*

$$E\left[V_{n+1}(X_{n+1}) - V_n(X_n)|\mathcal{F}_n\right] \leq -k(X_n) \leq 0,$$

where $k(\cdot)$ is continuous and positive for $x \neq 0$. Then $X_n \to 0$ with probability one.

The following extension of Theorem 5.2 will be useful in the stochastic approximation problem. In the theorem, $\{X_n\}$ is an $\mathbb{R}^r$-valued stochastic process, not necessarily a Markov process. Let $\{\mathcal{F}_n\}$ be a sequence of nondecreasing σ-algebras, with $\mathcal{F}_n$ measuring at least $\{X_i, i \leq n\}$, and let E_n denote the expectation conditioned on $\mathcal{F}_n$. If $X_n = \theta_n$, then the form (5.5) arises from the truncated Taylor expansion of $E[V(\theta_{n+1})|\theta_0, Y_i, i < n] - V(\theta_n)$ in stochastic approximation problems.

Theorem 5.3. *Let $V(\cdot)$ be a non-negative real-valued continuous function on $\mathbb{R}^r$ that is positive for $x \geq 0$, $V(0) = 0$, and with the property that for each $\epsilon > 0$ there is a $\delta > 0$ such that $V(x) \geq \delta$ for $|x| \geq \epsilon$. Suppose that $EV(X_0) < \infty$ and δ does not decrease as ϵ increases. Let there be random variables Y_n and a non-negative function $k(\cdot)$ such that for each $\epsilon > 0$ there is a $\delta > 0$ satisfying $k(x) \geq \delta$ for $|x| \geq \epsilon$. Let there be a $K_1 < \infty$ such that*

$$E_n V(X_{n+1}) - V(X_n) \leq -\epsilon_n k(X_n) + K_1 \epsilon_n^2 E_n |Y_n|^2, \tag{5.5}$$

where ϵ_n are positive $\mathcal{F}_n$-measurable random variables tending to zero with probability one and $\sum_n \epsilon_n = \infty$ with probability one. Suppose that $Ek(X_n) < \infty$ if $EV(X_n) < \infty$ and that there are $K_2 < \infty$ and $K < \infty$ such that

$$E_n |Y_n|^2 \leq K_2 k(X_n), \quad \text{when } |X_n| \geq K. \tag{5.6}$$

Let

$$E \sum_{i=0}^{\infty} \epsilon_i^2 |Y_i|^2 I_{\{|X_i| \leq K\}} < \infty. \tag{5.7}$$

Then $X_n \to 0$ with probability one.

Proof. The hypotheses imply that $EV(X_n) < \infty$ for all n; we leave the proof of this fact to the reader. Since $\epsilon_n \to 0$ with probability one, $\epsilon_n^2 \ll \epsilon_n$ for large n. Now, since $EV(X_n) < \infty$ for all n, by shifting the time origin we can suppose without loss of generality that $\epsilon_n^2 \ll \epsilon_n$ for all n. In particular, we suppose that $K_1 K_2 \epsilon_n^2 < \epsilon_n / 2$. Define

$$\delta V_n = K_1 E_n \sum_{i=n}^{\infty} \epsilon_i^2 |Y_i|^2 I_{\{|X_i| \leq K\}},$$

and the perturbed Liapunov function $V_n(X_n) = V(X_n) + \delta V_n$. Note that $V_n(X_n) \geq 0$ and

$$E_n \delta V_{n+1} - \delta V_n = -K_1 \epsilon_n^2 E_n |Y_n|^2 I_{\{|X_n| \leq K\}}.$$

This, together with (5.5) and (5.6), yields

$$E_n V_{n+1}(X_{n+1}) - V_n(X_n) \leq -\epsilon_n k(X_n)/2, \tag{5.8}$$

which implies that $\{V_n(X_n)\}$ is an $\mathcal{F}_n$-supermartingale sequence. By the supermartingale convergence theorem, there is a $\widetilde{V} \geq 0$ such that $V_n(X_n) \to \widetilde{V}$ with probability one. Since (5.7) implies that $\delta V_n \to 0$ with probability one, $V(X_n) \to \widetilde{V}$ with probability one.

For integers N and m, (5.8) yields

$$E_N V_{N+m}(X_{N+m}) - V_N(X_N) \leq - \sum_{i=N}^{N+m-1} E_N \epsilon_i k(X_i)/2. \tag{5.9}$$

As $m \to \infty$, the left side of (5.9) is bounded below by $-V_N(X_N)$. Suppose that $\widetilde{V} > 0$ with positive probability. Then, by the properties of $V(\cdot)$, X_N is asymptotically outside of some small neighborhood of the origin, with a positive probability. This and the fact that $\sum \epsilon_i = \infty$ with probability one and the properties of $k(\cdot)$ imply that the sum on the right side of (5.9) goes to infinity with a positive probability, leading to a contradiction. Thus $\widetilde{V} = 0$ with probability one. □

These theorems are concerned with convergence with probability one. The Liapunov function method is often used simply to prove recurrence. With the recurrence given, other methods might be used to prove convergence. Then the following special case of Theorem 5.3 is useful.

Theorem 5.4. *Let X_n, $\mathcal{F}_n$, and ϵ_n be as in Theorem 5.3. Let $V(x) \geq 0$ and suppose that there are $\delta > 0$ and compact $A \subset \mathbb{R}^r$ such that for all large n,*

$$E_n V(X_{n+1}) - V(X_n) \leq -\epsilon_n \delta < 0, \quad \text{for } x \notin A.$$

Then the set A is recurrent for $\{X_n\}$ in that $X_n \in A$ for infinitely many n with probability one.

5

Convergence with Probability One: Martingale Difference Noise

5.0 Outline of Chapter

Much of the classical work in stochastic approximation dealt with the situation where the "noise" in the observation Y_n is a martingale difference. That is, there is a function $g_n(\cdot)$ of θ such that $E[Y_n|Y_i, i < n, \theta_0] = g_n(\theta_n)$ [27, 65, 70, 72, 83, 110, 120, 182, 207, 215, 237, 250]. Then we can write $Y_n = g_n(\theta_n) + \delta M_n$, where δM_n is a martingale difference. This "martingale difference noise" model is still of considerable importance. It arises, for example, where Y_n has the form $Y_n = F_n(\theta_n, \psi_n)$, and where ψ_n are mutually independent. The convergence theory is relatively easy in this case, because the noise terms can be dealt with by well-known and relatively simple probability inequalities for martingale sequences. This chapter is devoted to this martingale difference noise case. Nevertheless, the ODE, compactness, and stability techniques to be introduced are of basic importance for stochastic approximation, and will be used in subsequent chapters.

A number of definitions that will be used throughout the book are introduced in Section 1. In particular, the general "ODE" techniques used in the rest of the book are based on the analysis of continuous-time interpolations of the stochastic approximation sequence. These interpolations are defined in Section 1. The general development in the book follows intuitively reasonable paths but cannot be readily understood unless the definitions of the interpolated processes are understood.

Section 2 gives a fundamental convergence theorem and shows how the stochastic approximation sequence is related to a "mean" ODE that characterizes the asymptotic behavior. The Arzelà–Ascoli theorem is crucial to getting the ODE since it guarantees that there will always be convergent subsequences of the set of interpolated processes. The limits of any of these subsequences will satisfy the "mean" ODE. The first theorem (Theorem 2.1) uses a simple constraint set to get a relatively simple proof and allows us to concentrate on the essential structure of the "ODE method"-type proofs. This constraint set is generalized in Theorem 2.3, where a method for characterizing the reflection terms is developed, which will be used throughout the book. All the results carry over to the case where the constraint set is a smooth manifold of any finite dimension.

The conditions used for the theorems in Section 2, for example, square summability of the step sizes ϵ_n, are more or less classical. The square summability, together with the martingale noise property and a stability argument, can be used to get a simpler proof if the algorithm is unconstrained. However, the type of proof given readily generalizes to one under much weaker conditions. The set to which the iterates converge is a limit or invariant set for the mean ODE. These limit or invariant sets might be too large in that the convergence can only be to a subset. Theorem 2.5 shows that the only points in the limit or invariant set that we need to consider are the "chain recurrent" points.

The conditions are weakened in Subsection 3.1, which presents the "final form" of the martingale difference noise case in terms of conditions that require the "asymptotic rates of change" of certain random sequences to be zero with probability one. These conditions are satisfied by the classical case of Section 2. They are phrased somewhat abstractly but are shown to hold under rather weak and easily verifiable conditions in Subsection 3.2. Indeed, these "rate of change" conditions seem to be nearly minimal for convergence, and they hold even for "very slowly" decreasing step sizes. The conditions have been proved to be necessary in certain cases. The techniques of this chapter owe much to the ideas in [135].

Stability and combined stability-ODE methods for getting convergence, when there are no *a priori* bounds on the iterates, are in Section 4. A stochastic Liapunov function method is used to prove recurrence of the iterates, and then the ODE method takes over in the final stage of the proof. This gives a more general result than one might obtain with a stability method alone and is more easily generalizable. Section 5 concerns "soft" constraints, where bounds on functionals of the iterate are introduced into the algorithm via a penalty function. The results in Section 6 on the random directions Kiefer–Wolfowitz method and on the minimization of convex functions are suggestive of additional applications. In order to illustrate the use of the theorems, in Section 7 we give the proof of convergence for the "lizard learning" problem of Section 2.1 and the pattern classification problem of Section 1.1.

When using stochastic approximation for function minimization, where the function has more than one local minimum, one would like to assure at least that convergence to other types of stationary points (such as local maxima or saddles) is impossible. One expects that the noise in the algorithm will destabilize the algorithm around these "undesirable" points. Such issues are discussed in Section 8.

5.1 Truncated Algorithms: Introduction

To develop the basic concepts behind the convergence theory in a reasonably intuitive way, we will first work with a relatively simple form and then systematically generalize it.

An important issue in applications of stochastic approximation concerns the procedure to follow if the iterates become too large. Practical algorithms tend to deal with this problem via appropriate adjustments to the basic algorithm, but these are often ignored in the mathematical developments, which tend to allow unbounded iterate sequences and put various "stability" conditions on the problem. However, even if these stability conditions do hold in practice, samples of the iterate sequence might get large enough to cause concern. The appropriate procedures to follow when the parameter value becomes large depends on the particular problem and the form of the algorithm that has been chosen, and it is unfortunate that there are no perfectly general rules to which one can appeal. Nevertheless, the useful parameter values in properly parameterized practical problems are usually confined by constraints of physics or economics to some compact set. This might be given by a hard physical constraint requiring that, say, a dosage be less than a certain number of milligrams or a temperature set point in a computer simulation of a chemical process be less than 200°C. In fact, there are implicit bounds in most problems. If θ_n is the set point temperature in a chemical processor and it reaches the temperature at the interior of the sun, or if the cost of setting the parameter at θ_n reaches the U.S. gross national product, then something is very likely wrong with either the model or algorithm or with both. The models used in simulations are often inaccurate representations of physical reality at excessive values of the parameter (or of the noise), and so a mathematical development that does not carefully account for the changes in the model as the parameter (and the noise) values go to infinity might well be assuming much more than is justified. The possibility of excessive values of θ_n is a problem unique to computer simulations, because any algorithm used on a physical process would be carefully controlled.

Excessively large values of θ_n might simply be a consequence of poor choices for the algorithmic structure. For example, instability can be caused by values of ϵ_n that are too large or values of finite difference intervals that are too small. The path must be checked for undesirable behavior, whether or not there are hard constraints. If the algorithm appears to be unstable, then one could reduce the step size and restart at an appropriate point or even reduce the size of the constraint set. The path behavior might suggest a better algorithm or a better way of estimating derivatives. Conversely, if the path moves too slowly, we might wish to increase the step sizes. If the problem is based on a simulation, one might need to use a cruder model, with perhaps fewer parameters and a more restricted constraint set, to get a rough estimate of the location of the important values of the parameters. Even hard constraints are often somewhat "flexible," in that they might be intended as rough guides of the bounds, so that if the iterate sequence "hugs" a bounding surface, one might try to slowly increase the bounds, or perhaps to test the behavior via another simulation. In practice, there is generally an upper bound, beyond which the user will not allow the iterate sequence to go. At this point, either the iterate will be truncated in some way by the rules of the algorithm or there will be external intervention.

Much of the book is concerned with projected or truncated algorithms, where the iterate θ_n is confined to some bounded set, because this is a common practice in applications. Allowing unboundedness can lead to needless mathematical complications because some sort of stability must be shown or otherwise assumed, with perhaps artificial assumptions introduced on the behavior at large parameter values, and it generally adds little to the understanding of practical algorithms.

Many practical variations of the constraints can be used if the user believes they will speed up convergence. For example, if the iterate leaves the constraint set, then the projection need not be done immediately. One can wait several iterates. Also, larger step sizes can be used near the boundary, if desired. When the iterate is not constrained to a compact set, the stability and stability-ODE methods of Section 5 can be used.

Throughout the book, the step size sequence will either be a constant or will satisfy the fundamental condition

$$\sum_{n=0}^{\infty} \epsilon_n = \infty, \ \epsilon_n \geq 0, \ \epsilon_n \to 0, \quad \text{for } n \geq 0; \quad \epsilon_n = 0, \ \text{for } n < 0. \tag{1.1}$$

When *random* ϵ_n are used, it will always be supposed that (1.1) holds *with probability one.* Let $Y_n = (Y_{n,1}, \ldots, Y_{n,r})$ denote the $\mathbb{R}^r$-valued "observation" at time n, with the real-valued components $Y_{n,i}$.

Many of the proofs in this and in the next chapter are based on the ideas in [135]. To facilitate understanding of these ideas, in Section 2 we start with conditions that are stronger than needed, and weaken them subsequently. The basic interpolations and time scalings will also be used in the subsequent chapters. In Theorem 2.1, we let the ith component of the state θ_n be confined to the interval $[a_i, b_i]$, where $-\infty < a_i < b_i < \infty$. Then the algorithm is

$$\theta_{n+1,i} = \Pi_{[a_i,b_i]} \left[\theta_{n,i} + \epsilon_n Y_{n,i} \right], \quad i = 1, \ldots, r. \tag{1.2}$$

We will write this in vector notation as

$$\theta_{n+1} = \Pi_H \left[\theta_n + \epsilon_n Y_n\right], \tag{1.3}$$

where Π_H is the projection onto the constraint set $H = \{\theta : a_i \leq \theta^i \leq b_i\}$. Define the *projection* or "correction" term Z_n by writing (1.3) as

$$\theta_{n+1} = \theta_n + \epsilon_n Y_n + \epsilon_n Z_n. \tag{1.4}$$

Thus $\epsilon_n Z_n = \theta_{n+1} - \theta_n - \epsilon_n Y_n$; it is the vector of shortest Euclidean length needed to take $\theta_n + \epsilon_n Y_n$ back to the constraint set H if it is not in H.

To get a geometric feeling for the Z_n terms, refer to Figures 1.1 and 1.2. In situations such as Figure 1.1, where only one component is being truncated, Z_n points inward and is orthogonal to the boundary at θ_{n+1}. If more than one component needs to be truncated, as in Figure 1.2, Z_n again points inward but toward the corner, and it is proportional to a convex combination of the inward normals at the faces that border on that corner. In both cases, $Z_n \in -C(\theta_{n+1})$, where the cone $C(\theta)$ determined by the outer normals to the active constraint at θ was defined in Section 4.3.

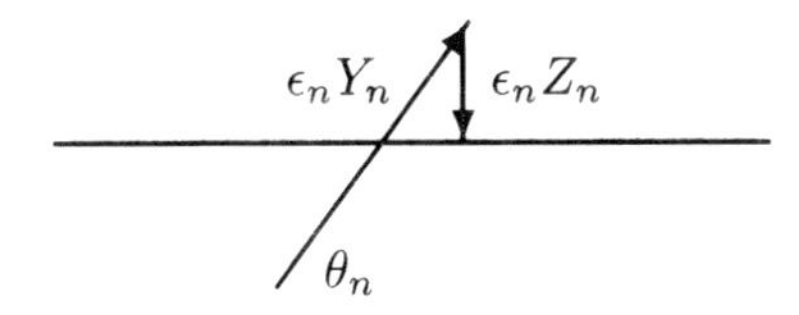

Figure 1.1. A projection with one violated constraint.

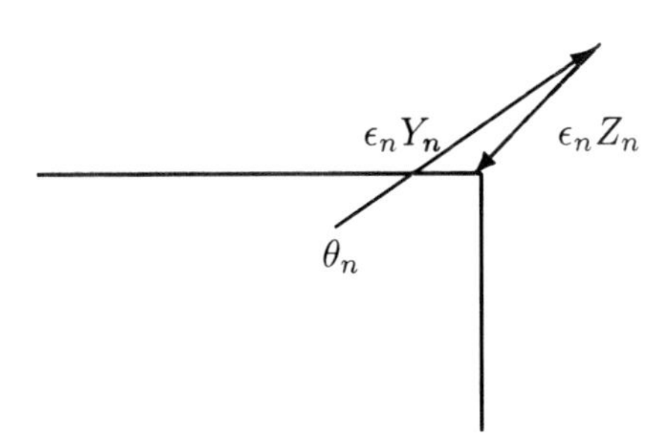

Figure 1.2. A projection with two violated constraints.

Martingale difference noise. In this chapter we will suppose that there are measurable functions $g_n(\cdot)$ of θ and random variables β_n such that Y_n can be decomposed as

$$Y_n = g_n(\theta_n) + \delta M_n + \beta_n, \quad \delta M_n = Y_n - E\left[Y_n|\theta_0, Y_i, i < n\right]. \tag{1.5}$$

The sequence $\{\beta_n\}$ will be "asymptotically negligible" in a sense to be defined. The sequence $\{\delta M_n\}$ is a martingale difference (with respect to the sequence of σ-algebras $\mathcal{F}_n$ generated by $\{\theta_0, Y_i, i < n\}$). The martingale difference assumption was used in the earliest work in stochastic approximation [27, 59, 65, 67, 70, 110, 182, 207]. Our proofs exploit the powerful ideas of the ODE methods stemming from the work of Ljung [164, 165] and Kushner [127, 135, 142]. In many of the applications of the Robbins–Monro or Kiefer–Wolfowitz algorithms, Y_n has the form $Y_n = F_n(\theta_n, \psi_n) + \beta_n$ where $\{\psi_n\}$ is a sequence of mutually independent random variables, $\{F_n(\cdot)\}$ is a sequence of measurable functions, $\beta_n \to 0$ and $E[F_n(\theta_n, \psi_n)|\theta_n = \theta] = g_n(\theta)$. For the Kiefer–Wolfowitz algorithm (see (1.2.1)–(1.2.4)), β_n represents the finite difference bias. The function $g_n(\cdot)$ might or might not depend on n. In the classical works on stochastic approximation, there was no n-dependence. The n-dependence occurs when the successive iterations are on different components of θ, the experimental procedure varies with n, or variance reduction methods are used, and so on. In the introductory result (Theorem 2.1), it will be supposed that $g_n(\cdot)$ is independent of n to simplify the development.

Definitions: Interpolated time scale and processes. The definitions and interpolations introduced in this section will be used heavily throughout the book. They are basic to the ODE method, and facilitate the effective exploitation of the time scale differences between the iterate process and the driving noise process. The ODE method uses a continuous-time interpolation of the $\{\theta_n\}$ sequence. A natural time scale for the interpolation is defined in terms of the step-size sequence. Define $t_0 = 0$ and $t_n = \sum_{i=0}^{n-1} \epsilon_i$. For $t \geq 0$, let $m(t)$ denote the unique value of n such that $t_n \leq t < t_{n+1}$. For $t < 0$, set $m(t) = 0$. Define the *continuous-time interpolation* $\theta^0(\cdot)$ on $(-\infty, \infty)$ by $\theta^0(t) = \theta_0$ for $t \leq 0$, and for $t \geq 0$,

$$\theta^0(t) = \theta_n, \qquad \text{for } t_n \leq t < t_{n+1}. \tag{1.6}$$

For later use, define the sequence of *shifted* processes $\theta^n(\cdot)$ by

$$\theta^n(t) = \theta^0(t_n + t), \quad t \in (-\infty, \infty). \tag{1.7}$$

Figures 1.3 and 1.4 illustrate the functions $m(\cdot)$, $m(t_n + \cdot)$, and interpolations $\theta^0(\cdot)$, and $\theta^n(\cdot)$.

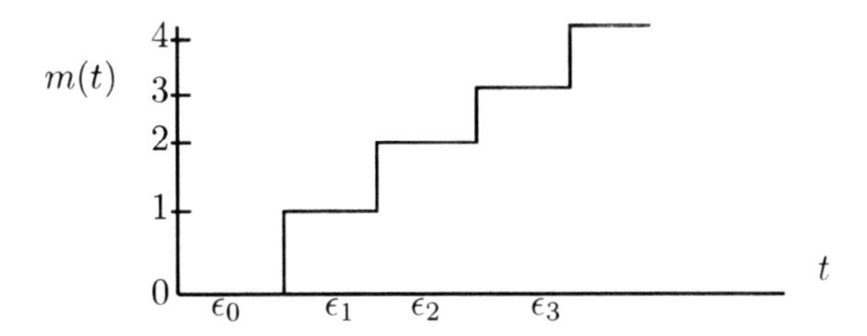

Figure 1.3a. The function $m(\cdot)$.

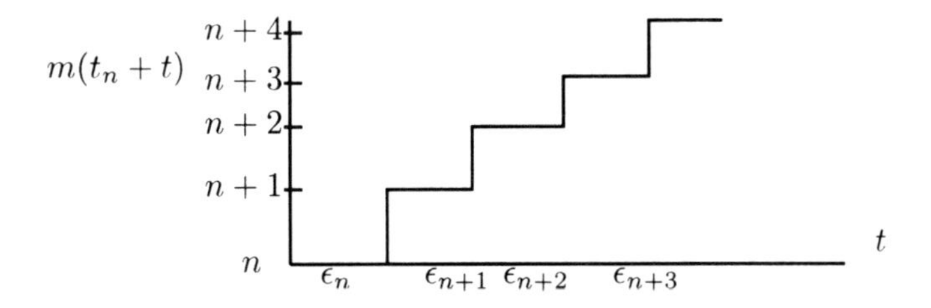

Figure 1.3b. The function $m(t_n + t)$.

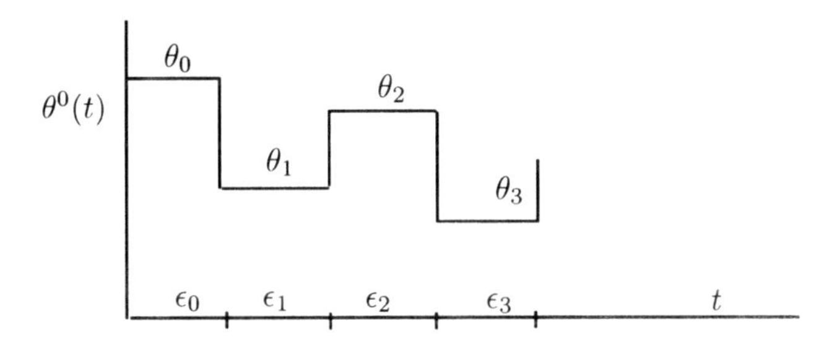

Figure 1.4a. The function $\theta^0(\cdot)$.

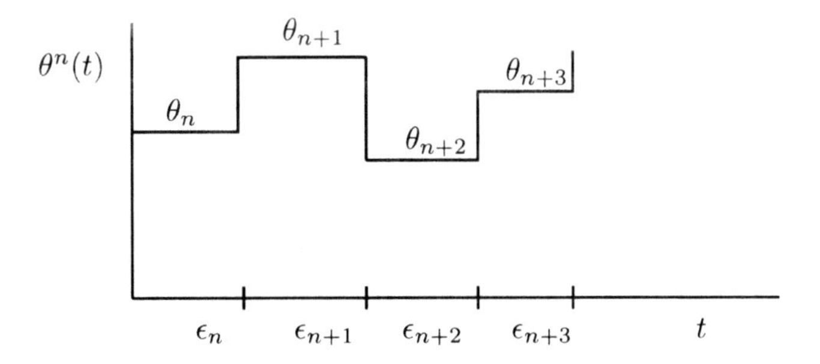

Figure 1.4b. The function $\theta^n(\cdot)$.

Let $Z_i = 0$ and $Y_i = 0$ for $i < 0$. Define $Z^0(t) = 0$ for $t \leq 0$ and

$$Z^0(t) = \sum_{i=0}^{m(t)-1} \epsilon_i Z_i, \quad t \geq 0.$$

Define $Z^n(\cdot)$ by

$$\begin{aligned} Z^n(t) &= Z^0(t_n + t) - Z^0(t_n) = \sum_{i=n}^{m(t_n+t)-1} \epsilon_i Z_i, \quad t \geq 0, \\ Z^n(t) &= -\sum_{i=m(t_n+t)}^{n-1} \epsilon_i Z_i, \quad t < 0. \end{aligned} \tag{1.8}$$

Define $Y^n(\cdot), M^n(\cdot)$, and $B^n(\cdot)$ analogously to $Z^n(\cdot)$ but using Y_i, δM_i, and β_i, resp., in lieu of Z_i. By the definitions (recall that $m(t_n) = n$)

$$\theta^n(t) = \theta_n + \sum_{i=n}^{m(t_n+t)-1} \epsilon_i [Y_i + Z_i] = \theta_n + Y^n(t) + Z^n(t), \quad t \geq 0, \tag{1.9a}$$

$$\theta^n(t) = \theta_n - \sum_{i=m(t_n+t)}^{n-1} \epsilon_i [Y_i + Z_i] = \theta_n + Y^n(t) + Z^n(t), \quad t < 0. \tag{1.9b}$$

Note on (1.9). *For simplicity, we always write the algorithm as* (1.9a), *whether t is positive or negative, with the understanding that it is to be interpreted as* (1.9b) *if* $t < 0$.

All the above interpolation formulas will be used heavily in the sequel. Note that the time origin of the "shifted" processes $\theta^n(\cdot)$ and $Z^n(\cdot)$ is time t_n for the original processes, the interpolated time at the nth iteration. The step sizes ϵ_n used in the interpolation are natural intervals for the continuous-time interpolation. Their use allows us to exploit the time scale differences between the mean and the noise terms under quite general conditions. We are concerned with the behavior of the tail of the sequence $\{\theta_n\}$. Since this is equivalent to the behavior of $\theta^n(\cdot)$ over any finite interval for large n, a very effective method (introduced in [135]) of dealing with the tails works with these shifted processes $\theta^n(\cdot)$.

Note on piecewise linear vs. piecewise constant interpolations. The basic stochastic approximation (1.3) is defined as a discrete-time process. We have defined the continuous-time interpolations $\theta^n(\cdot)$, $Z^n(\cdot)$ to be piecewise constant with interpolation intervals ϵ_n. We could have defined the interpolations to be piecewise linear in the obvious way by simply interpolating linearly between the "break" or "jump" points $\{t_n\}$. Nevertheless, there are some notational advantages to the piecewise constant interpolation. In the proofs in this chapter and Chapter 6, it is shown that for almost all sample paths the set $\{\theta^n(\omega, \cdot)\}$ is equi- (actually Lipschitz) continuous in the extended sense (see Theorem 4.2.2). Thus, the set of piecewise linear interpolations is also equicontinuous.

5.2 The ODE Method: A Basic Convergence Theorem

5.2.1 *Assumptions and the Main Convergence Theorem*

One way or another, all methods of analysis need to show that the "tail" effects of the noise vanish. This "tail" property is essentially due to the martingale difference property and the fact that the step sizes ϵ_n decrease to zero as $n \to \infty$.

Definition. Recall the definitions of stability, asymptotic stability, local stability, and local asymptotic stability, given in Section 4.2 for the projected ODE

$$\dot{\theta} = \bar{g}(\theta) + z, \quad z \in -C(\theta), \tag{2.1}$$

where z is the minimum force needed to keep the solution in H. Recall that L_H denotes the set of limit points of (2.1) in H, over all initial conditions. By invariant set, we always mean a two-sided invariant set; that is, if $x \in I$, an invariant set in H, then there is a path of the ODE in I on the time interval $(-\infty, \infty)$ that goes through x at time 0. If there is a constraint set, then the set of limit points might be smaller than the largest two-sided invariant set.

Let E_n denote the expectation conditioned on the σ-algebra $\mathcal{F}_n$, generated by $\{\theta_0, Y_i, i < n\}$. When it is needed, the definition will be changed to a larger σ-algebra.

Assumptions. The assumptions listed here, which will be used in Theorem 2.1, are more or less classical except for the generality of the possible limit set for the mean ODE and the use of a constraint set. All the conditions will be weakened in subsequent theorems. The proof is more complicated than the "minimal" convergence proof, since the algorithm is not necessarily of the gradient descent form and we do not require that there be a unique limit point, but allow the algorithm to have a possibly complicated asymptotic behavior. Also, the proof introduces decompositions and interpolations that will be used in the sequel, as well as the basic idea of the ODE method for the probability one convergence. Condition (A2.2) simply sets up the notation, where β_n satisfies (A2.5). Stronger results are obtained in Subsection 2.2, where it is shown that the process converges to the subset of the limit points consisting of "chain recurrent" points, a natural set.

(A2.1) $\sup_n E|Y_n|^2 < \infty$.

(A2.2) There is a measurable function $\bar{g}(\cdot)$ of θ and random variables β_n such that

$$E_n Y_n = E\left[Y_n | \theta_0, Y_i, i < n\right] = \bar{g}(\theta_n) + \beta_n.$$

(A2.3) $\bar{g}(\cdot)$ is continuous.

(A2.4) $\sum_i \epsilon_i^2 < \infty$.

(A2.5) $\sum_i \epsilon_i |\beta_i| < \infty$ w.p.1.

If there is a continuously differentiable real-valued function $f(\cdot)$ such that $\bar{g}(\cdot) = -f_\theta(\cdot)$, then the points in L_H are the stationary points, and are to be called S_H. They satisfy the stationarity condition

$$\bar{g}(\theta) + z = 0, \qquad \text{for some } z \in -C(\theta). \tag{2.2}$$

The set S_H can be divided into disjoint compact and connected subsets $S_i, i = 0, \ldots$ The following condition will sometimes be used.

(A2.6) $\bar{g}(\cdot) = -f_\theta(\cdot)$ for continuously differentiable real-valued $f(\cdot)$ and $f(\cdot)$ is constant on each S_i.

If $f(\cdot)$ and the $q_i(\cdot)$ in (A4.3.2) (which define the constraint set) are twice continuously differentiable, then (A2.6) holds.

Comment on equality constraints and smooth manifolds. The equality constrained problem and the case where the constraint set H is a smooth manifold in $\mathbb{R}^{r-1}$ are covered by the results of the book. A convenient alternative approach that works directly on the manifold and effectively avoids the reflection terms can be seen from the following comments. The reader can fill in the explicit conditions that are needed. Suppose that the constraint set H is a smooth manifold. The algorithm $\theta_{n+1} = \Pi_H(\theta_n + \epsilon_n Y_n)$ can be written as

$$\theta_{n+1} = \theta_n + \epsilon_n \gamma(\theta_n) Y_n + \epsilon_n \beta_n,$$

where $\gamma(\cdot)$ is a smooth function and $\epsilon_n \gamma(\theta_n) Y_n$ is the projection of $\epsilon_n Y_n$ onto the orthogonal complement of the normal to H at the point θ_n, and $\epsilon_n \beta_n$ represents the "error." Under reasonable conditions on the smoothness and on the sequence $\{Y_n\}$, the sequences $\{\gamma(\theta_n) Y_n, \beta_n\}$ will satisfy the conditions required on the $\{Y_n, \beta_n\}$ in the theorems. The mean ODE will be $\dot{\theta} = \gamma(\theta)\bar{g}(\theta)$. Similar comments hold when the ODE is replaced by a differential inclusion, for the correlated noise case of Chapter 6 and the various weak convergence cases of Chapters 7 and 8. The results can be extended to the case where H is the intersection of the $\mathbb{R}^{r-1}$-dimensional manifold defined by (A4.3.3) and a set satisfying (A4.3.2) or (A4.3.1).

Theorem 2.1. *Let* (1.1), (1.2), *and* (A2.1)–(A2.5) *hold for algorithm* (1.3). *Then there is a set* N *of probability zero such that for* $\omega \notin N$, *the set of functions* $\{\theta^n(\omega,\cdot), Z^n(\omega,\cdot), n<\infty\}$ *is equicontinuous. Let* $(\theta(\omega,\cdot), Z(\omega,\cdot))$ *denote the limit of some convergent subsequence. Then this pair satisfies the projected ODE* (2.1), *and* $\{\theta_n(\omega)\}$ *converges to some limit set of the ODE in* H.[1] *If the constraint set is dropped, but* $\{\theta_n\}$ *is bounded with probability one, then for almost all* ω, *the limits* $\theta(\omega,\cdot)$ *of convergent subsequences of* $\{\theta^n(\omega,\cdot)\}$ *are trajectories of*

$$\dot{\theta} = \bar{g}(\theta) \tag{2.3}$$

in some bounded invariant set and $\{\theta_n(\omega)\}$ *converges to this invariant set. Let* p_n *be integer-valued functions of* ω, *not necessarily being stopping times or even measurable, but that go to infinity with probability one. Then the conclusions concerning the limits of* $\{\theta^n(\cdot)\}$ *hold with* p_n *replacing* n. *If* $A \subset H$ *is locally asymptotically stable in the sense of Liapunov for* (2.1) *and* θ_n *is in some compact set in the domain of attraction of* A *infinitely often with probability* $\geq \rho$, *then* $\theta_n \to A$ *with at least probability* ρ. *Suppose that* (A2.6) *holds. Then, for almost all* ω, $\{\theta_n(\omega)\}$ *converges to a unique* S_i.

Remark. In many applications where $-\bar{g}(\cdot)$ is a gradient and the truncation bounds are large enough, there is only one stationary point of (2.1), and that is globally asymptotically stable. Then $\{\theta_n\}$ converges w.p.1 to that point. For simplicity, we use *equicontinuity* to mean "equicontinuity in the extended sense," as defined in the definition preceding Theorem 4.2.2.

Proof: Part 1. Convergence of the martingale and equicontinuity. Define $\delta M_n = Y_n - \bar{g}(\theta_n) - \beta_n$, and decompose the algorithm (1.3) as

$$\theta_{n+1} = \theta_n + \epsilon_n \bar{g}(\theta_n) + \epsilon_n Z_n + \epsilon_n \delta M_n + \epsilon_n \beta_n. \tag{2.4}$$

Then we can write

$$\begin{aligned}\theta^n(t) = \theta_n &+ \sum_{i=n}^{m(t+t_n)-1} \epsilon_i \bar{g}(\theta_i) + \sum_{i=n}^{m(t+t_n)-1} \epsilon_i Z_i \\ &+ \sum_{i=n}^{m(t+t_n)-1} \epsilon_i \delta M_i + \sum_{i=n}^{m(t+t_n)-1} \epsilon_i \beta_i.\end{aligned} \tag{2.5}$$

Define $M_n = \sum_{i=0}^{n-1} \epsilon_i \delta M_i$. This is a martingale sequence (with associated σ-algebras $\mathcal{F}_n$), since we have centered the summands about their condi-

[1] By convergence $\theta(t) \to A$ for a set A, we mean $\lim_{t\to\infty} \mathrm{dist}(\theta(t), A) = 0$, where $\mathrm{dist}(\theta, A) = \min_{x\in A} |\theta - x|$.

tional expectations, given the "past." By (4.1.4), for each $\mu > 0$,

$$P\left\{ \sup_{n \geq j \geq m} |M_j - M_m| \geq \mu \right\} \leq \frac{E\left|\sum_{i=m}^{n-1} \epsilon_i \delta M_i\right|^2}{\mu^2}.$$

By (A2.1), (A2.4), and the fact that $E\delta M_i \delta M_j' = 0$ for $i \neq j$, the right side is bounded above by $K \sum_{i=m}^{\infty} \epsilon_i^2$, for some constant K. Thus, for each $\mu > 0$,

$$\lim_m P\left\{ \sup_{j \geq m} |M_j - M_m| \geq \mu \right\} = 0. \tag{2.6}$$

Since $\theta^n(\cdot)$ is piecewise constant, we can rewrite (2.5) as

$$\theta^n(t) = \theta_n + \int_0^t \bar{g}(\theta^n(s))ds + Z^n(t) + M^n(t) + B^n(t) + \rho^n(t), \tag{2.7}$$

where $\rho^n(t)$ is due to the replacement of the first sum in (2.5) by an integral. $\rho^n(t) = 0$ at the times[2] $t = t_k - t_n, k > n$, at which the interpolated processes have jumps, and $\rho^n(t) \to 0$ uniformly in t as $n \to \infty$. By (2.6) and (A2.5), there is a null set N such that for $\omega \notin N$, $M^n(\omega, \cdot)$ and $B^n(\omega, \cdot)$ go to zero uniformly on each bounded interval in $(-\infty, \infty)$ as $n \to \infty$.

Let $\omega \notin N$. By the definition of N, the functions of t on the right side of (2.7) (except possibly for $Z^n(\cdot)$) are equicontinuous in n and the limits of $M^n(\cdot)$, $B^n(\cdot)$, and $\rho^n(\cdot)$ are zero. It will next be shown that the equicontinuity of $\{Z^n(\omega, \cdot), n < \infty\}$ is a consequence of the fact that

$$Z_n(\omega) \in -C(\theta_{n+1}(\omega)). \tag{2.8}$$

For $\omega \notin N$, $\theta_{n+1}(\omega) - \theta_n(\omega) \to 0$. If $Z^n(\omega, \cdot)$ is not equicontinuous, then there is a subsequence that has a jump asymptotically; that is, there are integers $\mu_k \to \infty$, uniformly bounded times s_k, $0 < \delta_k \to 0$ and $\rho > 0$ (all depending on ω) such that $|Z^{\mu_k}(\omega, s_k + \delta_k) - Z^{\mu_k}(\omega, s_k)| \geq \rho$. The changes of the terms other than $Z^n(\omega, t)$ on the right side of (2.7) go to zero on the intervals $[s_k, s_k + \delta_k]$. Furthermore $\epsilon_n Y_n(\omega) = \epsilon_n \bar{g}(\theta_n(\omega)) + \epsilon_n \delta M_n(\omega) + \epsilon_n \beta_n \to 0$ and $Z_n(\omega) = 0$ if $\theta_{n+1}(\omega) \in H^0$, the interior of H. Thus, this jump cannot force the iterate to the interior of the hyperrectangle H, and it cannot force a jump of the $\theta^n(\omega, \cdot)$ along the boundary either. Consequently, $\{Z^n(\omega, \cdot)\}$ is equicontinuous.

Part 2. Characterizing the limit of a convergent subsequence: Applying the Arzelà–Ascoli Theorem. Let $\omega \notin N$, let n_k denote a subsequence such that $\{\theta^{n_k}(\omega, \cdot), Z^{n_k}(\omega, \cdot)\}$ converges, and denote the limit by $(\theta(\omega, \cdot), Z(\omega, \cdot))$. Then

$$\theta(\omega, t) = \theta(\omega, 0) + \int_0^t \bar{g}(\theta(\omega, s))ds + Z(\omega, t). \tag{2.9}$$

[2] Recall that $t_n = \sum_{i=0}^{n-1} \epsilon_i$.

Note that $Z(\omega, 0) = 0$ and $\theta(\omega, t) \in H$ for all t. To characterize $Z(\omega, t)$, use (2.8) and the fact that $\theta_{n+1}(\omega) - \theta_n(\omega) \to 0$. These facts, together with the upper semicontinuity property (4.3.2) and the continuity of $\theta(\omega, \cdot)$, imply that (4.3.4) and (4.3.5) hold. In fact, it follows from the method of construction of $Z(\omega, \cdot)$ that the function simply serves to keep the dynamics $\bar{g}(\cdot)$ from forcing $\theta(\omega, \cdot)$ out of H. Thus, for $s > 0$, $|Z(\omega, t+s) - Z(\omega, t)| \leq \int_t^{t+s} |\bar{g}(\theta(\omega, u))| du$. Hence $Z(\omega, \cdot)$ is Lipschitz continuous, and $Z(\omega, t) = \int_0^t z(\omega, s) ds$, where $z(\omega, t) \in -C(\theta(\omega, t))$ for almost all t.

Recall the definition of the set A, and the definitions of the arbitrarily small positive numbers $\delta > \delta_1$ given in the definition of local stability in Subsection 4.2.2. Let $0 < \delta_2 < \delta_1$. Suppose that $\{\theta_n(\omega)\}$ has a limit point x_0 in a compact subset D of the domain of attraction of A. Then there is a subsequence m_k such that $\theta^{m_k}(\omega, \cdot)$ converges to a solution of (2.1) with initial condition x_0. Since the trajectory of (2.1) starting at x_0 ultimately goes to A, $\theta_n(\omega)$ must be in $N_{\delta_2}(A)$ infinitely often. It will be seen that escape from $N_\delta(A)$ infinitely often is impossible. Suppose that escape from $N_\delta(A)$ occurs infinitely often. Then, since $\theta_{n+1}(\omega) - \theta_n(\omega) \to 0$, there are integers n_k such that $\theta_{n_k}(\omega)$ converges to some point x_1 in $N_{\delta_1}(A)$, and the path $\theta^{n_k}(\omega, \cdot)$ converges to a solution of (2.1) starting at x_1 and that leaves $N_\delta(A)$. But, by the definition of local asymptotic stability of A, such a path cannot exist. This implies that $\theta_n(\omega)$ cannot exit $N_\delta(A)$ infinitely often.

Whether or not there is a constraint set H, if boundedness with probability one of the sequence $\{\theta_n\}$ is assumed, then the preceding arguments show that (with probability one) the limits of $\{\theta^n(\omega, \cdot)\}$ are bounded solutions to (2.1) (which is (2.3) if there is no constraint) on the time interval $(-\infty, \infty)$. Thus the entire trajectory of a limit $\theta(\omega, \cdot)$ must lie in a bounded invariant set of (2.1) by the definition of an invariant set. The fact that $\{\theta_n(\omega)\}$ converges to some invariant set of (2.1) then follows; otherwise there would be a limit of a convergent subsequence satisfying (2.1) on $(-\infty, \infty)$ but not lying entirely in an invariant set. To see that $\theta^n(\cdot)$ also converges to L_H with probability one, also work on the time interval $(-\infty, \infty)$. Then the limit of any convergent subsequence satisfies the ODE (with probability one) on $(-\infty, \infty)$. Given the "initial condition" $\theta(-T)$, the solution $\theta(0)$ must be arbitrarily close to L_H for large T, uniformly in $\theta(0)$.

These arguments do not depend on how the "sections" of $\theta^0(\omega, \cdot)$ are chosen. Any set of "sections" other than $\theta^n(\omega, \cdot)$ could have been used, as long as the initial times went to infinity. The statement of the theorem concerning $\{p_n\}$ then follows from what has been done.

Part 3. The case when $-\bar{g}(\cdot)$ is a gradient. Now assume (A2.6) and suppose that $\bar{g}(\cdot) = -f_\theta(\theta)$ for some continuously differentiable function $f(\cdot)$. As will be shown, the conclusion concerning the limits actually follows from what has been done.

We continue to work with $\omega \notin N$. Suppose for simplicity that there are only a finite number of S_i, namely, $S_0, \ldots, S_M$. In (2.1), $|z(t)| \le |\bar{g}(\theta(t))|$. Thus, if $\bar{g}(\cdot) = -f_\theta(\cdot)$, the derivative of $f(\theta(\cdot))$ along the solution of (2.1) at $\theta \in H$ is $f'_\theta(\theta)\left[-f_\theta(\theta) + z\right] \le 0$, and we see that all solutions of (2.1) tend to the set of stationary points defined by (2.2). For each c, the set $\{\theta : f(\theta) \le c\}$ is locally asymptotically stable in the sense of Liapunov, assuming that it is not empty. Then the previous part of the proof implies that $f(\theta_n(\omega))$ converges to some constant (perhaps depending on ω), and $\theta_n(\omega)$ converges to the set of stationary points.

It remains to be shown that $\{\theta_n(\omega)\}$ converges to a unique S_i. If the claimed convergence does not occur, the path will eventually oscillate back and forth among arbitrarily small neighborhoods of distinct S_i. This implies that there is a limit point outside the set of stationary points. □

An elaboration of the proof for the gradient descent case. For future use, and as an additional illustration of the ODE method, we will elaborate on the proof for the case where $\bar{g}(\cdot)$ is a gradient. The ideas are just those used in the previous proof. The details to be given are of more general applicability and will be used in Theorems 4.2 and 4.3 in combination with a Liapunov function technique.

We start by supposing that the path $\theta^0(\omega, t)$ oscillates back and forth between arbitrarily small neighborhoods of distinct S_i. This will be seen to contradict the "gradient descent" property of the ODE (2.1). The proof simply sets up the notation required to formalize this idea.

Since $\{\theta_n(\omega)\}$ converges to $S_H = \cup_i S_i$, there is a subsequence m_k such that $\theta_{m_k}(\omega)$ tends to some point $x_0 \in S_H$. Suppose that $x_0 \in S_0$. We will show that $\theta_n(\omega) \to S_0$. Suppose that $\theta_n(\omega) \not\to S_0$. Then there is an x_1 in some $S_i, i \neq 0$ (call it S_1 for specificity), and a subsequence $\{q_k\}$ such that $\theta_{q_k}(\omega) \to x_1$.

Continuing this process, let $S_0, \ldots, S_R, R > 0$, be all the sets that contain limit points of the sequence $\{\theta_n(\omega)\}$. Order the sets such that $f(S_R) = \liminf_n f(\theta_n(\omega))$, and suppose that S_R is the (assumed for simplicity) unique set on which the liminf is attained. The general (nonunique) case requires only a slight modification. Let $\delta > 0$ be such that $f(S_R) < f(S_i) - 2\delta, i \neq R$. For $\rho > 0$, define the ρ-neighborhood $N_\rho(S_R)$ of S_R by $N_\rho(S_R) = \{x : f(x) - f(S_R) < \rho\}$. By the definition of δ, $N_{2\delta}(S_R)$ contains no point in any $S_i, i \neq R$. By the hypothesis that more than one S_i contains limit points of $\{\theta_n(\omega)\}$, the neighborhood $N_\delta(S_R)$ of S_R is visited infinitely often and $N_{2\delta}(S_R)$ is exited infinitely often by $\{\theta_n(\omega)\}$. Thus there are $\nu_k \to \infty$ (depending on ω) such that $\theta_{\nu_k - 1}(\omega) \in N_\delta(S_R)$, $\theta_{\nu_k}(\omega) \notin N_\delta(S_R)$, and after time ν_k the path does not return to $N_\delta(S_R)$ until after it leaves $N_{2\delta}(S_R)$.

By the equicontinuity of $\{\theta^n(\omega, \cdot)\}$, there is a $T > 0$ such that $\theta_{\nu_k}(\omega) \to \partial N_\delta(S_R)$, the boundary of $N_\delta(S_R)$, and for large k, $\theta^{\nu_k}(\omega, t) \notin N_\delta(S_R)$ for $t \in [0, T]$.

There is a subsequence $\{\mu_m\}$ of $\{\nu_k\}$ such that $(\theta^{\mu_m}(\omega,\cdot), Z^{\mu_m}(\omega,\cdot))$ converges to some limit $(\bar{\theta}(\omega,\cdot), \bar{Z}(\omega,\cdot))$ that satisfies (2.1) with $\bar{\theta}(\omega,0) \in \partial N_\delta(S_R)$, the boundary of $N_\delta(S_R)$, and $\bar{\theta}(\omega,t) \notin N_\delta(S_R)$ for $t \le T$. This is a contradiction because, by the gradient descent property of (2.1) (with $\bar{g}(\cdot) = -f_\theta(\cdot)$) and the definitions of δ and $N_\delta(S_R)$, any solution to (2.1) starting on $\partial N_\delta(S_R)$ must stay in $N_\delta(S_R)$ for all $t > 0$. □

The preceding proof implies the following result.

Theorem 2.2. *Let (all with probability one) $\{\theta^n(\omega,\cdot), Z^n(\omega,\cdot)\}$ be equicontinuous, with all limits satisfying* (2.1), *where $\bar{g}(\cdot)$ is only measurable. Let the set A be locally asymptotically stable in the sense of Liapunov. Suppose that $\{\theta_n(\omega)\}$ visits a compact set in the domain of attraction of A infinitely often. Then $\theta_n(\omega) \to A$.*

Remark on the structure of the proof of Theorem 2.1. Let us review the structure of the proof. First, the increment was partitioned to get the convenient representation (2.7), with which we could work on one part at a time. Then it was shown that with probability one the martingale term $M^n(\cdot)$ converges with probability one to the "zero" process. Then the probability one convergence to zero of the bias $\{B^n(\cdot)\}$ was shown. The asymptotic continuity of $Z^n(\cdot)$ was obtained by a direct use of the properties of the Z_n as reflection terms. Then, by fixing ω not in some "bad" null set, and taking convergent subsequences of $\{\theta^n(\omega,\cdot), Z^n(\omega,\cdot)\}$, we were able to characterize the limit as a solution to the mean ODE. It then followed that the sequence $\{\theta_n(\omega)\}$ converges to some limit or invariant set of the ODE.

A more general constraint set. Using the same basic structure of the proof, Theorem 2.1 can be readily generalized in several useful directions with little extra work. (A2.1) and (A2.4) will be weakened in the next section. Appropriate dependence on n of $\bar{g}(\cdot)$ can be allowed, and the hyperrectangle H can be replaced by a more general constraint set. The techniques involved in the required minor alterations in the proofs will be important in the analysis of the "dependent" noise case in Chapter 6.

In Theorem 2.3, the constraint form (A4.3.2) or (A4.3.3) will be used, where (A4.3.2) includes (A4.3.1). For $\theta \in I\!R^r$, let $\Pi_H(\theta)$ denote the closest point in H to θ. If the closest point is not unique, select a closest point such that the function $\Pi_H(\cdot)$ is measurable. We will work with the algorithm

$$\theta_{n+1} = \Pi_H\left[\theta_n + \epsilon_n Y_n\right], \tag{2.10a}$$

that will be written as

$$\theta_{n+1} = \theta_n + \epsilon_n Y_n + \epsilon_n Z_n, \tag{2.10b}$$

where Z_n is the projection term. Recall the definition of $C(\theta)$ from Section 4.3. It follows from the calculus that $Z_n \in -C(\theta_{n+1})$ under (A4.3.3). Under

(A4.3.2), this is proved by applying the Kuhn–Tucker Theorem of nonlinear programming to the problem of minimizing $|x - (\theta_n + \epsilon_n Y_n)|^2$ subject to the constraints $q_i(x) \leq 0, i \leq p$, where $x = \theta_{n+1}$. That theorem says that there are $\lambda_i \geq 0$, with $\lambda_i = 0$ if $q_i(x) < 0$, such that

$$(x - (\theta_n + \epsilon_n Y_n)) + \sum_i \lambda_i q_{i,x}(x) = 0,$$

which implies that $Z_n \in -C(\theta_{n+1})$.

The following assumption generalizes (A2.2) and will in turn be relaxed in the next section.

(A2.7) There are functions $g_n(\cdot)$ of θ, which are continuous uniformly in n, a continuous function $\bar{g}(\cdot)$ and random variables β_n such that

$$E_n Y_n = g_n(\theta_n) + \beta_n, \tag{2.11}$$

and for each $\theta \in H$,

$$\lim_n \left| \sum_{i=n}^{m(t_n+t)} \epsilon_i \left[g_i(\theta) - \bar{g}(\theta) \right] \right| \to 0 \tag{2.12}$$

for each $t > 0$. (Thus $\bar{g}(\cdot)$ is a "local average" of the $g_n(\cdot)$.)

Dependence of $\bar{g}(\cdot)$ on the past. Note that in all the algorithmic forms, $g_n(\theta_n)$ can be replaced by dependence on the past of the form $g_n(\theta_n, \ldots, \theta_{n-K})$ provided that the continuity of $g_n(\cdot)$ is replaced by the continuity of $g_n(x_0, \ldots, x_K)$ on the "diagonal" set $x = x_0 = \cdots = x_K$, for each x, uniformly in n.

Theorem 2.3. *Assume the conditions of Theorem* 2.1 *but use the algorithm* $\theta_{n+1} = \Pi_H [\theta_n + \epsilon_n Y_n]$ *with any of the constraint set conditions* (A4.3.1), (A4.3.2), *or* (A4.3.3) *holding, and* (A2.7) *with* $\beta_n \to 0$ *with probability one replacing* (A2.2) *and* (A2.5). *Then the conclusions of Theorem* 2.1 *hold.*

Remark on the proof. The proof is essentially the same as that of Theorem 2.1, and we concentrate on the use of (A2.7) and the equicontinuity of $\{Z^n(\cdot)\}$ with probability one. The equicontinuity proof exploits the basic character of Z_n as *projection terms* to get the desired result, and the proof can readily be used for the general cases of Section 4 and Chapter 6.

Proof. Define

$$\bar{G}^n(t) = \sum_{i=n}^{m(t_n+t)-1} \epsilon_i \bar{g}(\theta_i), \quad \widetilde{G}^n(t) = \sum_{i=n}^{m(t_n+t)-1} \epsilon_i \left[g_i(\theta_i) - \bar{g}(\theta_i) \right].$$

For simplicity, we only work with $t \geq 0$. With these definitions,

$$\theta^n(t) = \theta_n + \bar{G}^n(t) + \widetilde{G}^n(t) + B^n(t) + M^n(t) + Z^n(t). \tag{2.13}$$

As in Theorem 2.1, (A2.1) and (A2.4) imply that $M^n(\cdot)$ converges to the "zero" process with probability one as $n \to \infty$. Since $\beta_n \to 0$ with probability one, the process $B^n(\cdot)$ also converges to zero with probability one. Since $g_n(\cdot)$ and $\bar{g}(\cdot)$ are uniformly bounded on H, the set $\{\bar{G}^n(\omega, \cdot), \widetilde{G}^n(\omega, \cdot)\}$ is equicontinuous for each ω. These bounds and convergences imply that the jumps in $\bar{G}^n(\cdot) + \widetilde{G}^n(\cdot) + M^n(\cdot) + B^n(\cdot)$ on any finite interval go to zero with probability one as $n \to \infty$. Consequently, with probability one the distance between $\theta_n + \epsilon_n Y_n$ and H goes to zero as $n \to \infty$.

Now fix attention on the case where H satisfies (A4.3.2). [The details under (A4.3.3) are left to the reader.] Then, if we were to ignore the effects of the terms $Z^n(\cdot)$, $\{\theta^n(\omega, \cdot)\}$ would be equicontinuous on $(-\infty, \infty)$ for ω not in a null set N, the set of nonconvergence to zero of $B^n(\omega, \cdot)$ or of $M^n(\omega, \cdot)$. Thus the only possible problem with equicontinuity would originate from Z_n. We now proceed to show that this is not possible.

Suppose that for some $\omega \notin N$, there are $\delta_1 > 0$, $m_k \to \infty$ and $0 < \Delta_k \to 0$ such that $|Z^{m_k}(\omega, \Delta_k)| \geq \delta_1$. Then the paths $Z^{m_k}(\omega, \cdot)$ will "asymptotically" have a jump of at least δ_1 at $t = 0$. This jump cannot take the path $\theta^{m_k}(\omega, \cdot)$ into the interior of H, since $Z_n(\omega) = 0$ if $\theta_{n+1}(\omega)$ is in the interior of H. Thus the effect of the assumed "asymptotic" jump in $Z^{m_k}(\omega, \cdot)$ is an "'asymptotic" jump of $\theta^{m_k}(\omega, \cdot)$ from one point on the boundary of H to another point on the boundary of H. But this contradicts the fact that $\epsilon_n Z_n(\omega)$ goes to zero and acts as either an inward normal at $\theta_{n+1}(\omega)$, if $\theta_{n+1}(\omega)$ is not on an edge or corner of H, or as a non-negative linear combination of the linearly independent set of the inward normals of the adjacent faces if $\theta_{n+1}(\omega)$ is on an edge or corner. The Lipschitz continuity follows from the same argument that was used in Theorem 2.1. A similar reasoning for the equicontinuity of $\{Z^n(\cdot)\}$ can be used when the constraint set is defined by (A4.3.3).

The rest of the reasoning is as in Theorem 2.1; we need only identify the limits of $\widetilde{G}^n(\omega, \cdot)$ and $\bar{G}^n(\omega, \cdot)$ along convergent subsequences and for ω not in N. Fix $\omega \notin N$, and let n_k index a convergent subsequence of $\{\theta^n(\omega, \cdot)\}$ with limit $\theta(\omega, \cdot)$. Then the equicontinuity of $\{\theta^n(\omega, \cdot)\}$, the uniform (in n) continuity of $g_n(\cdot)$, and (2.12) imply that $\widetilde{G}^{n_k}(\omega, t) \to 0$ and $\bar{G}^{n_k}(t) \to \int_0^t \bar{g}(\theta(\omega, s))ds$ for each t. □

Theorem 2.4. (Random ϵ_n.) *Let $\epsilon_n \geq 0$ be $\mathcal{F}_n$-measurable and satisfy*

$$\sum_n \epsilon_n^2 < \infty \text{ w.p.1.} \tag{2.14}$$

Then, under the other conditions of Theorem 2.3, the conclusions of the theorem remain valid.

Proof. The proof is essentially the same as that of Theorem 2.3. Modify the process on a set of arbitrarily small measure such that $E\sum \epsilon_n^2 < \infty$ and then prove convergence of the modified process. In particular, for $K > 0$ define $\epsilon_{n,K} = \epsilon_n$ until the first n that $\sum_{i=0}^{n-1} \epsilon_i^2 \geq K$, and set $\epsilon_{n,K} = 1/n$ at and after that time. The proof of Theorem 2.3 holds if the $\epsilon_{n,K}$ are used. Since $\lim_{K\to\infty} P\{\epsilon_n \neq \epsilon_{n,K},\ \text{any } n\} = 0$, the theorem follows. □

5.2.2 Convergence to Chain Recurrent Points

In the previous parts of this chapter and in the following sections, it is shown that θ_n and $\theta^n(\cdot)$ converge with probability one to a bounded invariant or limit set of the ODE. Sometimes the largest invariant or limit sets contain points to which convergence clearly cannot occur. It will now be seen that the possible limit points can be further restricted.

The paths of the stochastic approximation and chain recurrent points. As seen in Theorem 2.1, for almost all ω the path $\theta^n(\omega,\cdot)$ follows the solution to the ODE closely for a time that increases to infinity as $n \to \infty$. Let ω not be in the null set N of Theorem 2.1, and let $\Phi(t|\theta)$ denote the solution to the mean ODE with initial condition θ. Given $\delta > 0$, there are $T_0^n \to \infty$ such that $|\theta^n(\omega,t) - \Phi(t|\theta^n(\omega,0))| \leq \delta$ on $[0, T_0^n]$. Now repeat the procedure. There are times T_k^n such that $T_{k+1}^n - T_k^n \to \infty$ for each large n and

$$|\theta^n(\omega, T_k^n + t) - \Phi(t|\theta^n(\omega, T_k^n))| \leq \delta,\ t \leq T_{k+1}^n - T_k^n.$$

Thus, asymptotically, the path of the stochastic approximation will follow a path of the ODE which is restarted periodically at a value close to the value at the end of the previous section.

Consider a sequence of paths of the ODE on $[0, s_n]$, with initial condition x_n and $s_n \to \infty$. Then, for any $\mu > 0$, the fraction of time that the paths spend in $N_\mu(L_H)$ goes to one as $n \to \infty$.

It follows from these arguments that the path of the stochastic approximation spends an increasing amount of time (the fraction going to one) close to L_H as $n \to \infty$.

It also follows from these arguments and the definition of chain recurrence that if any small neighborhood of a point x is returned to infinitely often by $\theta_n(\omega)$, then that point will be chain recurrent. The point can be illustrated by the example of Figure 4.4.2. The paths of the stochastic approximation process will first be drawn to the center or to the boundary of the box. A path $\theta^n(\omega,\cdot)$ that is near the line $[a, b]$ will be drawn toward the point b, but if it is very slightly inside the box, as it gets close to b, it will be drawn towards c or to the center. The noise might eventually force it slightly outside the box, so that it will not necessarily end up at the point e. But if it does not go to e, it will stay close to the boundary of the box.

In this example, the process will eventually spend most of its time in an arbitrarily small neighborhood of the limit points $\{a, b, c, d, e\}$, It might visit any small neighborhood of some other chain recurrent point again and again, but the time intervals between such visits to a small neighborhood of such a point will go to infinity.

Comment. The proof of Theorem 2.5 implies that if any neighborhood of a point x is visited infinitely often for ω in some set Ω_x, and x and $\bar{x}$ are not chain connected, then any sufficiently small neighborhood of $\bar{x}$ can be visited only finitely often with probability one relative to Ω_x. Furthermore, there is a set that the path must enter infinitely often, which is disjoint from some neighborhood of $\bar{x}$, and which is locally asymptotically stable in the sense of Liapunov. In particular, there are arbitrarily small neighborhoods of this set such that the "flow is strictly inward" on the boundaries. This idea leads to the following theorem.

Theorem 2.5. *Let $\bar{g}(\cdot)$ be continuous and let the ODE be*

$$\dot{\theta} = \bar{g}(\theta) + z, \quad z \in -C(\theta) \tag{2.15}$$

where z is the reflection term. Suppose that there is a unique solution for each initial condition. Assume the other conditions of any of the Theorems 2.1 to 2.4. There is null set N such that if $\omega \notin N$ and if for points x and $\bar{x}$

$$\theta_n \in N_\delta(x),\ \theta_n \in N_\delta(\bar{x}),\ \text{infinitely often} \tag{2.16}$$

for all $\delta > 0$. then x and $\bar{x}$ are chain connected. [We allow $x = \bar{x}$.] Thus the assertions concerning convergence to an invariant or limit set of the mean ODE can be replaced by convergence to a set of chain recurrent points within that invariant or limit set.

Proof. For the constrained problem the neighborhoods are relative to the constraint set. Let N denote the "exceptional" set in Theorem 2.1. For $\omega \notin N$, $\{\theta^n(\cdot)\}$ is equicontinuous and the limits of any sequence of sections solves (2.15).

Let $R(A; T, \infty)$ denote the closure of the range of the solution of the ODE on the interval $[T, \infty)$ when the initial conditions are in the set A. Define $R(A) = \lim_{T\to\infty} R(A; T, \infty)$. Let $N_\delta(A)$ denote the δ-neighborhood of the set A. For $\delta > 0$, set $R_1^\delta(x) = R(N_\delta(x))$. For $n > 1$ define, recursively, $R_n^\delta(x) = R(N_\delta(R_{n-1}^\delta(x)))$, and let $R_\infty^\delta(x)$ be the closure of $\lim_n R_n^\delta(x)$. Note that $R_n^\delta(x) \subset R_{n+1}^\delta(x)$. For purposes of the proof, even without a constraint, we can suppose without loss of generality that all of the above sets are bounded. If x and $\bar{x}$ are not chain connected, then for small enough $\delta > 0$, either $\bar{x} \notin R_\infty^\delta(x)$ or $x \notin R_\infty^\delta(\bar{x})$. Suppose the former option, without loss of generality.

By the ODE method and the definition of $R_\infty^\delta(x)$, for $\omega \notin N$ each small neighborhood of $R_\infty^\delta(x)$ must be entered infinitely often. We need to show that if $\bar{x} \notin R^\delta_\infty(x)$, then there is some set which excludes a small neighborhood of $\bar{x}$ and which cannot be exited infinitely often. Since this will contradict (2.16), the theorem will be proved.

Let $\delta_i > 0$ be small enough such that for $\delta \leq \delta_1$, $N_{2\delta_2}(\bar{x}) \cap R^\delta_\infty(x) = \phi$, the empty set. The ν_i used below will be smaller than $\min\{\delta_1, \delta_2\}$, and δ will be less than δ_1. The δ and δ_i are fixed. Given $\nu_i > 0$, define the sets

$$\begin{aligned} S_1(\nu_1) &= N_{\nu_1}(R^\delta_\infty(x)), \\ S_2(\nu_2) &= \overline{N_{\nu_2}(R^\delta_\infty(x)) - R^\delta_\infty(x)}, \\ S_3(\nu_3) &= N_{\nu_3}(R^\delta_\infty(x)). \end{aligned}$$

We will show that, for any (small enough) $\nu_3 > 0$, there are $0 < \nu_1 < \nu_2 < \nu_3$ such that any solution to the ODE, that starts in $S_2(\nu_2) - S_1(\nu_1)$, cannot exit $S_3(\nu_3)$ and must return to $S_1(\nu_1)$ by a time $T(\nu_3)$, that is bounded in the initial condition in $S_2(\nu_2)$. This assertion can be used with the ODE method to prove the theorem, since the ODE method would then imply that exit infinitely often from any arbitrarily small neighborhood of $R^\delta_\infty(x)$ can occur only with probability zero.

We now prove the assertion. Let $\nu_2^n \to 0$, and suppose that for small $\nu_3 > 0$ there are $x_n \in S_2(\nu_2^n)$ and $T_n < \infty$ such that the trajectory $\Phi(t|x_n)$ first exits $S_3(\nu_3)$ at time T_n. Suppose that (take a subsequence, if necessary) $T_n \to T < \infty$. Then there is a point $y \in R^\delta_\infty(x)$ and a path $\Phi(t|y)$ that exits $S_3(\nu_3)$ by time T. By the continuity of the path of the ODE in the initial condition, the "δ-perturbation" method used for constructing the $R^\delta_n(x)$, the definition of $R^\delta_\infty(x)$, and the fact that any path starting at a point $y \in R^\delta_\infty(x)$ is in $R^\delta_\infty(x)$ for all t, this path would be contained in $R^\delta_\infty(x)$, a contradiction.

Now suppose that $T_n \to \infty$. Then by the "δ-perturbation" method which was used to construct the sets $R^\delta_n(x)$ and the definition of $R^\delta_\infty(x)$, we see that some point on the boundary of $S_3(\nu_3)$ would be in $R^\delta_\infty(x)$, a contradiction. We can conclude from this argument and from the definition of $R^\delta_\infty(x)$ that for small enough ν_3 there is a $\nu_2 > 0$ such that any solution which starts in $S_2(\nu_2)$ cannot exit $S_3(\nu_3)$ and also that it must eventually return to $S_1(\nu_1)$ for any $\nu_1 > 0$. We need only show that the time required to return to $S_1(\nu_1)$, $\nu_1 < \nu_2$, is bounded uniformly in the initial condition in $S_2(\nu_2)$ for small enough ν_2 and ν_3. We have shown that the paths starting in $S_2(\nu_2) - S_1(\nu_1)$ cannot exit $S_3(\nu_3)$ and must eventually converge to $S_1(\nu_1)$. Suppose that for each small $\nu_i, i \leq 3$, there is a sequence of initial conditions x_n in $S_2(\nu_2) - S_1(\nu_1)$ such that the time required for the path to reach $S_1(\nu_1)$ goes to infinity as $n \to \infty$. Then there is a path starting in $S_2(\nu_2) - S_1(\nu_1)$ and that stays in $S_3(\nu_3) - S_1(\nu_1)$ for an infinite amount of time, for small enough $\nu_i, i \leq 3$. But then the "δ-perturbation" definition of the $R^\delta_n(x)$ and the definition of $R^\delta_\infty(x)$ imply that such paths would be in $R^\delta_\infty(x)$ for small enough $\nu_i, i \leq 3$, a contradiction. □

5.3 A General Compactness Method

5.3.1 *The Basic Convergence Theorem*

The proofs of Theorems 2.1 to 2.3 used the classical summability condition (A2.4) to guarantee that the martingale M_n converges with probability one as $n \to \infty$. It was also supposed that $\beta_n \to 0$ with probability one. These were key points in the proofs that the sequence $\{\theta^n(\omega, \cdot)\}$ is equicontinuous with probability one, which allowed us to show that the limit points of $\theta^n(\cdot)$ are determined by the asymptotic behavior of the ODE determined by the "mean dynamics." An alternative approach, which was initiated in [135], starts with general conditions that guarantee the equicontinuity, and hence the limit theorem, and then proceeds to find specific and more verifiable conditions that guarantee the general conditions. Many such verifiable sets of conditions were given in [135]. The general conditions and the approach are very natural. They are of wide applicability and will be extended further in the next chapter. It has been shown that for certain classes of problems, the general conditions used are both necessary and sufficient [249].

We will continue to suppose (A2.7), namely,

$$E_n Y_n = g_n(\theta_n) + \beta_n, \tag{3.1}$$

and will work with algorithm (2.10).

Definition: Asymptotic "rate of change" conditions. Recall the definition

$$M^0(t) = \sum_{i=0}^{m(t)-1} \epsilon_i \delta M_i, \quad \delta M_n = Y_n - E_n Y_n,$$

and the analogous definition for $B^0(\cdot)$. Instead of using (A2.1) and (A2.4) (which implies the desired convergence of $\{M^n(\cdot)\}$), and the assumption (A2.5) to deal with the β_n effects, we will suppose that the *rates of change* of $M^0(\cdot)$ and $B^0(\cdot)$ go to zero with probability one as $t \to \infty$. By this, it is meant that for some positive number T,

$$\limsup_n \max_{j \ge n} \max_{0 \le t \le T} \left| M^0(jT + t) - M^0(jT) \right| = 0 \text{ w.p.1} \tag{3.2}$$

and

$$\limsup_n \max_{j \ge n} \max_{0 \le t \le T} \left| B^0(jT + t) - B^0(jT) \right| = 0 \text{ w.p.1.} \tag{3.3}$$

If (3.2) and (3.3) hold for some positive T, then they hold for all positive T. Note that (3.2) does not imply convergence of $\{M_n\}$. For example, the function $\log(t+1)$ for $t > 0$ satisfies (3.2) but does not converge. Condition (3.2) is guaranteed by (A2.1) and (A2.4), but we will show that it is much weaker. The conditions (3.2) and/or (3.3) will be referred to either by saying that the *asymptotic rate of change is zero with probability one* or that the *asymptotic rate of change goes to zero with probability one.*

Note that Theorem 3.1 does not require that Y_n be random, provided that there is some decomposition of the form $Y_n = g_n(\theta_n) + \delta M_n + \beta_n$ and (3.2) and (3.3) hold for whatever sequence $\{\delta M_n, \beta_n\}$ is used. Conditions (3.2) and (3.3) are equivalent to

$$\lim_n \sup_{|t|\leq T} |M^n(t)| = 0, \quad \lim_n \sup_{|t|\leq T} |B^n(t)| = 0. \tag{3.4}$$

With assumptions (3.2) and (3.3) used to eliminate many of the details, the proofs of Theorems 2.1 to 2.3 give us their conclusions, without the necessity of (A2.1), (A2.4), and (A2.5). We thus have the following theorem.

Theorem 3.1. *Suppose* (1.1) *and that* $E|Y_n| < \infty$ *for each* n. *Assume* (3.2), (3.3), (A2.7), *and any of the constraint set conditions* (A4.3.1), (A4.3.2), *or* (A4.3.3). *If* $\bar{g}(\cdot)$ *is a gradient, assume* (A2.6). *Then the conclusions of Theorems* 2.1 *to* 2.3 *hold.*

Under the additional conditions of Theorem 2.5, *for almost all* ω, $\theta_n(\omega)$ *converges to a set of chain recurrent points within the limit set.*

A sufficient condition for the asymptotic rate of change assumption (3.2). The main problem is the verification of (3.2) when (A2.4) fails to hold. The next theorem sets up a general framework for obtaining perhaps the weakest possible replacement for (A2.4), and this is illustrated by the examples in the next section. The general approach is reminiscent of the "large deviations" upper bounds.

Theorem 3.2. *Let* $E|Y_n| < \infty$ *for each* n. *For each* $\mu > 0$ *and some* $T > 0$, *suppose either that*

$$\lim_n P\left\{\sup_{j\geq n} \max_{0\leq t\leq T} \left| \sum_{i=m(jT)}^{m(jT+t)-1} \epsilon_i \delta M_i \right| \geq \mu \right\} = 0, \tag{3.5}$$

or

$$\sum_j q_j'(\mu) < \infty, \tag{3.6}$$

where $q_j'(\mu)$ *is defined by*

$$q_j'(\mu) = P\left\{\max_{0\leq t\leq T} \left| \sum_{i=m(jT)}^{m(jT+t)-1} \epsilon_i \delta M_i \right| \geq \mu \right\}. \tag{3.7}$$

Then (3.2) *holds for each* T.

Proof. If the conditions hold for some positive T, then they hold for all positive T. Equation (3.5) implies (3.2) for each T. Under (3.6), the Borel–Cantelli Lemma says that the event $\sup_{|t|\leq T} |M^n(t)| \geq \mu$ occurs only finitely often with probability one for each $\mu > 0$ and $T < \infty$. This implies (3.2) for each T. □

5.3.2 *Sufficient Conditions for the Rate of Change Condition*

Two classes of conditions will be used. First we consider the case where there are exponential moments, then only finite moments. Modifications for the Kiefer–Wolfowitz scheme will be given in the next subsection.

Exponential moments. The approach is reminiscent of the large deviations upper bounds.

(A3.1) For each $\mu > 0$,

$$\sum_n e^{-\mu/\epsilon_n} < \infty. \tag{3.8}$$

(A3.2) For *some* $T < \infty$, there is a $c_1(T) < \infty$ such that for all n,

$$\sup_{n\leq i\leq m(t_n+T)} \frac{\epsilon_i}{\epsilon_n} \leq c_1(T). \tag{3.9}$$

(A3.3) There is a real $K < \infty$ such that for small real γ, all n, and each component $\delta M_{n,j}$ of δM_n,

$$E_n e^{\gamma(\delta M_{n,j})} \leq e^{\gamma^2 K/2}. \tag{3.10}$$

(A3.2) is unrestrictive in applications. (A3.1) holds if $\epsilon_n \leq \gamma_n/\log n$, for any sequence $\gamma_n \to 0$. If γ_n does not tend to zero, then (excluding degenerate cases) it is not possible to get probability one convergence, since $\{\epsilon_n\}$ is in the "simulated annealing" range, where convergence is at best in the sense of convergence in probability as in Chapter 7. The sets of conditions (A2.1), (A2.4) and (A3.1), (A3.3) represent the extremes of the possibilities. In the intermediate cases, the speed at which the step sizes must go to zero for probability one convergence depends on the rate of growth of the moments $E|\delta M_n|^k$ of the noise as $k \to \infty$.

Theorem 3.3. (A3.1) *to* (A3.3) *imply* (3.6) *for real* ϵ_n. *The conclusion still holds if the* ϵ_n *are random, provided that* ϵ_n *is* $\mathcal{F}_n$*-measurable, that there are real* $\widetilde{\epsilon}_n$ *satisfying* (A3.1) *and* (A3.2), *and that* $\epsilon_n \leq \widetilde{\epsilon}_n$ *for all but a finite number of* n *with probability one.*

Proof. It is sufficient to work with one component of δM_i at a time, so we let δM_i be real-valued. The case of random ϵ_n is a straightforward extension of the nonrandom case, and we suppose that ϵ_n are nonrandom. To prove (3.6), it is enough to show that for some positive T there is a real $\alpha > 0$ (that can depend on j) such that for $\mu > 0$, $q_j(\mu)$ defined by

$$P\left\{\max_{0\le t\le T} \exp\left[\alpha \sum_{i=m(jT)}^{m(jT+t)-1} \epsilon_i \delta M_i\right] \ge e^{\alpha\mu}\right\} \equiv q_j(\mu) \tag{3.11}$$

is summable. [To deal with negative excursions, just replace δM_i with $-\delta M_i$.] Since M_n is a martingale and the exponential is a convex function, the martingale inequality (4.1.4) implies that

$$q_j(\mu) \le e^{-\alpha\mu} E\left\{\exp\left[\alpha \sum_{i=m(jT)}^{m(jT+T)-1} \epsilon_i \delta M_i\right]\right\}. \tag{3.12}$$

The summability of $q_j(\mu)$ will follow from (A3.1) to (A3.3) by evaluating (3.12) with an appropriate choice of α. By (3.10), for $m > n$,

$$\begin{aligned} E\left\{\exp\left[\alpha \sum_{i=n}^{m} \epsilon_i \delta M_i\right]\right\} &= E\left\{\exp\left[\alpha \sum_{i=n}^{m-1} \epsilon_i \delta M_i\right] E_m e^{\alpha \epsilon_m \delta M_m}\right\} \\ &\le E \exp\left[\alpha \sum_{i=n}^{m-1} \epsilon_i \delta M_i\right] e^{\alpha^2 \epsilon_m^2 K/2}. \end{aligned} \tag{3.13}$$

Iterating the procedure on the right side of (3.13) yields

$$q_j(\mu) \le \exp\left[K\alpha^2 \sum_{i=m(jT)}^{m(jT+T)-1} \epsilon_i^2/2\right] e^{-\alpha\mu}. \tag{3.14}$$

Minimizing the exponent in (3.14) with respect to α yields that

$$\alpha_{\min} = \mu \Big/ \left[K \sum_{i=m(jT)}^{m(jT+T)-1} \epsilon_i^2\right].$$

Thus

$$q_j(\mu) \le \exp\left[-\mu^2 \Big/ 2K \sum_{i=m(jT)}^{m(jT+T)-1} \epsilon_i^2\right].$$

By (A3.2), $\alpha_{\min} \ge \mu/[K\epsilon_{m(jT)} c_1(T) T] \equiv \alpha_0$. Using $\alpha = \alpha_0$ in (3.14) yields [using (A3.2) again]

$$q_j(\mu) \le \exp\left[\frac{K\mu^2 \displaystyle\sum_{i=m(jT)}^{m(jT+T)-1} \epsilon_i^2}{2K^2\epsilon_{m(jT)}^2 c_1^2(T)T^2} - \frac{\mu^2}{K\epsilon_{m(jT)}c_1(T)T}\right] \le \exp\left(\frac{-\mu^2}{2Kc_1(T)T\epsilon_{m(jT)}}\right) \equiv q_j^{''}(\mu). \tag{3.15}$$

The terms $q_j^{''}(\mu)$ are summable for each $\mu > 0$ by (A3.1). Note that it is enough for (3.10) to hold only for small γ because in (3.13) $\alpha_0\epsilon_i$ effectively replaces the γ in (3.10), and $\epsilon_i\alpha_0 = O(\mu)$, which is arbitrarily small. □

Examples of condition (A3.3). It is sufficient to work with real-valued random variables.

Example 1. Suppose that ξ_n are Gaussian, mutually independent, with mean zero and uniformly bounded variances σ_n^2. Let $\delta M_n = \nu_n(\theta_n)\xi_n$, where $\{\nu_n(\theta), \theta \in H\}$ is bounded. Then

$$E_n\left\{\exp\left[\gamma\nu_n(\theta_n)\xi_n\right]\right\} = \exp\left[\gamma^2\sigma_n^2\nu_n^2(\theta_n)/2\right],$$

and (3.10) holds.

Example 2. Let there be a $K_1 < \infty$ such that for all n, $k < \infty$, δM_n satisfies

$$E_n|\delta M_n|^{2k} \le K_1^k k!. \tag{3.16}$$

Then (3.10) holds. Without loss of generality, let $K_1 > 1$. Inequality (3.16) holds for the Gaussian distribution of Example 1, since $E|\xi_n|^{2k} = (2k-1)(2k-3)\cdots 3\cdot 1\cdot\sigma_n^{2k}$. Hence $K_1 = 2\sup_n \sigma_n^2$. Also, (3.16) holds if $\{\delta M_n\}$ is bounded. One canonical model takes the form $Y_n = g_n(\theta_n, \xi_n)$ with $\{\xi_n\}$ being mutually independent and independent of θ_0. Then (3.16) is essentially a condition on the moments of $g_n(\theta, \xi_n)$ for $\theta \in H$.

Comment on (3.16). Recall the discussion of robustness in Section 1.3.5. When truncation procedures are used to cull or truncate high values of Y_n to "robustify" the performance of the algorithm, (3.16) would generally be satisfied. It is not desirable to have the performance of the stochastic approximation procedure be too sensitive to the structure of the tails of the distribution functions of the noise terms.

To prove the sufficiency of (3.16), use $X = \delta M_n$ and

$$E_n\left\{\exp\left[\gamma X\right]\right\} \le 1 + \gamma E_n X + \frac{\gamma^2}{2}E_n X^2 + \sum_{k=3}^{\infty}\frac{\gamma^k}{k!}E_n|X|^k. \tag{3.17}$$

Since X is a martingale difference with respect to $\mathcal{F}_n$, $E_n X = 0$. By (3.16), there is a real K_2 such that

$$\frac{\gamma^{2k} E_n|X|^{2k}}{(2k)!} \leq \frac{\gamma^{2k} K_1^k k!}{(2k)!} \leq \frac{\gamma^{2k} K_2^k}{k!}, \quad k \geq 1.$$

For odd exponents ($k \geq 2$), Hölder's inequality yields

$$E_n|X|^{2k-1} \leq E_n^{(2k-1)/2k}|X|^{2k},$$

which implies that there is a $K_3 < \infty$ such that

$$\frac{\gamma^{2k-1} E_n|X|^{2k-1}}{(2k-1)!} \leq \frac{\gamma^{2k-1} K_3^k}{k!}, \quad k \geq 2.$$

Now writing $\gamma^{2k-1} \leq \gamma^{2k-2} + \gamma^{2k}$ and using upper bounds where needed yield that there is a $K < \infty$ such that

$$E_n\left\{\exp[\gamma X]\right\} \leq 1 + \sum_{k=1}^{\infty} \frac{\gamma^{2k} K^k}{2^k k!} = \exp\left[\gamma^2 K/2\right],$$

which yields (3.10).

Finite moments. For some even integer p, suppose that

$$\sum_n \epsilon_n^{p/2+1} < \infty, \quad \sup_n E|M_n|^p < \infty. \tag{3.18}$$

For large j, we have $\sum_{m(jT)}^{m(jT+T-1)} \epsilon_n \leq 1$. Then, Burkholder's inequality (4.1.12) and the bound (4.1.14) yield

$$\sum_j P\left\{\max_{0\leq t\leq T}\left|\sum_{i=m(jT)}^{m(jT+t)-1} \epsilon_i \delta M_i\right| \geq \mu\right\} < \infty,$$

which is (3.6). Thus, by Theorem 3.2, (3.2) holds.

5.3.3 *The Kiefer–Wolfowitz Algorithm*

Theorems 2.1–2.3 and 3.1–3.3 can be modified to hold for the Kiefer–Wolfowitz form of the projected algorithm (2.1). The few modifications required for Theorem 3.3 will be discussed next. The precise form of Y_n will depend on how one iterates among the coordinates. We will work with a form that includes the various cases discussed in Section 1.2, so that quite general choices of the coordinate(s) to be iterated on at each step are covered; see, for example, (1.2.4), (1.2.12), or the forms where one cycles among the coordinates. By modifying (A3.1)–(A3.3) appropriately, we will see that Theorem 3.1 continues to hold under general conditions.

Recall the discussion concerning variance reduction in connection with (1.2.6), where it was seen that if $\chi^+_{n,i} = \chi^-_{n,i}$ for all i, then the algorithm takes the form $\theta_{n+1} = \theta_n - \epsilon_n f_\theta(\theta_n) + \epsilon_n \widetilde{\chi}_n + \epsilon_n \beta_n$, where β_n is the finite difference bias and the noise term $\widetilde{\chi}_n$ is not proportional to $1/c_n$. In this case, the procedure is a Robbins-Monro method. If common random numbers cannot be used, then the noise term will be proportional to $1/c_n$. We will concentrate on this case.

Suppose that the observation can be written in the form

$$Y_n = g_n(\theta_n) + \beta_n + \frac{\delta M_n}{2c_n}, \tag{3.19}$$

where δM_n is a martingale difference. The term $\delta M_n/(2c_n)$ arises from the observation noise divided by the finite difference interval c_n. Redefine $M^0(\cdot)$ as

$$M^0(t) = \sum_{i=0}^{m(t)-1} \frac{\epsilon_i}{2c_i} \delta M_i. \tag{3.20}$$

Theorem 3.4. *Assume the conditions of Theorem* 3.1 *for the case where* $\bar{g}(\cdot)$ *is a gradient, with observations of the form* (3.19) *and the new definition* (3.20) *of* $M^0(\cdot)$ *used. Then the conclusions of Theorem* 3.1 *hold.*

Sufficient conditions for (3.2) with $M^0(\cdot)$ defined by (3.20). We next obtain a sufficient condition for (3.2) under the new definition (3.20). Assume the following.

(A3.4) For each $\mu > 0$,

$$\sum_n e^{-\mu c_n^2/\epsilon_n} < \infty. \tag{3.21}$$

(A3.5) For *some* $T < \infty$, there is a $c_2(T) < \infty$ such that for all n,

$$\sup_{n \le i \le m(t_n+T)} \frac{\epsilon_i/c_i^2}{\epsilon_n/c_n^2} \le c_2(T). \tag{3.22}$$

(A3.6) There is a real $K < \infty$ such that for small real γ, all n, and each component $\delta M_{n,j}$ of δM_n,

$$E_n e^{\gamma(\delta M_{n,j})} \le e^{\gamma^2 K/2}. \tag{3.23}$$

If ϵ_n and c_n are random, suppose that they are $\mathcal{F}_n$-measurable and that there are $\widetilde{\epsilon}_n$ and $\widetilde{c}_n$ satisfying (3.21) such that $\epsilon_n/c_n^2 \le \widetilde{\epsilon}_n/\widetilde{c}_n^2$ for all but a finite number of n with probability one. Then (3.2) holds for the new definition of $M^0(\cdot)$.

The proof is a repetition of the argument of Theorem 3.3, where ϵ_i/c_i replaces ϵ_i in (3.14). Thus, we need only show that

$$\sum_j \exp\left[\frac{-\mu^2}{2K \sum_{i=m(jT)}^{m(jT+T)-1} \epsilon_i^2/c_i^2}\right] < \infty \quad \text{for some } T < \infty \text{ and each } \mu > 0. \tag{3.24}$$

Using (3.22) yields the upper bound

$$\sum_{i=m(jT)}^{m(jT+T)-1} \frac{\epsilon_i^2}{c_i^2} \leq \frac{\epsilon_{m(jT)}}{c_{m(jT)}^2} \sum_{i=m(jT)}^{m(jT+T)-1} \epsilon_i c_2(T) \approx \frac{\epsilon_{m(jT)}}{c_{m(jT)}^2} c_2(T)T.$$

This and (3.21) yield (3.24).

Of course, M_n converges under the classical condition

$$\sum \epsilon_n^2/c_n^2 < \infty, \quad \sup_n E|\delta M_n|^2 < \infty. \tag{3.25}$$

5.4 Stability and Combined Stability–ODE Methods

Stability methods provide an alternative approach to proofs of convergence. They are most useful when the iterates are allowed to vary over an unbounded set and not confined to a compact set H. They can be used to prove convergence with probability one directly, but the conditions are generally weaker when they are used in combination with an ODE-type method. A stability method would be used to prove that the process is recurrent. That is, there is some compact set to which the stochastic approximation iterates return infinitely often with probability one. Then the ODE method takes over, starting at the recurrence times, and is used to show that (asymptotically) the iterates follow the path of the mean ODE, as in Sections 2 and 3. Indeed, if the paths are not constrained, there might be no alternative to starting the analysis with some sort of stability method.

Stability methods are generally based on a Liapunov function, and this Liapunov function is usually a small perturbation of one for the underlying ODE. The "combined" stability and ODE method is a powerful tool when the constraint set is unbounded or when there are no constraints. It can also be used for the state-dependent noise problem or where the algorithm is decentralized. In this section, we are still concerned with the martingale difference noise case. Extensions will be given in subsequent chapters.

In this regard, a discussion of stochastic stability for processes driven by nonwhite noise is in [127].

Recurrence and probability one convergence. Two types of theorems and proofs are presented. The first is a more or less classical approach via the use of a perturbed Liapunov function, as in Section 4.5. To construct the perturbation and assure that it is finite with probability one, a "local" square summability condition on ϵ_n is used. The step sizes ϵ_n are allowed to be random. The theorem will be used in Section 7 to prove probability one convergence for the lizard learning problem of Section 2.1, where ϵ_n are random. The perturbed Liapunov function-type argument is quite flexible as seen in [127]. A combination of a stability and a "local" ODE method yields convergence under weaker conditions. In particular, the square summability will be dropped. Such an approach is quite natural since the mean ODE characterizes the "flow" for large n. This "combined" method will be presented in Theorem 4.2. For notational simplicity, both Theorems 4.1 and 4.2 suppose that $\theta = 0$ is globally asymptotically stable in the sense of Liapunov for the mean ODE $\dot{\theta} = \bar{g}(\theta)$ and establish the convergence of $\{\theta_n\}$ to zero.

Theorem 4.3 gives a sufficient condition for recurrence, and then a "local" ODE is again used to get convergence. The theorems represent a few of the many possible variations. Starting with some canonical model, stability-type proofs are commonly tailored to the special application at hand. The statement of Theorem 4.2 is complicated because it is intended to cover cases where the mean ODE is an ordinary ODE, a differential inclusion, or where γ_n is an approximation to an element of a set of subgradients.

5.4.1 A Liapunov Function Method for Convergence

Theorem 4.1. *Assume* (1.1). *Let $V(\cdot)$ be a real-valued non-negative and continuous function on $I\!R^r$ with $V(0) = 0$, which is twice continuously differentiable with bounded mixed second partial derivatives. Suppose that for each $\epsilon > 0$, there is a $\delta > 0$ such that $V(\theta) \geq \delta$ for $|\theta| \geq \epsilon$, and δ does not decrease as ϵ increases. Let $\{\mathcal{F}_n\}$ be a sequence of nondecreasing σ-algebras, where $\mathcal{F}_n$ measures at least $\{\theta_0, Y_i, i < n\}$. Let $EV(\theta_0) < \infty$. Suppose that there is a function $\bar{g}(\cdot)$ such that $E_n Y_n = \bar{g}(\theta_n)$. For each $\epsilon > 0$ let there be a $\delta_1 > 0$ such that*

$$V_\theta'(\theta)\bar{g}(\theta) = -k(\theta) \leq -\delta_1$$

for $|\theta| \geq \epsilon$. Suppose that there are $K_2 < \infty$ and $K < \infty$ such that

$$E_n|Y_n|^2 \leq K_2 k(\theta_n), \quad \textit{when } |\theta_n| \geq K. \tag{4.1}$$

Let

$$E\sum_{i=1}^{\infty}\epsilon_i^2|Y_i|^2 I_{\{|\theta_i|\le K\}} < \infty. \tag{4.2}$$

Then $\theta_n \to 0$ with probability one.

Now, suppose that the step sizes ϵ_n are random. Let $\epsilon_n \to 0$ with probability one and be $\mathcal{F}_n$-measurable, with $\sum \epsilon_i = \infty$ with probability one. Let there be real positive $\widetilde{\epsilon}_n$ such that $\epsilon_n \le \widetilde{\epsilon}_n$ for all but a finite number of n with probability one. Suppose that (4.2) is replaced by

$$E\sum_{i=1}^{\infty}\widetilde{\epsilon}_i^2|Y_i|^2 I_{\{|\theta_i|\le K\}} < \infty. \tag{4.2$'$}$$

Then the conclusion continues to hold.

Comment on the proof. A truncated Taylor series expansion and the boundedness of the second partial derivatives yield that there is a constant K_1 such that

$$E_n V(\theta_{n+1}) - V(\theta_n) \le \epsilon_n V_\theta'(\theta_n) E_n Y_n + \epsilon_n^2 K_1 E_n |Y_n|^2.$$

By the hypotheses,

$$E_n V(\theta_{n+1}) - V(\theta_n) \le -\epsilon_n k(\theta_n) + \epsilon_n^2 K_1 E_n |Y_n|^2.$$

The statement of the part of the theorem that uses (4.2) is now the same as that of Theorem 4.5.3. The proof under (4.2$'$) is a simple modification; the details are left to the reader.

5.4.2 Combined Stability–ODE Methods

The next result extends Theorem 4.1 in several directions. It allows the conditional mean of Y_n to depend on n, and it covers the case where the γ_n are obtained as subgradients. It also uses either a "local" square summability condition as in Theorem 4.1, or a localized form of (3.2) if square summability cannot be assumed.

One often assumes a bound for $E_n|Y_n|^2$ in terms of the Liapunov function itself. This accounts for condition (4.5). Note that $\mathcal{F}_n$ and E_n are defined as in Theorem 4.1. The result will be presented in two parts; we first prove a lemma that will be needed in the theorem.

Lemma 4.1. *Consider the algorithm*

$$\theta_{n+1} = \theta_n + \epsilon_n Y_n. \tag{4.3}$$

Let ϵ_n be $\mathcal{F}_n$-measurable such that $\epsilon_n \to 0$ as $n \to \infty$ and $\sum_i \epsilon_i = \infty$, both with probability one. Let the real-valued continuous function $V(\cdot)$ have *continuous first and bounded second mixed partial derivatives. Suppose that* $V(0) = 0$, $V(\theta) > 0$ *for* $\theta \neq 0$, *and* $V(\theta) \to \infty$ *as* $|\theta| \to \infty$. *If* $E|Y_k| < \infty$, *then write*

$$E_k Y_k = \gamma_k. \tag{4.4}$$

Suppose that there are positive numbers K_1 *and* K_2 *such that*

$$E|Y_k|^2 + E|V_\theta'(\theta_k)\gamma_k| \leq K_2 V(\theta_k) + K_2. \tag{4.5}$$

Assume that $EV(\theta_0) < \infty$. *Then* $EV(\theta_n) < \infty$ *and* $E|Y_n|^2 < \infty$ *for each* n.

Remark. A common model has $V(\theta)$ growing at most as $O(|\theta|^2)$ and, if $\theta_n = \theta$, $|V_\theta(\theta)| + |\gamma_n|$ growing at most as $O(|\theta|)$ for large $|\theta|$.

Proof. The proof uses induction on n. By (4.5) and $EV(\theta_0) < \infty$, we have $E|Y_0|^2 < \infty$. Now suppose that $EV(\theta_n) < \infty$, for some n. Then $E|Y_n|^2 < \infty$ in view of (4.5). By a truncated Taylor series expansion and the boundedness of $V_{\theta\theta}(\cdot)$, we have

$$E_n V(\theta_{n+1}) - V(\theta_n) = \epsilon_n V_\theta'(\theta_n)\gamma_n + O(\epsilon_n^2) E_n |Y_n|^2. \tag{4.6}$$

The inequality (4.5) implies that there is a real K_3 such that the right side of (4.6) is bounded above by $\epsilon_n K_3[1 + V(\theta_n)]$. This, together with the induction hypothesis, implies that $EV(\theta_{n+1}) < \infty$. Hence (4.5) yields $E|Y_{n+1}|^2 < \infty$. Thus by induction, we have proven that $EV(\theta_n) < \infty$ and $E|Y_n|^2 < \infty$ for all n. □

Convergence to a unique limit.

Theorem 4.2. *Assume the conditions of Lemma 4.1 and suppose that* γ_n *has the following properties. For each real* K, $\gamma_n I_{\{|\theta_n| \leq K\}}$ *is bounded uniformly in* n. *There are convex and upper semicontinuous (see the definition (4.3.2)) sets* $G(\theta) \subset \mathbb{R}^r$ *with* $G(\theta)$ *being uniformly bounded on each bounded* θ*-set, such that for each real* K *and all* ω,

$$\min\{|\gamma_n - y| : y \in G(\theta_n)\} I_{\{|\theta_n| \leq K\}} = \text{distance}(\gamma_n, G(\theta_n)) I_{\{|\theta_n| \leq K\}} \to 0,$$

as $n \to \infty$. *Let* $c(\delta)$ *be a nondecreasing real-valued function with* $c(0) = 0$ *and* $c(\delta) > 0$ *for* $\delta > 0$ *such that for large* n *(that can depend on* δ*)*

$$V_\theta'(\theta_n)\gamma_n < -c(\delta), \quad \text{if } V(\theta_n) \geq \delta. \tag{4.7}$$

Suppose that there is a nondecreasing real-valued function $c_0(\delta)$ *with* $c_0(0) = 0$ *and* $c_0(\delta) > 0$ *for* $\delta > 0$ *such that*

$$V_\theta'(\theta)\gamma \leq -c_0(\delta), \quad \text{for all } \gamma \in G(\theta) \text{ if } V(\theta) > \delta. \tag{4.8}$$

Assume that (3.2) *holds with* δM_i *replaced by* $\delta M_i I_{\{|\theta_i| \leq K\}}$ *for each positive* K. *Then* $\theta_n \to 0$ *with probability one.*

Proof. For large n, (4.5) and (4.7) imply that the right side of (4.6) is negative outside of a neighborhood of $\theta = 0$, which decreases to the origin as $n \to \infty$. Thus, outside of this "decreasing" neighborhood, $V(\theta_n)$ has the supermartingale property. The supermartingale convergence theorem then implies that each neighborhood of $\theta = 0$ is recurrent, that is, θ_n returns to it infinitely often with probability one.

Completion of the proof under (A2.4). Once recurrence of each small neighborhood of the origin is shown, the rest of the proof uses a "local analysis." To illustrate the general idea in a simpler context, the proof will first be completed under the stronger conditions that (A2.4) holds and that $\sup_n E|Y_n|^p I_{\{|\theta_n| \leq K\}} < \infty$ for some positive K and even integer p. Define $\delta M_n = Y_n - E_n Y_n = Y_n - \gamma_n$. Fix δ and $\Delta > 2\delta > 0$ small, recall the definition $Q_\lambda = \{\theta : V(\theta) \leq \lambda\}$, and let Q_Δ^c denote the complement of the set Q_Δ. Let τ be a stopping time such that $\theta_\tau \in Q_\delta$. By (4.6), (4.7), and (A2.4), for large n the terms γ_n cannot force θ_n out of Q_Δ. The only way that $\{\theta_k, k \geq \tau\}$ can leave Q_Δ is by the effects of $\{\delta M_k, k \geq \tau\}$. But the convergence of

$$\sum_{i=0}^{m(t)} \epsilon_i \delta M_i I_{\{V(\theta_i) \leq \Delta\}}, \tag{4.9}$$

which is implied by the hypotheses (A2.4) being used in this paragraph, assures that these martingale difference terms cannot force the path from Q_δ to Q_Δ^c infinitely often. The convergence follows from this.

Completion of the proof under (3.2). Fix $\delta > 0$ and $\Delta > 2\delta$ as before. The proof is very similar to that of the gradient case assertions of Theorem 2.1. Let N denote the null set on which the asymptotic rate of change of (4.9) is not zero and let $\omega \notin N$. Suppose that there are infinitely many excursions of $\{\theta_n(\omega)\}$ from Q_δ to Q_Δ^c. Then there are $n_k \to \infty$ (depending on ω) such that n_{k-1} is the last index at which the iterate is in $Q_{2\delta}$ before exiting Q_Δ.

We now repeat the argument of Theorem 2.1. By selecting a subsequence if necessary, we can suppose that $\{\theta_{n_k}(\omega)\}$ converges to a point on $\partial Q_{2\delta}$, the boundary of $Q_{2\delta}$, and that $V(\theta_{n_k+i}(\omega)) \geq 2\delta$ until at least after the next time that $V(\theta_{n_k+i}(\omega)) \geq \Delta$. For $u \geq 0$,

$$\theta^{n_k}(u) - \theta_{n_k} - \sum_{i=n_k}^{m(t_{n_k}+u)-1} \epsilon_i \gamma_i = \sum_{i=n_k}^{m(t_{n_k}+u)-1} \epsilon_i \delta M_i. \tag{4.10}$$

For $\theta_i(\omega) \in Q_\Delta$, $\epsilon_i Y_i(\omega) \to 0$. This and the fact that the right side of (4.10) goes to zero if $\theta^{n_k}(\omega, s) \in Q_\Delta$ for $0 \leq s \leq u$ imply that there is a

$T > 0$ such that for large k, $\theta^{n_k}(\omega, t) \in Q_\Delta$ for $t \leq T$, that $\{\theta^{n_k}(\omega, \cdot)\}$ is equicontinuous for $t \leq T$, and that $V(\theta^{n_k}(\omega, t)) \geq 2\delta$ for $t \in (0, T]$.

Let $\theta(\omega, \cdot)$ be the limit of a convergent subsequence of $\{\theta^{n_k}(\omega, \cdot), t \leq T\}$. Write the first sum in (4.10) at ω in an obvious way as

$$\int_0^u g^k(\omega, s) ds.$$

By the hypothesis, the distance between γ_i and the set $G(\theta_i)$ goes to zero for the indices i involved in the first sum in (4.10). Using this and the convexity and upper semicontinuity properties of $G(\theta)$ in θ, it follows that the limit (along the convergent subsequence) of the integral has the representation

$$\int_0^u \gamma(s) ds,$$

where $\gamma(s) \in G(\theta(\omega, s))$ for almost all s. Thus, the limit $\theta(\omega, \cdot)$ of any convergent subsequence satisfies the differential inclusion

$$\dot{\theta} \in G(\theta), \tag{4.11}$$

with $V(\theta(\omega, t)) \geq 2\delta$, for $t \leq T$ and $V(\theta(\omega, 0)) = 2\delta$. But (4.8) implies that $V(\theta(\omega, \cdot))$ is strictly decreasing until it reaches the value zero, which contradicts the assertion of the previous sentence and (hence) the assertion that there are an infinite number of escapes from Q_δ. □

Recurrence and convergence to a limit set. A straightforward modification of the proof of Theorem 4.2 yields the following assertion, the proof of which is left to the reader. The theorem gives a very useful combination of the ODE and the stability methods. It first assures that some compact set is recurrent with probability one. Then, looking at the sequence of processes starting at the recurrence time, the ODE method takes over and the asymptotic stability of the mean ODE when starting in the recurrence set guarantees the convergence of $\{\theta_n\}$ with probability one. The ODE-type argument is like that in the proof of the gradient case assertions of Theorem 2.1.

Theorem 4.3. *Assume the conditions of Theorem 4.2 except let there be a λ_0 such that $c(\delta) > 0$ and $c_0(\delta) > 0$ for $\delta \geq \lambda_0$ and not necessarily otherwise. Then if $\lambda > \lambda_0$, Q_λ is a recurrence set for $\{\theta_n\}$. Let the asymptotic rate of change of (4.9) be zero with probability one for $\Delta = 2\lambda_0$. For almost all ω, the limit trajectories of $\{\theta^n(\omega, \cdot)\}$ are in an invariant set of (4.11) in Q_{λ_0}.*

Let $G(\theta)$ contain only the single point $\bar{g}(\theta)$ for each θ. If $\bar{g}(\cdot)$ is a gradient, let (A2.6) hold. Then, for each ω not in some null set, the convergence is to a single stationary set S_i. Otherwise, under the additional conditions of Theorem 2.5 or its corollary, the limit points for the algorithm are contained in the set of chain recurrent points, with probability one.

5.5 Soft Constraints

In Sections 2 and 3, we used hard constraints of either the form $a_i \leq \theta_{n,i} \leq b_i$ or where θ_n is confined to a compact set H defined in terms of the constraint functions $q_i(\cdot), i \leq p$. In these cases, the iterate is required to be in the set H for all n. Sometimes, the given constraint functions are merely a guide in that they should not be violated by much, but they can be violated. Then we say that the constraint is *soft*. Soft constraints (equivalently, penalty functions) can be added to the algorithm directly, and stability methods such as those introduced in the last section can be used to prove convergence. The idea is more or less obvious and will now be illustrated by a very simple example. The discussion to follow is merely a guide to possible variations of the basic stochastic approximation algorithms. The soft constraint might be combined with hard constraints.

In the example to follow, the soft constraint is the sphere S_0 in $\mathbb{R}^r$ with the origin as its center and with radius R_0. Define $q(\theta)$ to be the square of the distance of θ to S_0. Thus, $q(\theta) = (|\theta| - R_0)^2$ for $|\theta| \geq R_0$ and is zero for $|\theta| \leq R_0$. The gradient is $q_\theta(\theta) = 2\theta(1 - R_0/|\theta|)$ for $|\theta| \geq R_0$ and is zero otherwise.

Assume (1.1). The algorithm is

$$\theta_{n+1} = \theta_n + \epsilon_n Y_n - \epsilon_n K_0 q_\theta(\theta_n) \tag{5.1}$$

for sufficiently large positive K_0. The purpose of the $K_0 q_\theta(\cdot)$ term is to assure that the iterates do not wander too far from the sphere. Suppose that $Eq(\theta_0) < \infty$ and that there is a $K_1 < \infty$ (not depending on n) such that if $Eq(\theta_n) < \infty$, then $E_n|Y_n|^2 \leq K_1(q(\theta_n) + 1)$ and $E_n Y_n = \bar{g}(\theta_n) + \beta_n$, where $\beta_n \to 0$ with probability one and $\bar{g}(\cdot)$ is continuous. Suppose that for $K_0 > 0$ large enough there are $\alpha > 0$ and $C_0 \geq 0$ satisfying

$$q_\theta'(\theta)\left[\bar{g}(\theta) - K_0 q_\theta(\theta)\right] \leq -\alpha q(\theta) + C_0. \tag{5.2}$$

Recall the definition $\delta M_n = Y_n - E_n Y_n$. Suppose that the asymptotic rate of change of

$$\sum_i \epsilon_i \delta M_i I_{\{|\theta_i| \leq K\}}$$

is zero with probability one for each positive K. (See Section 3 for a useful criteria.) Then the conclusions of Theorem 2.1 continue to hold, with mean ODE

$$\dot{\theta} = \bar{g}(\theta) - K_0 q_\theta(\theta). \tag{5.3}$$

Note that if $\bar{g}(\theta) = -f_\theta(\theta)$ for a continuously differentiable real-valued function $f(\cdot)$, (5.1) is a gradient descent algorithm with the right side of (5.3) being $-[f_\theta(\theta) + K_0 q_\theta(\theta)]$. The proof is essentially that of Theorems 4.2 and 4.3 and will now be outlined. The essential point is the demonstration of recurrence. That is, some finite sphere is visited infinitely often by the sequence $\{\theta_n\}$ with probability one. As expected in such algorithms, the penalty function $q(\cdot)$ is used as the Liapunov function.

It will next be shown that

$$\sup_n Eq(\theta_n) < \infty. \tag{5.4}$$

A truncated Taylor series expansion yields

$$\begin{aligned} q(\theta_{n+1}) - q(\theta_n) = {} & \epsilon_n q_\theta'(\theta_n) \left[\bar{g}(\theta_n) + \beta_n - K_0 q_\theta(\theta_n)\right] \\ & + O(\epsilon_n^2) \left[|Y_n|^2 + K_0^2 |q_\theta(\theta_n)|^2\right] + \epsilon_n q_\theta'(\theta_n) \delta M_n. \end{aligned}$$

Using (5.2), the bound on $E_n|Y_n|^2$ in terms of $q(\theta_n)$, and the fact that $|q_\theta(\theta)|^2 \le O\left(q(\theta) + 1\right)$, for large n we have

$$\begin{aligned} q(\theta_{n+1}) \le {} & (1 - \epsilon_n \alpha/2) q(\theta_n) + O(\epsilon_n^2) \left[|Y_n|^2 - E_n|Y_n|^2\right] \\ & + \epsilon_n q_\theta'(\theta_n) \delta M_n + O(\epsilon_n^2) + \epsilon_n C_0 + o(\epsilon_n)|q_\theta(\theta_n)|. \end{aligned} \tag{5.5}$$

where α is defined above (5.2). Thus, for some $C_1 < \infty$,

$$E_n q(\theta_{n+1}) \le (1 - \epsilon_n \alpha/2) q(\theta_n) + O(\epsilon_n^2) + \epsilon_n C_1, \tag{5.6}$$

which implies (5.4).

The inequality (5.6) also implies that Q_λ is a recurrence set for $\{\theta_n\}$ for large enough λ (see Theorem 4.3 or Theorem 4.4.4). Now, Theorem 4.3 implies that the conclusions of Theorem 2.1 hold with the ODE being (5.3). In particular, for almost all ω the limit trajectories of $\theta^n(\omega, \cdot)$ satisfy the ODE (5.3) and are in a bounded invariant set for (5.3).

Finally, we remark that the development still holds if $E_n Y_n = g_n(\theta_n) + \beta_n$ and (5.2) holds for $g_n(\cdot)$ replacing $\bar{g}(\cdot)$. The "soft constraint" can be used in all subsequent chapters as well.

5.6 Random Directions, Subgradients, and Differential Inclusions

The random directions algorithm. Refer to the random directions algorithm (1.2.9), where additional references are given. Let $\mathcal{F}_n^d$ be the minimal σ-algebra that measures $\{\theta_0, Y_{i-1}, d_i, i \le n\}$. Adding the constraint set H yields the projected algorithm

$$\theta_{n+1} = \Pi_H \left[\theta_n - \epsilon_n d_n \frac{Y_n^+ - Y_n^-}{2c_n}\right], \tag{6.1}$$

that we write in the expanded form as

$$\theta_{n+1} = \theta_n - \epsilon_n f_\theta(\theta_n) + \epsilon_n d_n \beta_n + \epsilon_n d_n \frac{\delta M_n}{2c_n} + \epsilon_n \psi_n^d + \epsilon_n Z_n, \tag{6.2}$$

where we redefine

$$\delta M_n = (Y_n^- - E_{\mathcal{F}_n^d} Y_n^-) - (Y_n^+ - E_{\mathcal{F}_n^d} Y_n^+). \tag{6.3}$$

β_n is the finite difference bias and $\psi_n^d = [I - d_n d_n'] f_\theta(\theta_n)$ is the "random directions noise."

More generally, the random directions algorithm takes the following form. Let there be measurable functions $\gamma_n(\cdot)$ such that

$$E_{\mathcal{F}_n^d} Y_n^\pm = \gamma_n(\theta_n \pm c_n d_n). \tag{6.4}$$

Suppose that there are functions $g_n(\cdot)$ that are continuous in θ, uniformly in n, and random variables β_n such that

$$\frac{\gamma_n(\theta_n - c_n d_n) - \gamma_n(\theta_n + c_n d_n)}{2c_n} = d_n' g_n(\theta_n) + \beta_n.$$

Write the algorithm in the expanded form

$$\begin{aligned}\theta_{n+1} &= \theta_n + \epsilon_n d_n d_n' g_n(\theta_n) + \epsilon_n d_n \beta_n + \epsilon_n d_n \frac{\delta M_n}{2c_n} + \epsilon_n Z_n \\ &= \theta_n + \epsilon_n g_n(\theta_n) + \epsilon_n d_n \beta_n + \epsilon_n d_n \frac{\delta M_n}{2c_n} \\ &\quad + \epsilon_n \left(d_n d_n' - I\right) g_n(\theta_n) + \epsilon_n Z_n.\end{aligned} \tag{6.5}$$

When the general form (6.5) is used, we will require that the asymptotic rate of change of

$$\sum_{n=0}^{m(t)-1} \epsilon_n \left(d_n d_n' - I\right) g_n(\theta) \tag{6.6}$$

be zero for each θ.

The following theorem follows directly from Theorem 3.1. See also the comments on the Kiefer–Wolfowitz procedure at the end of Section 3 concerning the condition (3.2) with $d_n \delta M_n/(2c_n)$ and ψ_n^d replacing the δM_n there. If the constraint set H is dropped, then the stability theorems of Section 4 can be used, with $d_n \delta M_n/(2c_n)$ and ψ_n^d replacing the δM_n and $d_n \beta_n$ replacing β_n there.

Theorem 6.1. *Assume algorithm* (6.5), *the conditions of Theorem* 3.1 *for the gradient case* $\bar{g}(\cdot) = -f_\theta(\cdot)$, *with* $d_n \delta M_n/(2c_n)$ *replacing* δM_n, *and* $d_n \beta_n$ *replacing* β_n. *Assume that the asymptotic rate of change of* (6.6) *is zero with probability one. Then for almost all* ω, $\theta_n(\omega)$ *converges to a unique stationary set* S_i. *In particular if* $f(\cdot)$ *has a unique constrained stationary point* $\bar{\theta}$ *in* H, *then* θ_n *converges to* $\bar{\theta}$ *with probability one.*

Remark: Random directions and the Robbins–Monro procedure. The random directions idea can also be used for the Robbins–Monro procedure. This requires estimating the directional "increment." If the effort required to do this is commensurate with that required to get each component of Y_n and is also much less than what is required to get the full vector Y_n for high-dimensional problems, then it might be advantageous to use the random directions method for high dimensions; see Chapter 10.

Example of algorithm (6.5): n-dependent search. The use of $g_n(\cdot)$ rather than $-f_\theta(\cdot)$ in (6.5) arises if the direction of search is selected in a different subspace on successive iterations. For example, let $r = r_1 + r_2$, and suppose that we iterate in the subspace of the first r_1 (resp., the last r_2) components of θ on the even (resp., odd) iterations. Let $f_i(\cdot)$ denote the negative of the gradient of $f(\cdot)$ in the first r_1 (resp., second r_2) components. Then $g_{2n}(\cdot) = (f_1(\cdot), 0)$ and $g_{2n+1}(\cdot) = (0, f_2(\cdot))$ and $\bar{g}(\cdot) = (f_1(\cdot)+f_2(\cdot))/2$. On the even (resp., odd) numbered iterations, the last r_2 (resp., first r_1) components of the random direction vector are zero.

Convex function minimization and subgradients. Consider the constrained form of the algorithm (1.3.9), for the minimization of a convex function $f(\cdot)$ that is not necessarily continuously differentiable everywhere. Write the constrained form of (1.3.9) as

$$\theta_{n+1} = \theta_n - \epsilon_n \gamma_n + \epsilon_n \frac{\delta M_n}{2c_n} + \epsilon_n Z_n, \tag{6.7}$$

where Z_n is the reflection term, δM_n is the observation noise, and γ_n is a finite difference approximation to a subgradient of $f(\cdot)$ at θ_n. The required properties of γ_n were stated below (1.3.9). Again, the stability theorems of Section 4 can be applied if the constraint set H is dropped. Theorem 3.1 yields the following result.

Theorem 6.2. *Assume algorithm* (6.7), *where the* γ_n *are bounded and satisfy the condition below* (1.3.9). *Assume* (1.1), (3.2), *and* (3.3), *with* $\delta M_n/(2c_n)$ *replacing* δM_n *in the definition of* $M^0(\cdot)$ *in* (3.2). *Assume any of the constraint set conditions* (A4.3.1), (A4.3.2) *or* (A4.3.3), *Suppose that* $f(\cdot)$ *is not constant. Then the mean ODE is the differential inclusion*

$$\dot{\theta} \in -SG(\theta) + z, \quad z(t) \in -C(\theta(t)). \tag{6.8}$$

With probability one, all limit points of θ_n *are stationary points* [*i.e., points where* $0 \in -SG(\theta) + z, z \in -C(\theta)$]. *If there is a unique limit point* $\bar{\theta}$ *of the paths of* (6.8), *then* $\theta_n \to \bar{\theta}$ *with probability one.*

Differential inclusions. In some applications, (2.12) fails to hold owing to the nature of the variation in the distributions of the noise, but where nevertheless the local averages of $g_n(\cdot)$ are in a suitable set. The differential inclusion form (6.8) might then be useful. Write the algorithm as

$$\theta_{n+1} = \Pi_H \left[\theta_n + \epsilon_n \left(g_n(\theta_n) + \delta M_n\right)\right]. \tag{6.9}$$

The proof of the following useful result is like that of Theorem 3.1: the details are left to the reader. If the constraint set H is dropped, then the stability theorems of Section 4 can be used.

Theorem 6.3. *Assume the conditions in the first paragraph of Theorem 3.1 for the algorithm (6.9), except for (A2.7). Suppose that*

$$\lim_{\Delta\to 0}\limsup_{n}\sup_{m(t_n+\Delta)\geq i\geq n}\frac{|\epsilon_i-\epsilon_n|}{\epsilon_n}=0. \tag{6.10}$$

Let $g_n(\cdot)$ be continuous on H, uniformly in n, and suppose that for each θ, there is a $G(\theta)$ that is upper semicontinuous in the sense of (4.3.2) such that

$$\lim_{n,m\to\infty}\text{distance}\left[\frac{1}{m}\sum_{i=n}^{n+m-1}g_i(\theta),G(\theta)\right]=0. \tag{6.11}$$

Alternatively, replace the continuity of $g_n(\cdot)$ and (6.11) by the following: The $g_n(\theta_n)$ are bounded and for all α and α_i^n in H such that

$$\lim_{n,m\to\infty}\sup_{n\leq i\leq n+m}|\alpha_i^n-\alpha|=0, \tag{6.12}$$

we have

$$\lim_{n,m\to\infty}\text{distance}\left[\frac{1}{m}\sum_{i=n}^{n+m-1}g_i(\alpha_i^n),G(\alpha)\right]=0. \tag{6.13}$$

Then, for almost all ω, the limit points are contained in an invariant set of the differential inclusion

$$\dot{\theta}\in G(\theta)+z,\quad z(t)\in -C(\theta(t)). \tag{6.14}$$

If $A\subset H$ is locally asymptotically stable in the sense of Liapunov for (2.1) and θ_n is in some compact set in the domain of attraction of A infinitely often with probability $\geq\rho$, then $\theta_n\to A$ with at least probability ρ.

5.7 Convergence for Learning and Pattern Classification Problems

Theorems 2.1, 3.1, and 4.1 will be illustrated by proving convergence for two examples from Chapters 1 and 2.

5.7.1 *The Animal Learning Problem*

Using the stability Theorem 4.1, convergence with probability one will be proved for the lizard learning problem in Section 2.1. The algorithm is

(2.1.3) [equivalently, (2.1.4)], and the assumptions stated in that section will be used. In Theorem 4.1, $\bar{\theta} = 0$ was used for notational simplicity. Here $\theta - \bar{\theta}$ replaces the θ in that theorem.

Define the Liapunov function $V(\theta) = (\theta - \bar{\theta})^2$. Note that because there is a constant C such that $E_n[\tau_n^2 + r_n^2] \le C$ for all n and ω, we have

$$E_n \left[\theta_{n+1} - \theta_n\right]^2 = O(\epsilon_n^2) \left[1 + \bar{g}^2(\theta_n)\right].$$

Thus,

$$E_n V(\theta_{n+1}) - V(\theta_n) = 2\epsilon_n(\theta_n - \bar{\theta})\bar{g}(\theta_n) + O(\epsilon_n^2) \left[1 + \bar{g}^2(\theta_n)\right]. \tag{7.1}$$

By the properties of $\bar{g}(\cdot)$, the first term on the right side of (7.1) is bounded above by $-\epsilon_n \lambda |\theta_n - \bar{\theta}|^2$ for some positive number λ. Until further notice, suppose that $\epsilon_n \to 0$ with probability one. Then, using the fact that $|\bar{g}(\theta)| \le c_0 + c_1|\theta - \bar{\theta}|$ for some positive c_i, for large n we have

$$E_n V(\theta_{n+1}) - V(\theta_n) \le -\epsilon_n \lambda |\theta_n - \bar{\theta}|^2/2 + O(\epsilon_n^2).$$

It will next be shown that (4.2′) can be used with $\widetilde{\epsilon}_n = (2K + 2\bar{\theta})/(nE\tau_1)$ and $\theta_n - \bar{\theta}$ replacing θ_n. Recall that $\epsilon_n = 1/W_n$ and $\theta_n = T_n/W_n$. If $|\theta_n - \bar{\theta}| \le K$ for any real $K > 0$, then $1/W_n \le (K + \bar{\theta})/T_n$. Recall that $T_n \ge \sum_{i=0}^{n-1} \tau_i$, where τ_n are identically distributed and mutually independent with a positive mean value. Hence, the strong law of large numbers implies that with probability one we eventually have $T_n \ge nE\tau_1/2$. Hence, eventually, with probability one,

$$\epsilon_n I_{\{|\theta_n - \bar{\theta}| \le K\}} \le \frac{2K + 2\bar{\theta}}{nE\tau_1}.$$

Since the other conditions of Theorem 4.1 hold, it follows that $\theta_n \to \bar{\theta}$ with probability one.

It remains to be shown that $\epsilon_n \to 0$ with probability one. As was seen earlier, this will be true if $\theta_n \le K_1$ infinitely often for some real K_1. Thus, we need to consider what happens if $\theta_n \to \infty$ with a positive probability. Note that as θ_n increases to infinity the probability of pursuit increases to one. Thus, there are $\delta_0 > 0$ and $K_0 > 0$ sufficiently large such that for $\theta_n \ge K_0$, the probability of pursuit and capture (conditioned on the past) is greater than δ_0. Considering these n such that $\theta_n \ge K_0$, we have a (time varying) Bernoulli sequence with probability of success $\ge \delta_0$, and the strong law of large numbers guarantees that (except on a null set) there is a $c_2 > 0$ such that we eventually have

$$W_{n+1} \ge \sum_{i=1}^{n} w_i J_i I_i I_{\{\theta_i \ge K_0\}} \ge c_2 \sum_{i=1}^{n} I_{\{\theta_i \ge K_0\}}$$

on the set where the right side goes to infinity, where I_i (resp., J_i) is the indicator function of pursuit (resp., capture). Thus, $\epsilon_n \to 0$ with probability one. Finally, we note that these arguments imply that $\epsilon_n = O(1/n)$ with probability one.

5.7.2 The Pattern Classification Problem

An application of Theorem 3.1 yields the probability one convergence of the projected form of the algorithm (1.1.18). (The stability Theorem 4.1 can be used if the algorithm is untruncated.) The sequence $\{y_n, \phi_n\}$ is assumed to be mutually independent, but possibly nonstationary. Let H denote the box $\prod_{i=1}^r [a_i, b_i]$, where $-\infty < a_i < b_i < \infty$. The algorithm is

$$\theta_{n+1} = \Pi_H \left[\theta_n + \epsilon_n \phi_n \left(y_n - \phi_n' \theta_n\right)\right], \tag{7.2}$$

and it is supposed that (1.1) holds. Suppose that

$$\sup_n E\left[|y_n \phi_n| + |\phi_n|^2\right] < \infty, \tag{7.3}$$

and define $\bar{S}_n = E y_n \phi_n$ and $Q_n = E\phi_n \phi_n'$. Suppose that there are matrices $\bar{S}$ and $Q > 0$ such that for each $t > 0$,

$$\lim_n \sum_{i=n}^{m(t_n+t)} \epsilon_i \left[\bar{S}_i - \bar{S}\right] = 0, \tag{7.4a}$$

$$\lim_n \sum_{i=n}^{m(t_n+t)} \epsilon_i \left[Q_i - Q\right] = 0. \tag{7.4b}$$

Define the martingale difference

$$\delta M_n = \left(\phi_n y_n - \bar{S}_n\right) - \left(\phi_n \phi_n' - Q_n\right)\theta_n,$$

and suppose that it satisfies (3.2) or the sufficient conditions in Theorem 3.2. The algorithm can be written as

$$\theta_{n+1} = \theta_n + \epsilon_n \left[\bar{S}_n - Q_n \theta_n\right] + \epsilon_n \delta M_n + \epsilon_n Z_n. \tag{7.5}$$

Theorem 3.1 holds with the mean ODE being given by

$$\dot{\theta} = \bar{S} - Q\theta + z, \; z(t) \in -C(\theta(t)). \tag{7.6}$$

The limit points of $\{\theta_n\}$ are the stationary points of (7.6). If $\bar{\theta} = Q^{-1}\bar{S} \in H^0$, then this is the unique limit point. Otherwise it is the point θ on the boundary of H that satisfies $(\bar{S} - Q\theta) \in C(\theta)$. This will be the point in H closest to $\bar{\theta}$. As a practical matter, if the iterates cluster about some point on the boundary, one enlarges the box unless there is a reason not to. The averaging approach of Subsection 1.3.3 can be used to get an asymptotically optimal algorithm (see Chapters 9 and 11).

Now turn to the algorithm (1.1.16). Continue to assume the conditions listed above and that $n\Phi_n^{-1} \to Q^{-1}$ (equivalently, $\Phi_n/n \to Q$) with probability one. Define $\widetilde{\epsilon}_n = \Phi_n^{-1} Q$, and write the algorithm as

$$\theta_{n+1} = \Pi_H \left[\theta_n + \widetilde{\epsilon}_n Q^{-1} \phi_n \left(y_n - \phi_n' \theta_n\right)\right]. \tag{7.7}$$

By modifying the process on a set of arbitrarily small probability, it can be assumed (without loss of generality) that there are real $\delta_n \to 0$ such that $|n\widetilde{\epsilon}_n - I| \le \delta_n$. Then Theorem 3.1 continues to hold with the ODE

$$\dot{\theta} = Q^{-1}\bar{S} - \theta + z, \; z(t) \in -C(\theta(t)). \tag{7.8}$$

5.8 Non-Convergence to Unstable Points

Let $\bar{g}(\cdot) = -f_\theta(\cdot)$ for a real-valued continuously differentiable function $f(\cdot)$. Under appropriate conditions, the theorems in Sections 2 to 4 have shown that $\{\theta_n\}$ converges with probability one. If the algorithm is unconstrained, then the limit points are in the set S of stationary points, that is, the points satisfying $f_\theta(\theta) = 0$. If the algorithm is constrained, with constraint set H, the limit points are those $\theta \in H$ that satisfy $-f_\theta(\theta) + z = 0$, for some $z \in -C(\theta)$.

The set S (and similarly for the constrained problem) contains local minima and other types of stationary points such as local maxima and saddles. In practice, the local maxima and saddles are often seen to be unstable points for the algorithm. Yet this "practical" lack of convergence to the "bad" points does not follow from the given convergence theorems. The "bad" points tend to be unstable for the algorithm because of the interaction of the instability (or marginal stability) properties of the ODE near those points with the perturbing noise. In an intuitive sense, when the iterate is near a stationary point that is not a local minimum, the noise "shakes" the iterate sequence until it is "captured" by a path descending to a more stable point. This is similar in spirit to what happens with the simulated annealing procedure, except that we are not looking for a global minimum here. Under appropriate "directional nondegeneracy" conditions on the noise, instability theorems based on Liapunov function or large deviations methods can be used to prove the repelling property of local maxima and local saddles. A one dimensional problem was treated in [182, Chapter 5, Theorem 3.1], in which it was shown that convergence to a local maximum point is not possible under certain "nondegeneracy" conditions on the noise and the dynamics; see also [29, 30, 31, 189]. One result from [189] will be given at the end of the section. But, it is often hard to know whether the natural system noise is "sufficiently perturbing." The issue of possible convergence to an unstable point arises even if $\bar{g}(\cdot)$ is not a gradient.

The main problem in actually proving such an instability result is that little is known about this perturbing noise in general. This is particularly true for high dimensional problems, so that depending on the system noise to "robustify" the algorithm can be risky. Practical "robustification" might require an explicit introduction of carefully selected perturbations. This would reduce the theoretical rate of convergence, as defined in Chapter 10, but we would expect that added robustness has some associated price.

It can be shown that a slight perturbation of the basic algorithm guarantees convergence with probability one to a local minimum. The proof is still "asymptotic." We are not suggesting the use of such a perturbed algorithm. Indeed, the alteration is not a practical algorithm, although some sort of perturbation will no doubt be needed. In addition, even if an algorithm is constructed to ensure asymptotic escape from whatever local maximum or saddle the iterates might wander near, this is not necessarily a guarantee of good behavior during any run of "practical length." Keep in mind that if the iterate sequence is near a "bad" point for some large n, then it might stay near that point for a long time afterwards, for either the original or a perturbed form of the type discussed below, particularly when the step sizes are small. This can be a problem in practice. However, the results show the fundamental importance of the structure of the perturbing noise (or added perturbations) and the ODE. For practical robustness, one might require that neither the step sizes nor the perturbations go to zero.

One might wish to treat the problem of stationary points that are not local minima in the context of the problem where there are many local minima, and convergence to a reasonably good local minimum is desired. The basic methods suggested to date for such a problem involve either appropriate multiple stochastic approximations, as in [258], or some sort of crude perturbation as in [81, 130]; see also [108, 259, 269] and application to image segmentation and restoration [270]. The perturbation itself is equivalent to a "local" restarting, but without the step size sequence being reset. The issue of appropriate procedures for the general multi-stationary-point case is still open.

A perturbed algorithm. Suppose that there is no constraint set. Let the set of stationary points S consist of a finite number of points, define S_1 to be the set of (finite) local minima, and set $S_2 = S - S_1$. Let ν_k be a sequence of integers going to infinity, define $T_k = t_{\nu_{k+1}} - t_{\nu_k}$, and suppose that $T_k \to \infty$. Let $\{\chi_k\}$ be a sequence of mutually independent random variables uniformly distributed on the unit sphere in $I\!R^r$, and let $\{b_k\}$ be a sequence of positive numbers tending to zero. Also suppose that, for each k, χ_k is independent of $\{Y_i, i \le \nu_k - 1\}$. The perturbed algorithm is

$$\theta_n = \theta_{n-1} + \epsilon_{n-1} Y_{n-1}, \quad n \neq \nu_k, \text{ any } k,$$

$$\theta_n = \theta_{n-1} + \epsilon_{n-1} Y_{n-1} + b_k \chi_k, \quad n = \nu_k.$$

Note that ν_k are the perturbation times, and at each perturbation time ν_k, the iterate is changed by $b_k \chi_k$. Under appropriate conditions, the asymptotic rate of change of the right-continuous piecewise constant process

$$p(t) = \sum_{k:\nu_k \leq m(t)} b_k \chi_k$$

is zero, since $T_k \to \infty$ and $b_k \to 0$. Then, under appropriate conditions, the limiting ODE is just

$$\dot{\theta} = \bar{g}(\theta) = -f_\theta(\theta),$$

as for the unperturbed algorithm. and we have probability one convergence to S. Conditions on the b_k, ν_k which assure that the only possible limit points are in S_1 are not difficult to obtain.[3] But, there are always the issues of determining what the "best" values are, by balancing the considerations of destabilizing the algorithm about unstable points and rate of convergence. Additionally, we might want an algorithm that will converge to a global minimum with a high probability.

Nonconvergence to unstable points. Let $\widetilde{\theta}$ be a stationary point of the ODE $\dot{\theta} = \bar{g}(\theta)$, and suppose that some eigenvalue of $\bar{g}_\theta(\widetilde{\theta})$ has a positive real part. Then the point $\widetilde{\theta}$ is said to be *linearly unstable.* We are concerned with nonconvergence to $\widetilde{\theta}$ for the Robbins-Monro process, and assume that no constraints are active at $\widetilde{\theta}$. Consider the algorithm

$$\theta_{n+1} = \theta_n + \epsilon_n \left[\bar{g}(\theta_n) + \delta M_n\right],$$

where $\epsilon_n = 1/n$ and δM_n is a martingale difference in that $E[\delta M_n | \delta M_i, i < n, \theta_0] = 0$ with probability one.

Theorem 8.1. [189] *Let the mixed second-order partial derivatives of $\bar{g}(\cdot)$ be continuous in some neighborhood of $\widetilde{\theta}$, a linearly unstable point. Let $\{\delta M_n\}$ be bounded. Suppose that there is a constant $K > 0$ such that for all n and unit vectors x, we have $E \max\{0, x'\delta M_n\} \geq K$. Then the probability that θ_n converges to $\widetilde{\theta}$ is zero.*

---

[3]Such conditions and a proof of convergence were given in the first edition of this book. They are omitted here, since the proofs were intended only to be suggestive of possibilities.

6

Convergence with Probability One: Correlated Noise

6.0 Outline of Chapter

The most powerful method in Chapter 5 is the "general compactness method" of Section 5.3. The basic assumption for that method is that the asymptotic "rate of change" of the martingale $M^0(\cdot)$ is zero with probability one. It was shown in Subsection 5.3.2 that the rate of change condition is quite weak and yields convergence under what are probably close to the weakest conditions on both the noise and step-size sequences, when the underlying noise is of the martingale difference form. In this chapter it is shown that the same approach yields excellent results for correlated noise. General assumptions of the type used in Section 5.3 are stated in Subsection 1.1 and the main convergence theorem is proved in a straightforward way in Subsection 1.2. Sometimes, one can use a stability (or weak convergence) method to guarantee that some small neighborhood of a "stable" point $\bar{\theta}$ is entered infinitely often with probability $p > 0$. Then the ideas of Theorem 1.1 can be "localized" to give a proof of convergence to $\bar{\theta}$ with probability p, under "local" conditions, which is done in Theorem 1.2. In Section 2 it is shown that strong laws of large numbers imply the asymptotic rate of change condition.

An alternative but more complicated approach to verifying the asymptotic rate of change conditions can be stated in terms of a "perturbed state" process (as used in [127]). This latter method is particularly useful when the step sizes decrease very slowly so that the law of large numbers cannot be used. The general idea of the perturbed state method is introduced in

Section 3 for verifying the rate of change condition, and some examples are given in Section 4. Analogous perturbed state methods will frequently be used in the sequel to "average out" the noise effects in the course of proving convergence or stability when the noise is correlated. Section 5 gives the direct extension of the convergence results of Section 1 to the Kiefer–Wolfowitz algorithm and its random directions variant.

In Sections 1 to 5, the noise process $\{\xi_n\}$ is assumed to be "exogenous," in that its future evolution, given the past history of the iterates (or "state") and the noise, depends only on the noise. In many important applications, the evolution of the noise depends on the state in a Markovian way. Some examples of this were given in Sections 2.3–2.5. The basic outline of the state-dependent noise case is given in Section 6. General conditions for convergence are stated in terms of a "perturbed" state process. The method and general ideas are interesting and provide a very natural approach to the convergence theory for the state-dependent noise case. However, only a statement of the theorem and a rough outline of the proof and the general idea are given, without elaboration, because the weak convergence method of Chapters 7 and 8 provides a simpler approach to this problem. Section 7 deals with the stability method when there is no bounded constraint set H and the iterate sequence is not known to be bounded *a priori*. The use of perturbed Liapunov functions is crucial to the development. Since the noise is correlated, the perturbations are of the type to be introduced in Section 3 and not of the type used in Chapters 4 and 5. A result where the mean ODE is actually a differential inclusion is in Section 8, together with an application to the parameter identification problem. Some ideas from the theory of large deviations are introduced in Sections 9 and 10 to obtain "exponential" bounds on the path excursions. By *exponential bounds*, we mean bounds of the type: Given $\theta_n = x$ in a small neighborhood of an asymptotically stable point $\bar{\theta}$ of the ODE, the probability that some future iterate $\theta_i, i \geq n$, will leave some larger neighborhood of $\bar{\theta}$ is bounded above by e^{-c/ϵ_n}, for some $c > 0$. Under weaker conditions, the mean escape time is bounded below by e^{c/ϵ_n}, for some $c > 0$. These exponential bounds will also be used for the constant-step-size case with $\epsilon_n = \epsilon$, and will be useful in interpreting the weak convergence results in Chapters 7 and 8. The underlying theory of large deviations is quite technical; to simplify the development, the problem will be set up so that available results in the literature can be used.

6.1 A General Compactness Method

6.1.1 Introduction and General Assumptions

In Chapter 5, the noise process was a sequence of martingale differences. In Section 5.3, we discussed a method for proving convergence with probability one that used the condition that the asymptotic rate of change of the interpolated noise process $M^0(t) = \sum_{i=0}^{m(t)-1} \epsilon_i \delta M_i$ is zero with probability one. In this chapter, it will be shown that the basic idea of the asymptotic rate of change condition, as introduced and developed in [135], can be applied with great generality and relative ease to algorithms with correlated noise.

Until further notice, we will work with the projected or constrained algorithm (5.2.10), namely,

$$\theta_{n+1} = \Pi_H \left[\theta_n + \epsilon_n Y_n\right]. \tag{1.1}$$

In Chapter 5, we defined $\mathcal{F}_n$, the σ-algebra determined by the initial condition θ_0 and observations $Y_i, i < n$. When the driving noise is correlated, it is useful to keep track of the "effective memory" needed to compute the conditional expectation of the current observation given the past, and the following somewhat more general approach is useful. We introduce random variables ξ_n to represent the effective memory. Suppose that there are random variables ξ_n taking values in some topological space Ξ, a sequence of nondecreasing σ-algebras $\mathcal{F}_n$, measurable functions $g_n(\cdot)$, and "bias" random variables β_n such that $\{\theta_0, Y_{i-1}, \xi_i, i \le n\}$ is $\mathcal{F}_n$-measurable and if $E|Y_n| < \infty$, then

$$E_n Y_n = g_n(\theta_n, \xi_n) + \beta_n, \tag{1.2}$$

where E_n denotes the expectation conditioned on $\mathcal{F}_n$. Define $\delta M_n = Y_n - E_n Y_n$ and write the algorithm (1.1) in expanded form as

$$\theta_{n+1} = \theta_n + \epsilon_n \left[E_n Y_n + \delta M_n + Z_n\right]$$

or, equivalently,

$$\theta_{n+1} = \theta_n + \epsilon_n \left[g_n(\theta_n, \xi_n) + \delta M_n + \beta_n + Z_n\right], \tag{1.3}$$

where, as in Chapter 5, Z_n is the reflection term when there is a constraint set H.

The σ-algebra $\mathcal{F}_n$ would generally measure "details" of the experiments that led to $\{Y_i, i < n, \theta_0\}$. There is no pressing mathematical need to separate β_n from $g_n(\theta_n, \xi_n)$, but the form represents many applications, where β_n is a bias that is asymptotically unimportant and where it is natural to put conditions on $g_n(\cdot)$ and ξ_n.

Recall that when we say that the asymptotic "rate of change" of a process, say $M^0(\cdot)$, is zero with probability one (see Subsection 5.3.1), we mean that for ω not in some null set and for any positive T,

$$\limsup_n \max_{j \ge n} \max_{0 \le t \le T} |M^0(\omega, jT + t) - M^0(\omega, jT)| = 0. \tag{1.4}$$

Condition (1.4) does not imply that $M^0(\cdot)$ (or, equivalently, $M_n = \sum_{i=0}^{n-1} \epsilon_i \delta M_i$) converges. For example, $M^0(\omega, t) = \log t$ satisfies (1.4) but goes to infinity as $t \to \infty$. Conditions of this type, introduced and used heavily in [135], appear to be close to minimal and have been proved to be necessary for certain algorithms [249]. An analogous condition will be needed to deal with the β_n terms. The asymptotic rate of change conditions are a convenient way of getting to the heart of what needs to be done to get convergence. After proving a convergence theorem under such asymptotic rate of change conditions, one need only show that they can be verified in cases of interest. The approach is a convenient "division of labor."

In addition to eliminating the effects of the martingale differences $\{\delta M_n\}$ and the bias terms $\{\beta_n\}$ on the asymptotics, both of which were dealt with in Section 5.3 for the martingale noise case, one needs to "average" the sequence $\{g_n(\theta_n, \xi_n)\}$.

Suppose that $g_n(\cdot)$ does not depend on n and that for each θ,

$$\frac{1}{m} \sum_{i=n}^{m+n-1} g(\theta, \xi_i) \to \bar{g}(\theta) \text{ w.p.1 as } m, n \to \infty.$$

Then, since θ_n varies very slowly for large n, we might expect that a combination of the law of large numbers and a continuity condition could be used to approximate

$$\sum_{i=n}^{m(t_n+\Delta)-1} \epsilon_i g(\theta_i, \xi_i)/\Delta$$

by $\bar{g}(\theta_n)$ for small $\Delta > 0$ and that this approximation and the ODE method of Chapter 5 would yield the desired limit theorem. Of course, if $g_n(\cdot)$ depends on n, this dependence would have to be taken into account.

Thus, two issues arise: The first concerns the continuity of $g_n(\cdot, \xi)$ in θ, and the second an appropriate form of the law of large numbers or other averaging principle. The averaging principle we will use is weaker than the strong law of large numbers. (In the next two chapters, the requirements on the averaging principle are weakened even further.) A convenient condition on $g_n(\theta, \xi_n)$ uses a local averaging and asymptotic rate of change condition in the following form: There is a "mean" function $\bar{g}(\cdot)$ such that for any $T > 0$ and for each θ,

$$\limsup_n \max_{j \geq n} \max_{0 \leq t \leq T} \left| \sum_{i=m(jT)}^{m(jT+t)-1} \epsilon_i \left[g_i(\theta, \xi_i) - \bar{g}(\theta) \right] \right| = 0 \text{ w.p.1.} \tag{1.5}$$

To facilitate the verification, the conditions (1.4) and (1.5) will be written in the form of (A1.3)–(A1.5). In Section 2, it will be shown that the conditions are implied by strong laws of large numbers and other standard results. The following conditions will be used. [Keep in mind that we are working either with the constrained or the bounded θ_n. When the unconstrained algorithm is used (i.e., where θ_n might be unbounded), one first proves stability or recurrence, and then applies the results for the bounded case. Stability is dealt with in Section 7.]

(A1.1) $\sup_n E|Y_n| < \infty$.

(A1.2) $g_n(\theta, \xi)$ is continuous in θ for each ξ and n.

We require (A1.3)–(A1.5) to hold for each $\mu > 0$ and some $T > 0$.

(A1.3) There is a continuous function $\bar{g}(\cdot)$ such that for each θ,

$$\lim_{n\to\infty} P\left\{\sup_{j\geq n}\max_{0\leq t\leq T}\left|\sum_{i=m(jT)}^{m(jT+t)-1}\epsilon_i\left[g_i(\theta,\xi_i)-\bar{g}(\theta)\right]\right|\geq\mu\right\}=0.$$

(A1.4)
$$\lim_{n\to\infty} P\left\{\sup_{j\geq n}\max_{0\leq t\leq T}\left|\sum_{i=m(jT)}^{m(jT+t)-1}\epsilon_i\delta M_i\right|\geq\mu\right\}=0.$$

(A1.5)
$$\lim_{n\to\infty} P\left\{\sup_{j\geq n}\max_{0\leq t\leq T}\left|\sum_{i=m(jT)}^{m(jT+t)-1}\epsilon_i\beta_i\right|\geq\mu\right\}=0.$$

(A1.6) There are measurable and non-negative functions $\rho_3(\cdot)$ and $\rho_{n4}(\cdot)$ of θ and ξ, respectively, such that

$$|g_n(\theta,\xi)| \leq \rho_3(\theta)\rho_{n4}(\xi), \tag{1.6}$$

where $\rho_3(\cdot)$ is bounded on each bounded θ-set, and for each $\mu > 0$,

$$\lim_{\tau\to 0}\lim_{n} P\left\{\sup_{j\geq n}\sum_{i=m(j\tau)}^{m(j\tau+\tau)-1}\epsilon_i\rho_{i4}(\xi_i)\geq\mu\right\}=0. \tag{1.7}$$

(A1.7) There are non-negative measurable functions $\rho_1(\cdot)$, $\rho_{n2}(\cdot)$ of θ and ξ, respectively, such that $\rho_1(\cdot)$ is bounded on each bounded θ-set and

$$\begin{aligned}&|g_n(\theta,\xi)-g_n(y,\xi)| \leq \rho_1(\theta-y)\rho_{n2}(\xi)\\ &\text{where }\ \rho_1(\theta)\overset{\theta\to 0}{\longrightarrow}0\end{aligned} \tag{1.8}$$

and

$$P\left\{\limsup_{j}\sum_{i=j}^{m(t_j+\tau)}\epsilon_i\rho_{i2}(\xi_i)<\infty\right\}=1,\quad\text{for some }\tau>0. \tag{1.9}$$

(A1.8) If ϵ_n is random, let it be $\mathcal{F}_n$-measurable, with $\epsilon_n \geq 0$ and $\sum_n \epsilon_n = \infty$ with probability one.

Definition: Bounded rate of change condition. Replace $\rho_{i1}(\xi_i)$ in (1.7) by $\psi_i \geq 0$. Then, if (1.7) holds for each $\mu > 0$, we say that the *asymptotic rate of change* of the process

$$\sum_{i=0}^{m(t)-1} \epsilon_i \psi_i$$

is *bounded with probability one.*

Remarks on the conditions. Sufficient conditions for (A1.3) will be given in Section 2 in terms of laws of large numbers and in Sections 3 and 4 in terms of "perturbed state" criteria. Condition (A1.6) is used only in the proof that the set of interpolated paths is equicontinuous (always in the extended sense) with probability one and the condition holds if $g_n(\cdot)$ are uniformly bounded. (A1.7) holds if the functions $g_n(\cdot, \xi)$ are continuous in θ uniformly in n and ξ. If $\{\rho_{n2}(\cdot)\}$ satisfies the condition (1.7), then (1.9) holds. (1.7) and (1.9) can be reduced to a condition like (A1.3) as follows: Let $\sup_n E\rho_{n4}\,(\xi_n) < \infty$. For each $\mu > 0$ and $T < \infty$, suppose that

$$\lim_n P\left\{\sup_{j\geq n}\max_{0\leq t\leq T}\left|\sum_{i=m(jT)}^{m(jT+t)-1} \epsilon_i\left[\rho_{i4}(\xi_i) - E\rho_{i4}(\xi_i)\right]\right| \geq \mu\right\} = 0. \qquad (1.10)$$

Then (1.7) holds. Section 8 contains extensions when $\bar{g}(\cdot)$ is discontinuous or multivalued. But the weak convergence approach, to be presented in Chapter 8, is easier and requires weaker conditions.

6.1.2 The Basic Convergence Theorem

Remark on past state dependence in $g_n(\cdot)$. Suppose that $g_n(\theta_n, \xi_n)$ is replaced by $g_n(\theta_n, \ldots, \theta_{n-k}, \xi_n)$, $k \geq 0$. Define $\widehat{g}_n(\theta, \xi) = g_n(\theta, \ldots, \theta, \xi)$. Then the theorem holds if the continuity and bounding conditions (A1.6) and (A1.7) on $g_n(\theta, \xi)$ are replaced by analogous continuity and bounding conditions on $g_n(x_1, \ldots, x_{k+1}, \xi)$, and $\widehat{g}_n(\theta, \xi)$ replaces $g_n(\theta, \xi)$ in (A1.3).

Apart from the conditions, the next theorem is a combination of Theorem 5.2.1 and 5.3.1.

Theorem 1.1. *Assume* (5.1.1), *algorithm* (1.1), *and the conditions* (A1.1)–(A1.7), *with H satisfying any one of the conditions* (A4.3.1), (A4.3.2), *or* (A4.3.3). *If the ϵ_n are random, assume* (A1.8) *in lieu of* (5.1.1). *Then there is a null set N such that for $\omega \notin N$, the set of functions $\{\theta^n(\omega, \cdot), Z^n(\omega, \cdot), n < \infty\}$ is equicontinuous. Let $(\theta(\omega, \cdot), Z(\omega, \cdot))$ denote the limit of some convergent subsequence. Then this pair satisfies the projected ODE* (5.2.1), *and $\{\theta_n(\omega)\}$ converges to some limit set of the ODE* *in H. If the constraint set is dropped, but $\{\theta_n\}$ is bounded with probability one, then for almost all ω, the limits $\theta(\omega, \cdot)$ of convergent subsequences of $\{\theta^n(\omega, \cdot)\}$ are trajectories of*

$$\dot{\theta} = \bar{g}(\theta)$$

in some bounded invariant set and $\{\theta_n(\omega)\}$ converges to this invariant set. Let p_n be integer-valued functions of ω, not necessarily being stopping times or even measurable, but that go to infinity with probability one. Then the conclusions concerning the limits of $\{\theta^n(\cdot)\}$ hold with p_n replacing n. If $A \subset H$ is locally asymptotically stable in the sense of Liapunov for (5.2.1) and θ_n is in some compact set in the domain of attraction of A infinitely often with probability $\geq \rho$, then $\theta_n \to A$ with at least probability ρ. Suppose that (A5.2.6) holds. Then, for almost all ω, $\{\theta_n(\omega)\}$ converges to a unique S_i.

Under the additional conditions of Theorem 5.2.5 on the mean ODE, for almost all ω, $\theta^n(\omega, \cdot)$ and $\theta_n(\omega)$ converge to a set of chain recurrent points in the limit or invariant set.

Proof. Define the process $G^n(\cdot)$ on $-\infty < t < \infty$ by

$$G^n(t) = \sum_{i=n}^{m(t_n+t)-1} \epsilon_i g_i(\theta_i, \xi_i)$$

for $t \geq 0$ and use the definition analogous to the second line of (5.1.8) for $t < 0$. For simplicity, we henceforth write the definitions of the shifted interpolation processes for $t \geq 0$ only, with the obvious analogs of the second line of (5.1.8) used for $t < 0$. Rewrite the algorithm as (for $-\infty < t < \infty$)

$$\theta^n(t) = \theta_n + G^n(t) + M^n(t) + B^n(t) + Z^n(t). \tag{1.11}$$

The proof follows the basic lines of that of Theorem 5.2.3. First the (w.p.1) equicontinuity of processes on the right side of (1.11) is shown. Then it is shown that $M^n(\cdot)$ and $B^n(\cdot)$ go to zero w.p.1. The characterization of the limits of $G^n(\cdot)$ will be done by use of the continuity condition and the averaging condition (A1.3). The use of (A1.3) is possible since θ_n changes "slowly." This procedure yields the ODE that characterizes the asymptotic behavior.

First, we show that the sequence $\{G^n(\cdot)\}$ is equicontinuous with probability one. This is obviously true if the $g_i(\cdot)$ are uniformly bounded. Otherwise, proceed as follows. Since $\{\theta_n\}$ is bounded, by (A1.6) there is a $0 \leq K_1 < \infty$ such that

$$\sum_{i=n}^{m(t_n+t)-1} \epsilon_i |g_i(\theta_i, \xi_i)| \leq K_1 \sum_{i=n}^{m(t_n+t)-1} \epsilon_i \rho_{i4}(\xi_i) \equiv R_n(t).$$

Now (1.7) implies that there is a null set N_0 such that for $\omega \notin N_0$, the changes in the values of $R_n(\omega, \cdot)$ over small time intervals are small, uniformly in n for large n. Hence $\{R_n(\omega, \cdot)\}$ is equicontinuous.

By (A1.4) and (A1.5), $\{M^n(\omega, \cdot)\}$ and $\{B^n(\omega, \cdot)\}$ are equicontinuous for $\omega \notin N_1$, a null set, and all limits (as $n \to \infty$) are zero. As in the proof of Theorem 5.2.3, the equicontinuity of $\{G^n(\omega, \cdot) + M^n(\omega, \cdot) + B^n(\omega, \cdot)\}$ for $\omega \notin N_0 \cup N_1$ implies that of $\{Z^n(\omega, \cdot)\}$. Thus, for $\omega \notin N_0 \cup N_1$, $\{\theta^n(\omega, \cdot)\}$ is equicontinuous. We next identify the limits of convergent subsequences of $\theta^n(\omega, \cdot)$. This will be done using the continuity condition (A1.7) and the averaging condition (A1.3).

There is a countable dense set $H_0 \subset H$ and a null set N_2 such that for $\omega \notin N_2$ and each $\theta \in H_0$, (A1.3) implies that the set of centered functions $\{\bar{G}^n(\omega, \theta, \cdot)\}$ defined by

$$\bar{G}^n(\omega, \theta, t) = \sum_{i=n}^{m(t_n+t)-1} \epsilon_i \left[g_i(\theta, \xi_i(\omega)) - \bar{g}(\theta)\right]$$

is equicontinuous and all limits (as $n \to \infty$) are zero. Let N_3 denote the exceptional null set in (1.9) (i.e., where the limsup is infinite) and define $N = \cup_0^3 N_i$.

For $\omega \notin N$, extract a convergent subsequence of $\{\theta^n(\omega, \cdot), Z^n(\omega, \cdot)\}$, indexed by n_k, with limit denoted by $(\theta(\omega, \cdot), Z(\omega, \cdot))$. Henceforth we work with a fixed sample path for $\omega \notin N$, drop the ω argument, and use $t \geq 0$ for notational simplicity. Also, for notational simplicity, write n_k simply as k. Thus, we use t_k for t_{n_k}. Then, given small $\Delta > 0$ and supposing, without loss of generality, that t is an integral multiple of Δ, write

$$\theta^k(t) = \theta^k(0) + G_1^k(t) + G_2^k(t) + G_3^k(t) + M^k(t) + B^k(t) + Z^k(t), \quad (1.12)$$

where the $G_i^k(\cdot)$ terms are obtained by splitting $\sum_{i=k}^{m(t_k+t)-1} \epsilon_i g_i(\theta_i, \xi_i)$ as follows:

$$\begin{aligned}
G_1^k(t) &= \sum_{j=0}^{t/\Delta-1} \sum_{i=m(t_k+j\Delta)}^{m(t_k+j\Delta+\Delta)-1} \epsilon_i \bar{g}(\theta(j\Delta)), \\
G_2^k(t) &= \sum_{j=0}^{t/\Delta-1} \sum_{i=m(t_k+j\Delta)}^{m(t_k+j\Delta+\Delta)-1} \epsilon_i \left[g_i(\theta_i, \xi_i) - g_i(\theta(j\Delta), \xi_i))\right], \qquad (1.13)\\
G_3^k(t) &= \sum_{j=0}^{t/\Delta-1} \sum_{i=m(t_k+j\Delta)}^{m(t_k+j\Delta+\Delta)-1} \epsilon_i \left[g_i(\theta(j\Delta), \xi_i)) - \bar{g}(\theta(j\Delta))\right].
\end{aligned}$$

This splitting is a convenient way of exploiting the fact that the iterates change slowly to allow us to use the averaging condition (A1.3).

As $\Delta \to 0$,

$$G_1^k(t) \to \int_0^t \bar{g}(\theta(s))ds.$$

$G_2^k(\cdot)$ is handled by using (A1.7), which yields

$$|G_2^k(t)| \le \left[\max_{j\le t/\Delta}\ \max_{0\le t_i - j\Delta\le\Delta} \rho_1(\theta_i - \theta(j\Delta))\right] \sum_{i=m(t_k)}^{m(t_k+t)-1} \epsilon_i \rho_{i2}(\xi_i).$$

The right-hand side goes to zero as $\Delta \to 0$ and $k \to \infty$ by the convergence of $\theta^k(\cdot)$ to $\theta(\cdot)$ and the boundedness of $\limsup_k \sum_{i=m(t_k)}^{m(t_k+t)-1} \epsilon_i \rho_{i2}(\xi_i)$ (since $\omega \notin N$).

To show that $\lim_\Delta \limsup_k |G_3^k(t)| = 0$, use (A1.3) and (A1.7). It is enough to prove that for each small $\Delta > 0$,

$$\lim_k \sum_{i=m(t_k+j\Delta)}^{m(t_k+j\Delta+\Delta)-1} \epsilon_i \left[g_i(\theta(j\Delta),\xi_i) - \bar g(\theta(j\Delta))\right] = 0. \tag{1.14}$$

Given $\alpha > 0$, let $\{B_l^\alpha, l \le l_\alpha\}$ be a finite collection of disjoint sets of diameter smaller than α, whose union is H, and with some $x_l^\alpha \in H^0$ contained in B_l^α. Write the sum in (1.14) as the sum of the following three terms:

$$\sum_{l=1}^{l_\alpha} I_{\{\theta(j\Delta)\in B_l^\alpha\}} \sum_{i=m(t_k+j\Delta)}^{m(t_k+j\Delta+\Delta)-1} \epsilon_i \left[g_i(x_l^\alpha,\xi_i) - \bar g(x_l^\alpha)\right],$$

$$\sum_{l=1}^{l_\alpha} I_{\{\theta(j\Delta)\in B_l^\alpha\}} \sum_{i=m(t_k+j\Delta)}^{m(t_k+j\Delta+\Delta)-1} \epsilon_i \left[g_i(\theta(j\Delta),\xi_i) - g_i(x_l^\alpha,\xi_i)\right],$$

$$\sum_{l=1}^{l_\alpha} I_{\{\theta(j\Delta)\in B_l^\alpha\}} \sum_{i=m(t_k+j\Delta)}^{m(t_k+j\Delta+\Delta)-1} \epsilon_i \left[\bar g(x_l^\alpha) - \bar g(\theta(j\Delta))\right].$$

We need to show that these terms go to zero as $k \to \infty$ and then $\alpha \to 0$. The last term goes to zero uniformly in k as $\alpha \to 0$ by virtue of the continuity of $\bar g(\cdot)$. By (A1.3) and the fact that $\omega \notin N$, the first term goes to zero as $k \to \infty$. By (A1.7), the middle term goes to zero as $k \to \infty$ and then $\alpha \to 0$. Thus, (1.14) holds.

Putting the parts of the proof together and reintroducing the argument ω yields the representation, for $\omega \notin N$,

$$\theta(\omega,t) = \theta(\omega,0) + \int_0^t \bar g(\theta(\omega,s))ds + Z(\omega,t). \tag{1.15}$$

$Z(\omega,\cdot)$ is shown to be the reflection term as in the proof of Theorem 5.2.3, that is, $Z(\omega,t) = \int_0^t z(\omega,s)ds$, where $z(t) \in -C(\theta(\omega,t))$. The rest of the details are as in the proofs of Theorems 5.2.1 and 5.2.3. □

6.1.3 Local Convergence Results

It is sometimes convenient to prove convergence by the following two steps. First, a stability or weak convergence-type proof is used to show that each small neighborhood of a stable point $\bar{\theta}$ of the mean ODE is visited infinitely often with a nonzero probability. Then a "local" form of Theorem 1.1 is used, where the conditions are localized about $\bar{\theta}$. Such an approach is also useful when there is no constraint set.

The proof of Theorem 1.2, uses a "localized" form of the conditions of Theorem 1.1 and is very close to that of Theorem 1.1; the details are left to the reader. A stronger but more complicated result based on recurrence is in Theorem 7.1. Recall that $N_\delta(\theta)$ is used to denote the δ-neighborhood of θ.

Local conditions. The following localized forms of the conditions of Theorem 1.1 will also be used in the stability analysis of Section 7, where the bounded constraint set H will be dropped.

(A1.1′) $\sup_n E|Y_n|I_{\{|\theta_n|\le K\}} < \infty$ for each $K < \infty$.

(A1.4′) (A1.4) holds with δM_n replaced by $\delta M_n I_{\{|\theta_n|\le K\}}$ for each $K < \infty$.

(A1.5′) (A1.5) holds with β_n replaced by $\beta_n I_{\{|\theta_n|\le K\}}$ for each $K < \infty$.

(A1.6′) (A1.6) holds with ρ_{n4} replaced by $\rho_{n4} I_{\{|\theta_n|\le K\}}$ for each $K < \infty$.

(A1.7′) (A1.7) holds with ρ_{n2} replaced by $\rho_{n2} I_{\{|\theta_n|\le K\}}$ for each $K < \infty$.

Theorem 1.2. *Use algorithm* (1.1) *with H satisfying any of the constraint set conditions* (A4.3.1), (A4.3.2), *or* (A4.3.3), *or let $H = I\!R^r$. Let there be a $\bar{\theta}$ such that for each $\delta > 0$, $\theta_n \in N_\delta(\bar{\theta})$ infinitely often with probability one. Assume* (5.1.1) *and* (A1.1′), *and let* (A1.4′)–(A1.7′) *hold with $I_{\{|\theta_n - \bar{\theta}|\le K\}}$ replacing $I_{\{|\theta_n|\le K\}}$, where K is arbitrarily small. If ϵ_n is random, assume* (A1.8). *Let* (A1.2) *and* (A1.3) *hold for θ in an arbitrarily small neighborhood of $\bar{\theta}$. Let $\bar{\theta}$ be locally asymptotically stable in the sense of Liapunov for* (5.2.1). *Then $\theta_n \to \bar{\theta}$ with probability one. The result holds if $\bar{\theta}$ is replaced by a compact set A that is locally asymptotically stable in the sense of Liapunov.*

6.2 Sufficient Conditions for the Rate of Change Assumptions: Laws of Large Numbers

The assumptions (A1.3), (A1.4), (A1.6), and (A1.7) were stated in the current form since they are "minimal" conditions for the proof. Their forms suggest many ways of verifying them in applications. Since the verification of the conditions is a main issue in applications, two useful approaches to such verification will be discussed. In this section, a direct application of the strong law of large numbers and the "Mensov-Rademacher" moment estimates are used. When the strong law of large numbers holds for the sequence $\{g_n(\theta, \xi_n)\}$, then (A1.3) holds. The law of large numbers approach fails to work if the ϵ_n does not go to zero faster than the order of $1/\sqrt{n}$. Then the state perturbation approach of Section 3 is useful. Keep in mind that in the applications to date where $\epsilon_n \to 0$, it generally goes to zero faster than $O(1/\sqrt{n})$.

In what follows, ψ_n can represent $[g_n(\theta, \xi_n) - \bar{g}(\theta)]$, $[\rho_{n2}(\xi_n) - E\rho_{n2}(\xi_n)]$, or $[\rho_{n4}(\xi_n) - E\rho_{n4}(\xi_n)]$. Conditions will be given under which the asymptotic rate of change of

$$\Psi(t) = \sum_{i=0}^{m(t)-1} \epsilon_i \psi_i \tag{2.1}$$

is zero with probability one.

Examples: Strong laws of large numbers. A partial summation formula can be used to put (2.1) into a form where a strong law of large numbers can be applied. Define $S_m = \sum_{i=0}^{m-1} \psi_i$. Then

$$\sum_{i=n}^{m} \epsilon_i \psi_i = \epsilon_m \left[S_{m+1} - S_n\right] + \sum_{i=n}^{m-1} \left[S_{i+1} - S_n\right]\left[\epsilon_i - \epsilon_{i+1}\right]. \tag{2.2}$$

For $n = 0$ and $m = m(t) - 1$, (2.2) becomes

$$\Psi(t) = \epsilon_{m(t)-1} S_{m(t)} + \sum_{i=0}^{m(t)-2} S_{i+1} \frac{\epsilon_i - \epsilon_{i+1}}{\epsilon_i} \epsilon_i. \tag{2.3}$$

The representation (2.3) implies that if

$$\epsilon_n S_{n+1} \to 0 \text{ w.p.1} \tag{2.4a}$$

and

$$\frac{(\epsilon_n - \epsilon_{n+1})}{\epsilon_n} = O(\epsilon_n), \tag{2.4b}$$

then the asymptotic rate of change of $\Psi(\cdot)$ is zero.

Example 1. Let $\epsilon_n = n^{-\gamma}$, $\gamma \in (.5, 1]$. Then

$$\frac{\epsilon_n - \epsilon_{n+1}}{\epsilon_n} = O\left(\frac{1}{n}\right) = \epsilon_n O\left(\frac{1}{n^{1-\gamma}}\right).$$

Hence (2.4b) holds. Then if $\sum_{i=1}^{n} \psi_i / n^{\gamma}$ goes to zero with probability one (a strong law of large numbers), the asymptotic rate of change of $\Psi(\cdot)$ is zero with probability one. This implies that all we are asking for the averaging condition is that a law of large numbers holds.

Example 2. Random ϵ_n. Let $\epsilon_n = 1/\sum_{i=1}^{n} k_i$ for a sequence of positive random variables $\{k_n\}$. Then

$$\frac{\epsilon_n - \epsilon_{n+1}}{\epsilon_n} = \frac{k_{n+1}}{\sum_{i=1}^{n+1} k_i}.$$

If the following conditions are satisfied with probability one

$$\liminf_n \frac{1}{n}\sum_{i=0}^{n} k_i > 0, \quad \frac{1}{n}S_n \to 0, \quad \frac{1}{n}S_n k_n \to 0,$$

then (2.4a) is satisfied and (2.4b) holds.

Example 3. Mensov–Rademacher and moment estimates. The book [135, Chapter 2] gives some examples using other methods, and here are two of them. Suppose, without loss of generality, that ψ_n are real valued.

Suppose that ψ_n have mean zero and that there are $R(i)$ such that for all n, i, $|E\psi_{n+i}\psi_n| \le R(i)$, where $\sum_i R(i) < \infty$. Then in accordance with [135, Section 2.7], there is a real K such that

$$E \max_{n+M>m\ge n} \left|\sum_{i=n}^{m} \epsilon_i \psi_i\right|^2 \le K(\log_2 4M)^2 \sum_{i=n}^{M+n-1} \epsilon_i^2. \tag{2.5}$$

It follows from [135, Theorem 2.2.2] that $\Psi(\cdot)$ converges with probability one if

$$\sum_{i=1}^{\infty} [\epsilon_i \log_2 i]^2 < \infty. \tag{2.6}$$

For the Kiefer–Wolfowitz form, ϵ_n is replaced by ϵ_n/c_n in (2.5) and (2.6). The reference [135] contains applications of these estimates to processes $\{\psi_n\}$ defined by moving averages.

Recall that if the ψ_i are mutually independent with mean zero and for some even integer m

$$\sup_n E|\psi_n|^m < \infty, \quad \sum_n \epsilon_n^{m/2+1} < \infty,$$

then $\Psi(\cdot)$ converges with probability one (see (5.3.18) or [135, Example 6, Section 2.2]).

6.3 Perturbed State Criteria for the Rate of Change Assumptions

6.3.1 *Introduction to Perturbed Test Functions*

One of the powerful tools at our disposal is the method of perturbed test functions (sometimes called perturbed state, depending on the context).

which will be used heavily in the rest of the book. In Subsection 5.3.2, it was seen that if a process is composed of sums of martingale differences, then it is not too hard to get useful bounds on its asymptotic rate of change. In Section 5.4, the importance of the martingale or supermartingale process for proving stability was seen. Very often, in analyzing stochastic approximation processes, a process is close to a martingale or supermartingale in the sense that it would have that property when a small perturbation is added. A simple form of this perturbation technique was successfully used in the stability discussions in Sections 4.5 and 5.4. For additional motivation and in preparation for a more extensive use of the perturbation idea, we now review some of the previous discussion via a simple informal example.

In the discussion of stability in Sections 4.5 and 5.4, the Liapunov function was modified slightly so that the result $V_n(\theta_n) = V(\theta_n) + \delta V_n$ would be a supermartingale. In particular, we used

$$\delta V_n = E_n \sum_{i=n}^{\infty} \epsilon_i^2 |Y_i|^2 I_{\{|\theta_i| \leq K\}}.$$

Let $\{\psi_n\}$ be a sequence of real-valued random variables, and define $\Psi_n = \sum_{i=0}^{n-1} \epsilon_i \psi_i$. Let E_n denote the conditional expectation operator, given the past $\{\psi_i, i < n\}$. Then $E_n \Psi_{n+1} - \Psi_n = E_n \epsilon_n \psi_n$. If $E_n \psi_n$ does not equal zero with probability one for all n, then $\{\Psi_n\}$ is not a martingale. We would like to find a perturbation $\delta\Psi_n$ that goes to zero with probability one as $n \to 0$ such that $\Psi_n + \delta\Psi_n$ is a martingale and then use this martingale property to prove that the asymptotic rate of change of $\Psi_n + \delta\Psi_n$ is zero with probability one. Then the asymptotic rate of change of Ψ_n will also be zero with probability one. If the summands $\{\psi_n\}$ are "weakly" correlated and have mean value zero, it is reasonable to expect that $\sum_{i=n}^{n+m} \epsilon_i E_n \psi_i$ will be near zero for large n and m. The perturbations to be used exploit this idea.

In view of the fact that we use the conditional expectation E_n to define the martingale property at time n, define the perturbation $\delta\Psi_n = \sum_{i=n}^{\infty} \epsilon_i E_n \psi_i$. Note that the conditioning E_n is on the data up to time n only. This sum will be well defined if, loosely speaking, ψ_n have a "mixing" property, where the expectation of the future, given the present, goes to zero fast enough as the time difference goes to infinity. Thus, suppose that

$$\lim_n E \sum_{i=n}^{\infty} \epsilon_i |E_n \psi_i| = 0$$

and define $\widehat{\Psi}_n = \Psi_n + \delta\Psi_n$ (weaker conditions are used below). The process $\delta\Psi_n$ is constructed so that

$$E_n \delta\Psi_{n+1} - \delta\Psi_n = -\epsilon_n E_n \psi_n,$$

and hence

$$E_n \widehat{\Psi}_{n+1} - \widehat{\Psi}_n = 0.$$

This implies that $\widehat{\Psi}_{n+1} - \widehat{\Psi}_n$ is an $\mathcal{F}_n$-martingale difference.

Suppose that $\delta\Psi_n \to 0$ with probability one. Then the asymptotic rate of change of the martingale process

$$\sum_{i=0}^{m(t)-1} \left[\widehat{\Psi}_{n+1} - \widehat{\Psi}_n\right]$$

is that of

$$\sum_{i=0}^{m(t)-1} \epsilon_i \psi_i.$$

Nevertheless, the former sum, involving martingale differences, is often simpler to work with.

The perturbation $\delta\Psi_n$ can be extremely useful. It allows us to construct (slightly) perturbed state processes that have better properties than the original state process. It also allows us to construct Liapunov functions for stability proofs under broad conditions on the noise correlation, extending what was done in Sections 4.5 and 5.4.

Perturbed test function or perturbed state methods, to be used extensively in this book, are very powerful tools for proving approximation and limit theorems for sequences of random processes under quite broad conditions. They were used in [26] to obtain limits of two-time-scale Markov diffusion processes via PDE techniques. Perturbed test functions in the context of semigroup approximations were given in [117]. These approaches were extended considerably and applied to a large variety of approximation and limit problems for both Markovian and non-Markovian models in [124, 125, 127, 132]. These references developed a comprehensive theory and a set of practical techniques for the approximation of continuous-time systems driven by "wide-bandwidth noise" or with "singular" components, and discrete-time systems driven by correlated noise by diffusion or jump-diffusion type processes or solutions to ODEs. Many concrete examples of the use of the perturbation methods and their applications to problems arising in systems and communication theory are in [127]. The "combined first-order perturbation and direct averaging method" in [127] is perhaps the most useful for general problems where the limit processes are diffusions. For the problems of interest in this book, simple perturbations of the type $\delta\Psi_n$ will suffice. The general perturbed state or perturbed test function method will be used throughout the book to prove stability or for approximation theorems when the noise is correlated.

6.3.2 General Conditions for the Asymptotic Rate of Change

In this subsection, we use a perturbed state method to get simpler sufficient conditions for (A1.3). This will also give sufficient conditions for (1.7) and (1.9). Three types of perturbations will be used, namely, (3.2b), its discounted form (3.13), and (3.11), the latter two being variants on the former. For most problems they are of equal value and ease of use. The form (3.11) has advantages when the step sizes are random, and it is a little simpler to verify the conditions with (3.8). The following conditions will be used.

Assumptions and definitions.

(A3.1) $\sup_n E|g_n(\theta, \xi_n)| < \infty$, for each $\theta \in H$.

For each θ, define random variables $v_n(\theta)$ as follows: $v_0(\theta) = 0$ and for $n \geq 0$,

$$v_{n+1}(\theta) = v_n(\theta) + \epsilon_n[g_n(\theta, \xi_n) - \bar{g}(\theta)]. \tag{3.1}$$

(A3.2) For each θ,[1]

$$\lim_n E \sum_{i=n}^{\infty} \epsilon_i \left| E_n g_i(\theta, \xi_i) - \bar{g}(\theta) \right| = 0, \tag{3.2a}$$

and

$$\delta v_n(\theta) \equiv \sum_{i=n}^{\infty} \epsilon_i [E_n g_i(\theta, \xi_i) - \bar{g}(\theta)] \to 0 \text{ w.p.1 as } n \to \infty. \tag{3.2b}$$

Note that (3.2a) implies that, with probability one,

$$E_n \sum_{i=n+1}^{\infty} \epsilon_i E_{n+1}[g_i(\theta, \xi_i) - \bar{g}(\theta)] = \sum_{i=n+1}^{\infty} \epsilon_i E_n[g_i(\theta, \xi_i) - \bar{g}(\theta)]. \tag{3.2c}$$

Define the *perturbed state* $\widetilde{v}_n(\theta) = v_n(\theta) + \delta v_n(\theta)$ and the martingale difference

$$\begin{aligned} \delta N_n(\theta) &= \delta v_{n+1}(\theta) - E_n \delta v_{n+1}(\theta) \\ &= \sum_{i=n+1}^{\infty} \epsilon_i E_{n+1} \left[g_i(\theta, \xi_i) - \bar{g}(\theta) \right] - \sum_{i=n+1}^{\infty} \epsilon_i E_n \left[g_i(\theta, \xi_i) - \bar{g}(\theta) \right], \end{aligned}$$

[1] In fact, (3.2a) is used only to assure that the sum in (3.2b) is well defined, and that the interchange of summation and conditional expectation in $E_n \sum_{i=n+1}^{\infty} E_{n+1} g_i(\theta, \xi_i) = \sum_{i=n+1}^{\infty} E_n g_i(\theta, \xi_i)$ can be done, as in (3.2c). Any other condition assuring this, such as the discounted form introduced in the next subsection, can be used in lieu of (3.2a).

which we will write in the more compact form

$$\begin{aligned}\delta N_n(\theta) &= \sum_{i=n+1}^{\infty} \epsilon_i \left[E_{n+1} - E_n\right]\left[g_i(\theta,\xi_i) - \bar{g}(\theta)\right] \\ &= \sum_{i=n+1}^{\infty} \epsilon_i (E_{n+1} - E_n) g_i(\theta,\xi_i).\end{aligned} \tag{3.3}$$

(A3.3) For some $T > 0$ and each $\mu > 0$,

$$\lim_{n\to\infty} P\left\{\sup_{j\geq n}\max_{t\leq T}\left|\sum_{i=m(jT)}^{m(jT+t)-1} \delta N_i(\theta)\right| \geq \mu\right\} = 0.$$

In other words, the asymptotic rate of change of $\sum_{i=0}^{m(t)-1} \delta N_i(\theta)$ is zero with probability one.

Theorem 3.1. *Assume that ϵ_i are not random and that* (A3.1)–(A3.3) *hold. Then* (A1.3) *holds.*

Proof. By the definitions,

$$\begin{aligned}\widetilde{v}_{n+1}(\theta) - \widetilde{v}_n(\theta) &= \left[v_{n+1}(\theta) - v_n(\theta)\right] + \left[\delta v_{n+1}(\theta) - \delta v_n(\theta)\right] \\ &= \epsilon_n \left[g_n(\theta,\xi_n) - \bar{g}(\theta)\right] + \sum_{i=n+1}^{\infty} \epsilon_i E_{n+1}\left[g_i(\theta,\xi_i) - \bar{g}(\theta)\right] \\ &\quad - \sum_{i=n}^{\infty} \epsilon_i E_n \left[g_i(\theta,\xi_i) - \bar{g}(\theta)\right].\end{aligned} \tag{3.4}$$

Equation (3.4) can be rewritten as

$$\widetilde{v}_{n+1}(\theta) - \widetilde{v}_n(\theta) = \sum_{i=n+1}^{\infty} \epsilon_i \left[E_{n+1} - E_n\right]\left[g_i(\theta,\xi_i) - \bar{g}(\theta)\right] = \delta N_n(\theta). \tag{3.5}$$

By (A3.3), the asymptotic rate of change of

$$\sum_{i=0}^{m(t)-1} \delta N_i(\theta)$$

is zero with probability one. Then, by (3.2b) the asymptotic rate of change of $\sum_{i=0}^{m(t)-1} \epsilon_i[g_i(\theta,\xi_i) - \bar{g}(\theta)]$ is zero with probability one. □

Remarks on the assumptions. As noted in the previous subsection, the perturbation is constructed such that $\widetilde{v}_{n+1}(\theta) - \widetilde{v}_n(\theta)$ is a martingale difference. Note that (A3.3) is implied by

$$\sum_n E[\delta N_n(\theta)]^2 < \infty. \tag{3.6}$$

Some sufficient conditions for (A3.3) are provided in the next section; see also Subsection 5.3.2. Condition (3.2b) is implied by various "mixing-type" conditions on $\{\xi_n\}$. For very slowly decreasing step sizes, a set of conditions under which $\delta v_n(\theta)$ goes to zero with probability one can be obtained from the use of the Chebyshev's inequality applied to the exponential function, and the Borel–Cantelli Lemma, as used in Section 5.3.2, namely, $\delta v_n(\theta)$ goes to zero with probability one if there is a $0 \leq K < \infty$ such that for small α and all n,

$$Ee^{\alpha' \delta v_n(\theta)} \leq e^{K|\epsilon_n \alpha|^2},$$

and for each $\delta > 0$,

$$\sum_{n=0}^{\infty} e^{-\delta/\epsilon_n} < \infty;$$

see Section 4 for an illustration of this idea.

The idea of the theorem also applies to (1.7) and (1.9). Let $\sup_n E\rho_{n4}(\xi_i) < \infty$, and replace $g_n(\theta, \xi_n) - \bar{g}(\theta)$ by $\rho_{4n}(\xi_n) - E\rho_{n4}(\xi_n)$ in (A3.1)–(A3.3). Then (1.7) holds, with the analogous result for (1.9). Under "typical" conditions for stochastic approximation, the correlation of the ξ_n goes to zero as the time difference goes to infinity. If the conditional expectation $E_n[g_{n+i}(\theta, \xi_{n+i}) - \bar{g}(\theta)]$ goes to zero fast enough as $i \to \infty$, one expects that the variance of $\delta N_n(\theta)$ is of the order of ϵ_n^2, under broad conditions. Then the square summability of ϵ_n guarantees (3.6), but (A3.2) is weaker.

Other useful criteria for (A3.3) can be obtained from the discussion of the rate of change of $M^0(\cdot)$ in Chapter 5 and in the next section. The $\delta N_n(\theta)$ term would not appear in the analogous development under the weak convergence approach of the next chapter, and that approach is appreciably simpler than the probability one approach.

6.3.3 *Alternative Perturbations*

A discounted perturbation. The conditions (A3.2) and (A3.3) hold under broad mixing-type conditions on $\{\xi_n\}$, since in applications the "delayed" conditional expectation typically converges to zero fast enough as the delay goes to infinity. Solo [225] added a discount factor to the perturbation scheme developed in [127]. In applications of the type dealt with in this book, discounting adds little generality, but from an analytical point of view it reduces the concern that the perturbations are well defined and leads to a weakening of (A3.2).

For $i \geq n$, define the *discount factor*

$$\Pi(n, i) = \prod_{k=n}^{i} (1 - \epsilon_k), \tag{3.7}$$

where the empty product $\Pi(n+1,n)$ is defined to be unity. Then redefine the perturbation to take the discounted form

$$\delta v_n^d(\theta) = \sum_{i=n}^{\infty} \epsilon_i \Pi(n+1,i) E_n \left[g_i(\theta,\xi_i) - \bar{g}(\theta)\right]. \tag{3.8}$$

Define the *discounted* martingale difference as

$$\delta N_n^d(\theta) = \sum_{i=n+1}^{\infty} \epsilon_i \Pi(n+2,i) \left[E_{n+1} - E_n\right] \left[g_i(\theta,\xi_i) - \bar{g}(\theta)\right]. \tag{3.9}$$

In view of (A3.1) and the fact that

$$\sup_n \sum_{i=n}^{\infty} \epsilon_i \Pi(n+1,i) < \infty, \tag{3.10}$$

the sum in (3.8) is well defined and (3.2a) is not needed. There is an analogous extension for (1.7) and (1.9). Define the discounted perturbed test function

$$\widetilde{v}_n^d(\theta) = v_n(\theta) + \delta v_n^d(\theta).$$

Theorem 3.2. *Assume* (A3.1) *and* (A3.3), *except that the discounted forms* (3.8) *and* (3.9) *are used in lieu of* $\delta v_n(\theta)$ *and* $\delta N_n(\theta)$, *respectively, and let* $\delta v_n^d(\theta) \to 0$ *with probability one. Then* (A1.3) *holds.*

Proof. Proceeding as in the proof of Theorem 3.1 but using the "discounted" forms $\delta v_n^d(\theta)$ and $\delta N_n^d(\theta)$, we have

$$\begin{aligned} \widetilde{v}_{n+1}^d(\theta) - \widetilde{v}_n^d(\theta) &= \left[v_{n+1}(\theta) - v_n(\theta)\right] + \left[\delta v_{n+1}^d(\theta) - \delta v_n^d(\theta)\right] \\ &= \sum_{i=n+1}^{\infty} \epsilon_i \Pi(n+2,i) E_{n+1} \left[g_i(\theta,\xi_i) - \bar{g}(\theta)\right] \\ &\quad - \sum_{i=n+1}^{\infty} \epsilon_i \Pi(n+1,i) E_n \left[g_i(\theta,\xi_i) - \bar{g}(\theta)\right]. \end{aligned}$$

The right-hand side can be written as

$$\delta N_n^d(\theta) + \epsilon_{n+1}\gamma_{n+1},$$

where

$$\gamma_{n+1} = \sum_{i=n+1}^{\infty} \epsilon_i \Pi(n+2,i) E_n \left[g_i(x,\xi_i) - \bar{g}(x)\right] = E_n \delta v_{n+1}^d(\theta).$$

Since $\delta v_n^d(\theta) \to 0$ with probability one by the hypothesis, for the purposes of the proof and without loss of generality we can suppose that $\delta v_n^d(\theta)$ is bounded by some real number for large n and all ω. Then $E_n \delta v_n^d(\theta) \to 0$ with probability one, and the asymptotic rate of change of $\sum_{i=0}^{m(t)-1} \epsilon_i \gamma_i$ is zero with probability one. Then, using (A3.3) for $\delta N_n^d(\theta)$, the proof follows as in Theorem 3.1. □

The Poisson equation and an alternative perturbation. In lieu of the $\delta v_n(\theta)$ defined in (A3.2) or (3.8), one can use the perturbation defined by

$$\delta \widetilde{v}_n(\theta) = \epsilon_n q_n(\theta), \tag{3.11a}$$

where

$$q_n(\theta) = \sum_{i=n}^{\infty} E_n \left[g_i(\theta, \xi_i) - \bar{g}(\theta) \right]. \tag{3.11b}$$

Note that the step size ϵ_n in (3.11a) is outside of the sum. This form was used heavily in [16, 127] and is advantageous if the ϵ_n are random and $\mathcal{F}_n$-measurable. With the form (3.11) used in Theorem 3.1, $\delta N_n(\theta)$ is replaced by

$$\begin{aligned} \epsilon_{n+1} \sum_{i=n+1}^{\infty} & (E_{n+1} - E_n) \left[g_i(\theta, \xi_i) - \bar{g}(\theta) \right] \\ & + (\epsilon_{n+1} - \epsilon_n) \sum_{i=n+1}^{\infty} E_n \left[g_i(\theta, \xi_i) - \bar{g}(\theta) \right] \\ & \equiv \delta \widetilde{N}_n(\theta) + \epsilon_n \chi_n(\theta), \end{aligned} \tag{3.12}$$

where

$$\chi_n(\theta) = \frac{(\epsilon_{n+1} - \epsilon_n)}{\epsilon_n} \sum_{i=n+1}^{\infty} E_n \left[g_i(\theta, \xi_i) - \bar{g}(\theta) \right],$$

and

$$\delta \widetilde{N}_n(\theta) = \epsilon_{n+1} \sum_{i=n+1}^{\infty} (E_{n+1} - E_n) \left[g_i(\theta, \xi_i) - \bar{g}_i(\theta) \right]$$

is a martingale difference. The proof of Theorem 3.1 yields the following result.

Theorem 3.3. *Assume* (A1.8), (A3.1), *and* (A3.2) *with* ϵ_i *replaced by* ϵ_n *in* (3.2). *Suppose that* $\chi_n(\theta)$ *and* $\delta \widetilde{v}_n(\theta)$ *go to zero with probability one and that the asymptotic rate of change of*

$$\sum_{i=0}^{m(t)-1} \delta \widetilde{N}_i(\theta) \tag{3.13}$$

is zero with probability one. Then (A1.3) *holds.*

Comments on the discounted perturbation. Since $\chi_n(\theta)$ goes to zero with probability one, the asymptotic rate of change of $\sum_{i=0}^{m(t)-1} \epsilon_i \chi_i(\theta)$ is zero with probability one. The conditions needed with the use of the perturbation (3.11a) are not very different from those defined in (A3.2). A discounted form of the perturbation (3.11a) can also be used, namely the resolvent

$$\delta \tilde{v}_n^d(\theta) = \epsilon_n \sum_{i=n}^{\infty} (1 - \epsilon_n)^{i-n} E_n \left[g_i(\theta, \xi_i) - \bar{g}(\theta)\right]. \tag{3.14}$$

A possible advantage of the perturbation form (3.11) is that ϵ_n is outside of the summation.

Remark on the Poisson equation. Suppose that $g_n(\cdot)$ does not depend on n (write it as $g(\cdot)$) and that $\{\xi_n\}$ is a Markov chain with a time-independent one-step transition function. Now, let E_{ξ_n} denote the expectation given ξ_n. Suppose that $E_{\xi_0}|g(\theta, \xi_n)| < \infty$ for each n and initial condition ξ_0, and that there is a measurable function $q(\theta, \cdot)$ such that

$$\sum_{i=0}^{\infty} E_{\xi_0} \left[g(\theta, \xi_i) - \bar{g}(\theta)\right] = q(\theta, \xi_0),$$

where the sum is absolutely convergent for each value of ξ_0. If the order of summation and integration can be interchanged as

$$\sum_{i=1}^{\infty} E_{\xi_0} E\left[g(\theta, \xi_i) - \bar{g}(\theta) \middle| \xi_1\right] = E_{\xi_0} \sum_{i=1}^{\infty} E_{\xi_1} \left[g(\theta, \xi_i) - \bar{g}(\theta)\right],$$

then $q(\theta, \cdot)$ satisfies the *Poisson equation*

$$q(\theta, \xi_0) = \left[g(\theta, \xi_0) - \bar{g}(\theta)\right] + E_{\xi_0} q(\theta, \xi_1), \tag{3.15}$$

that was much used in the analysis in [16]. Of course, (3.11b) is always the solution to (3.15) and can be used even without the Markov assumption. This extra generality is important, because of the need for robustness in the algorithms and conditions. A practical procedure that converges for a particular Markov noise model should also converge for a non-Markov noise model, if the new model is "close" to the Markov model, and similarly if the basic model is non-Markov. The use of the perturbation forms (3.2b), (3.8), (3.11), or (3.14) facilitates dealing with this robustness concern. The equation (3.15) is useful mainly when information on its analytic form is available and can be exploited to get information concerning the solution.

6.4 Examples Using State Perturbation

In this section, the state perturbation idea of the last section will be applied to some concrete cases, providing additional sufficient conditions for (A3.1) that can be used even if $\epsilon_n \to 0$ very slowly. We will concentrate on ϵ_n of the order $O(1/\sqrt{n})$ or larger, but the methods can be useful no matter how fast $\epsilon_n \to 0$. We write $\psi_n = g_n(\theta, \xi_n) - \bar{g}(\theta)$, and the θ-argument will generally be dropped. If the ϵ_n sequence converges to zero no faster than $1/\sqrt{n}$, then the strong laws of large numbers are more difficult to apply via (2.4), since $\epsilon_n \sum_{i=1}^{n} \psi_i$ will not generally converge with probability one. For example, if $\epsilon_n = 1/\sqrt{n}$, we get a central limit theorem. But the state perturbation method of Section 3 remains useful and some examples will now be given. The key to the value of the state perturbation method is that the conditional expectation $E_n \psi_i$ often tends to zero quite rapidly as $i - n \to \infty$. In what follows, the ϵ_i are assumed to be not random.

Example 1. For an integer $M \geq 0$, the sequence ψ_n is said to be M-dependent if for each n the sets $\{\psi_i, i \leq n\}$ and $\{\psi_i, i \geq M+n\}$ are independent. Suppose that $\{\psi_n\}$ is M-dependent, $\sup_n E|\psi_n| < \infty$, $E\psi_n = 0$, and that

$$\sup_n \sup_{i \geq n} \frac{\epsilon_i}{\epsilon_n} < \infty. \tag{4.1}$$

By the M-dependence, the sums in (3.2) and (3.3) contain at most $M+1$ terms and there are random variables ρ_n and γ_n where the ρ_n are martingale differences and

$$\delta N_n(\theta) = \epsilon_n \rho_n, \quad \delta v_n(\theta) = \epsilon_n \gamma_n. \tag{4.2}$$

If $\{\psi_n\}$ is bounded, then so is $\{\rho_n, \gamma_n\}$. More generally, suppose that $E|\psi_n|^{2k} \leq C_k$ for all n, where $C_k \leq K_1^k k!$ for a finite K_1. Using the inequality [53, p. 45],

$$\left| \frac{1}{M+1} \sum_{i=1}^{M+1} x_i \right|^{2k} \leq \frac{1}{M+1} \sum_{i=1}^{M+1} |x_i|^{2k},$$

we have

$$E|\rho_n|^{2k} + E|\gamma_n|^{2k} \leq 2[(M+1)]^{2k} C_k.$$

Let

$$\epsilon_n \leq \alpha_n / \log n, \quad \alpha_n \to 0. \tag{4.3}$$

Now, split the process into two, with each being composed of alternating blocks (of the original) of length M; the blocks of each set are mutually independent. Now, using this independence, we see that in view of Example 2 in Section 5.3, the asymptotic rate of change of $\sum_{i=0}^{m(t)-1} \delta N_i$ is zero with probability one.

The proof that $\epsilon_n \gamma_n \to 0$ with probability one is easier and will now be given. It is sufficient to show that, for each $k \leq M$, $\epsilon_{n+k} |E_n \psi_{n+k}| \to 0$ with probability one as $n \to \infty$. Let $i = n + k$ and $\delta > 0$. Then, by the Chebyshev's inequality,

$$P\{\epsilon_i |E_n \psi_i| \geq \delta\} \leq e^{-\delta/\epsilon_i} E e^{|E_n \psi_i|}.$$

Under (4.3), the right side is summable over i for each δ. The Borel–Cantelli Lemma then yields that $\epsilon_{n+k}|E_n\psi_{n+k}|$ exceeds δ only finitely often with probability one. The conditions $\sup_n E|\psi_n|^2 < \infty$ and $\sum_n \epsilon_n^2 < \infty$ yield the same results.

Example 2. Let $q(\cdot)$ be a bounded and continuous real-valued function. Let $\{\xi_n\}$ be an aperiodic irreducible Markov chain with values in a compact metric state space, having a time-independent transition function satisfying Doeblin's condition [178, 203]. Then there is a unique invariant measure $\mu(\cdot)$. Let $\psi_n = q(\xi_n) - \int q(\xi)\mu(d\xi)$. Then $E_n\psi_{n+i} \to 0$ geometrically, and uniformly in ξ_n, as $i \to \infty$ [178, 203], which implies that the representations in (4.2) hold with the $\{\rho_n\}$ and $\{\gamma_n\}$ being bounded. Suppose that $g_n(\cdot)$ does not depend on n and is continuous. Then, (A3.2) holds if $\epsilon_n \to 0$, and (A3.3) holds under (4.1) and (4.3).

Example 3. This is an extension of Example 1. Suppose that $\{\psi_n\}$ is bounded and strongly mixing in the sense that there is a real $0 \le K_1$ such that

$$E\psi_n = 0 \text{ and } \sum_{i=n}^{\infty} |E_n\psi_i| \le K_1$$

for all n with probability one. Then (A3.2) holds if $\epsilon_n \to 0$, and (A3.3) holds under (4.1) and (4.3).

Example 4. Let $\{\psi_n\}$ be a sequence of real-valued and zero-mean Gaussian random variables with uniformly bounded variances. Then the conditional expectation $E[\psi_{n+m}|\psi_i, i \le n]$ is also Gaussian. If the correlation function $E\psi_n\psi_{n+m}$ goes to zero fast enough as $m \to \infty$, then (A3.2) and (A3.3) hold under (4.1) and (4.3). To prove such an assertion, it is convenient to have an explicit representation of the ψ_n in terms of a common sequence of independent random variables. Let $\{w_n\}$ be a sequence of mutually independent Gaussian random variables, with uniformly bounded variances and mean zero. Let the real numbers $\{a_n, n \ge 0\}$ satisfy

$$\sum_n |a_n| < \infty, \quad \sum_i \left[\sum_{j=i}^{\infty} |a_j| \right]^2 < \infty. \tag{4.4}$$

Condition (4.4) holds if $a_n = O(1/n^{3/2+\delta})$ for some $\delta > 0$. Suppose that ψ_i is defined by

$$\psi_i = \sum_{j=0}^{\infty} a_j w_{i-j} = \sum_{j=-\infty}^{i} a_{i-j} w_j. \tag{4.5}$$

Let E_n denote conditioning on $\{w_i, i \le n\}$. Then, for $i \ge n$,

$$E_n\psi_i = a_{i-n}w_n + a_{i-n+1}w_{n-1} + \cdots,$$

and the definition (3.2) yields

$$\begin{aligned}\delta v_n = \sum_{i=n}^{\infty} \epsilon_i E_n \psi_i &= w_n \left[\epsilon_n a_0 + \epsilon_{n+1} a_1 + \cdots\right] \\ &\quad + w_{n-1}\left[\epsilon_n a_1 + \epsilon_{n+1} a_2 + \cdots\right] + \cdots \\ &= \epsilon_n w_n \bar{\phi}_{n0} + \epsilon_n w_{n-1} \bar{\phi}_{n1} + \cdots,\end{aligned}$$

where

$$\bar{\phi}_{ni} = a_i + \frac{\epsilon_{n+1}}{\epsilon_n} a_{i+1} + \frac{\epsilon_{n+2}}{\epsilon_n} a_{i+2} + \cdots.$$

Since, by (4.1) and (4.4)

$$\sum_i \sup_n \bar{\phi}_{ni}^2 < \infty,$$

the sum $\delta v_n = \sum_{i=n}^{\infty} \epsilon_i E_n \psi_i$ is a zero mean Gaussian random variable with variance $\epsilon_n^2 A_n$, where $\{A_n\}$ is a bounded sequence. Now, Chebyshev's inequality leads to

$$P\{\delta v_n \geq \alpha\} \leq 2P\left\{e^{\delta v_n/\epsilon_n} \geq e^{\alpha/\epsilon_n}\right\} \leq 2e^{A_n/2} e^{-\alpha/\epsilon_n}.$$

Then (4.3) implies that the right side is summable. Hence, the Borel–Cantelli Lemma and (4.3) imply that $\delta v_n \to 0$ with probability one.

In this example,

$$E_{n+1}\psi_i - E_n \psi_i = a_{i-n-1} w_{n+1}.$$

Hence

$$\delta N_n = \sum_{i=n+1}^{\infty} \epsilon_i \left[E_{n+1} - E_n\right] \psi_i = \sum_{i=n+1}^{\infty} \epsilon_i a_{i-n-1} w_{n+1}.$$

Thus, δN_n is also Gaussian with mean zero and variance $O(\epsilon_n^2)$. Now, the argument used for Example 1 of Subsection 5.3.2 yields (A3.3).

6.5 Kiefer–Wolfowitz Algorithms

The Basic Algorithm. Following the approach taken in Section 5.3.3, the algorithm will be written in a general way that covers many special forms. Let us use algorithm (1.1) with the observation written as $Y_n = (Y_n^- - Y_n^+)/(2c_n)$. Suppose that there are random variables $\xi_n = (\xi_n^+, \xi_n^-)$, measurable functions $\gamma_n^{\pm}(\cdot)$, and nondecreasing σ-algebras $\mathcal{F}_n$, where $\mathcal{F}_n$ measures at least $\{\xi_i^{\pm}, Y_{i-1}, \theta_i, i \leq n\}$ such that

$$E_n Y_n^{\pm} = \gamma_n^{\pm}(\theta_n, c_n, \xi_n^{\pm}), \tag{5.1}$$

where E_n denotes the conditional expectation with respect to $\mathcal{F}_n$. Suppose there are measurable functions $g_n(\cdot)$ and random variables β_n such that

$$\frac{\gamma_n^-(\theta_n, c_n, \xi_n^-) - \gamma_n^+(\theta_n, c_n, \xi_n^+)}{2c_n} = g_n(\theta_n, \xi_n) + \beta_n. \tag{5.2}$$

The β_n will represent the finite difference bias, as usual. Redefine the martingale difference

$$\delta M_n = (Y_n^- - E_n Y_n^-) - (Y_n^+ - E_n Y_n^+),$$

and write the algorithm in the expanded form

$$\theta_{n+1} = \theta_n + \epsilon_n g_n(\theta_n, \xi_n) + \epsilon_n \beta_n + \epsilon_n \frac{\delta M_n}{2c_n} + \epsilon_n Z_n. \tag{5.3}$$

A more general form which adds a term $\widetilde{g}_n(\theta_n, \xi_n)/(2c_n)$ to (5.3) is dealt with in Theorem 8.3.3, where the development is via the simpler weak convergence approach. Recall the discussion concerning variance reduction and the use of common random numbers in connection with (1.2.6).

The proof of the following theorem is essentially that of Theorem 1.1 and the details are left to the reader.

Theorem 5.1. *Assume the conditions in the first paragraph of this section and the assumptions of Theorem* 1.1 *with* $\delta M_n/(2c_n)$ *replacing* δM_n. *Then the conclusions of Theorem* 1.1 *hold.*

The random directions algorithm. Recall the discussion of the random directions algorithm in connection with (1.2.9) and in Section 5.6. The development of the correlated noise case is virtually the same as that of the standard Kiefer–Wolfowitz algorithm, and is covered by it if β_n is defined appropriately. Use the notation in Theorem 5.1 and let the random direction vectors d_n be uniformly bounded. Let $\mathcal{F}_n^d$ be a sequence of nondecreasing σ-algebras such that $\mathcal{F}_n^d$ measures at least $\{\theta_i, d_i, Y_{i-1}, i \leq n\}$ and let there be measurable functions $\gamma_n(\cdot)$ and random variables $\xi_n^\pm$ such that

$$E_{\mathcal{F}_n^d} Y_n^\pm = \gamma_n(\theta_n \pm c_n d_n, \xi_n^\pm). \tag{5.4}$$

Suppose there are measurable functions $g_n(\cdot)$ and random variables β_n such that

$$\frac{\gamma_n(\theta_n - c_n d_n, \xi_n^-) - \gamma_n(\theta_n + c_n d_n, \xi_n^+)}{2c_n} = d_n' g_n(\theta_n, \xi_n) + \beta_n. \tag{5.5}$$

Redefine

$$\delta M_n = (Y_n^- - E_{\mathcal{F}_n^d} Y_n^-) - (Y_n^+ - E_{\mathcal{F}_n^d} Y_n^+). \tag{5.6}$$

and write the algorithm in the expanded form as

$$\begin{aligned}\theta_{n+1} &= \theta_n + \epsilon_n d_n d_n' g_n(\theta_n, \xi_n) + \epsilon_n d_n \beta_n + \epsilon_n d_n \frac{\delta M_n}{2c_n} + \epsilon_n Z_n \\ &= \theta_n + \epsilon_n g_n(\theta_n, \xi_n) + \epsilon_n d_n \beta_n + \epsilon_n d_n \frac{\delta M_n}{2c_n} \\ &\quad + \epsilon_n \left(d_n d_n' - I\right) g_n(\theta_n, \xi_n) + \epsilon_n Z_n.\end{aligned} \tag{5.7}$$

The proof of the following theorem is essentially that of Theorem 1.1.

Theorem 5.2. *Assume the preceding conditions and the assumptions of Theorem* 1.1 *with* $d_n \delta M_n/(2c_n)$ *replacing* δM_n, δM_n *defined by* (5.6), *and* $d_n \beta_n$ *replacing* β_n. *Suppose that for each* θ *the asymptotic rate of change of*

$$\sum_{n=0}^{m(t)-1} \epsilon_n \left(d_n d_n' - I\right) g_n(\theta, \xi_n) \tag{5.8}$$

is zero with probability one. Then the conclusions of Theorem 1.1 *hold.*

6.6 State-Dependent Noise

Perturbation methods were shown to be a very useful tool for proving stability in Chapters 4 and 5 and for verifying the rate of change conditions such as (A1.3) in Sections 3 and 4 of this chapter. In [127], closely related methods were developed and shown to be powerful tools for getting quite general results on convergence and approximation for stochastic processes. In this section, we briefly outline an approach based on the ideas developed in [127] for the types of "Markov" state-dependent noise introduced in the examples of Sections 2.3–2.5. Few details will be given and only the main steps are listed, so the reader can get a flavor of what needs to be done. The weak convergence approach and its required conditions are much simpler for this model and will be discussed in Chapter 8 for both the Markov and non-Markov state dependence. In the present context, by *noise* we mean the "conditional memory" random variable ξ_n used in the representations (1.2) and (1.3), and not the martingale difference term δM_n.

Assume algorithm (1.1), $\sup_n E|Y_n| < \infty$, and the representations (1.2) and (1.3). Define $\mathcal{F}_n$ as in Subsection 1.1. Up to this point in this chapter it was implicitly supposed that

$$P\left\{\xi_{n+1} \in \cdot \mid \xi_i, \theta_i, i \leq n\right\} = P\left\{\xi_{n+1} \in \cdot \mid \xi_i, i \leq n\right\}. \tag{6.1}$$

While this condition was never explicitly stated, conditions such as (A1.3) might not be meaningful unless (6.1) holds. Equation (6.1) says that the noise $\{\xi_n\}$ is *exogenous*; in other words, the law of the probabilistic evolution of the noise is not influenced by the evolution of the iterate sequence.

In the classes of noise models introduced in Sections 2.3–2.5, θ_n plays a crucial role in evaluating the left side of (6.1), and (6.1) does not hold. Indeed, if the algorithms in these sections were written in the expanded form (1.3) where the "memory" random variable ξ_n is used, we would have the Markov dependence

$$P\{\xi_{n+1} \in \cdot \mid \xi_i, \theta_i, i \leq n\} = P\{\xi_{n+1} \in \cdot \mid \xi_n, \theta_n\},$$

where θ_n cannot be eliminated on the right-hand side. [More detail is given in the examples at the end of the section.] This structure motivates the following general model, which was developed in [127, 143] for the weak convergence case. References [16, 57] contain an extensive development of the state perturbation ideas for probability one convergence.

For each θ, let $P(\cdot, \cdot|\theta)$ be a Markov transition function parameterized by θ such that $P(\cdot, A|\cdot)$ is Borel measurable for each Borel set A in the range space Ξ of ξ_n. Suppose that the law of evolution of the noise satisfies

$$P\{\xi_{n+1} \in \cdot \mid \xi_i, \theta_i, i \leq n\} = P(\xi_n, \cdot|\theta_n), \tag{6.2}$$

where $P(\xi, \cdot|\theta)$ denotes the one-step transition probability with starting point ξ and parameterized by θ.

Definition: Markov state-dependent noise. If (6.2) holds, the noise process $\{\xi_n\}$ is called *Markov state-dependent* or just state-dependent, where the state is the iterate. Owing to this dependence, (A1.3) cannot be used as stated, and the analysis becomes more difficult.

The fixed-θ process. For each fixed value of θ, $P(\cdot, \cdot|\theta)$ is the transition function of a Markov chain; we let $\xi_n(\theta)$ denote the random variables of that chain. This so-called *fixed-θ process* plays a fundamental role in the analysis and will be used again in Chapter 8. We expect that the probability law of this chain for given θ is close to the probability law of the true noise $\{\xi_n\}$ if θ_n varies slowly around θ, and hence that the mean ODE can be obtained in terms of this fixed-θ chain. This turns out to be the case. Of special interest is the fixed-θ processes $\{\xi_i(\theta_n), i \geq n\}$, defined for each n with initial condition $\xi_n(\theta_n) = \xi_n$. Thus, this process starts at value ξ_n at time n and then evolves as if the parameter were fixed at θ_n forever after.

Assumptions and the state perturbation.

(A6.1) There is a continuous function $\bar{g}(\cdot)$ such that for $\theta \in H$, the expression

$$v_n(\theta, \xi_n) = \sum_{i=n}^{\infty} \epsilon_i E_n \left[g_i(\theta, \xi_i(\theta)) - \bar{g}(\theta)\right] \tag{6.3}$$

is well defined when the initial condition for $\{\xi_i, i \geq n\}$ is $\xi_n(\theta) = \xi_n$, and

$$v_n(\theta_n, \xi_n) \to 0 \quad \text{w.p.1.} \tag{6.4}$$

The conditioning E_n in (6.3) is on $\xi_n(\theta) = \xi_n$ and the value of θ. If the ϵ_n are random or if desired, use one of the alternative perturbations (3.11) or (3.14), and assume (A1.8) and that $\epsilon_{n+1} - \epsilon_n = o(\epsilon_n)$.

The main conditions needed for convergence of $\{\theta_n\}$ are given in terms of $v_n(\theta, \xi)$. Define α_n and the $\mathcal{F}_n$-martingale difference δN_n by

$$\alpha_n = v_{n+1}(\theta_{n+1}, \xi_{n+1}) - v_{n+1}(\theta_n, \xi_{n+1}), \tag{6.5a}$$

$$\delta N_n = v_{n+1}(\theta_n, \xi_{n+1}) - E_n v_{n+1}(\theta_n, \xi_{n+1}). \tag{6.5b}$$

(A6.2) The asymptotic rates of change of the processes

$$A^0(t) = \sum_{i=0}^{m(t)-1} \alpha_i, \tag{6.6a}$$

$$N^0(t) = \sum_{i=0}^{m(t)-1} \delta N_i \tag{6.6b}$$

are zero with probability one.

If $v_n = O(\epsilon_n)$ and $v_n(\cdot, \xi)$ is Lipschitz continuous uniformly in n and ξ, then we would expect that $\alpha_n = O(\epsilon_n^2)$ and (6.6a) has zero asymptotic rate of change. The following localized forms will be used for the unconstrained algorithm in Chapter 7.

(A6.2′) (A6.1) holds if, for each $K < \infty$, α_i (resp., δN_i) is replaced by $\alpha_i I_{\{|\theta_i| \le K\}}$ (resp., by $\delta N_i I_{\{|\theta_i| \le K\}}$).

Theorem 6.1. *Assume* (5.1.1), (A1.1), (A1.2), (A1.4), (6.2), (A6.1), (A6.2), *and any of the constraint set conditions* (A4.3.1), (A4.3.2), *or* (A4.3.3). *Then the conclusions of Theorem* 1.1 *hold. Under the additional conditions of Theorem* 5.2.5 *on the mean ODE, for almost all* ω, *the limit is contained in the set of chain recurrent points.*

Comment on the proof. The proof uses a recursion for the perturbed state $\widetilde{\theta}_n = \theta_n + v_n(\theta_n, \xi_n)$ instead of for θ_n. By the definition of $v_n(\theta_n, \xi_n)$, we have the "perturbed-state algorithm"

$$\begin{aligned} \widetilde{\theta}_{n+1} = \; & \widetilde{\theta}_n + \epsilon_n \left[g_n(\theta_n, \xi_n) + \beta_n + \delta M_n + Z_n \right] \\ & + \left[v_{n+1}(\theta_{n+1}, \xi_{n+1}) - v_n(\theta_n, \xi_n) \right]. \end{aligned} \tag{6.7}$$

The last term in (6.7) needs to be expanded so the conditions can be applied. By the definitions,

$$
\begin{aligned}
v_{n+1}(\theta_{n+1},\xi_{n+1}) - v_n(\theta_n,\xi_n) &= -\epsilon_n\left[g_n(\theta_n,\xi_n) - \bar{g}(\theta_n)\right] \\
&\quad + \sum_{i=n+1}^{\infty} \epsilon_i E_{n+1}\left[g_i(\theta_{n+1},\xi_i(\theta_{n+1})) - \bar{g}(\theta_{n+1})\right] \\
&\quad - \sum_{i=n+1}^{\infty} \epsilon_i E_n\left[g_i(\theta_n,\xi_i(\theta_n)) - \bar{g}(\theta_n)\right] \\
&= -\epsilon_n\left[g_n(\theta_n,\xi_n) - \bar{g}(\theta_n)\right] + \alpha_n + \delta N_n.
\end{aligned}
\tag{6.8}
$$

Using (6.8), rewrite the perturbed state equation (6.7) as

$$
\widetilde{\theta}_{n+1} = \widetilde{\theta}_n + \epsilon_n\left[\bar{g}(\theta_n) + \beta_n + \delta M_n + Z_n\right] + \alpha_n + \delta N_n. \tag{6.9}
$$

The proof is completed by using the conditions with the method of Theorem 1.1.

On the role of the perturbation. The form (6.9) illustrates the value of the perturbed state approach. The term $v_n(\theta_n,\xi_n)$ can be shown to go to zero with probability one under reasonable "ergodic" hypotheses. The use of the perturbation removes $g_n(\theta_n,\xi_n)$ and replaces it by the average $\bar{g}(\theta_n)$. An "error" α_n (due to the replacement of θ_{n+1} by θ_n in $v_{n+1}(\theta_{n+1})$) and a new martingale difference term δN_n were introduced in the process. Owing to the rate of change conditions in (6.6), the sequence $\{\alpha_n,\delta N_n\}$ has no effect on the limit. In one way or another, in any particular application, the verification of these rate of change conditions and (6.4) is the focus of the analysis.

If $v_n(\theta,\xi_n)$ is redefined as in (3.11) such that the step sizes ϵ_i inside the summation are replaced by ϵ_n outside the summation and $g_n(\cdot)$ does not depend on n, then $v_n(\theta,\xi_n)$ becomes ϵ_n times the solution to the Poisson equation (see the remark on the Poisson equation below (3.14)), where $\xi_n(\theta)=\xi$, namely,

$$
\epsilon_n q_n(\theta,\xi) = \epsilon_n \sum_{i=n}^{\infty} E_n\left[g(\theta,\xi_i(\theta)) - \bar{g}(\theta)\right]. \tag{6.10}
$$

Then the condition given in (6.6a) can be written in terms of the smoothness of the solution of that equation. Perhaps more often, verification is done via conditions on the rate of convergence of the n-step transition probabilities to their limit and on their smoothness in θ, and one works directly with the form (6.3) rather than with the Poisson equation. The discounted forms of the perturbations that are analogous to (3.8) and (3.11) can also be used.

Examples. The conditions might look formidable. But they are often quite easy to verify. This will be illustrated via two examples from Chapter 2.

Consider the routing Example of Section 2.3, where $g(\theta, \xi)$ is defined above (2.3.4). The ξ_n is just the "occupancy" process. The function $g_n(\theta, \xi)$ does not depend on n. It is a simple polynomial in θ and is continuous in (θ, ξ), since ξ is finite-valued. The function $\bar{g}(\cdot)$ is defined below (2.3.4). The summands in the definition of $v_n(\theta, \xi)$ converge to zero at a geometric rate and all of the assumptions hold. In particular, δN_n is bounded and (A5.3.3) holds. Hence, under the minimal condition (A5.3.2), and $\epsilon_n \le a_n/\log n$ for any $a_n \to 0$, the process in (6.6b) has zero asymptotic rate of change. The α_n in (6.5a) is $O(\epsilon_n^2)$; hence $A^0(\cdot)$ also has zero asymptotic rate of change.

Now, consider the parametric optimization example of Subsection 2.5.3, where X_n is a finite-state controlled chain, and $\xi_n = (X_n, C_n)$. Suppose that $k_\theta(i, j, \theta), k(i, j, \theta)$ and $L(i, j, \theta)$ are bounded and continuous in θ for each value of (i, j). Let the chain be ergodic for each θ. Then $g_n(\theta_n, \xi_n)$ is the expectation of the coefficient of ϵ_n, ϵ_n', conditioned on X_n, C_n, and $g_n(\cdot)$ does not depend on n. The summands in the definition of $v_n(\theta, \xi)$ converge to zero at a geometric rate and, as for the example of the previous paragraph, all of the assumptions hold.

6.7 Stability and Stability ODE Methods for Probability One Convergence

Introduction. The stability proofs using ODE-type methods combined with local stability of the mean ODE, as in Theorems 5.4.2 and 5.4.3, are equally applicable to the correlated noise case and yield a versatile combination of tools. Recall the main idea: A stability theorem would first be used to show that the iterate sequence enters some compact set infinitely often with probability one. Then the ODE method is applied to the sequence of paths starting at the times that they enter that compact recurrence set, and it is proved that the limits of these paths are solutions to the mean ODE. Then the stability of the mean ODE is used to show that the iterates not only eventually stay in the compact recurrent set but that they converge to the limit set of the ODE in that recurrent set.

The proof of the main result will be done in several steps. First, in Theorem 7.1 it is *assumed* that some compact set is recurrent, and that the mean ODE satisfies a stability condition. Then the ODE method is used to show that $\{\theta_n\}$ is bounded with probability one. From this point on, the proof is just that of Theorem 1.1, but without the constraint. Thus, θ_n converges to a bounded invariant set or a subset of chain recurrent points. The advantage of this stability-ODE method type of proof is that the conditions on the step sizes and noise are quite weak. For example, as in Theorems 5.4.2 and 5.4.3, the condition $\sum_{i=0}^{\infty} \epsilon_i^2 < \infty$ is not needed.

Theorem 7.2 shows that boundedness in probability implies the recurrence needed in Theorem 7.1. In Theorem 7.3 a perturbed Liapunov function, whose form is suggested by that of the perturbed test function used in Theorem 3.1, is used to get a sufficient condition for recurrence. If an arbitrarily small neighborhood of a stable point $\bar{\theta}$ is visited infinitely often with probability one, then we need only verify the conditions near $\bar{\theta}$.

Note on soft constraints. Note that the idea of "soft constraint," as used in Chapter 5, is equally applicable here. Although we will not explicitly state a theorem, the reader should have no trouble in using the stability methods to that end.

Assumptions. We use the unconstrained algorithm

$$\theta_{n+1} = \theta_n + \epsilon_n Y_n$$

and assume (A1.1′) and the following conditions.

(A7.1) There are random variables $\{\xi_n\}$ and $\{\beta_n\}$, a sequence of nondecreasing σ-algebras $\{\mathcal{F}_n\}$, where $\mathcal{F}_n$ measures at least $\{\theta_i, \xi_i, Y_{i-1}, i \le n\}$, and measurable functions $g_n(\cdot)$, such that on each bounded θ_n-set,

$$E_n Y_n = g_n(\theta_n, \xi_n) + \beta_n,$$

where E_n is the expectation conditioned on $\mathcal{F}_n$.

By (A.1.1′) and (A7.1),

$$\delta M_n I_{\{|\theta_n| \le K\}} = [Y_n - E_n Y_n] I_{\{|\theta_n| \le K\}}$$

is an $\mathcal{F}_n$-martingale difference for each real K. Since we always assume that $|\theta_0| < \infty$ with probability one, an iterative use of (A1.1′) implies that Y_n and θ_n are finite with probability one for each n. Thus, we can write

$$\theta_{n+1} = \theta_n + \epsilon_n g_n(\theta_n, \xi_n) + \epsilon_n \delta M_n + \epsilon_n \beta_n \tag{7.1}$$

without ambiguity, even if we do not know *a priori* that $E|Y_n| < \infty$ for each n.

Theorem 7.1. *Assume the unconstrained algorithm* (7.1) *and conditions* (5.1.1), (A.1.1′), (A1.2), (A1.3), (A7.1), *and* (A1.4′)–(A1.7′). *For each* $1 > \rho > 0$, *let there be a compact set* R_ρ *such that* $\theta_n \in R_\rho$ *infinitely often with probability at least* ρ. *Let* $V(\cdot)$ *be a continuously differentiable Liapunov function for the ODE*

$$\dot{\theta} = \bar{g}(\theta) \tag{7.2}$$

in the following sense: $V(\theta) \ge 0$ *for each* θ, $V(\theta) \to \infty$ *as* $|\theta| \to \infty$, *and there are* $0 < \lambda_0 < \infty$ *and* $\lambda_1 > 0$ *such that if* $\theta(\cdot)$ *solves* (7.2) *and* $V(\theta(t)) \ge \lambda_0$, *then* $\dot{V}(\theta(t)) \le -\lambda_1$. *Then, with probability one,* θ_n *tends to the largest bounded invariant set of* (7.2) *in* $Q_{\lambda_0} = \{x : V(x) \le \lambda_0\}$. *Suppose that* (A5.2.6) *holds. Then, with probability one,* $\theta_n(\omega)$ *converges to a unique* S_i.

Under the additional conditions of Theorem 5.2.5, *for almost all* ω, $\theta_n(\omega)$ *converges to a set of chain recurrent points in the invariant set. If we use the conditions of Theorem* 6.1, *but localized in the same way, then the above conclusions hold for the state-dependent noise case.*

Proof. If it were shown that

$$\limsup_n |\theta_n| < \infty \text{ with probability one,} \tag{7.3}$$

then the proof for the constrained algorithm in Theorem 1.1 (but omitting the reflection terms) would yield the theorem. An "ODE method" will be used to prove (7.3).

Given $\rho > 0$, there is an $\alpha_\rho > \lambda_0$ and an ω-set Ω_ρ such that $P\{\Omega_\rho\} \ge \rho$ and for $\omega \in \Omega_\rho$, $\theta_n(\omega) \in Q^0_{\alpha_\rho}$ infinitely often. Let K in (A1.4′)–(A1.7′) be large enough that the K-sphere covers $Q_{2\alpha_\rho}$. Let N_0 denote the ω-set of nonconvergence to zero of the quantities in the brackets in (A1.4′)–(A1.7′) for the given value of K. Let D_{α_ρ} denote a countable dense set in $Q_{2\alpha_\rho}$, and N_1 the ω-set of nonconvergence to zero of the quantity in the brackets in (A1.3) for all $\theta \in D_{\alpha_\rho}$. Define $\Omega'_\rho = \Omega_\rho - (N_0 \cup N_1)$.

Henceforth, we work with a fixed $\omega \in \Omega'_\rho$ and show that

$$\limsup_n V(\theta_n(\omega)) \le \alpha_\rho, \tag{7.4}$$

which implies (7.3), since ρ is arbitrary. Suppose that (7.4) is false. Since by the hypotheses and the preceding paragraph, $\theta_n(\omega)$ enters $Q^0_{\alpha_\rho}$ infinitely often and (7.4) is false, there is a subsequence $\{n_k\}$ and a $\nu > \alpha_\rho$ such that, analogously to what was done in the proof of Theorem 5.4.2, the following holds: $n_k - 1$ is a last value of i for which $\theta_i(\omega) \in Q_{\alpha_\rho}$ prior to exiting Q_ν before it returns (if ever) to Q_{α_ρ}; $\{\theta_{n_k}(\omega)\}$ converges to some point on the boundary ∂Q_{α_ρ}; there is a $T > 0$ such that on $[0,T]$, $V(\theta^{n_k}(\omega,t)) \ge \alpha_\rho$ and $\{\theta^{n_k}(\omega,\cdot), t \le T\}$ is equicontinuous. By the argument of the proof in Theorem 1.1, the limit $\theta(\omega,\cdot)$ of any convergent subsequence on $[0,T]$ satisfies (7.2). But, by the stability hypothesis on (7.2), any solution of (7.2) with initial condition in Q_{α_ρ} remains strictly interior to Q_{α_ρ} for all $t > 0$. This contradicts the supposition that the $V(\theta^{n_k}(\omega,t)) \ge \alpha_\rho$ for $t \in (0,T]$. This contradiction implies (7.4). The details for the state-dependent noise case are similar and are omitted. □

Definition. Recall that a sequence of vector-valued random variables $\{X_n\}$ is said to be *bounded in probability* (*or tight*) if

$$\lim_{K\to\infty} \sup_n P\{|X_n| \ge K\} = 0. \tag{7.5}$$

The next theorem shows that tightness implies recurrence.

Theorem 7.2. *If*

$$\{\theta_n\} \text{ is bounded in probability} \tag{7.6}$$

and (5.1.1) *holds, the recurrence condition in Theorem* 7.1 *holds.*

Proof. Recurrence holds because (7.6) implies that there is a random $A < \infty$ with probability one such that $|\theta_n(\omega)| \le A(\omega)$ infinitely often. Equivalently, given any small $\rho > 0$, there is $\lambda_\rho < \infty$ such that

$$P\{|\theta_n| \le \lambda_\rho \text{ infinitely often}\} \ge 1 - \rho.$$

If this were not true, then for some $\rho > 0$ and all λ

$$P\{|\theta_n| \le \lambda \text{ only finitely often}\} \ge \rho,$$

which, with the arbitrariness of λ, contradicts (7.6). □

A sufficient condition for recurrence. We will use the perturbation $\delta v_n^d(\theta)$ defined in (3.8), (A1.1′), (A7.1), and the following conditions, where K_1 is a positive real number. The conditions are modeled on the needs of the example below. To simplify the development, the proof uses crude inequalities. A finer analysis under the conditions of a particular application should lead to weaker conditions.

(A7.2) $|V_\theta(\theta)|^2 \le K_1 \left(|V_\theta'(\theta)\bar{g}(\theta)| + 1\right)$.

(A7.3) $|V_\theta'(\theta)\bar{g}(\theta)| \le K_1 \left[V(\theta) + 1\right]$.

(A7.4) There are $\lambda_0 > 0$ and $\lambda_1 > 0$ such that for large n and $\theta_n \notin Q_{\lambda_0}^0 = \{\theta : V(\theta) < \lambda_0\}$,

$$V_\theta'(\theta_n)[\bar{g}(\theta_n) + \beta_n] \le -\lambda_1, \quad V_\theta'(\theta_n)\bar{g}(\theta_n) \le -2\lambda_1. \tag{7.7}$$

(A7.5) There are $k_n > 0$ such that $|Y_n|^2 \le k_n^2 \left[|V_\theta'(\theta)\bar{g}(\theta)| + 1\right]$, where $\epsilon_n k_n^2 \to 0$ with probability one.

(A7.6) There are $\widehat{k}_n > 0$ such that $\epsilon_n \widehat{k}_n^2 \to 0$ with probability one and

$$|\delta v_n^d(\theta)|^2 \le \epsilon_n^2 \widehat{k}_n^2 \left[|V_\theta'(\theta)\bar{g}(\theta)| + 1\right].$$

(A7.7) There are $1 \ge \gamma > 0$ and $\widetilde{k}_n > 0$ such that for all θ, y and n,

$$\left|\delta v_n^d(\theta + y) - \delta v_n^d(\theta)\right| \le \epsilon_n \widetilde{k}_n |y|^\gamma \tag{7.8}$$

and $\epsilon_n^{\gamma}\widetilde{k}_{n+1}\widehat{k}_n^{\gamma} \to 0$ with probability one.

Examples of the assumptions. The conditions (A7.6) and (A7.7) require a mixing property of the ξ_n. Now, recall the parameter identification problem in Section 3.1, with algorithm (1.1.18) used. Suppose that there are a vector S, a matrix $Q > 0$, and sequences $\{\widehat{\chi}_n, \widetilde{\chi}_n\}$ such that

$$\sum_{i=n}^{\infty} \epsilon_i \Pi(n+1,i) E_n \left[\phi_i \phi_i' - Q\right] = \epsilon_n \widehat{\chi}_n$$

and

$$\sum_{i=n}^{\infty} \epsilon_i \Pi(n+1,i) E_n \left[y_i \phi_i - S\right] = \epsilon_n \widetilde{\chi}_n.$$

Then $\bar{g}(\theta) = S - Q\theta$. Let $V(\theta) = |\theta|^2$. Then $v_n^d(\theta)$ takes the form

$$v_n^d(\theta) = \sum_{i=n}^{\infty} \epsilon_i \Pi(n+1,i) E_n \left[\phi_i y_i - \phi_i \phi_i' \theta - \bar{g}(\theta)\right].$$

If $\epsilon_n|\phi_n|^2 \to 0$, $\epsilon_n|\phi_n y_n| \to 0$, $\epsilon_n|\widehat{\chi}_n|^2 \to 0$, $\epsilon_n|\widetilde{\chi}_n|^2 \to 0$ with probability one, then (A7.2)–(A7.7) hold.

In the following discussion, $\bar{\theta} = 0$ is assumed for simplicity.

Theorem 7.3. *Assume algorithm* (7.1). *Let the real-valued function* $V(\cdot)$ *have continuous first and bounded mixed second partial derivatives with* $V(0) = 0$ *and* $V(\theta) \to \infty$ *as* $|\theta| \to \infty$. *Suppose that* $EV(\theta_0) < \infty$ *and that* (5.1.1), (4.1), (A1.1′), *and* (A7.1)–(A7.7) *hold. Then, for each positive integer* l,

$$\theta_n \in Q_{\lambda_0} \text{ infinitely often w.p.1,}$$

$$\lim_{K\to\infty} P\left\{ \sup_{t \in [l\text{th},\ (l+1)\text{st return to } Q_{\lambda_0}]} |\theta_n| \geq K \right\} = 0, \tag{7.9a}$$

and we can use $Q_{\lambda_0} = R_\rho$ *for all* $\rho \in (0,1]$ *in Theorem* 7.1. Finally,

$$\lim_{K\to\infty} \sup_n P\left\{|\theta_n| \geq K\right\} = 0. \tag{7.9b}$$

Proof. Suppose that $EV(\theta_n) < \infty$ and $E|v_n^d(\theta_n)| < \infty$. Then a truncated Taylor series expansion yields

$$E_n V(\theta_{n+1}) - V(\theta_n) = \epsilon_n V_\theta'(\theta_n)\left[g_n(\theta_n, \xi_n) + \beta_n\right] + O(\epsilon_n^2) E_n \left|Y_n\right|^2. \tag{7.10}$$

Since $\epsilon_n q_n \to 0$ with probability one, for q_n being either $k_n^2, \widehat{k}_n^2$ or $\widetilde{k}_n^2$, without loss of generality, we can assume that they are as small as we wish, and this will be assumed throughout the proof. Then by a slight modification of Lemma 5.4.1, it follows that, for purposes of the proof, we can suppose that $EV(\theta_{n+1}) < \infty$, $E|\delta v_{n+1}^d(\theta_{n+1})| < \infty$ for all n.

Define the perturbed Liapunov function

$$V_n(\theta_n) = V(\theta_n) + V_\theta'(\theta_n)\delta v_n^d(\theta_n).$$

Then

$$\begin{aligned} &E_n V_{n+1}(\theta_{n+1}) - V_n(\theta_n) \\ &\quad = \epsilon_n V_\theta'(\theta_n)\bar{g}(\theta_n) + \epsilon_n V_\theta'(\theta_n)\beta_n + O(\epsilon_n^2)E_n|Y_n|^2 \\ &\quad + E_n\left[V_\theta'(\theta_{n+1})\delta v_{n+1}^d(\theta_{n+1}) - V_\theta'(\theta_n)\delta v_{n+1}^d(\theta_n)\right] \\ &\quad + \epsilon_{n+1}V_\theta'(\theta_n)E_n\delta v_{n+1}^d(\theta_n). \end{aligned} \tag{7.11}$$

Thus, by (A7.2)–(A7.7), there is a $\lambda_2 < \lambda_1$ but arbitrarily close to it, such that

$$E_n V_{n+1}(\theta_{n+1}) - V_n(\theta_n) \leq -\epsilon_n\lambda_2 \tag{7.12}$$

for large n if $V(\theta_n) \geq \lambda_0$. Hence, $V_n(\theta_n)$ has the supermartingale property for large n if the process is stopped when θ_n first reaches Q_{λ_0}. Since $|\delta V_n^d| \leq \epsilon_n\widehat{k}_n[V(\theta_n) + 1]$, we have

$$V_n(\theta_n) = V(\theta_n) + O(\epsilon_n)\widehat{k}_n[V(\theta_n) + 1], \tag{7.13}$$

which (since $\epsilon_n\widehat{k}_n$ can be assumed to be arbitrarily small) guarantees that

$$\sup_n EV_n^-(\theta_n) < \infty,$$

where $V_n^-(\theta)$ is the negative part of $V_n(\theta)$. Now the supermartingale convergence theorem [see the statement below (4.1.6)] and (7.12) can be applied to $V_n(\theta_n)$, stopped on first entrance into Q_{λ_0}, to get that this set is again reached with probability one after each time it is exited. The proof of (7.9b) is similar to that of [127, Theorem 8, Chapter 6] and is omitted. □

Extensions. The following extensions are for the alternative perturbations, the Markov state-dependent noise cae and the constant $\epsilon_n = \epsilon$ case. The proofs are similar to that of Theorem 7.3 and are omitted. Note that the perturbation form (3.11) for the Markov state-dependent noise case is just the Poisson equation. The assumptions essentially require that ϵ times various noise terms are small for small ϵ. This is not unreasonable, since the step sizes in a well constructed algorithm should be small for small ϵ.

Theorem 7.4. *Assume the conditions of Theorem* 7.3, *except use either the perturbation* (3.2b), *or* (3.11) *with* $(\epsilon_{n+1} - \epsilon_n)/\epsilon_n = O(\epsilon_n)$. *For the state-dependent noise casc of Section* 6, *use either perturbation* (6.3), *or*

(6.10) with $(\epsilon_{n+1} - \epsilon_n)/\epsilon_n = O(\epsilon_n)$*, or the discounted forms. Then the conclusions of Theorem* 7.3 *hold.*

Theorem 7.5 (Constant step size). *Consider the case of constant-step-size* $\epsilon_n = \epsilon$*. Write the algorithm as* $\theta^\epsilon_{n+1} = \theta^\epsilon_n + \epsilon Y^\epsilon_n$*. Define the perturbations analogously to what was done in Theorems* 7.3 *or* 7.4*. With the replacements of* $k_n, \widehat{k}_n, \widetilde{k}_n$ *by* $k^\epsilon_n, \widehat{k}^\epsilon_n, \widetilde{k}^\epsilon_n$*, resp., in* (A7.5)–(A7.7)*, assume the conditions of Theorem* 7.3 *and that* $\epsilon[k^\epsilon_n]^2$*,* $\epsilon[\widehat{k}^\epsilon_n]^2$ *and* $\epsilon \widetilde{k}^\epsilon_{n+1}[\widehat{k}^\epsilon_n]^\gamma$ *all go to zero as* $\epsilon \to 0$ *and* $n \to \infty$*. Then, with* θ^ϵ_n *replacing* θ_n*, the conclusions of Theorems* 7.3 *and* 7.4 *hold for small* $\epsilon > 0$*.*

Remark. See Theorem 10.5.2 for the stability result

$$\limsup_n E|\theta_n|^2/\epsilon_n = O(1)$$

and an analogous result for the constant step size case.

6.8 Differential Inclusions Form of the ODE

The averaging methods of Theorem 1.1 required that $g_n(\theta, \xi_n)$ average out to a function $\bar{g}(\theta)$ in the sense of (A1.3). The use of differential inclusions in lieu of ODEs allows this to be generalized in the sense of Theorem 5.6.3. One practical use of the generalization will be illustrated by applying it to the parameter identification problem of Section 3.1.

The following theorem is an extension of Theorem 5.6.3 to the case of correlated noise; the details of the proof are left to the reader.

Theorem 8.1. *Assume condition* (5.6.10) *on the step sizes, algorithm* (1.1)*, and the conditions in the first paragraph of Theorem* 1.1 *except for* (A1.3)*. Suppose that there is a set-valued function* $G(\cdot)$ *that is upper semicontinuous in the sense of* (4.3.2) *and such that*

$$\lim_{n,m} \text{distance} \left[\frac{1}{m} \sum_{i=n}^{n+m-1} g_i(\theta, \xi_i), G(\theta) \right] = 0 \tag{8.1}$$

with probability one for each θ*. Then the conclusions of the first paragraph of Theorem* 1.1 *hold, with the ODE replaced by the differential inclusion*

$$\dot{\theta} \in G(\theta) + z, \quad z(t) \in -C(\theta(t)). \tag{8.2}$$

Alternatively to (A1.7) *and* (8.1)*, suppose that the* $g_i(\cdot)$ *are bounded and there is a null set* N *such that for* $\omega \notin N$ *and any sequence* α^n_i, α*, satisfying*

(5.6.12),

$$\lim_{n,m} \text{distance} \left[\frac{1}{m} \sum_{i=n}^{n+m-1} g_i(\alpha_i^n, \xi_i), G(\alpha) \right] = 0. \tag{8.1'}$$

Then the conclusions hold.

Application to the parameter identification problem. Return to the parameter identification problem of a linear system discussed in Section 3.1. Let the step sizes ϵ_n satisfy the conditions (5.1.1) and (5.6.10). The observation is $y_n = \phi_n' \bar{\theta} + \nu_n$, where $\phi_n, \bar{\theta}$ and θ_n have dimension r and y_n and ν_n are real-valued. Define $\widetilde{\theta}_n = \theta_n - \bar{\theta}$, and the constraint set $H = \prod_{i=1}^r [a_i, b_i]$, where $-\infty < a_i < b_i < \infty$. The constrained stochastic approximation form of the identification algorithm (1.1.18) is

$$\theta_{n+1} = \Pi_H \left[\theta_n + \epsilon_n \phi_n \left(y_n - \phi_n' \theta_n\right]\right) = \Pi_H \left[\theta_n + \epsilon_n \phi_n \left(\phi_n' \bar{\theta} + \nu_n - \phi_n' \theta_n\right)\right].$$

which can be written in terms of the $\widetilde{\theta}_n$ as

$$\widetilde{\theta}_{n+1} = \widetilde{\theta}_n - \epsilon_n \phi_n \phi_n' \widetilde{\theta}_n + \epsilon_n \phi_n \nu_n + \epsilon_n Z_n. \tag{8.3}$$

Suppose that the asymptotic rate of change of

$$\sum_{i=0}^{m(t)-1} \epsilon_n \phi_n \nu_n \tag{8.4}$$

is zero with probability one. [Otherwise there will be a bias term in the limit.] Suppose that

$$\sup_n E|\phi_n|^2 < \infty, \tag{8.5}$$

and define $Q_n = E\phi_n \phi_n'$. Let the asymptotic rate of change of

$$\sum_{i=0}^{m(t)-1} \epsilon_i \left[\phi_i \phi_i' - Q_i\right] \tag{8.6}$$

be zero with probability one. Suppose that there are positive definite matrices $\underline{Q} < \overline{Q}$ such that

$$\lim_{n,m} \text{distance} \left[\frac{1}{m} \sum_{i=n}^{n+m-1} Q_i, G \right] = 0 \tag{8.7}$$

with probability one, where G is the set of convex combinations of $\underline{Q}$ and $\overline{Q}$. Now, Theorem 8.1 holds with $G(\theta) = -G\theta$. If the box H is large enough, then there is convergence with probability one to $\widetilde{\theta} = 0$.

6.9 State Perturbation Methods for Convergence and Bounds on Escape Probabilities

Suppose that for some large n, the iterate is close to some locally asymptotically stable point of the mean ODE. Then we expect that it will stay there for a long time before (if ever) wandering away. In this and the next section, we will derive estimates of the probability that the iterate will stay within a small neighborhood of the stable point, either on a fixed finite time interval or on the infinite time interval, given that it is close to that stable point at the beginning of the interval. In fact, it will turn out that under broad conditions the probability of escape is "exponentially small," that is, of the form e^{-c/ϵ_n} for some $c > 0$.

The approaches in this and the next section are somewhat different. In this section, we use perturbed states of the type used in Sections 3 and 4, so that the essential requirements are on the conditional moments of the martingale differences that appear. The approach in the next section works with the original process (a perturbation is not added), and it is a more direct use of the methods of large deviations theory. The proofs of the basic results in this section are straightforward. But those in Section 10 require some of the methods of the theory of large deviations. Since those proofs are very technical, we will only set up the problem so that results from the literature can be used directly. The algorithm is (1.1), which we write in expanded form as

$$\theta_{n+1} = \theta_n + \epsilon_n Y_n + \epsilon_n Z_n = \theta_n + \epsilon_n \left[\delta M_n + g_n(\theta_n, \xi_n) + \beta_n + Z_n\right]. \quad (9.1)$$

The following assumptions will be used.

Assumptions and definitions. Use $N_\rho(x)$ to denote the open ρ-sphere about the point x. The terms ξ_n, $\mathcal{F}_n$, and E_n are as defined in Subsection 1.1. Let $P_{\mathcal{F}_n}$ denote the probability conditioned on $\mathcal{F}_n$. Recall that a *Hurwitz matrix* is a square matrix, all of whose eigenvalues have negative real parts. Let $J_n(\mu)$ denote the indicator function of the event that $|\theta_n - \bar{\theta}| \leq \mu$. Recall the definition of $\Pi(n, i)$ in (3.7).

(A9.1) $\bar{\theta} \in H^0$ is a locally asymptotically stable point (in the sense of Liapunov) of the ODE

$$\dot{\theta} = \bar{g}(\theta). \quad (9.2)$$

There is a Hurwitz matrix A such that $\bar{g}(\theta) = A(\theta - \bar{\theta}) + O(|\theta - \bar{\theta}|^2)$. For $\mu > 0$, let $\nu(\mu) > 0$ be such that if $|\theta(0) - \bar{\theta}| \leq \nu(\mu)$, then $|\theta(t) - \bar{\theta}| \leq \mu$ for all $t \geq 0$. [Clearly, $\mu \geq \nu(\mu)$.]

(A9.2) $g_n(\theta, \xi)$ can be written as

$$g_n(\theta, \xi) = g_n(\bar{\theta}, \xi) + g'_{n,\theta}(\bar{\theta}, \xi)(\theta - \bar{\theta}) + O(|\theta - \bar{\theta}|^2),$$

where $O(\cdot)$ is uniform in n and ξ.

(A9.3) $\sup_n E|g_n(\bar{\theta}, \xi_n)| < \infty$ and $\sup_n E|g_{n,\theta}(\bar{\theta}, \xi_n)| < \infty$.

By (A9.3), the random variables Γ_n (matrix-valued) and Λ_n (vector-valued) given by

$$\epsilon_n \Gamma_n = \sum_{i=n}^{\infty} \epsilon_i \Pi(n+1, i) E_n \left[g'_{i,\theta}(\bar{\theta}, \xi_i) - A \right], \tag{9.3}$$

$$\epsilon_n \Lambda_n = \sum_{i=n}^{\infty} \epsilon_i \Pi(n+1, i) E_n g_i(\bar{\theta}, \xi_i) \tag{9.4}$$

are well defined. Define the martingale differences

$$\begin{aligned} \delta M_n^{\Gamma} &= \Gamma_{n+1} - E_n \Gamma_{n+1}, \\ \delta M_n^{\Lambda} &= \Lambda_{n+1} - E_n \Lambda_{n+1}. \end{aligned}$$

(A9.4) There is a $0 \le K < \infty$ such that for small (in norm) α, and all n, with probability one

$$E_n e^{\alpha' \delta M_n} J_n(\mu) \le e^{K|\alpha|^2/2}, \tag{9.5a}$$

$$E_n e^{\alpha' \delta M_n^{\Gamma} x} \le e^{K|\alpha|^2/2}, \tag{9.5b}$$

$$E_n e^{\alpha' \delta M_n^{\Lambda}} \le e^{K|\alpha|^2/2}. \tag{9.5c}$$

(A9.5) There is a real-valued function $p(\cdot)$, continuous at zero, with $p(0) = 0$ such that for small $|\epsilon_n \alpha|$, small $\mu > 0$, and all n,

$$E e^{\epsilon_n \alpha' \Gamma_n} \le e^{p(\epsilon_n \alpha)}, \tag{9.6a}$$

$$E e^{\epsilon_n \alpha' \Lambda_n} \le e^{p(\epsilon_n \alpha)}, \tag{9.6b}$$

$$E e^{\epsilon_n \alpha' \Gamma_{n+1} Y_n J_n(\mu)} \le e^{p(\epsilon_n \alpha)}. \tag{9.6c}$$

(A9.6) For each $c > 0$,

$$\sum_{n=0}^{\infty} e^{-c/\epsilon_n} < \infty.$$

For each $c > 0$, there is a $c_1 > 0$ such that for large n

$$\sum_{i=n}^{\infty} e^{-c/\epsilon_i} \le e^{-c_1/\epsilon_n}.$$

(A9.7) $\limsup_n \sup_{i \ge n} \epsilon_i / \epsilon_n < \infty$.

Examples. Examples of (A9.4) can be based on those of Subsection 5.3.2. If $\{\xi_n\}$ is bounded and satisfies an appropriate mixing condition [127], then

(9.3) and (9.4) are bounded and (A9.3)–(A9.5) hold. Suppose that $\{\xi_n\}$ is Markov and (A9.3) holds. Then Γ_n and Λ_n are functions of ξ_n. (A9.5) is a condition on the growth rates of the moments of Λ_n, Γ_n and $\Gamma_{n+1} Y_n J_n(\mu)$, and it holds if there is a $0 \le K < \infty$ such that the kth moments of these random variables are bounded above by $K^k k!$.

Theorem 9.1. *Assume algorithm* (9.1), (5.1.1), *and* (A9.1)–(A9.7) *and let* $\beta_n \to 0$ *uniformly in* ω. *Suppose that there is a* $\zeta > 0$ *such that for each* $\rho > 0$, $\theta_n \in N_\rho(\bar\theta)$ *infinitely often with probability* ζ. *Then* $\theta_n \to \bar\theta$ *with probability* ζ. *Let* $\bar\theta$ *be at least a distance* $\mu_1 > 0$ *from* ∂H, *the boundary of* H. *Then there is a* $c > 0$ *such that for large* n *and small* $\mu \in (0, \mu_1)$,

$$P\left\{\theta_i \notin N_\mu(\bar\theta) \ \ \textit{for some } i \ge n \mid \theta_n \in N_{\nu(\mu/4)}(\bar\theta)\right\} \le e^{-c/\epsilon_n}. \tag{9.7}$$

Remark. Theorem 6.1 of [63] obtains estimates such as (9.7) even when $\bar\theta$ is on the boundary of the constraint set, under some minor additional conditions on the boundary. For ease of presentation here, we have assumed that $\bar\theta$ is interior to H.

Proof. Henceforth, $\mu > 0$ is small and in $(0, \mu_1)$. For simplicity and without loss of generality, we let $\zeta = 1$. Also, it can easily be shown that the effects of the β_n are strongly dominated by the effects of $A(\theta_n - \bar\theta)$, and for simplicity we drop the β_n term. Define the *perturbed state process*

$$\delta\theta_n = \epsilon_n \Gamma_n (\theta_n - \bar\theta) + \epsilon_n \Lambda_n, \tag{9.8a}$$

$$\widetilde{\theta}_n = (\theta_n - \bar\theta) + \delta\theta_n. \tag{9.8b}$$

The algorithm (9.1) can be written in the expanded form

$$\begin{aligned} \theta_{n+1} = \theta_n \ &+ \epsilon_n \left[\delta M_n + g_n(\bar\theta, \xi_n) + g'_{n,\theta}(\bar\theta, \xi_n)(\theta_n - \bar\theta) \right. \\ &\left. + O(|\theta_n - \bar\theta|^2) + Z_n\right]. \end{aligned} \tag{9.9}$$

Since we are concerned with the escape from a small neighborhood of $\bar\theta$, which is in the interior of the constraint set H, the terms Z_n do not appear in the analysis and will be dropped without loss of generality. Similarly, we can suppose without loss of generality that $\{\theta_n\}$ is bounded.

In preparation for rewriting the algorithm in terms of the perturbed state $\widetilde{\theta}_n$, note (3.7) and the expansions:

$$\begin{aligned} &\epsilon_{n+1}\Lambda_{n+1} - \epsilon_n \Lambda_n \\ &\quad = (\epsilon_{n+1} E_n \Lambda_{n+1} - \epsilon_n \Lambda_n) + \epsilon_{n+1}(\Lambda_{n+1} - E_n \Lambda_{n+1}) \\ &\quad = -\epsilon_n g_n(\bar\theta, \xi_n) + \epsilon_{n+1}^2 E_n \Lambda_{n+1} + \epsilon_{n+1}\left(\Lambda_{n+1} - E_n \Lambda_{n+1}\right), \end{aligned} \tag{9.10}$$

$$\begin{aligned}
&\epsilon_{n+1}\Gamma_{n+1}(\theta_{n+1}-\bar{\theta}) - \epsilon_n\Gamma_n(\theta_n-\bar{\theta}) \\
&\quad = \epsilon_{n+1}\Gamma_{n+1}(\theta_n-\bar{\theta}) - \epsilon_n\Gamma_n(\theta_n-\bar{\theta}) + \epsilon_{n+1}\Gamma_{n+1}(\theta_{n+1}-\theta_n) \\
&\quad = -\epsilon_n\left[g'_{n,\theta}(\bar{\theta},\xi_n) - A\right](\theta_n-\bar{\theta}) + \epsilon_{n+1}^2 E_n\Gamma_{n+1}(\theta_n-\bar{\theta}) \\
&\qquad + \epsilon_{n+1}\left[\Gamma_{n+1} - E_n\Gamma_{n+1}\right](\theta_n-\bar{\theta}) + \epsilon_{n+1}\Gamma_{n+1}(\theta_{n+1}-\theta_n).
\end{aligned} \tag{9.11}$$

Using (9.8)–(9.11) and noticing the cancellation of the $\pm g_n(\bar{\theta},\xi_n)$ and $\pm[g'_{n,\theta}(\bar{\theta},\xi_n) - A](\theta_n-\bar{\theta})$ terms, we arrive at

$$\begin{aligned}
\widetilde{\theta}_{n+1} = {} & \widetilde{\theta}_n + \epsilon_n A\widetilde{\theta}_n + \epsilon_n\delta M_n + \epsilon_n O(|\theta_n-\bar{\theta}|^2) \\
& + \epsilon_{n+1}\delta M_n^\Lambda + \epsilon_{n+1}\delta M_n^\Gamma(\theta_n-\bar{\theta}) + \epsilon_{n+1}^2 E_n\Gamma_{n+1}(\theta_n-\bar{\theta}) \\
& + \epsilon_{n+1}^2 E_n\Lambda_{n+1} + \epsilon_n\epsilon_{n+1}\Gamma_{n+1}Y_n - \epsilon_n A\delta\theta_n.
\end{aligned} \tag{9.12}$$

When estimating the following probabilities, for notational convenience and without loss of generality, we suppose that the random variables are real valued. For $k > 0$ and small $\alpha\epsilon_n > 0$, (A9.5) implies that

$$P\left\{\epsilon_n\Gamma_n \geq k\right\} = P\left\{e^{\alpha\epsilon_n\Gamma_n} \geq e^{\alpha k}\right\} \leq e^{p(\alpha\epsilon_n)}e^{-\alpha k}.$$

Choosing $\alpha = v/\epsilon_n$ for small $v > 0$, the right side becomes $e^{p(v)}e^{-vk/\epsilon_n}$. Repeating this for $-\epsilon_n\Gamma_n$ and for $\pm\epsilon_n\Lambda_n$ and using (9.6a), (9.6b), the first part of (A9.6) and the Borel–Cantelli Lemma, we see that $\delta\theta_n \to 0$ with probability one. Thus, if $\widetilde{\theta}_n \to 0$ with probability one, then so does $\theta_n - \bar{\theta}$. Furthermore, by the second part of (A9.6) for each small $\kappa > 0$ there is a $c_1 > 0$ such that $P\{\sup_{i\geq n}|\delta\theta_i| \geq \kappa\} \leq e^{-c_1/\epsilon_n}$. It follows that, for purposes of the proof we can suppose that $\widetilde{\theta}_n \in N_\rho(0)$ infinitely often with probability one for each $\rho > 0$, and it is enough to prove that $\widetilde{\theta}_n \to 0$ with probability one and that there is a $c > 0$ such that for large n,

$$P\left\{\widetilde{\theta}_i \notin N_{2\mu/3}(0) \text{ for some } i \geq n \;\middle|\; \widetilde{\theta}_n \in N_{\nu(\mu/3)}(0)\right\} \leq e^{-c/\epsilon_n}. \tag{9.13}$$

Note that the argument in the paragraph preceding (9.13) can be used to show that for each small $\kappa > 0$ there is a $c_1 > 0$ such that the maxima of the terms

$$\epsilon_i A\delta\theta_i, \quad \epsilon_{i+1}^2 E_i\Lambda_{i+1},$$

$$\epsilon_{i+1}^2 E_i\Gamma_{i+1}(\theta_i-\bar{\theta})J_i(\mu), \quad \epsilon_i\epsilon_{i+1}\Gamma_{i+1}Y_iJ_i(\mu)$$

over $i \in [n,\infty)$ are bounded by $\epsilon_n\kappa$ with probability greater than $1-e^{-c_1/\epsilon_n}$ for large n. This implies that the asymptotic rates of change of the sums of these terms are zero with probability one. By (A9.4), the asymptotic rates of change of the sums of the martingale difference terms

$$\epsilon_n\delta M_nJ_n(\mu), \quad \epsilon_{n+1}\delta M_n^\Gamma(\theta_n-\bar{\theta})J_n(\mu), \quad \epsilon_{n+1}\delta M_n^\Lambda$$

are zero with probability one.

Define $\widetilde{\theta}^0(\cdot)$ to be the *piecewise linear* interpolation of $\{\widetilde{\theta}_n\}$ with interpolation intervals $\{\epsilon_n\}$ and, as usual, define the shifted processes $\widetilde{\theta}^n(\cdot) = \widetilde{\theta}^0(t_n + \cdot)$. Now we localize the argument of Theorem 1.1 analogously to what was done in Theorem 7.1 by working with the segments of the interpolated process $\widetilde{\theta}^0(\cdot)$ on the sequence of time intervals starting at those exit times from $N_{\nu(\mu/3)}(\bar{\theta})$ after which it does not return to $N_{\nu(\mu/3)}(\bar{\theta})$ until at least the next time of exit from $N_{2\mu/3}(0)$, and ending at the next time of exit from $N_{2\mu/3}(0)$. Then the asymptotic rate of change properties cited earlier and the local asymptotic stability (in the sense of Liapunov) properties of the solution to (9.2) imply the convergence $\widetilde{\theta}_n \to 0$ with probability one, as in the proof of Theorem 7.1.

Proof of (9.13): Large deviations estimates of escape. Let $\epsilon_n v_n$ denote any one of the last four terms in (9.12). It was shown that for any small $\kappa > 0$ there is a $c_1 > 0$ such that the probability that $|v_i| > \kappa$ for some $i \in [n, \infty)$ is at most e^{-c_1/ϵ_n} for large n. Thus, for the purposes of proving (9.13), we can suppose without loss of generality that $|v_n| \leq \kappa$ for any small κ. It follows that for large n, these v_n terms are strongly dominated by $A\widetilde{\theta}_n + O(|\widetilde{\theta}_n|^2)$ when $\widetilde{\theta}_n \in N_{2\mu/3}(\bar{\theta}) - N_{\nu(\mu/3)}(\bar{\theta})$, and they can be ignored. Hence, to prove (9.13) we need only work with the modified algorithm

$$\begin{aligned}\widetilde{\theta}_{n+1} = {} & \widetilde{\theta}_n + \epsilon_n A\widetilde{\theta}_n + \epsilon_n O(|\theta_n - \bar{\theta}|^2) \\ & + \epsilon_n \delta M_n + \epsilon_{n+1}\delta M_n^{\Gamma}(\theta_n - \bar{\theta}) + \epsilon_{n+1}\delta M_n^{\Lambda}.\end{aligned} \tag{9.14}$$

Condition (A9.4), the last part of (A9.6), (A9.7), and the estimate (5.3.15) in Theorem 5.3.3 imply that there is $c_2 > 0$ such that for each $\rho > 0$

$$\begin{aligned}&P_{\mathcal{F}_n}\left\{\sup_{m\geq n}\left|\sum_{i=n}^{m}\left[\epsilon_i \delta M_i J_i(\mu) + \epsilon_{i+1}\delta M_i^{\Gamma}(\theta_i - \bar{\theta})J_i(\mu) + \epsilon_{i+1}\delta M_i^{\Lambda}\right]\right| \geq \rho\right\} \\ &\quad \leq e^{-c_2\rho^2/\epsilon_n}.\end{aligned} \tag{9.15}$$

This bound, the arbitrariness of ρ, and the form of the equation (9.14), imply that there is a $c > 0$ such that (9.13) holds. □

6.10 Large Deviations Estimates: Probability One Convergence and Escape Time Bounds

In Section 9, a state perturbation method, together with a linearization about the limit point $\bar{\theta}$, was used to get the large deviations-type estimates of escape (9.7). In this section, similar estimates will be obtained by a more direct appeal to the classical theory of large deviations, without the intervention of the perturbation function. The proofs are quite technical and will not be given. Instead, the problem will be set up so that the desired results can be taken from [61, 63], which contain many references to the general literature on the theory of large deviations. The main results come from the less technical paper [61] with minor changes. Much stronger results can be obtained from the ideas in [63]. The original applications of large deviations theory to the stochastic approximation problem can be found in [114, 128]. The last chapter of [127] contains an introduction to the subject of large deviations; a few other sources on the general theory are [42, 60, 75, 76, 220, 238].

6.10.1 Two-Sided Estimates

We will work with the following two forms of the algorithm (9.1):

$$\theta_{n+1} = \theta_n + \epsilon_n g_n(\theta_n, \xi_n) \tag{10.1}$$

and

$$\theta_{n+1} = \theta_n + \epsilon_n g_n(\theta_n, \xi_n) + \epsilon_n \delta M_n. \tag{10.2}$$

where δM_n is an $\mathcal{F}_n$-martingale difference and $\mathcal{F}_n$ is the minimal σ-algebra that measures $\{\theta_i, Y_{i-1}, \xi_i, i \leq n\}$. By an obvious definition of ξ_n, (10.2) is a special case of (10.1). But it is sometimes useful to distinguish between them when assumptions are to be verified.

Note the absence of the reflection term $\epsilon_n Z_n$ in (10.1) and (10.2). As in Section 9, the limit point $\bar{\theta}$ will be supposed to be strictly interior to the constraint set H. Since we will be concerned with escape from a small neighborhood of $\bar{\theta}$ when the path starts very near $\bar{\theta}$, the Z_n-terms can be dropped. The mean ODE will be

$$\dot{\theta} = \bar{g}(\theta). \tag{10.3}$$

Assumptions and definitions. For simplicity in citations to the literature, the observation terms Y_n will be assumed to be (uniformly) bounded when θ_n is close to $\bar{\theta}$. Some results for the unbounded case appear in [61] for the Gaussian noise and in [63, Section 7] if the conditional distribution of Y_n given the past of the iterate process has (uniformly) "exponential tails," which covers a quite general exogenous noise case. Whether mentioned explicitly or not, all the interpolations $\theta^n(\cdot)$ and $\theta^\epsilon(\cdot)$ are assumed to be *piecewise linear* and not piecewise constant, so that they are all continuous. We now list the assumptions. Note that the condition that the limit exists in (A10.4) will be weakened later; condition (A10.3), which was used only to get a slightly sharper bound in [61], will be dropped in Subsection 10.2.

(A10.1) Equation (10.3) has a unique solution for each initial condition, and $\bar{\theta} \in H^0$ is an asymptotically stable point in the sense of Liapunov. The $g_n(\cdot, \xi)$ are bounded and are Lipschitz continuous in θ uniformly in n and ξ.

(A10.2) The step sizes ϵ_n are nonincreasing, $\sum \epsilon_n = \infty$, and

$$\lim_n \sup_{i \geq n} \frac{\epsilon_i}{\epsilon_n} = 1.$$

(A10.3) There is a continuous and positive function $h_1(\cdot)$ on $[0, \infty)$ such that

$$\frac{\epsilon_{m(t_n+s+\delta)}}{\epsilon_n} \to h_1(s)$$

as $n \to \infty$ and then $\delta \to 0$. Define the function $h_0(\cdot) = 1/h_1(\cdot)$.

(A10.4) Let L and G be neighborhoods of the origin in $\mathbb{R}^r$ such that $\bar{\theta} + \bar{G} \equiv \{\bar{\theta} + y : y \in \bar{G}\} \in H$. The set L can be arbitrarily small. There is a real-valued function $H_0(\alpha, \psi)$ that is continuous in (α, ψ) in $L \times \bar{G}$ and whose α-derivative is continuous on L for each fixed $\psi \in \bar{G}$ such that the following limit holds: For any $T > 0$ and $\Delta > 0$ with T being an integral multiple of Δ, and any functions $(\alpha(\cdot), \psi(\cdot))$ taking values in $(L, \bar{G})$ and being constant on the intervals $[i\Delta, i\Delta + \Delta)$, $i\Delta < T$, we have

$$\begin{aligned} &\int_0^T H_0(\alpha(s), \psi(s))ds \\ &= \lim_{m,n} \frac{\Delta}{m} \log E \exp \left[\sum_{i=0}^{T/\Delta - 1} \alpha'(i\Delta) \right. \\ &\qquad \left. \times \sum_{j=im}^{im+m-1} \left[g_{n+j}(\bar{\theta} + \psi(i\Delta), \xi_{n+j}) - \bar{g}(\bar{\theta} + \psi(i\Delta)) \right] \right]. \end{aligned} \tag{10.4}$$

By $\lim_{n,m}$ we mean that the limit exists as $n \to \infty$ and $m \to \infty$ in any way at all. The limit in (10.4) exists if, for each α and ψ, it exists uniformly in ω (with probability one) in

$$\lim_{m,n} \frac{1}{m} \log E_n \exp \left[\alpha' \sum_{j=0}^{m-1} \left(g_{n+j}(\bar{\theta} + \psi, \xi_{n+j}) - \bar{g}(\bar{\theta} + \psi) \right) \right]. \tag{10.5}$$

Example of (A10.3). If $\epsilon_n = 1/n^\gamma$, for $0 < \gamma < 1$, then $h_1(s) = 1$. If $\epsilon_n = 1/n$, then $h_1(s) = e^{-s}$.

Definitions of the large deviations functionals. Define the functions (see Equation (2.10) in [61], where the ideas of this subsection originate)

$$H_1(\alpha, \psi, s) = h_0(s) H_0(h_1(s)\alpha, \psi) \tag{10.6}$$

and

$$L(\beta, \psi, s) = \sup_{\alpha} \left[\alpha' \left(\beta - \bar{g}(\bar{\theta} + \psi) \right) - H_1(\alpha, \psi, s) \right]. \tag{10.7}$$

Define the function $S(T, \phi)$ to be

$$S(T, \phi) = \int_0^T L(\dot{\phi}(s), \phi(s), s) ds \tag{10.8}$$

if $\phi(\cdot)$ is absolutely continuous and to take the value infinity otherwise. $S(T, \phi)$ is the usual action functional of the theory of large deviations.

Note that under (A10.3) and (A10.4), for each $\delta_1 > 0$ there is a $\delta_2 > 0$ such that

$$|\dot{\phi} - \bar{g}(\bar{\theta} + \phi)| \geq \delta_1 \text{ implies that } L(\dot{\phi}, \phi, t) \geq \delta_2, \ t \leq T. \tag{10.9}$$

In addition, $L(\dot{\phi}, \phi, t) = 0$ for $\dot{\phi} = \bar{g}(\bar{\theta} + \phi)$.

Comments on the limit (10.4) and its α-differentiability at $\alpha = 0$. The condition (A10.4) is a classical starting point for the theory of large deviations. It will be weakened later in various ways. The large deviations literature has devoted much attention to the verification of conditions such as (A10.4); see the examples below and in [61, 63].

For the following discussion, let α be real-valued and let ξ be a real-valued random variable such that $E\xi = 0$ and $Ee^{\alpha\xi} < \infty$ for some open α-interval containing the origin. Then the function $H(\alpha) = \log Ee^{\alpha\xi}$ is convex, nonnegative and takes the value zero at $\alpha = 0$. The differentiability of $H(\cdot)$ at $\alpha = 0$ is equivalent to there being a unique value of β (namely, $\beta = 0$) at which

$$L(\beta) = \sup_{\alpha} [\beta\alpha - H(\alpha)] = 0.$$

This is illustrated in Figures 10.1 and 10.2. In Figure 10.1, we have graphed a function $H(\alpha)$ that is not differentiable at $\alpha = 0$. It is piecewise linear, with slopes $-k_2$ and k_1, where $k_i > 0$.

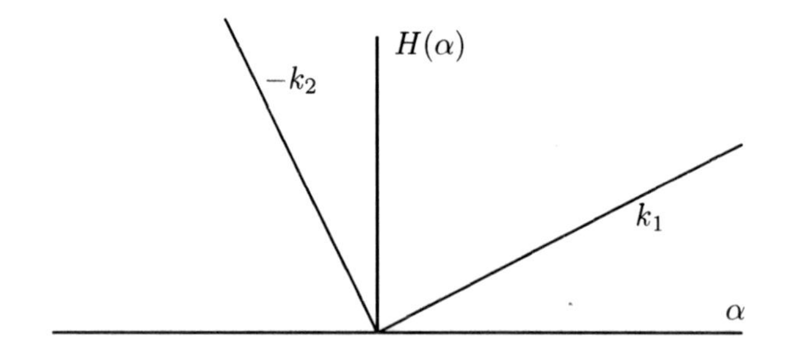

Figure 10.1. A nondifferentiable $H(\alpha)$.

Figure 10.2 plots $\beta\alpha$ and $-H(\alpha)$ as a function of α for some fixed

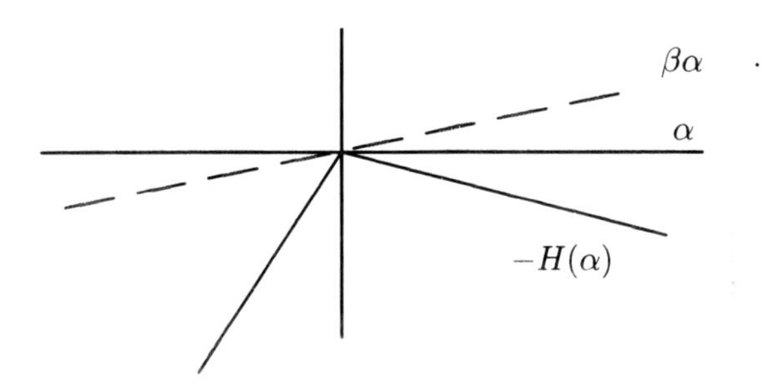

Figure 10.2. Plots of $\beta\alpha$ and $-H(\alpha)$.

It can be seen that $L(\beta) = 0$ for $\beta \in [-k_2, k_1]$. In the applications, the function $L(\cdot)$ plays the role of a cost rate, a penalty for the path to depart from the mean trajectory. In this case we have centered the random variable about the mean value. Adding the mean value, the form (10.7) would be $L(\beta - \bar{g}(\bar{\theta} + \psi))$ and $S(T, \phi)$ would be zero for any path $\phi(\cdot)$ that satisfies $-k_2 \leq \dot{\phi} - \bar{g}(\bar{\theta} + \phi) \leq k_1$ on $[0, T]$. Thus, it is likely that the noise effects would continually drive the path far from the mean. In any case, a noise process that would yield an H-functional of the type in the figure is quite unusual at the least.

Examples of α-differentiability. The α-differentiability and the characterization of its value at $\alpha = 0$ as a mean value is related to the validity of a law of large numbers, although we present only a "finite" heuristic argument for this.

Let $\{\xi_n\}$ be a sequence of real-valued random variables such that $E\sum_{i=0}^{n-1} \xi_i/n \to 0$ as $n \to \infty$. For each small real number α, define

$$H_n(\alpha) = \frac{1}{n} \log E \exp\left\{\alpha \sum_{i=0}^{n-1} \xi_i\right\},$$

and set $H(\alpha) = \lim_n H_n(\alpha)$. Differentiating $H_n(\alpha)$ yields

$$H_{n,\alpha}(\alpha) = \frac{1}{n} \frac{E\sum_{i=0}^{n-1} \xi_i \exp\left\{\alpha \sum_{i=0}^{n-1} \xi_i\right\}}{E \exp\left\{\alpha \sum_{i=0}^{n-1} \xi_i\right\}}.$$

Thus, $H_{n,\alpha}(0) = E\sum_{i=0}^{n-1} \xi_i/n$.

Let $\{\xi_i\}$ be mutually independent and satisfy

$$\frac{1}{m} E \sum_{i=n}^{n+m-1} \xi_i \to 0$$

as n and m go to infinity. Let $Ee^{\alpha\xi_n}$ be bounded uniformly in n for α in some neighborhood of the origin. Then, by a direct calculation it follows that the value defined by

$$H(\alpha) = \limsup_{n,m} \frac{1}{m} \log E \exp\left\{\alpha \sum_{i=n}^{n+m-1} \xi_i\right\}$$

exists and is continuous and differentiable in a neighborhood of the origin (with derivative zero at $\alpha = 0$). The limsup will replace the lim in the following subsections.

Suppose that the ξ_n are (vector-valued) Gaussian random variables with mean zero and correlation function $R(j) = E\xi_0\xi_j'$ and that the sum $\sum_{j=-\infty}^{\infty} R(j) = \bar{R}$, is absolutely convergent. Then, with α replaced by a row vector, α', $H(\alpha) = \alpha'\bar{R}\alpha/2$.

Now consider algorithm (10.1) with $g_n(\cdot) = g(\cdot)$ being bounded, continuous and not dependent on n. Suppose that $\{\xi_n\}$ is a Markov chain on a state space Ξ, with the one-step transition function denoted by $P(\xi, \cdot)$. The following fact is proved in [103] under a uniform recurrence condition on the chain. Assume that the $g(\cdot)$-function is "centered" such that $\int g(\theta, \xi)\mu(d\xi) = 0$, where $\mu(\cdot)$ is the invariant measure of the chain. Let $C(\Xi)$ denote the space of real-valued continuous functions on Ξ. For each θ and α, define the operator $\widehat{P}(\theta, \alpha)$ from $C(\Xi)$ to $C(\Xi)$ by

$$\widehat{P}(\theta, \alpha)(f)(\xi) = \int e^{\alpha' g(\theta, \xi_1)} f(\xi_1) P(\xi, d\xi_1).$$

The eigenvalue $\lambda(\theta, \alpha)$ of $\widehat{P}(\theta, \alpha)$ with maximum modulus is real, simple and larger than unity for $\alpha \neq 0$. Also, $H(\alpha, \theta) = \log \lambda(\theta, \alpha)$ is analytic in α for each θ and the limit

$$H(\theta, \alpha) = \lim_n \frac{1}{n} \log E_\xi \exp\left\{\alpha' \sum_{i=0}^{n-1} g(\theta, \xi_i)\right\}$$

exists uniformly in the initial condition ξ, where E_ξ is the expectation under the initial condition ξ.

Reference [63] contains other examples such as M-dependent and state-dependent noise and outlines a more abstract but more general approach to verifying the limit and differentiability conditions.

Definitions. For notational simplicity and without loss of generality, we suppose that $\bar{\theta} = 0$ in all subsequent results in this section, although we retain the argument $\bar{\theta}$. Let B_x be a set of continuous functions on $[0, T]$ taking values in the set $\bar{G}$, and with initial value x. Recall that B_x^0 denotes the interior of B_x, and $\bar{B}_x$ stands for the closure of B_x. We have the following theorem, in which we use $\theta^n(\cdot)$ to denote the *piecewise linear* interpolation with initial condition $\theta^n(0) = \theta_n$.

Theorem 10.1. [61, Theorem 3.2] *Assume algorithm* (10.1). *Under* (A10.1)–(A10.4)

$$
\begin{aligned}
-\inf_{\phi \in B_x^0} S(T,\phi) &\le \liminf_n \epsilon_n \log P_x^n \left\{\theta^n(\cdot) \in B_x\right\} \\
&\le \limsup_n \epsilon_n \log P_x^n \left\{\theta^n(\cdot) \in B_x\right\} \\
&\le -\inf_{\phi \in \bar{B}_x} S(T,\phi),
\end{aligned}
\tag{10.10}
$$

where P_x^n denotes the probability under the condition that $\theta^n(0) = \theta_n = x$.

Application of (10.10). The following estimate will be useful in the computation of the mean escape times. Recall the definition of $\nu(\mu)$ in (A9.1). Let μ be small enough that $\overline{N_\mu(0)} \subset G$. Define $B_{1,x}$ to be the set of continuous paths that start at $x \in N_{\nu(\mu)}(0)$ and either leave the set G on the interval $[0,T]$ or do not return to $N_{\nu(\mu)}(0)$ at time T. The inequality (10.9) says that there is a $\delta_3 > 0$ such that for all $x \in N_{\nu(\mu)}(0)$,

$$
S(T,\phi) \ge \delta_3 \quad \text{for } \phi(\cdot) \in B_{1,x}. \tag{10.11}
$$

Then Theorem 10.1 implies that there is a $c > 0$ such that for large n,

$$
P_x^n \left\{\theta^n(\cdot) \in B_{1,x}\right\} \le e^{-c/\epsilon_n}. \tag{10.12}
$$

Constant step size: $\epsilon_n = \epsilon$. Theorem 10.1 also holds for the constant step size ($\epsilon_n = \epsilon$) case, where the iterates are θ_n^ϵ, and we use $h_1(s) = 1$. Let $q_\epsilon \ge 0$ be a sequence of nondecreasing and non-negative integers. In the following theorems, $\theta^\epsilon(\cdot)$ is the *piecewise linear interpolation* of the process with values θ_n^ϵ on $[n\epsilon, n\epsilon + \epsilon)$. Theorem 10.1 becomes the following.

Theorem 10.2. *Assume algorithm* (10.1). *Let $\epsilon_n = \epsilon$. Under* (A10.1) *and* (A10.4),

$$
\begin{aligned}
-\inf_{\phi \in B_x^0} S(T,\phi) &\le \liminf_\epsilon \epsilon \log P_x^\epsilon \left\{\theta^\epsilon(\epsilon q_\epsilon + \cdot) \in B_x\right\} \\
&\le \limsup_\epsilon \epsilon \log P_x^\epsilon \left\{\theta^\epsilon(\epsilon q_\epsilon + \cdot) \in B_x\right\} \\
&\le -\inf_{\phi \in \bar{B}_x} S(T,\phi),
\end{aligned}
\tag{10.13}
$$

where P_x^ϵ denotes the probability under the condition that $\theta^\epsilon(\epsilon q_\epsilon) = \theta_{q_\epsilon}^\epsilon = x$. The inequality (10.12) *takes the form: There is a $c > 0$ such that for small $\epsilon > 0$ and large n,*

$$
P_x^\epsilon \left\{\theta^\epsilon(\epsilon q_\epsilon + \cdot) \in B_{1,x}\right\} \le e^{-c/\epsilon}.
$$

6.10.2 Upper Bounds and Weaker Conditions

The existence of the limits in (10.4) or (10.5) was needed to get the two-sided estimates in (10.10). But (10.12) is an application of only the upper bound in (10.10). Condition (A10.3) was used in [61] only to get a slightly sharper bound. If we need only an upper bound such as the second line of (10.10) or (10.12), then the limit in (10.4) or (10.5) can be replaced by a limsup, (A10.3) can be dropped, and (A10.4) can be replaced by the following much weaker condition.

(A10.5) There is a function $\bar{H}_1(\alpha, \psi)$ with the continuity and differentiability properties of $H_0(\alpha, \psi)$ in (A10.4) such that for any $\Delta > 0$ and $T > 0$ (with the functions $\alpha(\cdot)$ and $\psi(\cdot)$ being as in (A10.4), and T an integral multiple of Δ),

$$\begin{aligned} &\int_0^T \bar{H}_1(\alpha(s), \psi(s))ds \\ &\geq \limsup_{m,n} \frac{\Delta}{m} \log E \exp \left[\sum_{i=0}^{T/\Delta - 1} \alpha'(i\Delta) \right. \\ &\qquad \left. \times \sum_{j=im}^{im+m-1} \left[g_{n+j}(\bar{\theta} + \psi(i\Delta), \xi_{n+j}) - \bar{g}(\bar{\theta} + \psi(i\Delta)) \right] \right]. \end{aligned} \tag{10.14}$$

or that with probability one, for each α, ψ,

$$\bar{H}_1(\alpha, \psi) \geq \limsup_{m,n} \frac{1}{m} \log E_n \exp \left[\alpha' \sum_{j=0}^{m-1} \left[g_{n+j}(\bar{\theta} + \psi, \xi_{n+j}) - \bar{g}(\bar{\theta} + \psi) \right] \right], \tag{10.15}$$

where the limsup is attained uniformly in ω (for almost all ω). Define $\bar{H}(\alpha, \psi) = \bar{H}_1(\alpha, \psi)$ and set

$$\bar{L}(\beta, \psi) = \sup_{\alpha} \left[\alpha' \left(\beta - \bar{g}(\bar{\theta} + \psi) \right) - \bar{H}(\alpha, \psi) \right]. \tag{10.16}$$

Define $\bar{S}(T, \phi)$ as $S(T, \phi)$ was defined, but using $\bar{L}(\beta, \psi)$.

The continuity and α-differentiability properties of $\bar{H}(\alpha, \psi)$ imply that (10.9) holds for $\bar{L}(\beta, \psi)$ and (10.11) holds for $\bar{S}(T, \phi)$. Theorem 10.3 follows from the proof of the upper bound in [61, Theorem 3.2] or by the stronger results in [63, Theorem 3.2], whose proof is much more technical, but all of whose conditions hold by the assumptions here, as shown in [63, Sections 4 and 5].

Theorem 10.3. *Assume algorithm* (10.1) *and conditions* (A10.1), (A10.2), *and* (A10.5). *Then*

$$\limsup_n \epsilon_n \log P_x^n \left\{ \theta^n(\cdot) \in B_x \right\} \leq - \inf_{\phi \in B_x} \bar{S}(T, \phi), \tag{10.17}$$

and (10.12) *holds for some* $c > 0$. *The analog of the assertion holds for the constant step size* ($\epsilon = \epsilon_n$) *case.*

Algorithm (10.2). As previously noted, algorithms (10.1) and (10.2) are the same, as can be seen by letting δM_n be a component of ξ_n. But sometimes it is convenient to check the conditions separately on the sequences δM_n and ξ_n. Then the upper bounding idea of condition (A10.5) can be used and will be based on the following fact: By Hölder's inequality for real-valued random variables $X_1, \ldots, X_k$,

$$E\,|X_1 \cdots X_k| \leq E^{1/k}|X_1|^k \cdots E^{1/k}|X_k|^k.$$

Hence, for real-valued X_n,

$$\log E \exp \sum_{i=1}^{k} X_i \leq \frac{1}{k} \sum_{i=1}^{k} \log E e^{kX_i}. \tag{10.18}$$

We will need the following condition.

(A10.6) There is a function $\bar{H}_2(\alpha)$ that is continuous and differentiable in some small neighborhood L of the origin such that [for $\alpha(\cdot), T, \Delta$ as in (A10.4)],

$$\int_0^T \bar{H}_2(\alpha(s))ds \geq \limsup_{m,n} \frac{\Delta}{m} \log E \exp \left[\sum_{i=0}^{T/\Delta - 1} \alpha'(i\Delta) \sum_{j=im}^{im+m-1} \delta M_{n+j} \right] \tag{10.19}$$

or that with probability one,

$$\bar{H}_2(\alpha) \geq \limsup_{m,n} \frac{1}{m} \log E_n \exp \left[\alpha' \sum_{j=0}^{m-1} \delta M_{n+j} \right], \tag{10.20}$$

where the limsup is taken on uniformly in ω (for almost all ω).

Now with $k = 2$, use the bound in (10.18) to redefine

$$\bar{H}(\alpha, \psi) = \frac{1}{2}\bar{H}_1(2\alpha, \psi) + \frac{1}{2}\bar{H}_2(2\alpha),$$

and define $\bar{L}(\cdot)$ and $\bar{S}(\cdot)$ as before Theorem 10.3. The inequalities (10.9) and (10.11) still hold.

Theorem 10.4. *Assume algorithm* (10.2) *with bounded* $\{\delta M_n\}$ *and conditions* (A10.1), (A10.2), (A10.5), *and* (A10.6). *Then* (10.12) *and* (10.17) *hold. The analogous results hold for the constant step size* ($\epsilon_n = \epsilon$) *case.*

Linearization and local results. In Section 9, we worked with a linearization about $\bar{\theta}$. The same thing can be done here. Condition (A9.2) will be used with the following.

(A10.7) There is a function $\bar{H}_3(\alpha)$ that has the continuity and differentiability properties of $\bar{H}_2(\cdot)$ in (A10.6), and (10.19) or (10.20) holds for $\bar{H}_3(\alpha)$ when δM_n is replaced by $g_n(\bar{\theta}, \xi_n)$. There is a function $\bar{H}_4(\alpha, \psi)$ that has the continuity and differentiability properties of $\bar{H}_0(\alpha, \psi)$ in (A10.4), and (10.14) or (10.15) holds for $\bar{H}_4(\alpha, \psi)$ when $[g_{n+j}(\bar{\theta} + \psi, \xi_{n+j}) - \bar{g}(\bar{\theta} + \psi)]$ is replaced by $[g'_{n,\theta}(\bar{\theta}, \xi_{n+j}) - A]\psi$.

Using inequality (10.18) for $k = 3$, redefine $\bar{H}(\cdot)$ to be

$$\bar{H}(\alpha, \psi) = \frac{1}{3}\left[\bar{H}_2(3\alpha) + \bar{H}_3(3\alpha) + \bar{H}_4(3\alpha, \psi)\right]. \tag{10.21}$$

Redefine

$$\bar{L}(\beta, \psi) = \sup_{\alpha}\left[\alpha'\left(\beta - A\psi + O(|\psi|^2)\right) - \bar{H}(\alpha, \psi)\right]. \tag{10.22}$$

Redefine $\bar{S}(\cdot)$ using the $\bar{L}(\cdot)$ given in (10.22). Note that $A\psi$ strictly dominates $O(|\psi|^2)$ in some neighborhood of the origin. The proof of the following theorem is that of Theorem 10.3.

Theorem 10.5. *Assume algorithm* (10.2) *with bounded* $\{\delta M_n\}$ *and conditions* (A9.2), (A10.1), (A10.2), (A10.6), *and* (A10.7). *Then* (10.12) *and* (10.17) *hold. The analogous results hold for the constant step size* ($\epsilon_n = \epsilon$) *case.*

6.10.3 Bounds on Escape Times

We now show that if the iterate gets close to $\bar{\theta}$ at large n, it stays there for a very long time.

Definitions. For notational simplicity and without loss of generality, we retain the assumption that $\bar{\theta} = 0$. For $T > 0$, $j \geq 0$ and q_ϵ a sequence of non-negative integers, define the conditional probabilities

$$\begin{aligned}
&P^n_{x,jT}\left\{\theta^n(jT + \cdot) \in \cdot\right\} \\
&\qquad = P\left\{\theta^n(jT + \cdot) \in \cdot \,\middle|\, \theta^n(jT) = x, \mathcal{F}_{m(t_n+jT)}\right\}, \\
&P^\epsilon_{x,jT}\left\{\theta^\epsilon(\epsilon q_\epsilon + jT + \cdot) \in \cdot\right\} \\
&\qquad = P\left\{\theta^\epsilon(\epsilon q_\epsilon + jT + \cdot) \in \cdot \,\middle|\, \theta^\epsilon(\epsilon q_\epsilon + jT) = x, \mathcal{F}^\epsilon_{q_\epsilon + jT/\epsilon}\right\}.
\end{aligned}$$

Theorem 10.6. *Assume algorithm* (10.2) *with bounded* $\{\delta M_n\}$ *and conditions* (A10.1) *and* (A10.2). *Suppose that there are functions* $\bar{H}_1(\cdot)$ (*resp.,* $\bar{H}_2(\cdot)$) *satisfying the smoothness assumptions in* (A10.5) (*resp.,* (A10.6)) *such that* (10.15) (*resp.,* (10.20)) *holds. Let* $T > 0$ *and let* G *be an open set* *containing the origin. Then for small $\mu > 0$, there is a $c > 0$ such that for large n and all $x \in N_{\nu(\mu)}(\bar{\theta}) = N_{\nu(\mu)}(0)$,*

$$P_x^n \{\theta^n(t) \notin G \text{ for some } 0 \leq t \leq mT\} \leq \sum_{i=0}^{m-1} e^{-c/\epsilon_i}. \tag{10.23}$$

Let $\epsilon_n = \epsilon$ and assume the conditions with ξ_n^ϵ used in lieu of ξ_n. Then (10.23) holds for small ϵ and the processes $\theta^\epsilon(\epsilon q_\epsilon + \cdot)$ with the initial conditions $\theta^\epsilon(\epsilon q_\epsilon) = \theta^\epsilon_{q_\epsilon} = x$ replacing $\theta^n(\cdot)$. Let τ^ϵ denote the first (positive) escape time from G for $\theta^\epsilon(\epsilon q_\epsilon + \cdot)$. Then there is a $c > 0$ such that for small ϵ,

$$E^\epsilon_{x,0} \tau^\epsilon \geq e^{c/\epsilon}, \tag{10.24}$$

where $E^\epsilon_{x,0}$ denotes the expectation conditioned on $\theta^\epsilon(\epsilon q_\epsilon) = x$. All inequalities involving conditional expectations or probabilities hold with probability one.

Proof. For small μ, $\overline{N_\mu(0)} \subset G$. The fact that the limsups in (10.15) and (10.20) are taken on uniformly in ω (with probability one) and the proof cited for Theorems 10.3 and 10.4 implies that there are $c > 0$ and $\mu_0 > 0$ (with $\overline{N_{\mu_0}(0)} \subset G$) such that for $\mu \leq \mu_0$, all large n, and all $x \in N_{\nu(\mu)}(0)$,

$$P^n_{x,0} \left\{\theta^n(t) \notin G \text{ for some } 0 \leq t \leq T \text{ or } \theta^n(T) \notin N_{\nu(\mu)}(0)\right\} \leq e^{-c/\epsilon_n}. \tag{10.25}$$

For $j \geq 0$, define the ω-sets

$$C_j = \{\omega : j = \min\ [i : \theta^n(t) \notin G \text{ on } (iT, iT+T] \text{ or } \theta^n(iT+T) \notin N_{\nu(\mu)}(0)\ \}$$

and

$$\widehat{C}_j = \{\omega : \theta^n(\cdot) \text{ escapes from } G \text{ on } [0, jT]\}.$$

Then $\widehat{C}_m \subset \cup_{j=0}^{m-1} C_j$.

For $j > 0$, we can write

$$P^n_{x,0}\{C_j\} \leq E^n_{x,0} \sup_y P^n_{y,jT}\{C_j\} \leq \sup_\omega \sup_y P^n_{y,jT}\{C_j\}, \tag{10.26}$$

where the sup over ω is an almost sure sup. By the definition of C_j, $P^n_{y,jT}\{C_j\} = 0$ for $y \notin N_{\nu(\mu)}(\bar{\theta})$. Thus, by (10.25), for large n and $j \geq 0$,

$$P^n_{y,jT}\{C_j\} \leq e^{-c/\epsilon_j}. \tag{10.27}$$

This and (10.26) yield (10.23).

For the small constant step size ($\epsilon_n = \epsilon$) case, define C_j^ϵ as C_j was defined but with the process $\theta^\epsilon(\epsilon q_\epsilon + \cdot)$ used. Then (10.23) and (10.27) are, respectively,

$$P^\epsilon_{x,0} \{\theta^\epsilon(\epsilon q_\epsilon + t) \notin G \text{ for some } 0 \leq t \leq mT\} \leq m e^{-c/\epsilon}, \tag{10.28}$$

$$P^{\epsilon}_{y,jT}\left\{C^{\epsilon}_{j}\right\} \leq e^{-c/\epsilon}. \tag{10.29}$$

Turning to the proof of (10.24), for small $\epsilon > 0$, with probability one we have

$$\begin{aligned}
E^{\epsilon}_{x,0}\tau^{\epsilon} &\geq T\sum_{m=0}^{\infty} mP^{\epsilon}_{x}\left\{\text{first escape from } G \text{ is on } (mT, mT+T]\right\} \\
&= T\sum_{m=1}^{\infty} P^{\epsilon}_{x,0}\left\{\text{first escape is after time } mT\right\} \\
&= T\sum_{m=1}^{\infty} \left[1 - P^{\epsilon}_{x,0}\left\{\text{first escape is by time } mT\right\}\right] \\
&\geq T\sum_{m=1}^{e^{c/\epsilon}} \left[1 - me^{-c/\epsilon}\right] \geq TKe^{c/\epsilon},
\end{aligned}$$

where $K \approx .5$ for small ϵ. This yields (10.24) for a slightly smaller value of c. □

7

Weak Convergence: Introduction

7.0 Outline of Chapter

Up to now, we have concentrated on the convergence of $\{\theta_n\}$ or of $\{\theta^n(\cdot)\}$ to an appropriate limit set with probability one. In this chapter, we work with a weaker type of convergence. In practical applications, this weaker type of convergence most often yields the same information about the asymptotic behavior as the probability one methods. Yet the methods of proof are simpler (indeed, often substantially simpler), and the conditions are weaker and more easily verifiable. The weak convergence methods have considerable advantages when dealing with complicated problems, such as those involving correlated noise, state-dependent noise processes, decentralized or asynchronous algorithms, and discontinuities in the algorithms. If probability one convergence is still desired, starting with a weak convergence argument can allow one to "localize" the probability one proof, thereby simplifying both the argument and the conditions that are needed. For example, the weak convergence proof might tell us that the iterates spend the great bulk of the time very near some point. Then a "local" method such as that for the "linearized" algorithm in Theorem 6.1.2 can be used. The basic ideas have many applications to problems in process approximation and for getting limit theorems for sequences of random processes.

Mathematically, the basic idea of weak convergence concerns the characterization of the limits of the sequence of measures of the processes $\theta^n(\cdot)$ on the appropriate path space. In particular, one shows that the limit measures induce a process (on the path space) supported on some set of limit trajectories of the ODE $\dot{\theta} = \bar{g}(\theta) + z$, $z \in -C(\theta)$, (or $\dot{\theta} = \bar{g}(\theta)$ for the unconstrained case). Despite this abstract formulation, one does not work with the measures in either the proofs or the applications, but with the iterate sequence itself, and the entire process of proof and applications is actually simpler than what probability one methods require. The basic ideas are applications of only an elementary part of the theory of weak convergence of probability measures.

The main convergence results for stochastic approximations are in Chapter 8. This chapter provides an introduction to the subject. Section 1 motivates the importance and the role of weak convergence methods. The ideas and developments in Section 2 are intended to illustrate some of the ideas that underlie the theory of weak convergence and to provide a kind of "behind the scene" view. They do not require any of the machinery of the general theory. They need only be skimmed for the general ideas, because stronger results will be proved in Chapter 8 with the use of the general theory, without requiring any of the explicit constructions or methods that are used in the proofs in Section 2. Despite the fact that the statements of the theorems are more limited and the proofs require more details than those that use the general theory of Section 3, they are included since they relate the weak convergence method to what was done in Chapter 5, and illustrate the role of "tightness" and the minimal requirements on the step sizes ϵ_n and the moments of the martingale difference terms δM_n. It was seen in Section 6.10 that, under broad conditions and even for the constant-step-size algorithm, if the iterate is close to a stable point or set at time n, it will stay close to it for a time *at least* of the order of e^{c/ϵ_n} for some $c > 0$.

The general theory of weak convergence is introduced in Section 3. The theorems cited there and in Section 4 are all we will require for the convergence proofs in subsequent chapters. The reader need only understand the statements and need not know their proofs. Subsection 4.1 gives criteria for verifying that the "limit" is a martingale; this important idea will be used in the proofs of Chapter 8. Subsection 4.2 gives "martingale-type" criteria to verify that a given continuous-time martingale with continuous paths is a Wiener process. A very useful perturbed test function criterion for verifying tightness (relative compactness) is stated in Subsection 4.3. The latter two results will be used in the proofs of the rates of convergence in Chapter 10. The reference [241] contains a useful intuitive discussion of the advantages and nature of weak convergence, with many graphs illustrating the convergence.

7.1 Introduction

Introductory remarks on weak vs. probability one convergence. Chapters 5 and 6 were concerned with methods of proving the convergence of $\{\theta_n\}$ or of $\{\theta^n(\cdot)\}$ with probability one to an appropriate limit set. In the context of the actual way in which stochastic approximation algorithms are used in applications, an assertion of probability one convergence can be misleading. For example, there is usually some sort of stopping rule that tells us when to stop the iteration and to accept as the final value either the most recent iterate or some function of the iterates that were taken "shortly" before stopping. The stopping rule might simply be a limit on the number of iterations allowed, or it might be a more sophisticated rule based on an estimate of the improvement of the mean performance over the recent past, or perhaps on the "randomness" of the behavior of the recent iterations (the more "random," the more likely that the iterate is in a neighborhood of a stationary point). Generally, at the stopping time all we know about the closeness to a limit point or set is information of a distributional type.

If the application of stochastic approximation is done via a simulation, then one can control the model so that it does not change over time (but even then there is a stopping rule). Nevertheless, the situation is different when the stochastic approximation is used to optimize a system on-line, since convergence with probability one implies that we can iterate essentially forever, and the system will remain unchanged however long the procedure is. In practical on-line applications, the step size ϵ_n is often not allowed to decrease to zero, due to considerations concerning robustness and to allow some tracking of the desired parameter as the system changes slowly over time. Then probability one convergence does not apply. Indeed, it is the general practice in signal processing applications to keep the step size bounded away from zero. In the Kiefer–Wolfowitz procedure for the minimization of a function via a "noisy" finite difference-based algorithm, the difference interval is often not allowed to decrease to zero. This creates a bias in the limit, but this bias might be preferred to the otherwise slower convergence and "noisier" behavior when the variance of the effective noise is inversely proportional to the square of a difference interval that goes to zero. Thus, even under a probability one convergence result, the iterates might converge to a point close to the minimum, but not to the minimum itself. Such biases reduce the value of a probability one convergence result.

The proofs of probability one results tend to be quite technical. They might not be too difficult when the noise terms are martingale differences, but they can be very hard for multiscale, state-dependent-noise cases or decentralized/asynchronous algorithms. To handle the technical difficulties in an application where one wishes to prove probability one convergence, one might be forced to introduce assumptions that are not called for (such as modifications of the algorithm) or that are hard to verify.

These concerns do not eliminate the value of convergence with probability one. Convergence theorems are a guide to behavior. Although no algorithm is carried to infinity, it is still comforting to know that if the iterations are allowed to continue forever in the specified ideal environment, they will assuredly converge. However, the concerns that have been raised emphasize that methods for probability one convergence might offer less than what appears at first sight, and that methods with slightly more limited convergence goals can be just as useful, particularly if they give a lot of insight into the entire process, are technically easier, require weaker conditions, and are no less informative under the conditions that prevail in applications.

This and the next chapter will focus on convergence in a weak or distributional sense. It will turn out that the proofs are easier and conditions weaker and that we can learn nearly as much about where the iterate sequence spends its time as with probability one methods. For complicated algorithms, the proofs are substantially simpler. The methods are the natural ones if the step sizes do not decrease to zero, where probability one convergence is not pertinent. When the step sizes do go to zero, weak convergence does not preclude convergence with probability one. In fact, first proving weak convergence can simplify the ultimate proof of probability one convergence, since it allows a "localization" of the proof. Recall that the general approach has been to get the mean ODE determined by the "average dynamics," show that the solution to the ODE tends to an appropriate limit set (or a set of stationary points if the algorithm is of the gradient descent type), and then show that the chosen limit points of the solution to the ODE are the limit points of $\{\theta_n\}$. The mean ODE is easier to derive in that there are weaker conditions and simpler proofs when weak convergence methods are used. The process $\{\theta_n\}$ can still be shown to spend nearly all of its time arbitrarily close to the same limit point or set. For example, suppose that the limit set is a just a unique asymptotically stable (in the sense of Liapunov) point $\bar{\theta}$ of the ODE. Then, once we know, via a weak convergence analysis, how to characterize the path to $\bar{\theta}$ and that $\{\theta_n\}$ spends nearly all of its time (asymptotically) in any arbitrarily small neighborhood of $\bar{\theta}$, one can use a *local analysis* to get convergence with probability one, under weaker conditions (due to the local nature of the proof) than what would be needed by a pure probability one technique. For example, the methods of Chapters 5 and 6 can be used locally, or the local large deviations methods of [63] can be used. Whether or not one follows a weak convergence proof with a probability one convergence proof, under broad conditions it can be shown that if the error $|\theta_n - \bar{\theta}|$ is small, it stays small afterwards for an average time of at least the order of e^{c/ϵ_n} for some $c > 0$.

Some basic ideas and facts from the theory of weak convergence will be discussed in the next section. The theory is a widely used tool for obtaining approximation and limit theorems for sequences of stochastic processes. There is only a small amount of machinery to be learned, and this machinery has applications well beyond the needs of this book. Before discussing the ideas of the theory of weak convergence in detail, we return to the model of Chapter 5, where the noise terms are martingale differences and prove a convergence theorem under weaker conditions than used there. The proof that will be given is of a "weak convergence nature" and gives some of the flavor of weak convergence. Owing to the martingale difference property, it is quite straightforward and does not require any of the general machinery of weak convergence analysis of Sections 3 and 4. The proofs and the statements of the theorems are intended to be illustrative of some of the ideas underlying the theory of weak convergence. They are more awkward than necessary, since the tools of the general theory are avoided. However, the constructions used are of independent interest and play an important role in relating the general theory to what has been done for the probability one case, although they will not be used in applications of that general theory in the following chapters. Since more general results will be obtained in Chapter 8, the results and ideas of the next section should be skimmed for their intuitive content and insights into the types of approximations that can be used for "distributional-sense" approximation and limit theorems, and what might be required if the general theory were not available.

7.2 Martingale Difference Noise: Simple Alternative Approaches

Introductory comments and definitions. Convergence results for two simple algorithmic forms will be given in this section. The theorems are quite similar, although different methods are used for the proofs. In the second problem, there is no constraint set H, and it is assumed that $\{\theta_n\}$ is bounded in probability. These models are chosen for illustrative purposes only. The methods to be used avoid the explicit use of the machinery of weak convergence theory, but they illustrate some of the concepts. The explicit constructions used in these theorems are not necessary when the general theory is used.

The first result (Theorem 2.1 and its corollary) depends on the fact that if a sequence of random variables converges in probability, there is always a subsequence that converges with probability one to the same limit. The second result (Theorem 2.2) depends on the fact that any sequence of random variables which is bounded in probability has a subsequence that converges in distribution to some random variable. These basic facts provide simple connections between the convergence of the sequence $\{\theta^n(\cdot)\}$ with probability one and in the weak or distributional sense. In both cases, the technique of proof depends on the choice of appropriate subsequences. In the first case, it is shown that for any sequence $\{\theta^n(\cdot), Z^n(\cdot)\}$, there is always a subsequence to which the methods of Theorems 5.2.1 and 5.2.3 can be applied. The second method works with convergence in distribution directly and leads to a "functional" limit theorem. The reader should keep in mind that the assumptions are selected for convenience in exposing the basic ideas and that stronger results are to be obtained in Chapter 8.

Let $\{\mathcal{F}_n\}$ denote a sequence of nondecreasing σ-algebras, where $\mathcal{F}_n$ measures at least $\{\theta_i, Y_{i-1}, i \leq n\}$, and let E_n denote the expectation conditioned on $\mathcal{F}_n$. Suppose that we can write $E_n Y_n = \bar{g}(\theta_n) + \beta_n$, where β_n is a small bias term. First, we work with the constrained algorithm

$$\theta_{n+1} = \Pi_H(\theta_n + \epsilon_n Y_n) = \theta_n + \epsilon_n Y_n + \epsilon_n Z_n. \tag{2.1}$$

It will be shown that the mean ODE

$$\dot{\theta} = \bar{g}(\theta) + z, \quad z(t) \in -C(\theta(t)) \tag{2.2}$$

continues to characterize the asymptotic behavior. Recall the definitions $t_n = \sum_{i=0}^{n-1} \epsilon_i$, $\theta^0(t) = \theta_n$ on $[t_n, t_{n+1})$, and $\theta^n(t) = \theta^0(t + t_n)$, where $\theta^n(t) = \theta_0$ for $t \leq -t_n$. As usual, decompose the interpolated process $\theta^n(\cdot)$ as

$$\theta^n(t) = \theta_n + \bar{G}^n(t) + M^n(t) + B^n(t) + Z^n(t), \tag{2.3}$$

where we recall that, for $t \geq 0$,

$$\begin{aligned} \bar{G}^n(t) &= \sum_{i=n}^{m(t_n+t)-1} \epsilon_i \bar{g}(\theta_i), & M^n(t) &= \sum_{i=n}^{m(t_n+t)-1} \epsilon_i \delta M_i, \\ B^n(t) &= \sum_{i=n}^{m(t_n+t)-1} \epsilon_i \beta_i, & Z^n(t) &= \sum_{i=n}^{m(t_n+t)-1} \epsilon_i Z_i, \end{aligned}$$

where $\delta M_n = Y_n - E_n Y_n$.

Theorem 2.1. *Assume the step-size condition* (5.1.1), (A5.2.1)–(A5.2.3), *and any of the constraint set conditions* (A4.3.1), (A4.3.2), *or* (A4.3.3). *Suppose that* $E|\beta_n| \to 0$. *Then, for each subsequence of* $\{\theta^n(\cdot), Z^n(\cdot)\}$, *there is a further subsequence (indexed by* n_k*) such that* $\{\theta^{n_k}(\cdot), Z^{n_k}(\cdot)\}$ *is equicontinuous in the extended sense with probability one (on a sequence of intervals going to infinity), and whose limits satisfy the ODE* (2.2). *Let there be a unique limit point* $\bar{\theta}$ *of* (2.2), *which is asymptotically stable in the sense of Liapunov. Then, for each* $\mu > 0, T > 0$,,

$$\lim_n P\left\{\sup_{t \leq T} \left|\theta^n(t) - \bar{\theta}\right| > \mu\right\} = 0. \tag{2.4a}$$

More generally, there are $\mu_n \to 0, T_n \to \infty$ *such that*

$$\lim_n P\left\{\sup_{t \leq T_n} \text{distance}\left[\theta^n(t), L_H\right] \geq \mu_n\right\} = 0. \tag{2.4b}$$

In the sense of convergence in probability, the fraction of time $\theta^n(\cdot)$ spends in a δ neighborhood of L_H goes to one as $n \to \infty$.

Remarks on the theorem. According to the estimates in Sections 6.9 and 6.10, under broad conditions one can use $T_n = O(e^{c/\epsilon_n})$ in (2.4) for some $c > 0$. It will be seen that the behavior proved by both the weak convergence and the probability one methods are similar. They both show that the path essentially follows the solution to the ODE, for large n. Suppose that the path enters the domain of attraction of an asymptotically stable point $\bar{\theta}$ infinitely often. Then (ignoring some null set of paths), the probability one methods show that it will eventually converge to $\bar{\theta}$. Under the weaker conditions used for the weak convergence proofs we might not be able to prove that it will never escape. But this escape, if it ever occurs, will be a "large deviations" phenomena; i.e., it will be very rare, perhaps too rare to be of concern.

Note that we do not need a summability condition of the type $\sum_n \epsilon_n^{1+\gamma} < \infty$ for some $\gamma > 0$; only $\epsilon_n \to 0$ is needed. The corollary given after the proof shows that uniform square integrability of $\{Y_n\}$ can be replaced by uniform integrability. Further comments on the nature of the convergence results appear after Theorems 2.2 and 8.2.1.

Proof. The proof is modeled on that of Theorem 5.2.1. The main idea is the careful choice of subsequence. By the fact that there is a $0 \le K_1 < \infty$ such that $\sup_n E|Y_n|^2 \le K_1$, the inequality (4.1.4) implies that for $T > 0$ and $\mu > 0$,

$$P\left\{\sup_{t\le T}|M^n(t)| \ge \mu\right\} \le \frac{E\left|\sum_{i=n}^{m(t_n+T)-1}\epsilon_i \delta M_i\right|^2}{\mu^2} \le \frac{K_1\sum_{i=n}^{m(t_n+T)-1}\epsilon_i^2}{\mu^2}. \tag{2.5}$$

Next, it will be shown that, for any $T < \infty$ and $\mu > 0$,

$$\lim_n P\left\{\sup_{t\le T}|y^n(t)| \ge \mu\right\} = 0 \tag{2.6}$$

for $y^n(\cdot)$ being either $M^n(\cdot)$ or $B^n(\cdot)$. Since $\epsilon_n \to 0$, we have

$$\lim_n \sum_{i=n}^{m(t_n+T)} \epsilon_i^2 = 0 \tag{2.7}$$

for each $T > 0$, which yields (2.6) for $M^n(\cdot)$. Since $E|\beta_n| \to 0$,

$$E\sum_{i=n}^{m(t_n+T)} \epsilon_i|\beta_i| \to 0,$$

which implies that (2.6) also holds for $B^n(\cdot)$.

By (2.6), for $M^n(\cdot)$ and $B^n(\cdot)$ (or by the proof of the following corollary in the case where uniform integrability of $\{Y_n\}$ replaces uniform square integrability),

$$\theta^n(t) = \theta_n + \sum_{i=n}^{m(t_n+t)-1} \epsilon_i \bar{g}(\theta_i) + Z^n(t) + \kappa^n(t), \tag{2.8}$$

where $\kappa^n(t) = M^n(t) + B^n(t)$ and for any $\mu > 0$,

$$\lim_n P\left\{\sup_{t\le T} |\kappa^n(t)| \ge \mu\right\} = 0.$$

By the fact that $M^n(\cdot)$ and $B^n(\cdot)$ satisfy (2.6), there are $m_k \to \infty$ and $T_k \to \infty$ such that

$$P\left\{\sup_{t\le T_k} |\kappa^n(t)| \ge 2^{-k}\right\} \le 2^{-k}, \quad n \ge m_k. \tag{2.9}$$

Now, (2.9) and the Borel–Cantelli Lemma imply that for any sequence $n_k \ge m_k$,

$$\lim_k \sup_{t\le T_k} |\kappa^{n_k}(t)| = 0, \tag{2.10}$$

with probability one. From this point on, the proof concerning equicontinuity and the mean ODE follows that of Theorems 5.2.1 or 5.2.3.

The proof of (2.4) uses a contradiction argument. Assuming that it is false, extract a suitable subsequence for which the liminf is positive and use the previous conclusions to get a contradiction. Proceed as follows. Let $T > 0$ and (extracting another subsequence if necessary) work with the left shifted sequence $\{\theta^{n_k}(-T+\cdot), Z^{n_k}(-T+\cdot)$. Then use the fact that for any $\delta > 0$ the time required for the solution of (2.2) to reach and remain in $N_\delta(L_H)$ is bounded in the initial condition in H. Since T is arbitrary, this yields that the solution to the ODE on $[0,\infty)$ is in $N_\delta(L_H)$ for each $\delta > 0$. The few remaining details are left to the reader. □

Remark on equicontinuity. In Theorem 5.2.1, the original sequence $\{\theta^n(\cdot), Z^n(\cdot)\}$ was equicontinuous in the extended sense with probability one and $\lim_n \sup_{t\le T} |\kappa^n(t)| = 0$ with probability one. Thus, we were able to examine the convergent subsequences of $\{\theta^n(\omega,\cdot), Z^n(\omega,\cdot)\}$ for (almost all) fixed ω, with the "errors" $\kappa^n(\omega,\cdot)$ vanishing as $n \to \infty$. In the current case, we know only that each sequence $\{\theta^n(\cdot), Z^n(\cdot)\}$ has a further subsequence that is equicontinuous in the extended sense with probability one (on a sequence of time intervals increasing to the entire real line), and the errors vanish (for almost all ω) only along that subsequence. Hence, under only the conditions of Theorem 2.1, we cannot expect that for almost all ω, any subsequence of $\{\theta^n(\omega,\cdot), Z^n(\omega,\cdot)\}$ will always have a further subsequence that converges to a solution to the mean ODE.

Definition. A sequence $\{Y_n\}$ of vector-valued random variables is said to be *uniformly integrable* if

$$\sup_n E|Y_n| I_B \to 0 \quad \text{as } P\{B\} \to 0,$$

where B is a measurable set. This is equivalent to the property

$$\lim_{K\to\infty} \sup_n E|Y_n| I_{\{|Y_n|\geq K\}} = 0.$$

Remark on uniform integrability. A nice aspect of the weak convergence approach is that the uniform integrability of $\{Y_n\}$ is enough to assure that the limit processes are continuous without using the "reflection" character of the Z_n terms, as required in the proof of Theorem 5.2.3.

Corollary. *The conclusions of the theorem continue to hold if the square integrability of* $\{Y_n\}$ *is replaced by uniform integrability.*

Proof of the corollary. Assume the uniform integrability condition in lieu of square integrability. The only problem is the verification of (2.6) for $y^n(\cdot) = M^n(\cdot)$. For $K > 0$, let $I_K(v)$ denote the indicator function of the set $\{v \in I\!R^r : |v| \geq K\}$. Define the truncated sequence $\{Y_{n,K}\}$ by $Y_{n,K} = Y_n(1 - I_K(Y_n))$. Then $Y_n = Y_{n,K} + Y_n I_K(Y_n)$. Define $\delta M_{n,K}$ and $\delta\kappa_{n,K}$ by

$$\begin{aligned} \delta M_n &= (Y_n - E_n Y_n) = [Y_{n,K} - E_n Y_{n,K}] + [Y_n I_K(Y_n) - E_n Y_n I_K(Y_n)] \\ &\equiv \delta M_{n,K} + \delta\kappa_{n,K}. \end{aligned}$$

The uniform integrability of $\{Y_n\}$ and Jensen's inequality (4.1.11) imply that

$$\lim_{K\to\infty} \sup_n E\left[|Y_n I_K(Y_n)| + |E_n Y_n I_K(Y_n)|\right] \leq 2 \lim_{K\to\infty} \sup_n E|Y_n| I_K(Y_n) = 0. \tag{2.11}$$

Equation (2.11) and the definition of $\delta\kappa_{i,K}$ imply that

$$\lim_{K\to\infty} \sup_n E \sum_{i=n}^{m(t_n+T)} \epsilon_i |\delta\kappa_{i,K}| = 0. \tag{2.12}$$

For $\mu > 0$ and $T > 0$, we can write

$$\begin{aligned}\limsup_n P\left\{\sup_{t\leq T}|M^n(t)| \geq \mu\right\} \\ \leq \limsup_n P\left\{\sup_{t\leq T}\left|\sum_{i=n}^{m(t_n+t)-1}\epsilon_i \delta M_{i,K}\right| \geq \mu/2\right\} \\ + \limsup_n P\left\{\sup_{t\leq T}\left|\sum_{i=n}^{m(t_n+t)-1}\epsilon_i \delta \kappa_{i,K}\right| \geq \mu/2\right\}.\end{aligned} \tag{2.13}$$

Now, given $\nu > 0$, there is a $0 \leq K < \infty$ such that (2.12) implies that the last term on the right side in (2.13) is less than ν. Next, (2.5) holds with $\delta M_{n,K}$ (that is bounded by $2K$) replacing δM_n, where $K_1 = 4K^2$. These facts imply (2.6) for $y^n(\cdot) = M^n(\cdot)$. The rest of the details are as in the theorem. □

Remarks on extensions to correlated noise. The conditions required in Theorem 2.1 and its corollary showed some of the possibilities inherent in the weak convergence method, since we required only $\epsilon_n \to 0$, uniform integrability of $\{Y_n\}$, and $E|\beta_n| \to 0$ (the latter condition will be weakened in Chapter 8). For more general algorithms, where there are noise processes such as the sequence $\{\xi_n\}$ appearing in Chapter 6, some additional averaging is needed. In Theorem 6.1.1, the condition (A6.1.3) was used to average out the noise. To use (A6.1.3), it was necessary to show that the paths of $\theta^n(\cdot)$ were asymptotically continuous with probability one, so that the time varying θ_n could be replaced by a fixed value of θ over small time intervals. The analog of that approach in the present weak convergence context involves showing that for each positive T and μ,

$$\lim_{\Delta\to 0}\limsup_n P\left\{\max_{j\Delta\leq T}\max_{0\leq t\leq\Delta}|\theta^n(j\Delta+t) - \theta^n(j\Delta)| \geq \mu\right\} = 0. \tag{2.14}$$

This condition does not imply asymptotic continuity of $\{\theta^n(\cdot)\}$ with probability one, and hence it is weaker than what was needed in Chapter 6. Indeed, (2.14) is implied by the uniform integrability of $\{Y_n\}$. Thus, for the correlated noise case and under uniform integrability, one could redo Theorem 2.1 by replacing condition (A6.1.3) by the weaker condition obtained by deleting the $\sup_{j\geq n}$ inside the probability there. With analogous adjustments to assumptions (A6.1.4)–(A6.1.7), such an approach can be carried out. But the method to be used in the next chapter is preferable in general because it is simpler to use, requires weaker conditions, and is more versatile.

An alternative approach. In the next theorem, $\widetilde{\theta}^0(\cdot)$ denotes the *piecewise linear* interpolation of $\{\theta_n\}$ with interpolation interval $\{[t_n, t_{n+1})\}$, and $\widetilde{\theta}^n(\cdot)$ the left shift by t_n. Let $C^r[0,\infty)$ denote the space of continuous $\mathbb{R}^r$-valued functions on the time interval $[0,\infty)$ with the local sup norm metric (i.e., a sequence converges if it converges uniformly on each bounded time interval). The convergence assertion (2.15) is in terms of the convergence of a sequence of expectations of bounded and continuous functionals. This is actually equivalent to the types of convergence assertions given in Theorem 2.1, as can be seen by suitable choices of the function $F(\cdot)$. The form (2.15) of the convergence assertion is typical of the conclusions of weak convergence theory. The methods to be used in Chapter 8 work with the original piecewise constant interpolated processes $\theta^n(\cdot)$ and do not require the piecewise linear interpolation.

Theorem 2.2. *Assume the step-size condition* (5.1.1) *and that* $\{Y_n\}$ *is uniformly integrable. Drop the constraint set* H *and let* $\{\theta_n\}$ *be bounded in probability. Let* $E_n Y_n = \bar{g}(\theta_n) + \beta_n$, *where* $E|\beta_n| \to 0$ *and* $\bar{g}(\cdot)$ *is bounded and continuous. Suppose that the solution to the ODE is unique (going either forward or backward in time) for each initial condition, and that the limit set* L, *over all initial conditions, is bounded. Then, for each subsequence of* $\{\widetilde{\theta}^n(\cdot)\}$ *there is a further subsequence (indexed by* n_k*) and a process* $\theta(\cdot)$ *that satisfies* $\dot{\theta} = \bar{g}(\theta)$ *such that* $\{\widetilde{\theta}^{n_k}(\cdot)\}$ *converges in distribution to* $\theta(\cdot)$ *in the sense that for any bounded and continuous real-valued function* $F(\cdot)$ *on* $C^r[0,\infty)$

$$EF(\widetilde{\theta}^{n_k}(\cdot)) \xrightarrow{k} EF(\theta(\cdot)). \tag{2.15}$$

For almost all ω, $\theta(t,\omega)$ *takes values in an invariant set of the ODE. Also,* (2.4b) *holds when* L_H *is replaced by the invariant set. If the limit set is simply a point,* $\bar{\theta}$, *then* $EF(\widetilde{\theta}^n(\cdot)) \to F(\bar{\theta}(\cdot))$, *where* $\bar{\theta}(t) = \bar{\theta}$.

Remark on the convergence assertion and the limit points. Analogously to the conclusions of Theorem 2.1, the theorem says that for large n, the paths of $\widetilde{\theta}^n(\cdot)$ are essentially concentrated on the set of limit trajectories of the ODE $\dot{\theta} = \bar{g}(\theta)$. This can be seen as follows. Let L denote the largest bounded invariant set of the ODE. For $y(\cdot) \in C^r[0,\infty)$ and any positive T, define the function

$$\widetilde{F}_T(y(\cdot)) = \sup_{t \le T} \text{distance}[y(t), L],$$

where $\text{distance}[y, L] = \min_{u\in L} |y - u|$. The function $\widetilde{F}_T(\cdot)$ is continuous on $C^r[0,\infty)$. Then the theorem says that for each subsequence there is a further subsequence (indexed by n_k) such that $E\widetilde{F}_T(\widetilde{\theta}^{n_k}(\cdot)) \to E\widetilde{F}_T(\theta(\cdot)) = 0$, where the limit is zero since the value of $\widetilde{F}_T(\cdot)$ on the paths of the limit process is zero with probability one. Thus, the sup over $t \in [0,T]$ of the distance between the original sequence $\widetilde{\theta}^n(t)$ and L goes to zero in probability as $n \to \infty$. Indeed, the same result holds if T is replaced by $T_n \to \infty$ slowly enough. Thus, there are $T_n \to \infty$ such that for any $\mu > 0$,

$$\lim_n P\left\{ \sup_{t \le T_n} \text{distance}[\widetilde{\theta}^n(t), L] \ge \mu \right\} = 0.$$

We note the key role to be played by the estimate (2.14). This estimate implies the "tightness" condition, which will be basic to the results of Chapter 8 and is guaranteed by the uniform integrability of $\{Y_n\}$.

Proof. The theorem remains true if, for any $T > 0$, $F(\cdot)$ depends on the values of its argument only at times $t \le T$. Both $F(\cdot)$ and T will be fixed henceforth. Recall that a compact set in $C^r[0,T]$ is a set of equicontinuous functions. Let $y(\cdot)$ denote the canonical element of $C^r[0,T]$. For any $\nu > 0$ and compact set $C_0 \subset C^r[0,T]$, there is a $\Delta > 0$ and a real-valued continuous function $F_\Delta(\cdot)$ on $C^r[0,T]$ that depends on $y(\cdot)$ only at times $\{i\Delta, i\Delta \le T\}$ such that

$$|F(y(\cdot)) - F_\Delta(y(\cdot))| \le \nu, \quad y(\cdot) \in C_0.$$

We can write

$$\widetilde{\theta}^n(t) = \theta_n + \sum_{i=n}^{m(t_n+t)-1} \epsilon_i \bar{g}(\theta_i) + \kappa^n(t), \tag{2.16}$$

where $\kappa^n(t) = M^n(t) + B^n(t)$. Since $E|\beta_n| \to 0$, $\lim_n E \max_{t \le T} |B^n(t)| = 0$. By the martingale property, the uniform integrability, (4.1.5) (applied to the sums of $\epsilon_i \delta M_{i,K}$ in the corollary to Theorem 2.1) and the estimates (2.13), for each $T > 0$ we have

$$\lim_n E \max_{t \le T} |M^n(t)| = 0.$$

Since for each $\mu > 0$,

$$\lim_n P\left\{ \sup_{t \le T} |\kappa^n(t)| \ge \mu \right\} = 0, \tag{2.17a}$$

we have

$$\lim_{K \to \infty} \sup_n P\left\{ \sup_{t \le T} \left| \widetilde{\theta}^n(t) \right| \ge K \right\} = 0. \tag{2.17b}$$

Equation (2.17a) and the representation (2.16) imply that (2.14) holds.

Equations (2.14) and (2.17b) imply that for each $\nu > 0$ there is a compact set $C_\nu \subset C^r[0,T]$ such that for each n, $\widetilde{\theta}^n(\cdot) \in C_\nu$ with probability greater than $1 - \nu$. [This is also implied directly by the uniform integrability of $\{Y_n\}$.] Thus we need only show that there is a subsequence n_k and a process $\theta(\cdot)$ satisfying $\dot{\theta} = \bar{g}(\theta)$ and taking values in the largest bounded invariant set of this ODE such that for any $\Delta > 0$,

$$EF_\Delta(\widetilde{\theta}^{n_k}(\cdot)) \to EF_\Delta(\theta(\cdot)),$$

where the bounded and continuous real-valued function $F_\Delta(\cdot)$ depends only on the values of the argument at times $\{i\Delta, i\Delta \le T\}$. [We note that a key point in the general theory is to show that, with a high probability (not depending on n), the paths of $\widetilde{\theta}^n(\cdot)$ are confined to a compact set in the path space. The general theory also uses limits taken along convergent subsequences to help characterize the limits of the original sequence.]

By the fact that $\{\theta_n\}$ is bounded in probability, there is a subsequence n_k and a random variable $\theta(0)$ (on some probability space) such that $\{\theta_{n_k}\}$ converges in distribution to $\theta(0)$. Let $\mathcal{T}$ denote the positive rational numbers. By (2.17b) and the diagonal method, we can take a further subsequence $\{m_k\}$ such that $\{\widetilde{\theta}^{m_k}(t), t \in \mathcal{T}\}$ converges in distribution, and denote the limit (on some probability space) by $\{\theta(t), t \in \mathcal{T}\}$. By the representation (2.16), the boundedness of $\bar{g}(\cdot)$ and the fact that (2.17a) holds for each T, there is a version of the limit that is continuous on $\mathcal{T}$ with probability one. Hence, we can suppose that $\theta(\cdot)$ is defined for all $t \ge 0$ and is continuous.

By (2.14), (2.16), and (2.17a), we can write

$$\theta(n\delta) = \theta(0) + \sum_{i=0}^{n-1} \delta \bar{g}(\theta(i\delta)) + \kappa_\delta(n\delta), \tag{2.18}$$

where $\lim_\delta P\{\sup_{t \le T} |\kappa_\delta(t)| \ge \mu\} = 0$ for each $\mu > 0, T > 0$. By the continuity of $\theta(\cdot)$ we see that it must satisfy the ODE $\dot{\theta} = \bar{g}(\theta)$. By the uniqueness to the solution of the ODE for each initial condition and the boundedness of $\bar{g}(\cdot)$, the limit does not depend on the chosen further subsequence $\{m_k\}$ and the original subsequence can be used. Now (2.15) clearly holds for $F_\Delta(\cdot)$.

We need only show that with probability one the paths of $\theta(\cdot)$ take values in the largest bounded invariant set of the ODE, and we will sketch the details. We have worked on the time interval $[0, \infty)$. Follow the same procedure on the interval $(-\infty, \infty)$ by replacing $\mathcal{T}$ by $\mathcal{T}_1 = \mathcal{T} \cup (-\mathcal{T})$. Then (extracting a further subsequence $\{m_k\}$ if necessary), $\{\widetilde{\theta}^{m_k}(t), t \in \mathcal{T}_1\}$ converges in distribution to a process $\theta(t), t \in \mathcal{T}_1$. As above, it can be assumed that $\theta(t)$ is defined and continuous on $(-\infty, \infty)$ and satisfies the ODE. Again, by the uniqueness of the solution to the ODE for each initial condition, the further subsequence indexed by m_k is not needed, and one can use the original subsequence.

Next, note that the boundedness in probability of the sequence $\{\theta_n\}$ implies that for any $\mu > 0$, there is a $K_\mu < \infty$ such that if θ is any limit in distribution of a subsequence of $\{\theta_n\}$, then

$$P\{|\theta| \ge K_\mu\} \le \mu. \tag{2.19}$$

Thus, for each $\mu > 0$, we can suppose that $|\theta(t)| \le K_\mu$ for each t with probability $\ge 1 - \mu$. Now, this fact and the stability property of the limit set L implies that the solution is bounded and lies in L for all t. □

7.3 Weak Convergence

7.3.1 Definitions

Convergence in distribution. Let $\{A_n\}$ be a sequence of $I\!R^k$-valued random variables on a common probability space $(\Omega, P, \mathcal{F})$, with $(a_{n,i}, i = 1, \ldots, k)$ being the real-valued components of A_n. Let P_n denote the measures on the Borel sets of $I\!R^k$ determined by A_n, and let $x = (x_1, \ldots, x_k)$ denote the canonical variable in $I\!R^k$. If there is an $I\!R^k$-valued random variable A with real-valued components $(a_1, \ldots, a_k)$ such that

$$P\{a_{n,1} < \alpha_1, \ldots, a_{n,k} < \alpha_k\} \to P\{a_1 < \alpha_1, \ldots, a_k < \alpha_k\} \tag{3.1}$$

for each $\alpha = (\alpha_1, \ldots, \alpha_k) \in I\!R^k$ at which the right side of (3.1) is continuous, then we say that A_n *converges to* A *in distribution*. Let P_A denote the measure on the Borel sets of $I\!R^k$ determined by A. An equivalent definition [34] is that

$$EF(A_n) = \int F(x)dP_n(x) \to EF(A) = \int F(x)dP_A(x) \tag{3.2}$$

for each bounded and continuous real-valued function $F(\cdot)$ on $I\!R^k$. We say that the sequence $\{P_n\}$ is *tight* or *bounded in probability* if

$$\lim_{K\to\infty} \sup_n P_n\{(-\infty, -K] \cup [K, \infty)\} = \lim_{K\to\infty} \sup_n P\{|A_n| \geq K\} = 0. \tag{3.3a}$$

For real- or vector-valued random variables, the term *mass preserving* is sometimes used in lieu of *tight*. An equivalent definition of boundedness in probability is: Let $|A_n| < \infty$ with probability one for each n and for each small $\mu > 0$, let there be finite M_μ and K_μ such that

$$P\{|A_n| \geq K_\mu\} \leq \mu, \quad \text{for } n \geq M_\mu. \tag{3.3b}$$

Given a sequence of random variables $\{A_n\}$ with values in $I\!R^k$ (or more generally, in any complete separable metric space), tightness is a necessary and sufficient condition that any subsequence has a further subsequence that converges in distribution [34]. Convergence in distribution is also called *weak convergence.*

The notion of convergence in distribution extends, via the general theory of weak convergence, to sequences of random variables that take values in more abstract spaces than $I\!R^k$. The extension provides a powerful methodology for the approximation of random processes and for obtaining useful limit theorems for sequences of random processes, such as our $\theta^n(\cdot)$.

The following example is one of the classical illustrations of weak convergence. Let $\{\xi_n\}$ be a sequence of real-valued random variables that are mutually independent and identically distributed, with mean zero and unit variance. Then, by the classical central limit theorem $\sum_{i=1}^{n} \xi_i/\sqrt{n}$ converges in distribution to a normally distributed random variable with zero mean and unit variance. Now, define $q(t) = \max\{i : i/n \leq t\}$ and define the process with piecewise constant paths

$$W_n(t) = \sum_{i=0}^{q(t)-1} \xi_i/\sqrt{n}. \tag{3.4}$$

Then the central limit theorem tells us that $W_n(t)$ converges in distribution to a normally distributed random variable with mean zero and variance t. For an integer k, let $0 = t_0 < t_1, \ldots, t_{k+1}$ be real numbers, and let $W(\cdot)$ be a real-valued Wiener process with unit variance parameter. Then, by the multivariate central limit theorem [34], the set $\{W_n(t_{i+1}) - W_n(t_i), i \leq k\}$ converges in distribution to $\{W(t_{i+1}) - W(t_i), i \leq k\}$. It is natural to ask whether $W_n(\cdot)$ converges to $W(\cdot)$ in a stronger sense. For example, will the distribution of the first passage time for $W_n(\cdot)$ defined by $\min\{t : W_n(t) \geq 1\}$ converge in distribution to the first passage time for $W(\cdot)$ defined by $\min\{t : W(t) \geq 1\}$? Will the maximum $\max\{W_n(t) : t \leq 1\}$ converge in distribution to $\max\{W(t) : t \leq 1\}$ and similarly for other useful functionals? In general, we would like to know the class of functionals $F(\cdot)$ for which $F(W_n(\cdot))$ converges in distribution to $F(W(\cdot))$. Donsker's Theorem states that this convergence occurs for a large class of functionals [25, 68].

Now, let us consider the following extension, where the ξ_n are as given above. For given real-valued $U(0)$ and $\Delta > 0$, define real-valued random variables U_n^Δ by $U_0^\Delta = U(0)$ and for $n \geq 0$,

$$U_{n+1}^\Delta = U_n^\Delta + \Delta g(U_n^\Delta) + \sqrt{\Delta}\xi_n,$$

where $g(\cdot)$ is a continuous function. Define the interpolated process $U^\Delta(\cdot)$ by: $U^\Delta(t) = U_n^\Delta$ on $[n\Delta, n\Delta + \Delta)$. Then in what sense will $U^\Delta(\cdot)$ converge to the process defined by the stochastic differential equation

$$dU = g(U)dt + dW?$$

We expect that this stochastic differential equation is the "natural" limit of $U^\Delta(\cdot)$.

More challenging questions arise when the random variables ξ_n are correlated. The central limit theorem and the laws of large numbers are very useful for the approximation of random variables, which are the "sums" of many small effects whose mutual dependence is "local," by the simpler normally distributed random variable or by some constant, respectively. The theory of weak convergence is concerned with analogous questions when the random variables are replaced by random processes as in the above examples. There are two main steps analogous to what is done for proving the central limit theorem: First show that there are appropriately convergent subsequences and then identify the limits. The condition (3.3) says that, neglecting a set of small probability for each n (small, uniformly in n), the values of random variables A_n are confined to some compact set. There will be an analogous condition when random processes replace random variables.

The path spaces $D(-\infty, \infty)$ and $D[0, \infty)$. The processes $\theta^n(\cdot), M^n(\cdot)$, and $Z^n(\cdot)$ used in Chapters 5 and 6 and Section 1 were piecewise constant and had small discontinuities (of the order of the step size) at the "jump times." However, when using the Arzelà–Ascoli Theorem to justify the extraction of subsequences that converge uniformly on bounded time intervals to continuous limits, we checked equicontinuity by looking at the processes at the jump times. Equivalently, we used the extended definition of equicontinuity (4.2.2) and the extended Arzelá–Ascoli Theorem 4.2.2. This procedure is obviously equivalent to working with the piecewise linear interpolations. We could continue to work with piecewise linear interpolations or the extended definition of equicontinuity. However, from a technical point of view it turns out to be easier to use a path space that allows discontinuities, in particular because the verification of the extension of the concept of tightness will be simpler.

The statement of Theorem 2.2 used the path space $C^r[0, \infty)$ of the piecewise linear interpolations of θ_n. The applications of weak convergence theory commonly use the space of paths that are right continuous and have limits from the left, with a topology known as the *Skorohod topology.* This topology is weaker than the topology of uniform convergence on bounded time intervals. The key advantage of this weaker topology is that it is easier to prove the *functional* analog of the tightness condition (3.3) for the various processes of interest. This will be more apparent in Chapters 10 and 11, where a more sophisticated form of the theory is used to get the rates of convergence. Since the limit processes $\theta(\cdot)$ will have continuous paths in our applications, the strength of the assertions of the theorem is the same no matter which topology is used.

Let $D(-\infty, \infty)$ (resp., $D[0, \infty)$) denote the space of real-valued functions on the interval $(-\infty, \infty)$ (resp., on $[0, \infty)$) that are right continuous and have left-hand limits, with the Skorohod topology used, and $D^k(-\infty, \infty)$ (resp., $D^k[0, \infty)$) its k-fold product. The exact definition of the Skorohod topology is somewhat technical and not essential for our purposes. It is given at the end of the chapter, but is not explicitly used in subsequent chapters. Full descriptions and treatments can be found in [25, 68]. We note the following properties here. Let $f_n(\cdot)$ be a sequence in $D(-\infty, \infty)$. Then the convergence of $f_n(\cdot)$ to a continuous function $f(\cdot)$ in the Skorohod topology is equivalent to *convergence uniformly on each bounded time interval.* Under the Skorohod topology, $D(-\infty, \infty)$ is a complete and separable metric space. Since we will later use the Skorohod topology to prove rate of convergence results, for consistency we will use it from this point on. Loosely speaking, the Skorohod topology is an extension of the topology of uniform convergence on bounded time intervals in the sense that a "local" small (n, t)-dependent stretching or contraction of the time scale is allowed, the purpose of which is to facilitate dealing with "nice" discontinuities that do not disappear in the limit.

Definition of weak convergence. Let B be a metric space. In our applications, it will be either $\mathbb{R}^k$ or one of the product path spaces $D^k(-\infty, \infty)$ or $D^k[0, \infty)$ for an appropriate integer k. Let $\mathcal{B}$ denote the minimal σ-algebra induced on B by the topology. Let $\{A_n, n < \infty\}$ and A be B-valued random variables defined on a probability space $(\Omega, P, \mathcal{F})$, and suppose that P_n and P_A are the probability measures on $(B, \mathcal{B})$ determined by A_n and A, respectively. We say that P_n *converges weakly* to P_A if (3.2) holds for all bounded, real-valued, and continuous functions on B, and write the convergence as $P_n \Rightarrow P_A$. Equivalently, with a convenient abuse of terminology, we say that A_n *converges weakly* to A or that A is the *weak limit* or *weak sense limit* of $\{A_n\}$, and write $A_n \Rightarrow A$. These ways of expressing weak convergence will be used interchangeably.

A set $\{A_n\}$ of random variables with values in B is said to be *tight* if for each $\delta > 0$ there is a compact set $B_\delta \subset \mathcal{B}$ such that

$$\sup_n P\{A_n \notin B_\delta\} \leq \delta. \tag{3.5}$$

To prove tightness of a sequence of $\mathbb{R}^k$-valued processes, it is enough to prove tightness of the sequence of each of the k components. A set $\{A_n\}$ of B-valued random variables is said to be *relatively compact* if each subsequence contains a further subsequence that converges weakly.

7.3.2 Basic Convergence Theorems

A basic result, Prohorov's Theorem, is given next.

Theorem 3.1. [25, Theorems 6.1 and 6.2]. *If $\{A_n\}$ is tight, then it is relatively compact (i.e, it contains a weakly convergent subsequence). If B is complete and separable, tightness is equivalent to relative compactness.*

Theorem 3.2. [25, Theorem 5.1] *Let $A_n \Rightarrow A$. Let $F(\cdot)$ be a real-valued bounded and measurable function on B that is continuous with probability one under the measure P_A. Then $EF(A_n) \to EF(A)$.*

Tightness in $D(-\infty, \infty)$ and $D[0, \infty)$. An advantage to working with the path space $D(-\infty, \infty)$ in lieu of $C(-\infty, \infty)$ or $C[0, \infty)$ is that it is easier to prove tightness in $D(-\infty, \infty)$. Let $A_n(\cdot)$ be processes with paths in $D(-\infty, \infty)$. The following criteria for tightness will be easy to apply to our problems. Let $\mathcal{F}_t^n$ denote the σ-algebra generated by $\{A_n(s), s \leq t\}$, and let τ denote an $\mathcal{F}_t^n$-stopping time.

Theorem 3.3. [[68, Theorem 8.6, Chapter 3], [118, Theorem 2.7b]] *Let $\{A_n(\cdot)\}$ be a sequence of processes that have paths in $D(-\infty, \infty)$. Suppose that for each $\delta > 0$ and each t in a dense set in $(-\infty, \infty)$, there is a compact set $K_{\delta,t}$ in $I\!R$ such that*

$$\inf_n P\{A_n(t) \in K_{\delta,t}\} \geq 1 - \delta \tag{3.6}$$

and for each positive T,

$$\lim_{\delta \downarrow 0} \limsup_n \sup_{|\tau| \leq T} \sup_{s \leq \delta} E \min[|A_n(\tau + s) - A_n(\tau)|, 1] = 0. \tag{3.7}$$

Then $\{A_n(\cdot)\}$ is tight in $D(-\infty, \infty)$. If the interval $[0, \infty)$ is used, tightness holds if $|\tau| \leq T$ is replaced by $0 \leq \tau \leq T$.

Remarks on tightness and the limit process. Let the piecewise constant interpolations $\theta^n(\cdot)$ and $Z^n(\cdot)$ be defined on $(-\infty, \infty)$ until further notice. Note that representation (2.16) and estimate (2.17) imply the tightness of $\{\theta^n(\cdot)\}$ in $D^r(-\infty, \infty)$. If a compact constraint set H is used, then (3.6) holds. For the problems in Chapters 5 and 6, the fact that the extended Arzelà–Ascoli Theorem was applicable (with probability one) to $\{\theta^n(\cdot), Z^n(\cdot)\}$ implies (3.7) for these processes. Thus the tightness criterion is always satisfied under the conditions used in Chapters 5 and 6. It is clear that (3.7) does not imply the continuity of the paths of either $A_n(\cdot)$ or any weak sense limit $A(\cdot)$. Indeed, (3.7) holds if $\{A_n(\cdot)\}$ is a sequence of continuous-time Markov chains on a compact state space S with uniformly bounded and time independent transition rates. Then it can be shown that any weak sense limit process is also a continuous-time Markov chain with values in S and time independent transition functions.

Suppose that a sequence of processes $\{A^n(\cdot)\}$ is tight in $D^r(-\infty, \infty)$ and that on each interval $[-T, T]$ the size of the maximum discontinuity goes to zero in probability as $n \to \infty$. Then any weak sense limit process must have continuous paths with probability one.

Suppose that $\{\theta^n(\cdot), Z^n(\cdot)\}$ (or, otherwise said $\{P_n\}$) is tight. Let $(\theta(\cdot), Z(\cdot))$ denote the weak sense limit of a weakly convergent subsequence. The next question concerns the characterization of the process $(\theta(\cdot), Z(\cdot))$. It will be shown that the weak sense limit process is characterized as solutions to the mean ODE. In other words, any limit measure is concentrated on a set of paths that satisfy the ODE, with $Z(\cdot)$ being the reflection term. In particular, the limit measure is concentrated on a set of paths that are limit trajectories of the ODE, as $t \to \infty$.

Skorohod representation and "probability one" convergence. In the general discussion of weak convergence and in Theorems 3.1–3.3, the sequence $\{A_n\}$ was defined on some given probability space $(\Omega, P, \mathcal{F})$. Since weak convergence works with measures P_n induced on the range space of the sequence $\{A_n\}$, the actual probability space itself is unimportant, and one can select it for convenience. For purely analytical purposes, it is often helpful to be able to suppose that the convergence is with probability one rather than in the weak sense, since it enables us to work with paths directly and simplifies parts of the proofs. It turns out that the probability space can be chosen such that the weak convergence "implies" convergence with probability one. This basic result is known as the *Skorohod representation.*

Theorem 3.4. [[68, Chapter 3, Theorem 1.8], [222, Theorem 3.1]] *Let B be a complete and separable metric space with metric $d(\cdot,\cdot)$, and let $A_n \Rightarrow A$ for B-valued random variables A_n and A. Then there is a probability space $(\widetilde{\Omega}, \widetilde{\mathcal{B}}, \widetilde{P})$ with associated B-valued random variables $\widetilde{A}_n$ and $\widetilde{A}$ defined on it such that for each set $D \in \mathcal{B}$,*

$$\widetilde{P}\{\widetilde{A}_n \in D\} = P\{A_n \in D\}, \quad \widetilde{P}\{\widetilde{A} \in D\} = P\{A \in D\}, \tag{3.8}$$

and

$$d(\widetilde{A}_n, \widetilde{A}) \to 0 \quad \textit{with probability one.} \tag{3.9}$$

The choice of the probability space in the theorem is known as the *Skorohod representation.* Its use facilitates proofs without changing the distributions of the quantities of interest. In the rest of the book, it will be supposed where convenient in the proofs, and without loss of generality, that the probability space has been selected so that weak convergence is "equivalent to" convergence with probability one.

Note that we have started with a range space B with a σ-algebra $\mathcal{B}$, and measures P_n, P_A defined on it, but with only weak convergence $P_n \Rightarrow P_A$. The Skorohod representation constructs a single probability space with B-valued random variables $\widetilde{A}_n$ and $\widetilde{A}$ defined on it, where $\widetilde{A}_n$ (resp., $\widetilde{A}$) determine the measure $\widetilde{P}_n$ (resp., $\widetilde{P}_A$), on the range space $(B, \mathcal{B})$, and where the convergence is with probability one. For notational simplicity, when the Skorohod representation is used in the sequel, the *tilde* notation will generally be omitted.

Define $A_n = (\theta^n(\cdot), Z^n(\cdot))$, which takes values in $D^{2r}(-\infty,\infty)$. If it has been shown that the sequence $(\theta^n(\cdot), Z^n(\cdot))$ converges weakly to some $D^{2r}(-\infty,\infty)$-valued random variable $(\theta(\cdot), Z(\cdot))$, where $(\theta(\cdot), Z(\cdot))$ have continuous paths with probability one, then by the Skorohod representation (see Theorem 3.4), it can be supposed in the proof of the characterization of $(\theta(\cdot), Z(\cdot))$ that the convergence is with probability one uniformly on bounded time intervals, provided that the conclusions of the theorem remain in terms of weak convergence. In particular, the use of the Skorohod representation itself does not imply that the original sequence θ_n (or $\theta^n(\cdot)$) converges with probability one.

A simple example of Skorohod representation. Let $\{Y_n\}$ be a sequence of real-valued random variables that converges in distribution to a random variable Y, and let $F_n(\cdot)$ (resp., $F(\cdot)$) be the distribution function of Y_n (resp., of Y). Suppose for simplicity that each of the distribution functions is strictly monotonically increasing. The sequence $\{Y_n\}$ might not converge to Y with probability one. In fact, Y_n might not even be defined on the same probability space. But there are random variables $\widetilde{Y}_n$ and $\widetilde{Y}$ such that each of the pairs Y_n and $\widetilde{Y}_n$ (as well as Y and $\widetilde{Y}$) have the same distribution, and $\widetilde{Y}_n$ converges to $\widetilde{Y}$ with probability one.

The construction is as follows. Let the probability space be $(\widetilde{\Omega}, \widetilde{\mathcal{B}}, \widetilde{P})$ where $\widetilde{\Omega} = [0,1]$, $\widetilde{\mathcal{B}}$ is the collection of Borel sets on $[0,1]$, and $\widetilde{P}$ is the Lebesgue measure. For $\widetilde{\omega} \in [0,1]$ define $\widetilde{Y}_n(\widetilde{\omega}) = F_n^{-1}(\widetilde{\omega})$ and $\widetilde{Y}(\widetilde{\omega}) = F^{-1}(\widetilde{\omega})$. By the construction and the uniform distribution on $[0,1]$, $P\{\widetilde{Y}_n \leq a\} = F_n(a)$ for all a. Thus $\widetilde{Y}_n$ (resp., $\widetilde{Y}$) has the distribution function $F_n(\cdot)$ (resp., $F(\cdot)$). Furthermore the uniform convergence of $F_n(\cdot)$ to $F(\cdot)$ and the strict monotonicity imply that $F_n^{-1}(\cdot)$ also converges pointwise to $F^{-1}(\cdot)$. This is equivalent to the convergence of $\widetilde{Y}_n \to \widetilde{Y}$ for all $\widetilde{\omega}$. This is an easy example. In the more general case, where Y_n is replaced by a random process and $\{Y_n\}$ is tight, the limit of any weakly convergent subsequence is not so easily characterized. Then the Skorohod representation can be quite helpful in the analysis.

Return to the central limit theorem discussed in connection with (3.4). The theory of weak convergence tells us that the *process* $W_n(\cdot)$ constructed in (3.4) converges weakly to the Wiener process with unit variance parameter. This result gives us more information on the distributions of real-valued functionals of the paths of $W_n(\cdot)$ for large n than can be obtained by the classical central limit theorem alone, which is confined to working with values at a finite number of fixed points and not with the entire process: see [25, 68] for the details and a full development of the general theory and other examples. For the basic background, effective methods for dealing with wide-bandwidth noise-driven processes or discrete time processes with correlated driving noise, including many applications of the theory to approximation and limit problems arising in applications to control, communication and signal processing theory, as well as to various stochastic approximation-type problems, consult [127].

Some auxiliary results. The following theorems will simplify the analysis. Theorem 3.5 shows the fundamental role of uniform integrability in establishing the Lipschitz continuity of the paths of the weak sense limit processes and generalizes the corollary to Theorem 2.1. Further details of the proof are in Theorem 8.2.1. Sometimes one can show that a sequence of processes can be approximated in some sense by one that can be shown to be tight and for which the weak sense limit can be exhibited. This is dealt with in Theorem 3.6. The proof of Theorem 3.6 follows from the definition of weak convergence; the details are left to the reader.

Theorem 3.5. *Let $\{Y_i^n; n \geq 0, i \geq 0\}$ be a sequence of real-valued random variables that is uniformly integrable, and let ϵ_i^n be non-negative numbers that satisfy*

$$\sum_{i=0}^{\infty} \epsilon_i^n = \infty, \quad \text{for all } n \text{ and}$$

$$\lim_n \sup_i \epsilon_i^n = 0.$$

Define $\tau_k^n = \sum_{i=0}^{k-1} \epsilon_i^n$ and the processes $X^n(\cdot)$ on $D^r[0,\infty)$ by

$$X^n(t) = \sum_{i=0}^{k-1} \epsilon_i^n Y_i^n \quad \text{on } [\tau_k^n, \tau_{k+1}^n).$$

Then $\{X^n(\cdot)\}$ is tight, and all weak sense limit processes have Lipschitz continuous paths with probability one. If $E|Y_i^n| \to 0$ as n and i go to infinity, then the weak sense limit process is identically zero. The analogous results hold for $D^r(-\infty,\infty)$.

Theorem 3.6. *Let the processes $X^n(\cdot)$ have paths in $D^r[0,\infty)$ with probability one. Suppose that for each $1 > \rho > 0$ and $T > 0$ there is a process $X^{n,\rho,T}(\cdot)$ with paths in $D^r[0,\infty)$ with probability one such that*

$$P\left\{\sup_{t \leq T} \left|X^{n,\rho,T}(t) - X^n(t)\right| \geq \rho\right\} \leq \rho.$$

If $\{X^{n,\rho,T}(\cdot), n \geq 0\}$ is tight for each ρ and T, then $\{X^n(\cdot)\}$ is tight. If $\{X^{n,\rho,T}(\cdot), n \geq 0\}$ converges weakly to a process $X(\cdot)$ that does not depend on (ρ, T), then the original sequence converges weakly to $X(\cdot)$. Suppose that for each $1 > \rho > 0$ and $T > 0$, $\{X^{n,\rho,T}(\cdot), n \geq 0\}$ converges weakly to a process $X^{\rho,T}(\cdot)$, and that there is a process $X(\cdot)$ such that the measures of $X^{\rho,T}(\cdot)$ and $X(\cdot)$ on the interval $[0,T]$ are equal, except on a set whose probability goes to zero as $\rho \to 0$. Then $\{X^n(\cdot)\}$ converges weakly to $X(\cdot)$. The analogous result holds for processes with paths in $D^r(-\infty,\infty)$.

7.4 Martingale Limit Processes and the Wiener Process

7.4.1 Verifying that a Process Is a Martingale

The criteria for tightness in Theorem 3.3 will enable us to show that for any subsequence of the shifted stochastic approximation processes $\{\theta^n(\cdot), Z^n(\cdot)\}$, there is always a further subsequence that converges weakly.

The next step will be to identify the limit process; in particular to show that it is a solution to the desired mean ODE, and to do this without excessive effort and under weak conditions. If the noise is not of the martingale difference type, then this step requires an averaging of the noise effects so that the "mean dynamics" appear. Suppose that $(\theta(\cdot), Z(\cdot))$ is the weak sense limit of a weakly convergent subsequence. A particularly useful way of doing both the averaging under weak conditions and identifying the limit process involves showing that $\theta(t)-\theta(0)-\int_0^t \bar{g}(\theta(s))ds - Z(t)$ is a martingale with Lipschitz continuous paths. Recall the fact (Section 4.1) that any continuous-time martingale with Lipschitz continuous paths (with probability one) is a constant (with probability one). The Lipschitz continuity will be easy to prove. Then the martingale property implies that the expression is a constant. Since it takes the value zero at $t = 0$, the limit process satisfies the desired ODE. A convenient criterion for showing that a process is a martingale is needed, and a useful approach is suggested by the definition of a martingale in terms of conditional expectations.

Let Y be a random variable with $E|Y| < \infty$, and let $\{V(s), 0 \le s < \infty\}$, be an arbitrary sequence of random variables. Suppose that for fixed real $t > 0$, each integer p and each set of real numbers $0 \le s_i \le t, i = 1, \ldots, p$, and each bounded and continuous real-valued function $h(\cdot)$, we have

$$Eh(V(s_i), i \le p)Y = 0.$$

This and the arbitrariness of $h(\cdot)$ imply that $E[Y|V(s_i), i \le p] = 0$. The arbitrariness of p and $\{s_i, i \le p\}$ now imply that

$$E[Y|V(s), s \le t] = 0$$

with probability one [34]. To extend this idea, let $U(\cdot)$ be a random process with paths in $D^r[0,\infty)$ such that for all p, $h(\cdot)$, $s_i \le t, i \le p$, as given above and a given real $\tau > 0$,

$$Eh(U(s_i), i \le p) \left[U(t+\tau) - U(t)\right] = 0. \tag{4.1}$$

Then $E[U(t+\tau) - U(t)|U(s), s \le t] = 0$. If this holds for all t and $\tau > 0$ then, by the definition of a martingale, $U(\cdot)$ is a martingale. Sometimes it is convenient to work with the following more general format whose proof follows from the preceding argument. The suggested approach is a standard and effective method for verifying that a process is a martingale.

Theorem 4.1. *Let $U(\cdot)$ be a random process with paths in $D^r[0,\infty)$, where $U(t)$ is measurable on the σ-algebra $\mathcal{F}_t^V$ determined by $\{V(s), s \le t\}$ for some given process $V(\cdot)$ and let $E|U(t)| < \infty$ for each t. Suppose that for each real $t \ge 0$ and $\tau \ge 0$, each integer p and each set of real numbers $s_i \le t$, $i = 1, \ldots, p$, and each bounded and continuous real-valued function $h(\cdot)$,*

$$Eh(V(s_i), i \le p) \left[U(t+\tau) - U(t)\right] = 0, \tag{4.2}$$

then $U(t)$ is an $\mathcal{F}_t^V$-martingale.

7.4.2 The Wiener Process

One of the most important martingales in applications is the Wiener process. Theorem 4.1.2 gave a criterion for verifying that a process is a Wiener process, and we now repeat and elaborate it. Let $W(\cdot)$ be an $I\!R^r$-valued process with continuous paths such that $W(0) = 0$, $EW(t) = 0$, for any set of increasing real numbers $\{t_i\}$, the set $\{W(t_{i+1}) - W(t_i)\}$ is mutually independent, and the distribution of $W(t+s) - W(t)$, $s > 0$ does not depend on t. Then $W(\cdot)$ is called a vector-valued *Wiener process* or *Brownian motion*. There is a matrix Σ, called the covariance, such that $EW(t)W'(t) = \Sigma t$, and the increments are normally distributed [34].

There are other equivalent definitions; one will now be given. Let the $I\!R^r$-valued process $W(\cdot)$ have continuous paths and satisfy $W(0) = 0$ w.p.1. Let $\mathcal{F}_t$ be a sequence of nondecreasing σ-algebras such that $W(t)$ is $\mathcal{F}_t$-measurable and let $E_{\mathcal{F}_t}[W(t+s) - W(t)] = 0$ with probability one for each t and each $s \geq 0$. Let there be a non-negative definite matrix Σ such that for each t and each $s \geq 0$

$$E_{\mathcal{F}_t}\left[W(t+s) - W(t)\right]\left[W(t+s) - W(t)\right]' = \Sigma s \text{ w.p.1.}$$

Then $W(\cdot)$ is a vector-valued Wiener process with covariance Σ, also called an $\mathcal{F}_t$-Wiener process [163, Volume 1, Theorem 4.1].

The criterion of Theorem 4.1 for verifying that a process is a martingale can be adapted to verify that it is a vector-valued $\mathcal{F}_t$-Wiener process for appropriate $\mathcal{F}_t$. Suppose that $W(\cdot)$ is a continuous vector-valued process with $E|W(t)|^2 < \infty$ for each t. Let $V(\cdot)$ be a random process and let $\mathcal{F}_t^V$ be the smallest σ-algebra that measures $\{V(s), W(s), s \leq t\}$. Let $h(\cdot), p, t, \tau > 0, s_i \leq t$ be arbitrary but satisfy the conditions put on these quantities in Theorem 4.1. Suppose that

$$Eh(V(s_i), W(s_i), i \leq p)\left[W(t+\tau) - W(t)\right] = 0 \tag{4.3}$$

and that there is a non-negative definite matrix Σ such that

$$\begin{aligned} &Eh(V(s_i), W(s_i), i \leq p) \\ &\quad\times \left[\left[W(t+\tau) - W(t)\right]\left[W(t+\tau) - W(t)\right]' - \Sigma\tau\right] = 0. \end{aligned} \tag{4.4}$$

Then $W(\cdot)$ is an $\mathcal{F}_t^V$-Wiener process, with covariance Σ.

Proving that (4.4) holds for the weak sense limit of a sequence of processes $\{W^n(t)\}$ might require showing that $\{|W^n(t)|^2\}$ is uniformly integrable. This can be avoided by using the following equivalent characterization.

For a matrix $\Sigma = \{\sigma_{ij}\}$, let A_Σ denote the operator acting on twice continuously differentiable real-valued functions $F(\cdot)$ on $I\!R^r$:

$$A_\Sigma F(w) = \frac{1}{2}\sum_{i,j}\sigma_{ij}\frac{\partial^2 F(w)}{\partial w_i \partial w_j}. \tag{4.5}$$

Theorem 4.2. *Let $F(\cdot)$ be an arbitrary continuous real-valued function on $\mathbb{R}^r$ with compact support and whose mixed partial derivatives up to order three are continuous and bounded. Let $V(\cdot)$ be a random process. Let the $\mathbb{R}^r$-valued process $W(\cdot)$ have continuous paths with probability one and $\Sigma = \{\sigma_{ij}\}$ a non-negative definite symmetric matrix. Suppose that for each real $t \geq 0$ and $\tau \geq 0$, each integer p and each set of real numbers $s_i \leq t, i = 1, \ldots, m$, each bounded and continuous real-valued function $h(\cdot)$,*

$$
\begin{aligned}
&Eh\left(V(s_i), W(s_i), i \leq p\right) \\
&\quad \times \left[F(W(t+\tau)) - F(W(t)) - \int_t^{t+\tau} A_\Sigma F(W(u))du\right] = 0.
\end{aligned}
\tag{4.6}
$$

Then $W(\cdot)$ is an $\mathcal{F}_t^V$-Wiener process with covariance Σ, where $\mathcal{F}_t^V$ is the smallest σ-algebra that measures $\{V(s), W(s), s \leq t\}$.

7.4.3 A Perturbed Test Function Method for Verifying Tightness and Verifying the Wiener Process

In Chapters 5 and 6, we have seen the usefulness of perturbed Liapunov functions and perturbed state methods for proving stability or for averaging correlated noise. Perturbed test function methods are also very useful for proving tightness or characterizing the limit of a weakly convergent sequence. The original perturbed test function ideas stem from the work of Blankenship and Papanicolaou [26], Papanicolaou, Stroock, and Varadhan [187], and Kurtz [117]. Kushner extended them to cover quite general non-Markovian situations and developed powerful techniques for their construction, exploitation and applications to diverse problems; see for example, [127, 132]; see also the remarks on perturbations in Subsection 6.3.1.

In the following perturbed test function theorems, ϵ_i^n are positive real numbers and $\tau_i^n = \sum_{j=0}^{i-1} \epsilon_j^n$. The $\mathbb{R}^r$-valued processes $X^n(\cdot)$ are constant on the intervals $[\tau_i^n, \tau_{i+1}^n)$ and are right continuous. Define $m^n(t) = \max\{i : \tau_i^n \leq t\}$. For each n, let $\mathcal{F}_i^n$ be a sequence of nondecreasing σ-algebras such that $\mathcal{F}_i^n$ measures at least $\{X^n(\tau_j^n), j \leq i\}$, and let E_i^n denote the expectation conditioned on $\mathcal{F}_i^n$. Let $\mathcal{D}^n$ denote the class of right continuous, real-valued random functions $F(\cdot)$ that are constant on the intervals $[\tau_i^n, \tau_{i+1}^n)$, with bounded expectation for each t, and that $F(\tau_i^n)$ is $\mathcal{F}_i^n$-measurable. Define the operator $\widehat{A}^n$ acting on random functions $F(\cdot)$ in $\mathcal{D}^n$ by

$$
\widehat{A}^n F(\tau_i^n) = \frac{E_i^n F(\tau_{i+1}^n) - F(\tau_i^n)}{\epsilon_i^n}. \tag{4.7}
$$

The next theorem is an extension of Theorem 3.5.

Theorem 4.3. [127, Theorems 4 and 8; Chapter 3] *For each real-valued function $F(\cdot)$ on $\mathbb{R}^r$ with compact support and whose mixed partial deriva-*

tives up to second order are continuous, let there be a sequence of processes $F^n(\cdot) \in \mathcal{D}^n$ such that for each $\alpha > 0$ and $T > 0$,

$$\lim_n P\left\{\sup_{s\le T} |F^n(s) - F(X^n(s))| \ge \alpha\right\} = 0, \tag{4.8}$$

and suppose that

$$\lim_{N\to\infty} \sup_n P\left\{\sup_{t\le T} |X^n(t)| \ge N\right\} = 0. \tag{4.9}$$

Define γ_i^n by

$$E_i^n F^n(\tau_{i+1}^n) - F^n(\tau_i^n) = \epsilon_i^n \gamma_i^n = \epsilon_i^n \widehat{A}^n F^n(\tau_i^n). \tag{4.10}$$

If $\{\gamma_i^n; n, i : \tau_i^n \le T\}$ is uniformly integrable for each T and $F(\cdot)$, then $\{F(X^n(\cdot)\}$ is tight in $D[0,\infty)$, and $\{X^n(\cdot)\}$ is tight in $D^r[0,\infty)$. The analogous result holds on $D^r(-\infty,\infty)$. If, in addition, for each $T > 0$, $E|\gamma_i^n| \to 0$ uniformly in $\{i : \tau_i^n \le T\}$ as $n \to \infty$, then the weak sense limit is the "zero" process.

Theorem 4.4. *Let $X^n(0) = 0$, and suppose that $\{X^n(\cdot)\}$ is tight in $D^r[0,\infty)$ and that each of the weak sense limit processes has continuous paths with probability one. Let*

$$\lim_n \sup_i \epsilon_i^n = 0. \tag{4.11a}$$

Suppose that there are integers μ_i^n such that $\lim_n \inf_i \mu_i^n = \infty$ with the following properties:

(a)

$$\lim_n \sup_i \sum_{j=i}^{i+\mu_i^n-1} \epsilon_i^n = 0, \quad \lim_n \sup_{i+\mu_i^n\ge j\ge i} \left|\frac{\epsilon_j^n - \epsilon_i^n}{\epsilon_i^n}\right| = 0; \tag{4.11b}$$

(b) for each continuous real-valued function $F(\cdot)$ on $I\!R^r$ with compact support and whose mixed partial derivatives up to order three are continuous, and for each $T > 0$, there is an $F^n(\cdot)$ in $\mathcal{D}^n$ such that

$$\lim_n E\left|F^n(t) - F(X^n(t))\right| = 0, \quad t \le T, \tag{4.12}$$

$$\sup_{t\le T} E|\widehat{A}^n F^n(t)| < \infty, \quad \text{each } n; \tag{4.13}$$

(c) for the non-negative definite matrix $\Sigma = \{\sigma_{ij}\}$

$$\lim_n E\left|\frac{1}{\mu_i^n}\sum_{j=i}^{i+\mu_i^n-1} E_i^n\left[\widehat{A}^n F^n(\tau_j^n) - A_\Sigma F(X^n(\tau_j^n))\right]\right| = 0, \tag{4.14}$$

where the limit is taken on uniformly in i for $\tau_i^n \le T$, and each $T > 0$, and A_Σ is defined in (4.5).

Then $X^n(\cdot)$ converges weakly in $D^r[0,\infty)$ to the Wiener process with covariance Σ.

Let the set $\{V^n(\cdot)\}$ also be tight in $D^r[0,\infty)$, where $V^n(\cdot)$ is constant on the intervals $[\tau_i^n, \tau_{i+1}^n)$, and $\{V^n(\cdot), X^n(\cdot)\}$ converges weakly to $(V(\cdot), X(\cdot))$. Let (4.12)–(4.14) continue to hold, where $\mathcal{F}_i^n$ measures at least $\{X^n(\tau_j^n), V^n(\tau_j^n), j \leq i\}$. If $\mathcal{F}_t^V$ is the smallest σ-algebra that measures $\{V(s), X(s),\ s \leq t\}$, then $X(\cdot)$ is an $\mathcal{F}_t^V$-Wiener process. The analogous result holds on $(-\infty, \infty)$.

Let there be a continuous function $\bar{g}(\cdot)$ such that the conditions hold with $A_\Sigma F(x)$ replaced by $A_\Sigma F(x) + F_x'(x)\bar{g}(x)$. Then the limit $X(\cdot)$ of any weakly convergent subsequence of $\{X^n(\cdot)\}$ can be characterized as follows: There is a Wiener process $W(\cdot)$ with covariance matrix Σ such that $X(\cdot)$ satisfies the stochastic differential equation

$$X(t) = X(0) + \int_0^t \bar{g}(X(s))ds + W(t), \tag{4.15}$$

where for each t, $\{X(s), s \leq t\}$ and $\{W(s) - W(t), s \geq t\}$ are independent.

Note on the proof. This is actually an adaptation of [127, Theorem 8, Chapter 5] to the decreasing-step-size case. In the proof, one needs to show that for each $t \geq 0$ and small $\tau > 0$, with $t + \tau \leq T$,

$$E_{m^n(t)}^n \sum_{j=m^n(t)}^{m^n(t+\tau)-1} \epsilon_j^n \left[\widehat{A}^n F^n(\tau_j^n) - A_\Sigma F(X^n(\tau_j^n))\right] \to 0$$

in mean as $n \to \infty$. By the conditions on ϵ_i^n, this is implied by (4.14).

The Skorohod topology. Let Λ_T denote the space of continuous and strictly increasing functions from the interval $[0,T]$ onto the interval $[0,T]$. The functions in this set will be "allowable time scale distortions" for the functions in $D[0,T]$. Define the metric $d_T(\cdot)$ by

$$\begin{aligned} d_T(f(\cdot), g(\cdot)) = \inf \Big\{\mu : & \sup_{0 \leq s \leq T} |s - \lambda(s)| \leq \mu \text{ and} \\ & \sup_{0 \leq s \leq T} |f(s) - g(\lambda(s))| \leq \mu \text{ for some } \lambda(\cdot) \in \Lambda_T\Big\}. \end{aligned}$$

If there are $\eta_n \to 0$ such that the discontinuities of $f_n(\cdot)$ are less than η_n in magnitude and if $f_n(\cdot) \to f(\cdot)$ in $d_T(\cdot)$, then the convergence is uniform on $[0,T]$ and $f(\cdot)$ must be continuous. Because of the "time scale distortion" involved in the definition of the metric $d_T(\cdot)$, we can have (loosely speaking) convergence of a sequence of discontinuous functions, where there are only a finite number of discontinuities, where both the locations and the values of the discontinuities converge, and a type of "equicontinuity" condition holds between the discontinuities. For example, let $T > 1$ and define $f(\cdot)$

by: $f(t) = 1$ for $t < 1$ and $f(t) = 0$ for $t \geq 1$. Define the function $f_n(\cdot)$ by $f_n(t) = 1$ for $t < 1 + 1/n$ and $f_n(t) = 0$ for $t \geq 1 + 1/n$. Then $f_n(\cdot)$ converges to $f(\cdot)$ in the Skorohod topology, but not in the sup norm, and $d_T(f(\cdot), f_n(\cdot)) = 1/n$. The minimizing time scale distortion is illustrated in Figures 4.1 and 4.2.

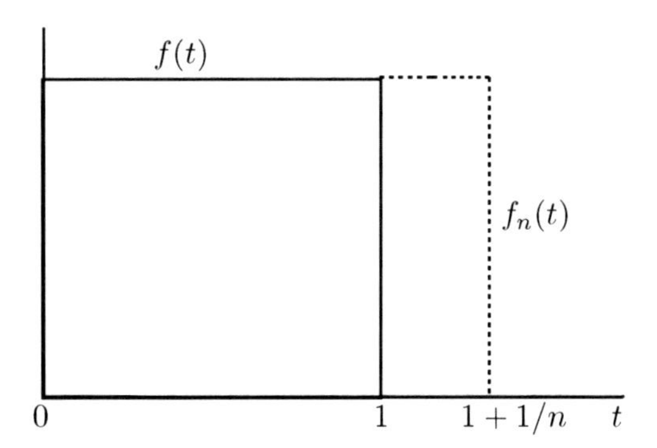

Figure 4.1. The functions $f(\cdot)$ and $f_n(\cdot)$.

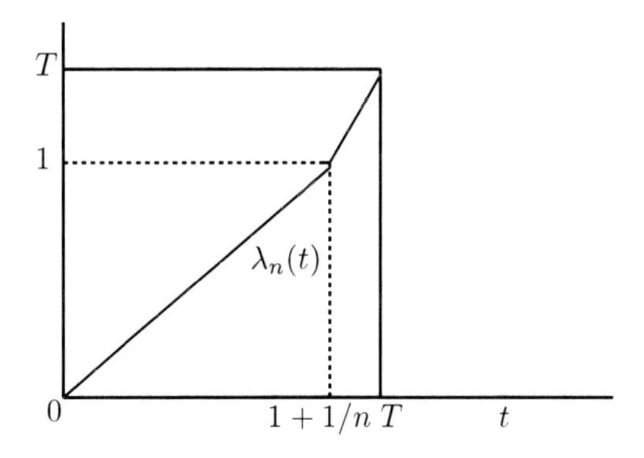

Figure 4.2. The time scale distortion $\lambda_n(\cdot)$.

Under $d_T(\cdot)$, the metric space $D[0, T]$ is separable but not complete [25, 68]. There is an equivalent metric $\widehat{d}_T(\cdot)$ under which the space is both complete and separable. The $\widehat{d}_T(\cdot)$ weights the "derivative" of the time scale changes $\lambda(t)$ and its deviation from t. For $\lambda(\cdot) \in \Lambda_T$, define

$$|\lambda| = \sup_{s<t<T} \left| \log \left\{ \frac{\lambda(t) - \lambda(s)}{t - s} \right\} \right|.$$

The metric $\widehat{d}_T(\cdot)$ is defined by

$$\widehat{d}_T(f(\cdot), g(\cdot)) = \inf \Big\{ \mu : |\lambda| \le \mu \text{ and } \sup_{0 \le s \le T} |f(s) - g(\lambda(s))| \le \mu, \\ \text{for some } \lambda(\cdot) \in \Lambda_T \Big\}.$$

Both $d_T(\cdot)$ and $\widehat{d}_T(\cdot)$ are referred to as *Skorohod metrics*. The topology under $\widehat{d}_T(\cdot)$ is called the *Skorohod topology*.

On the space $D[0, \infty)$, the metric for the Skorohod topology is defined by

$$\widehat{d}(f(\cdot), g(\cdot)) = \int_0^\infty e^{-t} \min(1, \widehat{d}_t\,(f(\cdot), g(\cdot))\, dt,$$

and analogously on $D(-\infty, \infty)$. The metrics on the product spaces $D^r[0, \infty)$ and $D^r(-\infty, \infty)$ can be taken to be the sum of the metrics on the component spaces.

8
Weak Convergence Methods for General Algorithms

8.0 Outline of Chapter

The main results for the weak convergence of stochastic approximation algorithms are given in this chapter, using the ideas of Sections 7.3 and 7.4. The relative simplicity of the proofs and weakness of the required conditions in comparison with the probability one method will be seen. Section 1 lists the main conditions used for both the martingale difference noise case and the correlated "exogenous" noise case when the step size does not change with time. For a single process with fixed step size $\epsilon_n = \epsilon$, there can only be convergence in distribution as $n \to \infty$. With an arbitrarily high probability, for small enough ϵ the limit process is concentrated in an arbitrarily small neighborhood of some limit set of the mean ODE.

When the noise is not of the martingale difference type, the term *exogenous* noise is used in a loose way to indicate that the conditional distribution of the "future of the noise," given its "past," does not change if we also condition on the past of the iterate sequence. This is to distinguish it from the "state-dependent" noise model, which was introduced in the examples of Sections 2.3–2.5 and discussed briefly in Section 6.6. The precise meaning of *exogenous noise* is that assumption (A1.9) holds. Section 2, which contains the basic convergence results, starts by using the theory of weak convergence to redo the martingale difference noise problem of Section 7.2 in a more general and efficient way. The special constructions used in Section 7.2 are not needed, and the results are stronger due to the use of the general theory of weak convergence. Then the exogenous correlated noise is treated. The minor alterations needed when the step size goes to zero are given in Subsection 2.3.

The proof for the martingale difference noise case in Theorem 2.1 lays out the general "martingale" method for proving convergence and will be used for all subsequent cases as well. This is a very useful and widely used way of exploiting the theory of weak convergence for proving limit results for sequences of stochastic processes. In the proof for the correlated exogenous noise case in Theorem 2.2, the efficiency of the martingale method for the averaging of the correlated noise is seen. These proofs are then easily extended to the decreasing and random step-size problems, where it is seen that little is required for the step sizes other than that they go to zero. When the underlying data change with time on a scale that is commensurate with that determined by the step sizes, then the mean ODE might depend on time. Such issues arise in tracking time-varying systems. This is dealt with in Theorem 2.6.

Several forms of the Kiefer–Wolfowitz algorithm are covered in Section 3. If the noise is of the martingale difference type, then the proof is a minor modification of that of Theorem 2.1, and the essential condition on (ϵ_n, c_n) is that $\epsilon_n/c_n^2 \to 0$. When the noise is correlated, the analysis of the Kiefer–Wolfowitz algorithm depends on whether or not there is a correlated noise term that is inversely proportional to the finite difference interval c_n in the expansion of the finite difference estimator. Depending on how the algorithm is constructed, there might or might not be such a term. Two forms are covered in Section 3. In the first case, when the finite difference expansion is done and the martingale difference term separated out, the remaining term, which contains the correlated noise, is not divided by c_n. Then the proof is very close to that of the Robbins–Monro algorithm with correlated noise. When there is a term containing correlated noise that is divided by c_n, the problem has a "singular" character. But it can be effectively treated by a perturbed test function method such as that introduced in Theorem 7.4.3. One such class will be dealt with in Theorem 3.3 to illustrate some of the underlying ideas. The proof of Theorem 3.3 introduces the use of truncated processes to prove tightness and convergence when we are not *a priori* certain that the sequence of processes of concern is suitably bounded in probability.

The Markov (and non-Markov) state-dependent noise models are treated in Section 4. For the Markovian form of the model, the (effective noise memory, state) pair has a Markovian structure. This models effectively many problems of current interest in manufacturing, communications, and queue control, among others, as noted in Sections 2.3–2.5. A powerful and widely used general method of proof is the "invariant measure" approach, which originated in [127, 142] and generalized in [145]. This and two alternative approaches for such state-dependent noise problems will be discussed, each being practical and having its own advantages. A full appreciation of their value and the relative simplicity of their use can be seen by comparing them with the much more complicated proofs and more limited conclusions obtained by alternative currently used approaches, as noted in Section 4. Except for some special considerations due to the state dependence of the noise, the proofs are direct extensions of what was done for the exogenous noise case. Loosely speaking, the main additional condition is a weak continuity of the Markov transition function. The model is quite general, since the state space of the Markov chain can include the entire past.

The method of proof of Theorem 4.1 is based on the "weak continuity" idea in Appendix 2 of [145]. It is simpler than what is required for the invariant measure method, in both the assumptions and the proof. The invariant measure method is dealt with in Theorem 4.4 and an extension for non-Markov state-dependent noise processes is in Theorem 4.5. Theorem 4.6 is modeled on Theorem 4.1 but uses a more direct approach. Its conditions are implied by those of Theorem 4.1. It differs from Theorem 4.1 in that it essentially assumes that a certain approximation holds. Nevertheless, this can often be proved to be the case in rather complex applications, as seen in the queue control problem in Chapter 9. Most of the analysis is in terms of the conditional mean functions $g_n^\epsilon(\cdot)$ or $g_n(\cdot)$, since it typically has a smoother dependence on θ than do Y_n^ϵ or Y_n. But sometimes it is most convenient to work directly in terms of the observations Y_n^ϵ or Y_n. This requires only some notational changes and is dealt with in Theorem 4.7.

Section 5 deals with unconstrained algorithms. The basic idea is to first show either recurrence or boundedness in probability of the iterate sequence, and then to "locally" adapt the methods used earlier in the chapter for the constrained problem. Although the proof and statement of the theorem are confined to the unconstrained form of Theorem 2.1, the same idea works for all of the cases dealt with in this chapter. The use of soft constraints, as in Section 5.5, can be dealt with by the same stability and localization methods. A Liapunov function method for proving tightness of the iterate sequence is in Section 6.7. The development of the two time scale problem in Section 6 requires only minor modifications.

The assumptions on the noise process do not imply stationarity. Nevertheless, for the most part, they imply that the dynamics "average out" to a function $\bar{g}(\theta)$ of θ that does not depend on time. Otherwise, the mean ODE might be time dependent or replaced by a differential inclusion. An example of the required modification of the conditions when the mean ODE depends on time is given in Subsection 2.6 for the basic algorithm with constant step size. There are obvious analogies for all the cases dealt with.

8.1 Exogenous Noise: Problem Formulation and Constant Step Size

The primary development will be for the constant-step-size case, where $\epsilon_n = \epsilon$ and ϵ is small. The proof for the decreasing-step-size case $\epsilon_n \to 0$ is essentially the same, and the minor adjustments that are required will be given in Subsection 2.3.

Definitions and interpolated processes. With fixed step size ϵ, the constrained algorithm is

$$\theta_{n+1}^\epsilon = \Pi_H(\theta_n^\epsilon + \epsilon Y_n^\epsilon), \tag{1.1}$$

where $\{Y_n^\epsilon\}$ is the observation sequence. As usual, define the reflection term Z_n^ϵ by rewriting (1.1) as

$$\theta_{n+1}^\epsilon = \theta_n^\epsilon + \epsilon Y_n^\epsilon + \epsilon Z_n^\epsilon. \tag{1.2}$$

Analogously to what was done for the decreasing-step-size cases in Chapters 5 and 6, where $\epsilon_n \to 0$, we work with a sequence of continuous-time interpolations. It is natural that the interpolation interval be the constant step size ϵ. Define the interpolation $\theta^\epsilon(\cdot)$ as follows. For $t < 0$, set $\theta^\epsilon(t) = \theta_0^\epsilon$. For $t \geq 0$, set $\theta^\epsilon(t) = \theta_n^\epsilon$ on the time interval $[n\epsilon, n\epsilon + \epsilon)$. Define the interpolation $Z^\epsilon(t) = \epsilon \sum_{i=0}^{t/\epsilon - 1} Z_i^\epsilon$, where $t \geq 0$, and set $Z^\epsilon(t) = 0$ for $t < 0$. The *integer part* of t/ϵ is always used in the limit of summation when such sums are defined. The integer part of a real number τ is defined to be the largest (positive, negative, or zero) integer less than or equal to τ. Define $Y^\epsilon(\cdot)$ analogously but using Y_n^ϵ in lieu of Z_n^ϵ.

When proving the asymptotic results for the decreasing-step-size case in Chapters 5 and 6, it was convenient to work with the sequence of shifted process $\{\theta^n(\cdot)\}$. The limits of pathwise convergent subsequences of these shifted processes satisfy the mean ODE and characterize the asymptotic properties of $\{\theta_n\}$. An analogous method will be used with the weak convergence approach.

Let $\{q_\epsilon\}$ be a sequence of nondecreasing non-negative integers, and define $Z_n^\epsilon = Y_n^\epsilon = 0$ for $n < 0$. Analogously to the definition (5.1.8), define the process $Z^{\epsilon,q}(\cdot)$ by

$$Z^{\epsilon,q}(t) = \epsilon \sum_{i=q_\epsilon}^{t/\epsilon + q_\epsilon - 1} Z_i^\epsilon, \quad \text{for } t \geq 0, \tag{1.3a}$$

and

$$Z^{\epsilon,q}(t) = -\epsilon \sum_{i=t/\epsilon + q_\epsilon}^{q_\epsilon - 1} Z_i^\epsilon, \quad \text{for } t < 0. \tag{1.3b}$$

Define the processes $Y^{\epsilon,q}(\cdot)$ analogously, using Y_n^ϵ in lieu of Z_n^ϵ. Then, for all t,

$$\theta^\epsilon(\epsilon q_\epsilon + t) = \theta^\epsilon_{q_\epsilon} + Y^{\epsilon,q}(t) + Z^{\epsilon,q}(t). \tag{1.4}$$

There are two cases that are usually of interest. In the first, we do not shift at all, so that $q_\epsilon = 0$ and ϵ is small. In this case, one is concerned with the behavior of the θ_n^ϵ for small ϵ and moderate values of $n\epsilon$. On the other hand, to study the problem for large time (the behavior of θ_n^ϵ for large n and small ϵ, where $n\epsilon \to \infty$), one works with $\theta^\epsilon(\epsilon q_\epsilon + \cdot)$ where $\epsilon q_\epsilon \to \infty$ and $\epsilon \to 0$. This is analogous to what was done with the sequence $\theta^n(\cdot)$ in Chapters 5 and 6 to get the tail behavior of $\{\theta_n\}$.

Assumptions: Martingale difference noise. Let $\{\mathcal{F}_n^\epsilon\}$ be a sequence of nondecreasing σ-algebras, where $\mathcal{F}_n^\epsilon$ measures at least $\{\theta_j^\epsilon, Y_{j-1}^\epsilon, j \le n\}$, and let E_n^ϵ denote the expectation conditioned on $\mathcal{F}_n^\epsilon$. The following assumptions will be used in the next section for the martingale difference noise case. In Section 5, which deals with the unconstrained case, the conditions will be "localized."

(A1.1) $\{Y_n^\epsilon; \epsilon, n\}$ is uniformly integrable.

(A1.2) There are functions $g_n^\epsilon(\cdot)$ that are continuous uniformly in n and ϵ and random variables β_n^ϵ such that

$$E_n^\epsilon Y_n^\epsilon = g_n^\epsilon(\theta_n^\epsilon) + \beta_n^\epsilon. \tag{1.5}$$

(A1.3) There is a continuous function $\bar{g}(\cdot)$ such that for each $\theta \in H$,

$$\lim_{m,n,\epsilon} \frac{1}{m} \sum_{i=n}^{n+m-1} \left[g_i^\epsilon(\theta) - \bar{g}(\theta)\right] = 0. \tag{1.6}$$

(A1.4)
$$\lim_{m,n,\epsilon} \frac{1}{m} \sum_{i=n}^{n+m-1} E_n^\epsilon \beta_i^\epsilon = 0 \quad \text{in mean.}$$

The $\lim_{m,n,\epsilon}$ means that the limit is taken as $m \to \infty$, $n \to \infty$, and $\epsilon \to 0$ simultaneously in any way at all.

Assumptions for the correlated noise case. Following the correlated noise model used in Chapter 6, let ξ_n^ϵ be "memory" random variables taking values in some complete and separable metric space Ξ, and let $\mathcal{F}_n^\epsilon$ measure at least $\{\theta_j^\epsilon, Y_{j-1}^\epsilon, \xi_j^\epsilon, j \le n\}$. In Theorem 2.2, the following conditions will be used in addition to (A1.1) and (A1.4). In what follows A is an arbitrary compact set in Ξ. Examples are given in Chapter 9.

(A1.5) There are measurable functions $g_n^\epsilon(\cdot)$ and random variables β_n^ϵ such that

$$E_n^\epsilon Y_n^\epsilon = g_n^\epsilon(\theta_n^\epsilon, \xi_n^\epsilon) + \beta_n^\epsilon. \tag{1.7}$$

(A1.6)

$$g_n^\epsilon(\cdot, \xi) \text{ is continuous in } \theta \tag{1.8}$$

uniformly in n, ϵ and in $\xi \in A$.

(A1.7) For each $\delta > 0$ there is a compact set $A_\delta \subset \Xi$ such that

$$\inf_{n,\epsilon} P\{\xi_n^\epsilon \in A_\delta\} \geq 1 - \delta. \tag{1.9}$$

(A1.8) For each θ, the sequences

$$\{g_n^\epsilon(\theta_n^\epsilon, \xi_n^\epsilon); \epsilon, n\}, \quad \{g_n^\epsilon(\theta, \xi_n^\epsilon); \epsilon, n\} \tag{1.10}$$

are uniformly integrable.

(A1.9) There is a continuous function $\bar{g}(\cdot)$ such that for each $\theta \in H$,

$$\lim_{m,n,\epsilon} \frac{1}{m} \sum_{i=n}^{n+m-1} E_n^\epsilon \left[g_i^\epsilon(\theta, \xi_i^\epsilon) - \bar{g}(\theta)\right] I_{\{\xi_n^\epsilon \in A\}} = 0 \text{ in probability.} \tag{1.11}$$

The following condition will sometimes be used when the constraint set H is dropped.

(A1.10) $\{\theta_n^\epsilon\}$ is tight.

Liapunov function based methods for proving the tightness in (A1.10) are in Section 6.7. The uniform continuity in (A1.6) implies that the $\bar{g}(\cdot)$ in (A1.9) is continuous, and also that the convergence in (A1.9) is uniform in θ in each compact set. One can get somewhat weaker conditions by assuming the uniformity in (A1.9) first, and the following set of conditions (A1.11)–(A1.13) can replace (A1.6) and (A1.9). It allows us to take advantage of the "smoothing effects"of the randomness over several iterates, and the fact that θ_n^ϵ and θ_n change very slowly. Throughout the rest of the chapter expressions such as $E_n g_l(\theta, \xi_l^\epsilon), n \leq l$, will appear. To simplify the statements of the assumptions, it is always supposed that such integrals are well defined.

(A1.11) There is a continuous function $\bar{g}(\cdot)$ such that as $m \to \infty$

$$\frac{1}{m} E \left| \sum_{l=n}^{n+m-1} E_n^\epsilon g_l^\epsilon(\theta, \xi_l^\epsilon) - \bar{g}(\theta) \right| I_{\{\xi_n^\epsilon \in A\}} \to 0 \tag{1.12}$$

uniformly in (large) n, (small) ϵ, and in θ in any compact set.

(A1.12) For each large integer m,

$$\frac{1}{m} E \left| \sum_{l=n}^{n+m-1} \left[E_n^\epsilon g_l^\epsilon(\theta_n^\epsilon, \xi_l^\epsilon) - E_n^\epsilon g_l^\epsilon(\theta_l^\epsilon, \xi_l^\epsilon)\right] \right| I_{\{\xi_n^\epsilon \in A\}} \to 0, \tag{1.13}$$

uniformly in n, as $\epsilon \to 0$.

(A1.13) For each large integer m and as $|\widehat{\theta} - \widetilde{\theta}| \to 0$,

$$\frac{1}{m} E \left| \sum_{l=n}^{n+m-1} \left[E_n^\epsilon g_l^\epsilon(\widehat{\theta}, \xi_i^\epsilon) - E_n^\epsilon g_l^\epsilon(\widetilde{\theta}, \xi_i^\epsilon) \right] \right| I_{\{\xi_n^\epsilon \in A\}} \to 0, \tag{1.14}$$

uniformly in n and ϵ, where the vectors $\widetilde{\theta}$ and $\widehat{\theta}$ are in any compact set.

Remarks on the conditions. The conditions are phrased so that they fit a wide variety of applications and are relatively easy to use in the proofs. The triple $(\xi_n^\epsilon, \beta_n^\epsilon, g_n^\epsilon(\cdot))$ is not unique, but the averaging conditions (A1.3) and (A1.4) or (A1.9) are in reference to them, and this should be kept in mind when the ξ_n^ϵ are chosen. However, the best choice is often obvious. The β_n^ϵ are presumed to represent a "bias" that is asymptotically unimportant, although the only property of the β_n^ϵ that we use is (A1.4).

Note that the use of the *conditional expectation* E_n^ϵ in (A1.4) and (A1.9) gives a weaker condition than that given by the weak law of large numbers that one would have without it. The conditions are probably not far from necessary. If the continuity in (A1.6) is uniform in $\xi \in \Xi$, then we can choose the topology on the range space of ξ_n^ϵ so that it is compact. Then (A1.7) automatically holds.

Notice that it is not necessary that $Eg_n^\epsilon(\theta, \xi_n^\epsilon) = \bar{g}(\theta)$. The term $Eg_n^\epsilon(\theta, \xi_n^\epsilon)$ can be periodic or "locally average" to $\bar{g}(\theta)$. Under any of these conditions, (A1.9) holds under simple mixing conditions, M-dependence, or ergodicity type conditions on $\{\xi_n^\epsilon\}$. Under M-dependence of $\{\xi_n^\epsilon\}$, $E_n^\epsilon g_i^\epsilon(\theta, \xi_i^\epsilon) = Eg_i^\epsilon(\theta, \xi_i^\epsilon)$ for $i - n > M$. Consider the special case in which $\xi_n^\epsilon = \sum_{k=0}^\infty b_k \psi_{n-k}$, where ψ_i are mutually independent and identically distributed with mean zero and finite variance, and $\{b_k\}$ is square summable. Suppose that, in (A1.6), $g_n^\epsilon(\cdot) = g(\cdot)$ (it does not depend on (n, ϵ)) and is continuous in (θ, ξ). Then the uniform integrability (A1.1) implies that $E_n^\epsilon g(\theta, \xi_i) \to Eg(\theta, \xi)$ in the mean as $i - n \to \infty$, where the expectation is taken with respect to the stationary distribution of ξ_n^ϵ. By the uniform integrability in (1.10b), the convergence in probability in (1.11) assures convergence in the mean.

8.2 Convergence under Exogenous Noise

8.2.1 Constant Step Size: Martingale Difference Noise

The proof of the following theorem proceeds by a series of natural approximations, each one simplifying the process a little more until we get what is needed. The proof is fundamental in that all subsequent proofs of weak convergence of the stochastic approximation iterates are variations of it, and all the basic techniques are illustrated in it. More general dependence on the past $\theta_j^\epsilon, j \le n$, in the functions $g_n^\epsilon(\cdot)$ can be allowed; see the comments after the theorem for extensions and interpretations.

Theorem 2.1. *Assume* (A1.1)–(A1.4), *and any of the constraint set conditions* (A4.3.1), (A4.3.2), *or* (A4.3.3). *Then, for any nondecreasing sequence of integers* q_ϵ, *for each subsequence of* $\{\theta^\epsilon(\epsilon q_\epsilon + \cdot), Z^\epsilon(\epsilon q_\epsilon + \cdot), \epsilon > 0\}$, *there exist a further subsequence and a process* $(\theta(\cdot), Z(\cdot))$ *such that*

$$(\theta^\epsilon(\epsilon q_\epsilon + \cdot), Z^\epsilon(\epsilon q_\epsilon + \cdot)) \Rightarrow (\theta(\cdot), Z(\cdot))$$

as $\epsilon \to 0$ *through the convergent subsequence, where*

$$\theta(t) = \theta(0) + \int_0^t \bar{g}(\theta(s))ds + Z(t). \tag{2.1}$$

In addition, there is an integrable $z(\cdot)$ *such that*

$$Z(t) = \int_0^t z(s)ds, \text{ where } z(t) \in -C(\theta(t))$$

for almost all ω, t. *Furthermore, for any bounded and continuous function* $F(\cdot)$ *on* $D^r(-\infty, \infty)$,

$$EF(\theta^\epsilon(\epsilon q_\epsilon + \cdot)) \to EF(\theta(\cdot)), \tag{2.2}$$

as $\epsilon \to 0$ *through the convergent subsequence. For any* $\delta > 0$, *the fraction of time that* $\theta^\epsilon(\epsilon q_\epsilon + \cdot)$ *spends in* $N_\delta(L_H)$ *on* $[0, T]$ *goes to one (in probability) as* $\epsilon \to 0$ *and* $T \to \infty$. *Suppose that the solution to* (2.1) *is unique for each initial condition. Then* L_H *can be replaced by the set of chain recurrent points.*

Let $\epsilon q_\epsilon \to \infty$ *as* $\epsilon \to 0$. *Then for almost all* ω, *the path* $\theta(\omega, \cdot)$ *lies in a limit set of* (2.1). *There are* $T_\epsilon \to \infty$ *such that for* $\delta > 0$,

$$\limsup_\epsilon P\left\{\theta^\epsilon(\epsilon q_\epsilon + t) \notin N_\delta(L_H), \text{ some } t \le T_\epsilon\right\} = 0.$$

Now, drop the constraint set H *and assume* (A1.10). *Then the conclusions continue to hold with* $Z(t) = 0$ *and* L_H *replaced by a bounded invariant set of* $\dot{\theta} = \bar{g}(\theta)$.

Remarks on the support of the limit processes and the convergence assertions. For a fixed step size $\epsilon_n = \epsilon$ or if $\epsilon_n \to 0$ very slowly, there can only be convergence in some distributional sense as $n \to \infty$. Fix $\epsilon_n = \epsilon$. Let $q_\epsilon = 0$ and $\theta_0^\epsilon = \theta_0$, so that we are concerned with $\theta^\epsilon(\cdot)$ on an interval that is arbitrarily large and can even go to infinity appropriately as $\epsilon \to 0$. The weak convergence assertion in the theorem implies that the path $\theta^\epsilon(\cdot)$ will closely follow the solution to the ODE on any finite interval, with an arbitrarily high probability (uniformly in the initial condition) as $\epsilon \to 0$. As the length of the time interval increases and $\epsilon \to 0$, the fraction of time that the path of the ODE must eventually spend in a small neighborhood of L_H goes to one. Thus, $\theta^\epsilon(\cdot)$ will eventually (with an arbitrarily high probability) spend nearly all of its time in a small neighborhood of L_H as well.

The time interval over which the path of the stochastic approximation follows the path of the ODE goes to infinity as $\epsilon \to 0$. The speed depends on the details of the algorithm, particularly on the local dynamics, the value of ϵ, and the noise terms.

Now, let $\epsilon q_\epsilon \to \infty$. Then, the weak sense limit paths $\theta(\cdot)$ lie in L_H; for small ϵ, and with an arbitrarily high probability, $\theta^\epsilon(\epsilon q_\epsilon + \cdot)$ will spend nearly all of its time in an arbitrarily small neighborhood of L_H.

Suppose that there is a set A that is asymptotically stable in the sense of Liapunov and that for small $\delta > 0$, $\limsup_n P\{\theta_n^\epsilon \in N_\delta(A)\} > \mu$ for some $\mu > 0$ and small ϵ. Then, the problem of escape from some larger neighborhood of A is one in the theory of large deviations. It will be a rare event for small ϵ. As noted in Section 7.2, by the estimates in Sections 6.9 and 6.10, the order of the escape time (if it ever occurs) is often of the form $e^{c/\epsilon}$ for some $c > 0$. These comments also apply to all of the theorems in the rest of the chapter.

A comment on chain recurrence. We do not have a complete analog of the result concerning chain recurrence in Section 5.2.2 and its extensions elsewhere in Chapters 5 and 6, since that requires working on the infinite time interval with individual sample paths. But the same intuition is helpful. Suppose that the conditions on the ODE of Theorem 5.2.5 holds. Let $\delta > 0$ be small. Suppose that x is not chain recurrent and that for each $\delta > 0$ there are $\mu > 0$ and $T > 0$ such that

$$\limsup_{\epsilon \to 0} \limsup_{s \to \infty} P\left\{\theta^\epsilon(s+t) \in N_\delta(x) \text{ for some } t \leq T\right\} \geq \mu.$$

The fact that x is not chain recurrent implies that, for small $\delta > 0$, $x \notin R_\infty^\delta(x)$ (see Section 5.2.2 for the definition). Thus, for small enough $\delta_1 < \delta$ the solution of the ODE (starting in $N_{\delta_1}(x)$ and δ-perturbed as in the definition in Section 5.2.2) will eventually not return to $N_{\delta_1}(x)$. Also, it will enter a neighborhood of $R_\infty^\delta(x)$ that is attracting to $R_\infty^\delta(x)$.

Let $\Phi(t|x)$ denote the solution of the ODE at time t with initial condition x. Given $q_\epsilon, \delta > 0$ and $\mu_1 > 0$, there are $T_\epsilon \to \infty$ such that

$$\sup_{t \leq T_\epsilon} |\theta^\epsilon(\epsilon q_\epsilon + t) - \Phi(t|\theta^\epsilon(\epsilon q_\epsilon))| \leq \delta,$$

and

$$\sup_{t \leq T_\epsilon} \left|\theta^\epsilon(\epsilon q_\epsilon + kT_\epsilon + t) - \Phi(t\middle|\theta^\epsilon(\epsilon q_\epsilon + kT_\epsilon))\right| \leq \delta, \; k \geq 1,$$

each with probability $\geq 1 - \mu_1$. Thus, as $\epsilon \to 0$, if the path starts in and eventually returns to a small neighborhood of a point (with positive probability, which does not go to zero as $\epsilon \to 0$) on an arbitrarily large (but finite) time interval, then that point has to be chain recurrent.

Although we cannot work with individual paths as in the case of Chapters 5 and 6, the above facts give us related information. The differences between successive return times to a small neighborhood of a non-chain-recurrent point x will eventually (as $\epsilon \to 0$) go to infinity in probability; indeed under broad conditions the "return time intervals" to $N_{\delta_1}(x)$ will be at least of the order of $e^{c/\epsilon}$ for some $c > 0$.

Unstable points in the limit set. The theorem asserts that the process spends nearly all of its time in a small neighborhood of the limit or invariant set (or of the subset of chain recurrent points of such a set). Consider the case where $\bar{g}(\cdot)$ is the negative of a gradient. Then the chain recurrent points in the limit set consists of the stationary points. The theorem does not rule out the possibility that convergence will be to an unstable or saddle point, with some positive probability. Under the given conditions, it is always possible (although very unlikely in typical examples) that some noise value will push the path to a local maximum or saddle, after which time, there will be no noise. This issue was addressed briefly in Section 5.8. There, in the context of probability one convergence and martingale difference noise, it was seen that nonconvergence to the "bad" points could be assured if the noise "acts in all directions" in a non-degenerate way. Alternatively, one could add occasional perturbing noise to the algorithm. As noted in Section 5.8, the effects of the perturbing noise that assure nonconvergence to the bad points might be hard to see in a practical algorithm, since they are asymptotic, and the algorithm might spend a lot of time in a neighborhood of such a point before moving away.

In the weak convergence context, the situation with either martingale difference or correlated noise is similar. One could deal with the problem using the methods of large deviations theory. Let $\epsilon_n = \epsilon$. Under broad conditions, the asymptotic behavior can be characterized as follows. Let $K_i, i \leq m$, denote the set of connected compact invariant or limit sets for the ODE. The path might wander from a small neighborhood of one to a small neighborhood of another, with relatively little time spent in transit. The motion between such sets behaves, asymptotically, like a Markov chain. The amount of time that the path spends in a neighborhood of some K_i before moving out depends on the ability of the noise to drive it out, and on the stabilizing or unstabilizing role of the dynamics. If the noise acts "in all directions" in a non-degenerate way, then the time spent in a small neighborhood of an asymptotically stable K_i (if it ever got there) would be very large, of the order of $e^{c/\epsilon}$ for some $c > 0$. The time spent in a small neighborhood of an unstable K_i (or only marginally stable, such as a saddle point) would not be of exponential order. Thus, asymptotically, the path spends the bulk (perhaps all) of its time in a small neighborhood of the asymptotically stable K_i. It might even converge to one of them. Precise formulas can be worked out for the transition probabilities and the time. Details of a related large deviations problem in stochastic approximation are in [130], where the algorithm was constructed with the aim that it would never get stuck in any of the K_i, so that the entire space could be searched for a global minimum. See also [76, Chapter 6] where the basic ideas for the asymptotic chain were developed in the context of diffusion processes.

Proof. For notational simplicity, we restrict our attention to the case $q_\epsilon = 0$. Before proceeding further, the basic ideas of the proof will be outlined so that its essential structure is clear.

Part 1. Outline of the basic steps of the proof. Define the martingale difference $\delta M_n^\epsilon = Y_n^\epsilon - g_n^\epsilon(\theta_n^\epsilon) - \beta_n^\epsilon$. Recall the definition of $Z^\epsilon(\cdot)$ given below (1.2) and define the sums

$$M^\epsilon(t) = \epsilon \sum_{i=0}^{t/\epsilon - 1} \delta M_i^\epsilon, \quad B^\epsilon(t) = \epsilon \sum_{i=0}^{t/\epsilon - 1} \beta_i^\epsilon, \quad Y^\epsilon(t) = \epsilon \sum_{i=0}^{t/\epsilon - 1} Y_i^\epsilon, \tag{2.3a}$$

$$\bar{G}^\epsilon(t) = \epsilon \sum_{i=0}^{t/\epsilon - 1} \bar{g}(\theta_i^\epsilon), \quad \widetilde{G}^\epsilon(t) = \epsilon \sum_{i=0}^{t/\epsilon - 1} \left[g_i^\epsilon(\theta_i^\epsilon) - \bar{g}(\theta_i^\epsilon) \right], \tag{2.3b}$$

where by t/ϵ in the limit of the sum we mean the integer part as usual. Rewrite the algorithm as

$$\theta^\epsilon(t) = \theta_0^\epsilon + \bar{G}^\epsilon(t) + \widetilde{G}^\epsilon(t) + M^\epsilon(t) + B^\epsilon(t) + Z^\epsilon(t). \tag{2.4}$$

Define the process $W^\epsilon(\cdot)$ by

$$W^\epsilon(t) = \theta^\epsilon(t) - \theta_0^\epsilon - \bar{G}^\epsilon(t) - Z^\epsilon(t) = \widetilde{G}^\epsilon(t) + M^\epsilon(t) + B^\epsilon(t).$$

In Part 2 of the proof, it will be shown that the uniform integrability (A1.1) implies tightness of $\{\theta^\epsilon(\cdot), Z^\epsilon(\cdot)\}$ and the Lipschitz continuity of the paths of any weak sense limit process. It then follows that there is a subsequence $\epsilon(k) \to 0$ as $k \to \infty$ and a pair of processes $(\theta(\cdot), Z(\cdot))$ with Lipschitz continuous paths such that

$$(\theta^{\epsilon(k)}(\cdot), Z^{\epsilon(k)}(\cdot)) \Rightarrow (\theta(\cdot), Z(\cdot)). \tag{2.5}$$

To simplify the notation, we write ϵ for $\epsilon(k)$ in what follows. The limit process will be characterized by using the method of Theorem 7.4.1 to show that a process is a martingale. This martingale will have Lipschitz continuous paths, and hence (by virtue of Theorem 4.1.1) it is a constant. This will yield the mean ODE.

Fix t and, for any integer p, let $s_i \leq t,\ i \leq p$, and $\tau > 0$. Let $h(\cdot)$ be a bounded, continuous, and real-valued, function of its arguments. Then, by the definition of $W^\epsilon(\cdot)$ and the representation (2.4),

$$\begin{aligned}&Eh(\theta^\epsilon(s_i), Z^\epsilon(s_i), i \leq p)\left[W^\epsilon(t+\tau) - W^\epsilon(t)\right]\\ &-Eh(\theta^\epsilon(s_i), Z^\epsilon(s_i), i \leq p)\left[\widetilde{G}^\epsilon(t+\tau) - \widetilde{G}^\epsilon(t)\right]\\ &-Eh(\theta^\epsilon(s_i), Z^\epsilon(s_i), i \leq p)\left[M^\epsilon(t+\tau) - M^\epsilon(t)\right]\\ &-Eh(\theta^\epsilon(s_i), Z^\epsilon(s_i), i \leq p)\left[B^\epsilon(t+\tau) - B^\epsilon(t)\right] = 0.\end{aligned} \tag{2.6}$$

Note that $\mathcal{F}^\epsilon_{t/\epsilon}$ measures $\{M^\epsilon(s), s \leq t\}$ and that the process $M^\epsilon(\cdot)$ is an $\mathcal{F}^\epsilon_{t/\epsilon}$-martingale. The next to last term in (2.6) equals zero by this martingale property, as is seen from the following evaluation of the nested conditional means, that is, by writing the expectation as an expectation of a conditional expectation:

$$Eh(\theta^\epsilon(s_i), Z^\epsilon(s_i), i \leq p)E\left[M^\epsilon(t+\tau) - M^\epsilon(t) | \theta^\epsilon_j, Y^\epsilon_j, j \leq t/\epsilon\right] = 0. \tag{2.7}$$

It will be shown in Part 3 of the proof that the second and fourth lines of (2.6) also go to zero as $\epsilon \to 0$. The form of the second line of (2.6) is particularly useful for use of the "averaging condition" (A1.3) (or (A1.9) for the correlated noise case), which will be used to show that the term goes to zero as $\epsilon \to 0$. Condition (A1.4) will be used to show that the last line of (2.6) goes to zero.

Until the end of this outline, let us suppose the convergences to zero of the second and fourth lines of (2.6), which were claimed in the last paragraph. To complete the characterization of the limit process, define

$$W(t) = \theta(t) - \theta(0) - \int_0^t \bar{g}(\theta(s))ds - Z(t). \tag{2.8}$$

Then the assumed convergence of the terms in (2.6) and the assumed weak convergence imply that

$$Eh(\theta(s_i), Z(s_i), i \leq p)\left[W(t+\tau) - W(t)\right] = 0. \tag{2.9}$$

By the definition (2.8), $W(t)$ is a function of $\{\theta(s), Z(s), s \leq t\}$. Using this, the arbitrariness of the choices of $h(\cdot)$, p, t, $s_i \leq t$, and τ, and Theorem 7.4.1 yield that $W(\cdot)$ is a martingale. In particular,

$$E\left[W(t+\tau) - W(t) | \theta(s), Z(s), s \leq t\right] = 0. \tag{2.10}$$

The fact that $\theta(\cdot)$ and $Z(\cdot)$ have Lipschitz continuous paths with probability one implies that $W(\cdot)$ is a constant with probability one (Theorem 4.1.1). Since $W(0) = 0$, $W(t) = 0$ for all t. Thus (2.1) holds. $Z(\cdot)$ will be shown to be the reflection term in the last paragraph of the proof. The details will now be filled in.

Part 2. Tightness and Lipschitz continuity. [This part provides some of the details of the proof of Theorem 7.3.5.] If the Y_n^ϵ were uniformly bounded, then tightness would be obvious from the criteria (7.3.6) and (7.3.7). The uniform integrability (A1.1) effectively allows us to suppose that the Y_n^ϵ are uniformly bounded, as can be seen from the following argument. In the rest of the proof, T and δ will be fixed positive numbers, with T being arbitrarily large and δ arbitrarily small.

For each $K > 0$, define $Y_{n,K}^\epsilon$ to equal Y_n^ϵ on the set where $|Y_n^\epsilon| \leq K$ and zero elsewhere. Define $\delta Y_{n,K}^\epsilon = Y_n^\epsilon - Y_{n,K}^\epsilon$. The uniform integrability (A1.1) implies that $\sup_{i,\epsilon} E|\delta Y_{i,K}^\epsilon| \to 0$ as $K \to \infty$. Hence,

$$\epsilon \sum_{i=0}^{T/\epsilon - 1} E|\delta Y_{i,K}^\epsilon| \leq \epsilon \frac{T}{\epsilon} \sup_{\epsilon, i \leq T/\epsilon} E|\delta Y_{i,K}^\epsilon| \to 0 \tag{2.11}$$

as $K \to \infty$. Also, the function defined by

$$\epsilon \sum_{i=0}^{t/\epsilon - 1} |Y_{i,K}^\epsilon| \tag{2.12}$$

can change values only at integral multiples of ϵ (and at each of those points the value can change by at most $K\epsilon$). Consequently,

$$\lim_{\alpha \to 0} \limsup_{\epsilon} P\left\{ \sup_{s \leq \alpha, t \leq T} |Y^\epsilon(t+s) - Y^\epsilon(t)| \geq \delta \right\} = 0. \tag{2.13}$$

Since for $s > 0$,

$$|Z^\epsilon(t+s) - Z^\epsilon(t)| \leq \sum_{i=t/\epsilon}^{(t+s)/\epsilon} \epsilon |Y_i^\epsilon|, \tag{2.14}$$

(2.13) also holds for $\{Z^\epsilon(\cdot)\}$, hence for $\{\theta^\epsilon(\cdot)\}$. Indeed,

$$\lim_{\alpha \to 0} \limsup_{\epsilon} E \sup_{s \leq \alpha, t \leq T} |\theta^\epsilon(t+s) - \theta^\epsilon(t)| = 0. \tag{2.15}$$

Now the fact that (2.13) holds for the processes $(\theta^\epsilon(\cdot), Z^\epsilon(\cdot))$ together with the fact that $\theta_0^\epsilon = \theta^\epsilon(0)$ is bounded and $Z^\epsilon(0) = 0$, yields the tightness of

$$\{\theta^\epsilon(\cdot), Z^\epsilon(\cdot)\}$$

via the criterion (7.3.6) and (7.3.7). Similarly, (2.11) and the (uniform in ϵ) Lipschitz continuity (at the jump points) of the process defined by (2.12) (for which the Lipschitz constant is $\leq K$) imply that any limit processes $(\theta(\cdot), Z(\cdot))$ must have Lipschitz continuous paths with probability one.

Part 3. Eliminating the $\widetilde{G}^\epsilon(\cdot)$ and $B^\epsilon(\cdot)$ terms in (2.6). Next, the continuity in condition (A1.2) and the averaging condition (A1.3) will be used to show that the second line of (2.6) goes to zero as $\epsilon \to 0$. Let $\Delta > 0$. By (2.15) and the continuity part of (A1.2), replacing the bracketed term in the second line of (2.6) by

$$\sum_{j=t/\Delta}^{(t+\tau)/\Delta-1} \Delta \left[\frac{\epsilon}{\Delta} \sum_{i=j\Delta/\epsilon}^{(j\Delta+\Delta)/\epsilon-1} \left[g_i^\epsilon(\theta^\epsilon(j\Delta)) - \bar{g}(\theta^\epsilon(j\Delta))\right] \right] \tag{2.16}$$

and letting $\epsilon \to 0$ and then $\Delta \to 0$ yields the same as that of the original terms as $\epsilon \to 0$. By (A1.3) and the continuity part of (A1.2), for each fixed $\Delta > 0$ the limit as $\epsilon \to 0$ of the bracketed term in (2.16) is zero. By (A1.4), the limit (as $\epsilon \to 0$) of the last term in (2.6) is zero.

Part 4. Completion. By the tightness of $\{\theta^\epsilon(\cdot), Z^\epsilon(\cdot)\}$, there is a weakly convergent subsequence. Let us work with such a subsequence and (with a slight abuse of notation) index it by ϵ for notational simplicity. Let $(\theta(\cdot), Z(\cdot))$ denote the weak sense limit process. Then, by the weak convergence and the uniform integrability of $\{W^\epsilon(t)\}$ for each t, the first line in (2.6) converges to the left side of (2.9). Since the other terms in (2.6) go to zero, (2.9) holds. Now, (2.9) and Theorem 7.4.1 imply that $W(\cdot)$ is a martingale. The paths are Lipschitz continuous and $W(0) = 0$. Hence, by Theorem 4.1.1, $W(t) = W(0) = 0$.

By the Lipschitz continuity of $Z(\cdot)$, we can write it as an integral $Z(t) = \int_0^t z(s)ds$, and it remains only to characterize $z(\cdot)$. At this point suppose, without loss of generality, that the probability space is chosen according to the Skorohod representation of Theorem 7.3.4, so that we can suppose that the convergence is in the sense of probability one in the topology of $D^{2r}[0,\infty)$. Since the limit process is continuous, this means that (under the Skorohod representation) with probability one, $(\theta^\epsilon(\cdot), Z^\epsilon(\cdot))$ converges to $(\theta(\cdot), Z(\cdot))$ uniformly on each bounded time interval. Now, analogously to the situation in Theorem 5.2.3, under (A4.3.1) or (A4.3.2), $Z_n^\epsilon \in -C(\theta_{n+1}^\epsilon)$, and $Z_n^\epsilon = 0$ if $\theta_{n+1}^\epsilon \in H^0$, the interior of H. Under (A4.3.3), Z_n^ϵ is orthogonal to the surface at θ_{n+1}^ϵ. These facts, the upper semicontinuity (4.3.2), and the convergence imply that for almost all (t, ω), $z(t) \in -C(\theta(t))$. The rest of the details are left to the reader. □

Details of use of the Skorohod representation. In the proof of the theorem, we have shown that

$$Eh(\theta^{\epsilon(k)}(s_i), Z^{\epsilon(k)}(s_i), i \le p) \left[W^{\epsilon(k)}(t+\tau) - W^{\epsilon(k)}(t)\right] \to 0, \tag{$*$}$$

and that $\{W^{\epsilon(k)}(t)\}$ is uniformly integrable for each t, where we now use $\epsilon(k)$ to index the weakly convergent subsequence. If we can assert that $(\theta^{\epsilon(k)}(\cdot), Z^{\epsilon(k)}(\cdot), W^{\epsilon(k)}(\cdot))$ converges to the limit $(\theta(\cdot), Z(\cdot), W(\cdot))$ uniformly on each interval $[0, T]$, then (2.9) will hold. Since the processes in

(2.9) appear only via the expectation, the underlying probability space is irrelevant.

By Theorem 7.3.4, the weak convergence implies that there is a probability space $(\widetilde{\Omega}, \widetilde{\mathcal{F}}, \widetilde{P})$ with processes $(\widetilde{\theta}^{\epsilon(k)}(\cdot), \widetilde{Z}^{\epsilon(k)}(\cdot), \widetilde{W}^{\epsilon(k)}(\cdot)), k = 0, 1, \ldots,$ and $(\widetilde{\theta}(\cdot), \widetilde{Z}(\cdot), \widetilde{W}(\cdot))$ defined on it, which have the same distribution as the processes without the tilde, and $(\widetilde{\theta}^{\epsilon(k)}(\cdot), \widetilde{Z}^{\epsilon(k)}(\cdot), \widetilde{W}^{\epsilon(k)}(\cdot)) \rightarrow (\widetilde{\theta}(\cdot), \widetilde{Z}(\cdot), \widetilde{W}(\cdot))$ with probability one in the topology of $D^{3r}[0, \infty)$. Since the weak sense limit processes have continuous paths with probability one, so do the processes $(\widetilde{\theta}(\cdot), \widetilde{Z}(\cdot), \widetilde{W}(\cdot))$. Due to this continuity of the paths of the limit process, the Skorohod topology has the property that the convergence of the tilde processes is uniform on each bounded time interval with probability one. Then, because (*) holds for the tilde processes, (2.9) holds for the tilde processes. Hence, (2.9) holds as stated. In applications of Skorohod representation in the sequel, the tilde notation will not be used.

8.2.2 Correlated Noise

The following theorem concerns the case of correlated noise; see also the comments after Theorem 2.1.

Theorem 2.2. *Assume the conditions of Theorem* 2.1 *except with* (A1.5)–(A1.9) *replacing* (A1.2) *and* (A1.3). *Then the conclusions of Theorem* 2.1 *hold. The conclusions continue to hold if* (A1.11)-(A1.13) *replaces* (A1.6) *and* (A1.9).

Note on (A1.6) and (A1.11). The proof (here and in the sequel) will be done under (A1.6) and (A1.9) since the notation is a little simpler. But, the way that conditional expectations are used in (2.21)–(2.25) allows the use of (A1.11)-(A1.13) instead of (A1.6) and (A1.9).

Proof. Redefine $\widetilde{G}^\epsilon(\cdot)$ as

$$\widetilde{G}^\epsilon(t) = \epsilon \sum_{i=0}^{t/\epsilon - 1} \left[g_i^\epsilon(\theta_i^\epsilon, \xi_i^\epsilon) - \bar{g}(\theta_i^\epsilon) \right]. \tag{2.17}$$

Then the proof is the same as that of Theorem 2.1 except for the method of dealing with the second line in (2.6). The following paragraph provides the main ideas.

Eliminating the term $\widetilde{G}^\epsilon(\cdot)$ in (2.6). The assumptions (A1.6)–(A1.9) will be used to show that the second line in (2.6) goes to zero as $\epsilon \rightarrow 0$. The idea is simple and requires two basic steps, which will be outlined before the formal details are given.

First, by (2.15), with a "high probability," the process $\theta^\epsilon(\cdot)$ changes "little" on small time intervals, so that it can be approximated uniformly by piecewise constant functions whose intervals of constancy do not depend on ϵ (for small ϵ). The approximation need only be in the sense of probability or in the mean because (2.6) involves only expectations. Second, after having made the piecewise constant approximation, we rewrite the expectation as a suitable expectation of a conditional expectation and use (A1.9).

Now returning to the proof, let $\Delta > 0$ and $T > 0$, and let ϵ be small enough so that Δ/ϵ is large. Define the index set $I_j^{\epsilon,\Delta} = \{i : j\Delta/\epsilon \le i < (j+1)\Delta/\epsilon\}$. For any positive number K, let $g_{n,K}^\epsilon(\theta,\xi)$ denote the function $g_n^\epsilon(\theta,\xi)$ whose components are truncated at $\pm K$. It will next be shown that

$$\sup_{j\le T/\Delta}\sup_{i\in I_j^{\epsilon,\Delta}} E|g_{i,K}^\epsilon(\theta_i^\epsilon,\xi_i^\epsilon) - g_{i,K}^\epsilon(\theta^\epsilon(j\Delta),\xi_i^\epsilon)| \to 0 \tag{2.18}$$

as $\epsilon \to 0$ and then $\Delta \to 0$. The assertion (2.18) is a consequence of the following facts: (a) the tightness condition (A1.7), which allows us to suppose that the values of the $\{\xi_n^\epsilon\}$ are confined to a compact set when evaluating the expectation in (2.18); (b) the θ-continuity (A1.6) of $g_i^\epsilon(\cdot,\xi)$, uniformly in (ϵ, i) on each compact ξ-set; (c) the continuity result (2.15), which holds here.

By (2.15),

$$\lim_\Delta \limsup_\epsilon E\sum_{j=0}^{T/\Delta-1}\epsilon\sum_{i\in I_j^{\epsilon,\Delta}}|\bar g(\theta_i^\epsilon) - \bar g(\theta^\epsilon(j\Delta))| = 0. \tag{2.19}$$

Let $s_i \le t, i \le p$, and $\tau > 0$. By (2.18), (2.19), and the uniform integrability of the set in (1.10), the set of possible limits (as $\epsilon \to 0$) of

$$Eh(\theta^\epsilon(s_i), Z^\epsilon(s_i), i\le p)\left[\widetilde G^\epsilon(t+\tau) - \widetilde G^\epsilon(t)\right] \tag{2.20}$$

is the same (as $\epsilon \to 0$, then $\Delta \to 0$ and then $K \to \infty$) as that of

$$Eh(\theta^\epsilon(s_i), Z^\epsilon(s_i), i\le p)\left[\sum_{j=t/\Delta}^{(t+\tau)/\Delta-1}\epsilon\sum_{i\in I_j^{\epsilon,\Delta}}\left[g_{i,K}^\epsilon(\theta^\epsilon(j\Delta),\xi_i^\epsilon) - \bar g(\theta^\epsilon(j\Delta))\right]\right]. \tag{2.21}$$

The expression (2.21) equals the expression obtained when the bracketed term is replaced by

$$\left[\sum_{j=t/\Delta}^{(t+\tau)/\Delta-1}\epsilon E_{j\Delta/\epsilon}^\epsilon\sum_{i\in I_j^{\epsilon,\Delta}}\left[g_{i,K}^\epsilon(\theta^\epsilon(j\Delta),\xi_i^\epsilon) - \bar g(\theta^\epsilon(j\Delta))\right]\right], \tag{2.22}$$

where the conditional expectation can be used since $j\Delta \ge t$ and the data in the argument of $h(\cdot)$ in (2.21) are at time points no greater than t.

Thus, to show that the limit of (2.21) is zero, it is only necessary to prove that the bracketed term in (2.22) goes to zero in mean as $\epsilon \to 0$, then $\Delta \to 0$, and then $K \to \infty$, and (A1.9) is used for this following an argument similar to the one used in the proof of Theorem 6.1.1. For small $\alpha > 0$, let $\{B_k^\alpha, k \le l_\alpha\}$ be a finite partition of H where each of the sets B_k^α has diameter less than α. Let x_k^α be any point in B_k^α. Now write the sum in (2.22) in the partitioned form

$$\sum_{j=t/\Delta}^{(t+\tau)/\Delta-1} \Delta \frac{\epsilon}{\Delta} \sum_{k=1}^{l_\alpha} \sum_{i \in I_j^{\epsilon,\Delta}} \left[E_{j\Delta/\epsilon}^\epsilon g_{i,K}^\epsilon(\theta^\epsilon(j\Delta), \xi_i^\epsilon) - \bar{g}(\theta^\epsilon(j\Delta)) \right] I_{\{\theta^\epsilon(j\Delta) \in B_k^\alpha\}}. \tag{2.23}$$

Similarly to what was done to get (2.18), the continuity (A1.6), the boundedness of $g_{i,K}^\epsilon(\cdot)$, the tightness (A1.7), and the fact that α is arbitrarily small imply that it is enough to show for each j and small $\Delta > 0$,

$$\sum_{k=1}^{l_\alpha} \epsilon \sum_{i \in I_j^{\epsilon,\Delta}} \left[E_{j\Delta/\epsilon}^\epsilon g_{i,K}^\epsilon(x_k^\alpha, \xi_i^\epsilon) - \bar{g}(x_k^\alpha) \right] I_{\{\theta^\epsilon(j\Delta) \in B_k^\alpha\}} \tag{2.24}$$

goes to zero in mean as $\epsilon \to 0$ and then $K \to \infty$. By the uniform integrability of the set in (1.10) for each θ, K can be dropped in the $g_{i,K}^\epsilon(\cdot)$ in (2.24). Define $n_\epsilon = \Delta/\epsilon$. Since $\{g_i^\epsilon(\theta, \xi_i^\epsilon)\}$ is uniformly integrable for each θ by (A1.8), the expression (2.24) goes to zero in mean as desired if

$$\lim_\epsilon \frac{1}{n_\epsilon} \sum_{i=jn_\epsilon}^{jn_\epsilon+n_\epsilon-1} \left[E_{jn_\epsilon}^\epsilon g_i^\epsilon(\theta, \xi_i^\epsilon) - \bar{g}(\theta) \right] I_{\{\xi_{jn_\epsilon}^\epsilon \in A\}} = 0 \tag{2.25}$$

in mean, uniformly in j for each $\theta \in H$, where A is a compact set. This last condition is just (A1.9). The rest of the details are as in the proof of Theorem 2.1. □

Comment on the grouping of the iterates in the averaging argument. In (2.21) the summands (indices in the inner sum) are grouped as

$$[0, \Delta/\epsilon - 1], [\Delta/\epsilon, 2\Delta/\epsilon - 1], \ldots$$

for the purposes of averaging via use of (2.25) or (A1.9). Once grouped, the summands within each group are approximated as in (2.24) and then (A1.9) or (2.25) is applied to each group as $n_\epsilon \to \infty$ or $\epsilon \to 0$.

There is considerable flexibility in the grouping; other groupings will be used in the next two subsections and in Theorem 3.1. One such grouping method takes the following form. Let n_ϵ be any sequence of integers going to infinity such that $\Delta_\epsilon = \epsilon n_\epsilon \to 0$. Note that the proof of Theorem 2.2 will still work if we group the iterates in lots of size n_ϵ. Then (2.23) is replaced by

$$\sum_{j=t/\Delta_\epsilon}^{(t+\tau)/\Delta_\epsilon-1} \epsilon \sum_{k=1}^{l_\alpha} \sum_{i=jn_\epsilon}^{jn_\epsilon+n_\epsilon-1} \left[E^\epsilon_{j\Delta_\epsilon/\epsilon} g^\epsilon_{i,K}(\theta^\epsilon(j\Delta_\epsilon), \xi^\epsilon_i) - \bar{g}(\theta^\epsilon(j\Delta_\epsilon))\right] I_{\{\theta^\epsilon(j\Delta_\epsilon)\in B^\alpha_k\}}. \tag{2.26}$$

Then one replaces $\theta^\epsilon(j\Delta_\epsilon)$ in the arguments of the functions by x^α_k as in (2.24) and applies (A1.9) with the new definition of n_ϵ.

8.2.3 Step Size $\epsilon_n \to 0$

The development for the decreasing-step-size case requires only minor modifications of the proofs for the constant-step-size case. Return to the algorithm

$$\theta_{n+1} = \Pi_H \left[\theta_n + \epsilon_n Y_n\right]$$

of (5.2.10) and (6.1.1). Recall the definitions of $\mathcal{F}_n$ and E_n given above (6.1.2); namely, ξ_n are the "memory" random variables, taking values in some topological space Ξ, $\{\mathcal{F}_n\}$ is a sequence of nondecreasing σ-algebras, and $\{\theta_0, Y_{i-1}, \xi_i, i \le n\}$ is $\mathcal{F}_n$-measurable. E_n is the expectation conditioned on $\mathcal{F}_n$.

The following modifications of the conditions in Section 1 will be used. The martingale difference and the correlated noise cases will be dealt with simultaneously.

(A2.1) $\{Y_n\}$ is uniformly integrable.

(A2.2) There are measurable functions $g_n(\cdot)$ and random variables β_n such that

$$E_n Y_n = g_n(\theta_n, \xi_n) + \beta_n.$$

(A2.3) For each compact set $A \subset \Xi$,

$$g_n(\cdot, \xi_n) \text{ is continuous in } \theta$$

uniformly in n and in $\xi_n \in A$.

(A2.4) For each $\delta > 0$, there is a compact set $A_\delta \subset \Xi$ such that

$$\inf_n P\{\xi_n \in A_\delta\} \ge 1 - \delta. \tag{2.27}$$

(A2.5) The following sets

$$\{g_n(\theta_n, \xi_n)\}, \quad \{g_n(\theta, \xi_n)\} \text{ for each } \theta, \tag{2.28}$$

are uniformly integrable.

(A2.6) Let

$$\lim_{n,m} \frac{1}{m} \sum_{i=n}^{n+m-1} E_n \beta_i = 0, \tag{2.29}$$

where the limit is in mean.

(A2.7) There is a continuous function $\bar{g}(\cdot)$ such that for each $\theta \in H$ and compact A,

$$\lim_{n,m} \frac{1}{m} \sum_{i=n}^{n+m-1} E_n \left[g_i(\theta, \xi_i) - \bar{g}(\theta) \right] I_{\{\xi_n \in A\}} = 0, \tag{2.30}$$

where the limit is in mean.

The $\lim_{n,m}$ means that the limit is taken as $n \to \infty$ and $m \to \infty$ simultaneously in any way at all.

(A2.8) Suppose that ϵ_n changes slowly in the sense that there is a sequence of integers $\alpha_n \to \infty$ such that

$$\lim_{n} \sup_{0 \le i \le \alpha_n} \left| \frac{\epsilon_{n+i}}{\epsilon_n} - 1 \right| = 0. \tag{2.31}$$

We are in a position to state Theorem 2.3.

Theorem 2.3. *Assume the algorithm*

$$\theta_{n+1} = \Pi_H \left[\theta_n + \epsilon_n Y_n \right],$$

(5.1.1), (A2.1)–(A2.8), *and any of the constraint conditions* (A4.3.1), (A4.3.2), *or* (A4.3.3). *Then for each subsequence of* $\{\theta^n(\cdot), Z^n(\cdot)\}$ *there is a further subsequence, which will be indexed by* n_k, *and a process* $(\theta(\cdot), Z(\cdot))$ *such that*

$$(\theta^{n_k}(\cdot), Z^{n_k}(\cdot)) \Rightarrow (\theta(\cdot), Z(\cdot)),$$

where

$$\theta(t) = \theta(0) + \int_0^t \bar{g}(\theta(s))ds + Z(t). \tag{2.32}$$

The reflection term $Z(\cdot)$ *satisfies*

$$Z(t) = \int_0^t z(s)ds, \quad z(t) \in -C(\theta(t)) \tag{2.33}$$

for almost all ω, t. *The rest of the conclusions of Theorem* 2.1 *hold, with* $\theta^n(\cdot)$ *replacing* $\theta^\epsilon(\cdot)$.

Comment on the proof. The proof is the same as that of Theorem 2.2. The only difference concerns the grouping of the terms to get the analog of the approximation (2.23) or (2.26) via the use of (A2.7) and (A2.8). This grouping can be done as follows. Let $\nu_n \to \infty$ be a sequence of integers. For $j \geq n$, define $q_j^n = \sum_{i=n}^{j-1} \nu_i$ and

$$t_j^n = \sum_{i=n}^{n+q_j^n-1} \epsilon_i. \tag{2.34a}$$

Suppose that $\nu_n \leq \alpha_n$ (defined in (A2.8)) and that

$$\Delta_j^n = t_{j+1}^n - t_j^n \to 0 \tag{2.34b}$$

as $j \to \infty$. Then group the summands as

$$[n, n+\nu_n - 1], [n+\nu_n, n+\nu_n+\nu_{n+1}-1], \ldots,$$

and replace (2.23) (or (2.26)) by

$$\sum_{j:t \leq t_j^n < t+\tau} \sum_{k=1}^{l_\alpha} \sum_{i=n+q_j^n}^{n+q_{j+1}^n-1} \epsilon_i \left[E_{n+q_j^n} g_{i,K}(\theta^n(t_j^n), \xi_i^\epsilon) - \bar{g}(\theta^n(t_j^n)) \right] I_{\{\theta^n(t_j^n) \in B_k^\alpha\}}. \tag{2.35}$$

Figures 2.1 and 2.2 illustrate the connections among the terms.

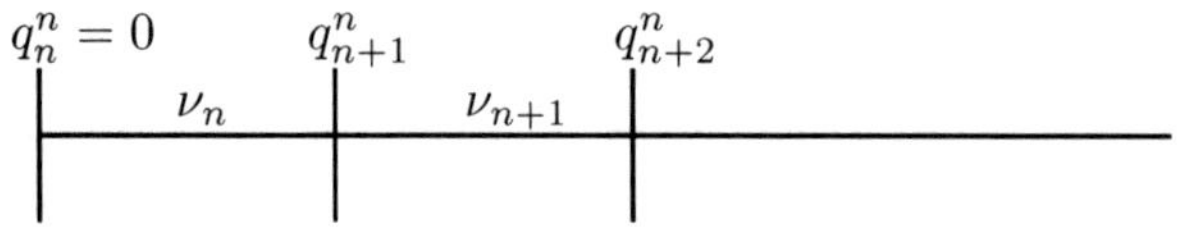

$q_n^n = 0$ $\quad \nu_n \quad$ q_{n+1}^n $\quad \nu_{n+1} \quad$ q_{n+2}^n

Figure 2.1. Illustration of the grouping terms: Number of indices.

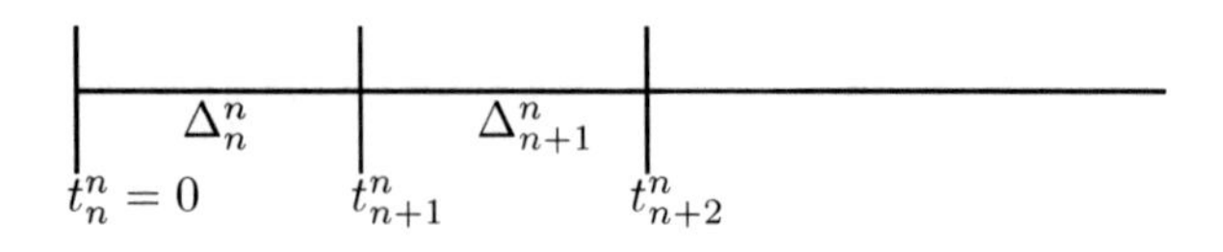

Figure 2.2. Illustration of the grouping terms: Interpolated time.

Note that $\theta^n(t_j^n) = \theta_{n+q_j^n}$. By (A2.3), (A2.4), and (A2.8), approximate

(2.35) by

$$\sum_{j:t\leq t_j^n<t+\tau} \sum_{k=1}^{l_\alpha} \Delta_j^n \frac{1}{\nu_j} \sum_{i=n+q_j^n}^{n+q_{j+1}^n-1} \left[E_{n+q_j^n} g_{i,K}(x_k^\alpha, \xi_i^\epsilon) - \bar{g}(x_k^\alpha)\right] I_{\{\theta^n(t_j^n)\in B_k^\alpha\}}. \tag{2.36}$$

Now extract a weakly convergent subsequence of $\{\theta^n(\cdot), Z^n(\cdot)\}$. The rest of the details are as in the proof of Theorem 2.2 and are left to the reader.

8.2.4 Random ϵ_n

If the ϵ_n are random such that $\epsilon_n \to 0$ with probability one, then the proof is essentially the same as that of Theorem 2.3. The uniform integrability condition (A2.1) implies the tightness. Although the t_j^n are now random, the proof remains the same.

Theorem 2.4. *Assume* (A6.1.8) *and the conditions of Theorem* 2.3, *but replace* (A2.8) *with the following: ϵ_n is $\mathcal{F}_n$-measurable and there is a sequence of integers $\alpha_n \to \infty$ such that for each $\delta > 0$*

$$\lim_n P\left\{\sup_{0\leq i\leq\alpha_n}\left|\frac{\epsilon_{n+i}}{\epsilon_n}-1\right|\geq\delta\right\}=0.$$

Then the conclusions of Theorem 2.3 *hold.*

Remark. Suppose that there are bounded and non-negative random variables k_n, such that $\{k_i, i \leq n\}$ is $\mathcal{F}_n$-measurable and $\epsilon_n = [\sum_{i=0}^n k_i]^{-1}$. Then the condition on ϵ_n requires that

$$\lim_n \frac{\sum_{i=n+1}^{n+\alpha_n} k_i}{\sum_{i=0}^n k_i} = 0,$$

where the convergence is in the sense of probability.

8.2.5 Differential Inclusions

Recall Theorem 6.8.1, where the ODE characterizing the limit behavior was replaced by a differential inclusion, and an application to the parameter identification problem was given. There is an analogous result for the weak convergence case. The proof of the following result is close to those of Theorems 2.1 and 2.2 and is omitted.

Theorem 2.5. *Assume the conditions of Theorem* 2.2, *but replace the limit condition* (A1.9) *with the following. There are sets $G(\cdot)$ that are upper semi-continuous in the sense of* (4.3.2) *such that for each $\theta \in H$ and compact* *set A,*

$$\text{distance}\left[\frac{1}{m}\sum_{i=n}^{n+m-1} E_n^\epsilon g_i^\epsilon(\theta,\xi_i^\epsilon), G(\theta)\right] I_{\{\xi_n^\epsilon\in A\}} \to 0 \tag{2.37}$$

in probability as m and n go to infinity. If $\epsilon_n \to 0$, assume the conditions of Theorem 2.3 but replace the limit (2.30) with the following:

$$\text{distance}\left[\frac{1}{m}\sum_{i=n}^{n+m-1} E_n g_i(\theta,\xi_i), G(\theta)\right] I_{\{\xi_n\in A\}} \to 0 \tag{2.38}$$

in probability as m and n go to infinity. Alternatively to (A1.6) and (A1.9), suppose that for any sequence α_i^n, α, satisfying (5.6.12),

$$\lim_{n,m}\text{distance}\left[\frac{1}{m}\sum_{i=n}^{n+m-1} E_n^\epsilon g_i(\alpha_i^n,\xi_i), G(\alpha)\right] I_{\{\xi_n\in A\}} = 0 \tag{2.37'}$$

in probability (or the analogous form for $\epsilon_n \to 0$). Then the conclusions of Theorem 2.2 hold, with the ODE replaced by the differential inclusion

$$\dot\theta \in G(\theta) + z,\ z(t) \in -C(\theta(t)) \ \text{ for almost all } t,\omega. \tag{2.39}$$

Example. As an example of $g_n(\cdot)$ satisfying the above conditions, let $g_n^\epsilon(\theta_n^\epsilon,\xi_n^\epsilon) = g_n^\epsilon(\theta_n^\epsilon) + h_n^\epsilon(\theta_n^\epsilon,\xi_n^\epsilon)$, where $g_n^\epsilon(\cdot)$ satisfies (2.37) and $h_n^\epsilon(\cdot)$ satisfies the conditions on $g_n^\epsilon(\cdot)$ of Theorem 2.1 and averages to zero. For example, in convex optimization, we might observe a subgradient plus additive noise, which does not depend on the particular subgradient that is selected.

8.2.6 Time-Dependent ODEs

Averaging conditions such as (A1.9) are based on the supposition that the "local" averages of the conditional means do not depend on time. While this is rarely the case in an absolute sense, it is an appropriate working hypothesis for the bulk of applications to date. Some applications are explicitly concerned with tracking the effects of a process that varies on a time scale that is commensurate with that implied by the step size ϵ. For example, recall the adaptive antenna model of Section 3.5, where the dominant issue is the tracking of the effects of the time-varying Doppler frequency. It is still important to understand the behavior of the adaptive process for the case where the observed data is a stationary process that evolves on a time scale that is faster than that implied by the step size ϵ, where conditions such as (A1.9) can be used. Such models can also be used for tracking analysis. But, it is also useful to derive the mean ODE directly when the time scales are comparable. Condition (A1.9) will be replaced by the following:

(A2.9) There is a sequence $m_\epsilon \to \infty$ such that $\epsilon m_\epsilon \to 0$, and a vector-valued process $v(\cdot)$ whose paths are right continuous and have left hand limits such that for each $\theta \in H$ and $t < \infty$, and as $\epsilon \to 0$ and $n\epsilon \to t$,

$$\frac{1}{m_\epsilon} \sum_{l=n}^{n+m_\epsilon-1} E_n \left[g_l(\theta, \xi_l) - \bar{g}(\theta, v(t)) \right] \to 0,$$

in the mean uniformly in t on each bounded time interval. Let $\mathcal{T}$ be a set of time points that contains at most a finite number on each bounded interval. Then the convergence need only be for $t \neq \mathcal{T}$, and the uniformity can exclude an arbitrarily small neighborhood of $\mathcal{T}$.

Theorem 2.6. *Assume the conditions of Theorem* 2.2, *but with* (A2.9) *replacing* (A1.9) *and an analogous replacement for* (A1.11). *Then* $\{\theta^\epsilon(\cdot)\}$ *is tight and the limit of any weakly convergent subsequence satisfies the mean ODE*

$$\dot{\theta}(t) = \bar{g}(\theta(t), v(t)). \tag{2.40}$$

Comment on the time scales. The mean ODE (2.40) is in terms of the time scale implied by the step size ϵ. Suppose that

$$\lim_{\epsilon \to 0} \epsilon \,[\text{Number of iterates per unit time}] \to K < \infty.$$

Then, in the real time-scale, the mean ODE is

$$\dot{\theta}(t) = K\bar{g}(\theta(t), v(t)). \tag{2.41}$$

The factor K is important when analyzing the speed of tracking in real time.

8.3 The Kiefer–Wolfowitz Algorithm

Subsection 5.3.3 (resp., Section 6.5) dealt with the Kiefer–Wolfowitz procedure, where the algorithm took the particular form (5.3.19) (resp., (6.5.3)). In practice, one rarely lets the difference intervals go to zero. Sometimes the difference interval is simply a small constant. Sometimes one starts a problem with a relatively large difference interval, with the aim of converging to a "reasonable" neighborhood of the solution while keeping the effects of the noise (in the estimates) small (trading bias for small noise), and it eventually allows the interval to decrease, but not to zero. With these approaches, the Kiefer–Wolfowitz procedure reduces to the Robbins–Monro algorithm and is covered by Theorems 2.1 and 2.2. We will now give some results if the difference interval c_n does go to zero. (Then, of course, we must have $\epsilon_n \to 0$.)

8.3.1 Martingale Difference Noise

The martingale difference noise case arises in sequential monte carlo optimization, when the "driving noises" are independent from run to run. The algorithm is (3.1), which is the scheme used in Subsection 5.3.3 and Section 6.5, but with $g_n(\theta_n, \xi_n)$ replaced by $g_n(\theta_n)$. The following definitions and assumptions will be used.

Assumptions and definitions. Let $\{\mathcal{F}_n\}$ be a sequence of nondecreasing σ-algebras, where $\mathcal{F}_n$ measures at least $\{\theta_0, Y_{i-1}^{\pm}, i \leq n\}$, with E_n the associated conditional expectation. Recall that the terms $Y_{n,i}^{\pm}$ are the observations at parameter value $\theta_n \pm e_i c_n$, where e_i is the unit vector in the ith coordinate direction and $Y_n^{\pm} = \{Y_{n,i}^{\pm}, i \leq r\}$.

(A3.1) $\sup_n E|Y_n^{\pm}|^2 < \infty$.

(A3.2) There are functions $g_n(\cdot)$ that are continuous uniformly in n and random variables β_n such that $E_n[Y_n^- - Y_n^+]/(2c_n) = g_n(\theta_n) + \beta_n$.

(A3.3) There is a continuous function $\bar{g}(\cdot)$ such that for each $\theta \in H$,

$$\lim_{n,m} \frac{1}{m} \sum_{i=n}^{n+m-1} [g_i(\theta) - \bar{g}(\theta)] = 0,$$

where the limit is as n and m go to infinity in any way.

(A3.4) $\lim_n \epsilon_n / c_n^2 = 0$.

As in Subsection 5.3.3, we suppose that the algorithm can be written as

$$\begin{aligned} \theta_{n+1} &= \Pi_H \left[\theta_n + \epsilon_n \frac{Y_n^- - Y_n^+}{2c_n} \right] \\ &= \theta_n + \epsilon_n g_n(\theta_n) + \epsilon_n \beta_n + \frac{\epsilon_n}{2c_n} \delta M_n + \epsilon_n Z_n, \end{aligned} \tag{3.1}$$

where $\delta M_n = [Y_n^- - Y_n^+] - E_n[Y_n^- - Y_n^+]$. Theorem 3.1 presents the convergence of the algorithm. The remarks on extensions below Theorem 2.1 also hold here.

Theorem 3.1. *Assume* (5.1.1), (A2.6), (A2.8), (A3.1)–(A3.4) *and any one of the constraint set conditions* (A4.3.1), (A4.3.2), *or* (A4.3.3). *Then the conclusions of Theorem 2.3 hold. If there is a unique stationary point $\bar{\theta}$ and it is a minimum, then the limit process $\theta(\cdot)$ is concentrated at $\bar{\theta}$.*

For the random directions Kiefer–Wolfowitz algorithm, let $\mathcal{F}_n$ measure d_n as well, and assume that it can be expanded in the form (5.6.2), *where $\{d_n\}$ is bounded and*

$$\frac{1}{m} \sum_{i=n}^{n+m-1} E_n[d_i d_i' - I] g_i(\theta) \to 0 \tag{†}$$

in the mean for each θ as n and m go to infinity. Then the conclusions of Theorem 2.3 continue to hold. If the constraint set H is dropped, then the conclusions continue to hold under (A1.10), *where a bounded invariant set replaces the limit set.*

Comment on the role of $g_i(\cdot)$ in (†). If the successive d_n are selected at random in proper subspaces of $I\!R^r$ and not in $I\!R^r$ itself, then the term $g_i(\cdot)$ is needed. For example, we could partition the vector θ into q subvectors of dimensions $r_i, i \leq q$, with $r = \sum_i r_i$, and work with the subvectors in sequence. In applications, $\bar{g}(\cdot)$ is usually the negative of the gradient of a function that we wish to minimize.

Proof. In view of the proof of Theorem 2.1 and the alterations required for Theorem 2.3, we need only show that the sequence of martingale processes $\bar{M}^n(\cdot)$ defined by

$$\bar{M}^n(t) = \sum_{i=n}^{m(t_n+t)-1} \frac{\epsilon_i}{c_i} \delta M_i$$

is tight and converges weakly to the process that is identically zero as $n \to \infty$. Note that (A3.1) implies that $\sup_n E|\delta M_n|^2 < \infty$. The tightness and zero limit follow from this, (A3.4), the limit

$$\lim_n E|\bar{M}^n(t)|^2 = \lim_n \sum_{i=n}^{m(t_n+t)-1} \epsilon_i \frac{\epsilon_i}{c_i^2} E|\delta M_i|^2 = 0,$$

and, for each $\delta > 0$ and $T > 0$, the martingale inequality (4.1.4), which yields

$$P\left\{\sup_{t \leq T} |\bar{M}^n(t)| \geq \delta\right\} \leq \frac{E|\bar{M}^n(T)|^2}{\delta^2} \to 0 \text{ as } n \to \infty. \qquad \square$$

8.3.2 Correlated Noise

In applications of the Kiefer–Wolfowitz procedure, correlated noise might arise if the entire optimization were done using one long continuous run of a physical process or simulation. We will work with two canonical forms. The first is that defined in expanded form by (6.5.3). The algorithmic form (6.5.3) is covered by a combination of Theorems 2.3 and 3.1; we can state the following result.

Theorem 3.2. *Assume the algorithmic form* (6.5.3)*, the conditions of Theorem* 2.3 (*except for* (A2.1)), *with* $Y_n = [Y_n^- - Y_n^+]/(2c_n)$ *in* (A2.2), *and assume* (A3.1) *and* (A3.4). *Let* $\{\beta_n\}$ *be uniformly integrable. Then the conclusions of Theorem* 2.3 *hold.*

Correlated noise inversely proportional to c_n. Recall the forms used in (6.5.1) and on the left side of (6.5.2). Suppose that there are functions $F_i(\cdot)$ and random variables $\chi_n^{\pm}$ such that $\chi_n^+ \neq \chi_n^-$ and $Y_{n,i}^{\pm} = F_i(\theta_n \pm c_n e_i, \chi_n^{\pm})$. Formally expanding

$$\frac{Y_{n,i}^- - Y_{n,i}^+}{2c_n} = \frac{F_i(\theta_n, \chi_n^-) - F_i(\theta_n, \chi_n^+)}{2c_n} \\ -\frac{1}{2}\left[F_{i,\theta^i}(\theta_n, \chi_n^+) + F_{i,\theta^i}(\theta_n, \chi_n^-)\right] + \text{ finite difference bias.}$$

This motivates a more general form of the Kiefer–Wolfowitz algorithm with correlated noise, which is written in the expanded form

$$\begin{aligned} \theta_{n+1} &= \Pi_H \left[\theta_n + \frac{\epsilon_n}{2c_n}\left(Y_n^- - Y_n^+\right)\right] \\ &= \theta_n + \epsilon_n g_n(\theta_n, \xi_n) + \epsilon_n \beta_n + \frac{\epsilon_n}{2c_n}\widetilde{g}_n(\theta_n, \xi_n) \\ &+ \frac{\epsilon_n}{2c_n}\delta M_n + \epsilon_n Z_n. \end{aligned} \tag{3.2}$$

The main additional feature is the $\epsilon_n \widetilde{g}_n(\theta_n, \xi_n)/2c_n$ term, because it complicates both the proof of tightness and the characterization of the limit. The assumptions to be used guarantee that the effects of this term average to zero. The difficulty will be handled by a very useful technique, known as the *perturbed test function method.* The idea is fully developed in [127] for a large variety of problems, and many examples were given. That book did not deal explicitly with stochastic approximation, but the translation into stochastic approximation terminology is straightforward. The perturbations are small and have the effect of "smoothing" or "averaging" the "conditional differences," as will be seen. We have seen applications of the idea in Subsection 7.4.3 where it was used to get a sufficient condition for the Wiener process, and in Chapter 6, where it was used in the form of a state perturbation or a Liapunov function perturbation. The next result is an application of Theorem 7.4.3.

In preparation for the convergence theorem, define the following processes. Let T be an arbitrary positive number, and define

$$N_n = \min\left\{i : t_{n+i} - t_n \geq T\right\}.$$

The conditions on (3.3) and (3.4) are of the "mixing type." It is required that the conditional expectations go to zero fast enough so that the sums are well defined. The perturbed test functions (3.6) in the proof are defined so that when taking conditional differences, the dominant "bad" terms will cancel. This is, of course, the usual motivation for the perturbations. Define (the sum of the troublesome terms)

$$\Lambda_i^n = \sum_{j=n}^{n+i} \frac{\epsilon_j}{2c_j}\widetilde{g}_j(\theta_j, \xi_j), \tag{3.3}$$

$$\Gamma_i^n = \sum_{j=n+i}^{n+N_n} \frac{\epsilon_j}{2c_j} E_{n+i} \widetilde{g}_j(\theta_{n+i}, \xi_j), \quad i \le N_n. \tag{3.4}$$

Γ_i^n will be the perturbation in the iterate. Define the process $\Lambda^n(\cdot)$ by $\Lambda^n(t) = \Lambda_i^n$ on $[t_{n+i}, t_{n+i+1})$. Define γ_i^n by

$$\epsilon_{n+i}\gamma_i^n = \sum_{j=n+i}^{n+N_n} \frac{\epsilon_j}{2c_j} E_{n+i} \left[\widetilde{g}_j(\theta_{n+i+1}, \xi_j) - \widetilde{g}_j(\theta_{n+i}, \xi_j)\right], \quad i \le N_n. \tag{3.5}$$

We will use the following assumptions.

(A3.5) $\sup_n E|\widetilde{g}_n(\theta_n, \xi_n)|^2 < \infty$.

(A3.6) $\limsup_n \sup_{0 \le i \le N_n} \left[(\epsilon_{n+i}/c_{n+i}^2)/(\epsilon_n/c_n^2) + c_{n+i}/c_n\right] < \infty$.

(A3.7) $\limsup_n \sup_{0 \le i \le N_n} \epsilon_{n+i}/\epsilon_n < \infty$.

(A3.8) $\lim_n \sup_{0 \le i \le N_n} E|\gamma_i^n| = 0$.

(A3.9) $\sup_n \sum_{j=n}^{n+N_n} E^{1/2} \left|E_n \widetilde{g}_j(\theta_n, \xi_j)\right|^2 < \infty$.

Comment on the assumptions. Condition (A3.9) is used only to assure that the processes $C_{2,i}^n$ to be given in the proof are well defined and go to zero as $n \to \infty$. Condition (A3.8) holds, for example, if $E_n \widetilde{g}(\theta_n, \xi_i)$ goes to zero fast enough as $i - n \to \infty$ and $\widetilde{g}(\cdot, \xi)$ is smooth in θ, uniformly in ξ on any compact set.

Theorem 3.3. *Assume that the algorithm can be written in the expanded form* (3.2), *where* δM_n *is defined below* (3.1). *Assume the conditions of Theorem* 2.3 (*except* (A2.1)), (A3.1), (A3.4), (A3.5), *and the uniform integrability of* $\{\beta_n, g_n(\theta_n, \xi_n)\}$. *For each positive* T, *assume* (A3.6)–(A3.9). *Then the conclusions of Theorem* 2.3 *hold.*

Proof. The fact that $\sup_n E|Y_n^{\pm}|^2 < \infty$ implies that $\sup_n E|\delta M_n|^2 < \infty$. Hence, as in the proof of Theorem 3.1, (A3.1) and (A3.4) imply that the sequence $\{\bar{M}^n(\cdot)\}$ defined there converges weakly to the process that is identically zero.

If $\{\theta^n(\cdot)\}$ can be shown to be tight, then all weak sense limit processes have continuous paths with probability one, because the jumps of the $\theta^n(\cdot)$ go to zero as $n \to \infty$. The main problem in proving tightness is due to the term $\epsilon_n \widetilde{g}_n(\theta_n, \xi_n)/(2c_n)$; more particularly, in proving tightness of the interpolations of its sum $\Lambda^n(\cdot)$. If it can be shown that $\{\Lambda^n(\cdot)\}$ converges weakly to the process that is identically zero, then the proof outlined for Theorem 2.3 yields this theorem. Thus, we need only prove the tightness and "zero" limit property for $\{\Lambda^n(\cdot)\}$. This will be done by using Theorem

7.4.3. For notational simplicity and without loss of generality, suppose that θ_n is real valued.

A potentially serious problem for the proof is that, *a priori*, we do not know whether $\sup_{0\leq i\leq N_n} |\Lambda_i^n|$ is bounded in probability uniformly in n, as required by condition (7.4.9). To deal with this problem, we use a truncation method. This truncation method is of wide applicability in related problems of approximation and convergence and will be used again in Section 5.

For $\nu > 0$, define $\tau_\nu^n = n + \min\{i : |\Lambda_i^n| > \nu\}$. We will show that for each $\nu > 0$, the truncated sequence $\{\Lambda_\nu^n(\cdot)\}$ converges weakly to the "zero process" as $n \to \infty$, where we define

$$\Lambda_\nu^n(t) = \sum_{j=n}^{m(t_n+t)\wedge\tau_\nu^n} \frac{\epsilon_j}{2c_j}\widetilde{g}_j(\theta_j, \xi_j).$$

By this (to be proved) convergence to the "zero process," with a probability arbitrarily close to one and large ν, the level ν is not reached (asymptotically) on any interval $[0, T]$ by $\Lambda_\nu^n(\cdot)$. Thus, asymptotically with a probability arbitrarily close to one, for large n the processes $\Lambda_\nu^n(\cdot)$ are not truncated on any finite interval, which implies that the convergence assertion holds for the original sequence of untruncated processes.

Define

$$\Lambda_{i,\nu}^n = \sum_{j=n}^{(n+i-1)\wedge\tau_\nu^n} \frac{\epsilon_j}{2c_j}\widetilde{g}_j(\theta_j, \xi_j).$$

The term $\sup_{0\leq i\leq N_n} |\Lambda_{i,\nu}^n|$ is bounded by ν plus the maximum value of $\epsilon_{n+j}|\widetilde{g}_{n+j}(\theta_{n+j}, \xi_{n+j})|/(2c_{n+j})$ for $0 \leq j \leq N_n$. For any $\delta > 0$,

$$\begin{aligned} P&\left\{\sup_{0\leq j\leq N_n} \frac{\epsilon_{n+j}}{c_{n+j}}|\widetilde{g}_{n+j}(\theta_{n+j}, \xi_{n+j})| \geq \delta\right\} \\ &\leq \sum_{j=0}^{N_n} P\left\{\frac{\epsilon_{n+j}}{c_{n+j}}|\widetilde{g}_{n+j}(\theta_{n+j}, \xi_{n+j})| \geq \delta\right\}. \end{aligned}$$

By Chebyshev's inequality, the above expression is bounded above by

$$\frac{1}{\delta^2}\sum_{j=0}^{N_n} \frac{\epsilon_{n+j}^2}{c_{n+j}^2} E|\widetilde{g}_{n+j}(\theta_{n+j}, \xi_{n+j})|^2,$$

which goes to zero as $n \to \infty$ by (A3.4)–(A3.6).

Now that the boundedness is shown, we apply the perturbed test function method of Theorem 7.4.3. Let $F(\cdot)$ be a real-valued function with compact support whose partial derivatives up to the second order are continuous. For $n + i < \tau_\nu^n$, evaluate

$$\begin{aligned} &E_{n+i}F(\Lambda_{i+1,\nu}^n) - F(\Lambda_{i,\nu}^n) \\ &\quad = F_\theta'(\Lambda_{i,\nu}^n)\frac{\epsilon_{n+i}}{2c_{n+i}}\widetilde{g}_{n+i}(\theta_{n+i}, \xi_{n+i}) + O(1)\frac{\epsilon_{n+i}^2}{c_{n+i}^2}\left|\widetilde{g}_{n+i}(\theta_{n+i}, \xi_{n+i})\right|^2. \end{aligned}$$

The left-hand side is zero for $n+i \geq \tau_\nu^n$. In all of the following expressions, $n+i < \tau_\nu^n$. Define the perturbed test function

$$F_i^n = F(\Lambda_{i,\nu}^n) + F_\theta'(\Lambda_{i,\nu}^n)\Gamma_i^n. \tag{3.6}$$

Note that $\sup_{i \leq N_n} |\Gamma_i^n| \to 0$ in probability by (A3.4), (A3.6), and (A3.9). By the assumptions, we can write

$$E_{n+i}F_{i+1}^n - F_i^n = \epsilon_{n+i}\left[C_{1,i}^n + C_{2,i}^n + C_{3,i}^n\right],$$

where

$$\begin{aligned}
C_{1,i}^n &= O(1)\frac{\epsilon_{n+i}}{c_{n+i}^2}|\widetilde{g}_{n+i}(\theta_{n+i},\xi_{n+i})|^2,\\
C_{2,i}^n &= O(1)\left|\sum_{j=n+i+1}^{n+N_n}\frac{\epsilon_j}{c_j}E_{n+i+1}\widetilde{g}_j(\theta_{n+i+1},\xi_j)\right|\frac{1}{c_{n+i}}|\widetilde{g}_{n+i}(\theta_{n+i},\xi_{n+i})|,\\
C_{3,i}^n &= F_\theta'(\Lambda_{i,\nu}^n)\gamma_i^n.
\end{aligned}$$

By (A3.5) and the fact that $\epsilon_n/c_n^2 \to 0$,

$$\max_{0\leq i\leq N_n} E|C_{1,i}^n| \to 0,$$

and by (A3.8),

$$\max_{0\leq i\leq N_n} E|C_{3,i}^n| \to 0.$$

By (A3.1), (A3.4), (A3.5), and (A3.9), the term $\max_{0\leq i\leq N_n} EC_{2,i}^n$ is bounded by $O(1)\epsilon_n/c_n^2 \to 0$. Thus, Theorem 7.4.3 implies that the ν-truncated sequence $\{\Lambda^n(\cdot)\}$ converges weakly to the "zero process." Hence the truncation is not needed, and the original untruncated sequence also converges to the "zero process." □

8.4 State-Dependent Noise: Markovian and Non-Markovian

In this section, we treat the "state-dependent" noise model introduced in Sections 2.3–2.5. As noted there, the (effective noise memory, state) pair has a Markov structure for many problems of current interest, and the approach to be discussed is powerful and practical for such problems. A full appreciation of its value and the relative simplicity of its use can be seen by comparing it with the greater difficulty of proofs and the more limited conclusions obtained by alternative current approaches. For example, for the controlled queue example of Section 2.5, compare the proof in Chapter 9 with those of [52, 161]; see also [145] and Section 2.5 for additional comments. The Markovian assumption enables the "memory" variable to depend on the entire past. The forms used in Subsection 4.4 avoid the use of a transition probability, and do not require the Markov assumption.

The method of proof is based on the idea in Appendix 2 of [145]. The proof is simpler than that for the original "invariant measure" method of [127, 142] and its generalizations in [145], which in one form or another are widely used.

8.4.1 Constant Step Size

Continuing to assume (A1.1), (A1.4), (A1.5), (A1.7), and the terminology of Sections 1 and 2 for the correlated noise case, we will also use the following assumptions. By $n \to \infty$ and $\epsilon \to 0$, we mean, as usual, that they can go to their limits in any way at all.

Assumptions and definitions.

(A4.1) For each ϵ, θ, and n, there is a transition function $P_n^\epsilon(\cdot, \cdot|\theta)$ such that $P_n^\epsilon(\cdot, A|\cdot)$ is measurable for each Borel set A in the range space Ξ of ξ_n^ϵ and

$$P\left\{\xi_{n+1}^\epsilon \in \cdot|\mathcal{F}_n^\epsilon\right\} = P_n^\epsilon(\xi_n^\epsilon, \cdot|\theta_n^\epsilon). \tag{4.1}$$

(A4.2) For each fixed θ, there is a transition function $P(\xi, \cdot|\theta)$ such that

$$P_n^\epsilon(\xi, \cdot|\theta) \Rightarrow P(\xi, \cdot|\theta), \text{ as } n \to \infty \text{ and } \epsilon \to 0, \tag{4.2}$$

where the limit is uniform on each compact (θ, ξ) set. That is, for each bounded and continuous real-valued function $F(\cdot)$ on Ξ,

$$\int F(\tilde{\xi}) P_n^\epsilon(\xi, d\tilde{\xi}|\theta) \to \int F(\tilde{\xi}) P(\xi, d\tilde{\xi}|\theta) \tag{4.3}$$

uniformly on each compact (θ, ξ)-set, as $n \to \infty$ and $\epsilon \to 0$.

(A4.3) $P(\xi, \cdot|\theta)$ is weakly continuous in (θ, ξ). That is, the right side of (4.3) is continuous in (θ, ξ) for each bounded and continuous real-valued $F(\cdot)$.

Definition. Fixed-θ chain. For each fixed θ, $P(\cdot, \cdot|\theta)$ is the transition function of a Markov chain with state space Ξ. This chain is referred to as the *fixed-θ chain*, and the random variables of the chain will be denoted by $\{\xi_n(\theta)\}$. Unless otherwise specified, when using the expression $E_n^\epsilon F(\xi_{n+j}(\theta))$ for $j \geq 0$, the conditional expectation is for the fixed-θ chain

$$\{\xi_{n+j}(\theta), j \geq 0;\ \xi_n(\theta) = \xi_n^\epsilon\}, \tag{4.4}$$

which starts at time n with initial condition $\xi_n(\theta) = \xi_n^\epsilon$ and given θ. These fixed-θ chains play a major role in analyzing Markov state-dependent noise for stochastic approximation.

(A4.4) $g_n^\epsilon(\cdot)$ is continuous on each compact (θ, ξ)-set, uniformly in ϵ and n.

(A4.5) The set

$$\{g_n^\epsilon(\theta_n^\epsilon, \xi_n^\epsilon); \epsilon, n\} \text{ is uniformly integrable,} \tag{4.5a}$$

and for any compact set $A \subset \Xi$ and each $\theta \in H$,

$$\{g_{n+j}^\epsilon(\theta, \xi_{n+j}(\theta)), j \geq 0, \text{ all } \xi_n(\theta) \in A;\ n \geq 0\} \tag{4.5b}$$

is uniformly integrable.

(A4.6) There is a continuous function $\bar{g}(\cdot)$ such that for any sequence of integers $n_\epsilon \to \infty$ satisfying $\epsilon n_\epsilon \to 0$ as $\epsilon \to 0$ and each compact set $A \subset \Xi$,

$$\frac{1}{n_\epsilon} \sum_{i=jn_\epsilon}^{jn_\epsilon+n_\epsilon-1} E_{jn_\epsilon}^\epsilon \left[g_i^\epsilon(\theta, \xi_i(\theta)) - \bar{g}(\theta)\right] I_{\{\xi_{jn_\epsilon}^\epsilon \in A\}} \to 0 \tag{4.6}$$

in the mean for each θ, as $j \to \infty$ and $\epsilon \to 0$.

(A4.7) For each compact set $A \subset \Xi$ and each $\delta > 0$, there is a compact set $A_\delta \subset \Xi$ such that

$$P\left\{\xi_{n+j}(\theta) \in A_\delta \middle| \xi_n(\theta)\right\} \geq 1 - \delta,$$

for all $\theta \in H, j > 0, n \geq 0$ and $\xi_n(\theta) \in A$.

We are now in a position to state the following theorem. The remarks on extensions below Theorem 2.1 also hold here.

Theorem 4.1. *Assume algorithm* (1.1), (A1.1), (A1.4), (A1.5), (A1.7), (A4.1)–(A4.7), *and any one of the constraint set conditions* (A4.3.1), (A4.3.2), *or* (A4.3.3). *Then the conclusions of Theorem* 2.1 *hold. If the constraint set H is dropped, then the conclusions continue to hold under* (A1.10), *where a bounded invariant set replaces the limit set. The extensions concerning chain recurrence hold if the corresponding conditions are satisfied.*

Proof. We follow the proof of Theorem 2.2 as closely as possible, but with the use of the n_ϵ from (A4.6) in lieu of Δ/ϵ. Referring to the proof of Theorem 2.2, let $n_\epsilon \to \infty$ such that $\Delta_\epsilon \equiv \epsilon n_\epsilon \to 0$. Define $g_{i,K}(\cdot), B_k^\alpha$, and x_k^α as in the proof of Theorem 2.2. By the uniform integrability of the set (4.5a), for large enough K we can approximate (in the mean and uniformly in j, ϵ)

$$\frac{1}{n_\epsilon} \sum_{i=jn_\epsilon}^{jn_\epsilon+n_\epsilon-1} E_{jn_\epsilon}^\epsilon g_i^\epsilon(\theta_i^\epsilon, \xi_i^\epsilon)$$

by

$$\frac{1}{n_\epsilon}\sum_{i=jn_\epsilon}^{jn_\epsilon+n_\epsilon-1} E_{jn_\epsilon}^\epsilon g_{i,K}^\epsilon(\theta_i^\epsilon,\xi_i^\epsilon). \tag{4.7}$$

Suppose that we can approximate (4.7) by

$$\frac{1}{n_\epsilon}\sum_{i=jn_\epsilon}^{jn_\epsilon+n_\epsilon-1} E_{jn_\epsilon}^\epsilon g_{i,K}^\epsilon(\theta_{jn_\epsilon}^\epsilon,\xi_i(\theta_{jn_\epsilon}^\epsilon)), \tag{4.8}$$

in the sense that the difference converges to zero in the mean as $\epsilon \to 0$ and $j \to \infty$. Suppose also that we can approximate (4.8) by

$$\sum_{k=1}^{l_\alpha}\frac{1}{n_\epsilon}\sum_{i=jn_\epsilon}^{jn_\epsilon+n_\epsilon-1} E_{jn_\epsilon}^\epsilon g_{i,K}^\epsilon(x_k^\alpha,\xi_i(x_k^\alpha))I_{\{\theta_{jn_\epsilon}^\epsilon\in B_k^\alpha\}} \tag{4.9}$$

in the sense that the difference converges to zero in the mean as $\epsilon \to 0$, $j \to \infty$, and then $\alpha \to 0$ when the initial conditions $\{\xi_{jn_\epsilon}^\epsilon\}$ are confined to a compact set A. Now, (A1.7) implies that for a large enough compact set $A \subset \Xi$, the limit of (4.9) as $\epsilon \to 0$ changes little if the sum is multiplied by $I_{\{\xi_{jn_\epsilon}^\epsilon\in A\}}$. In other words, in evaluating (4.9) we can suppose, without loss of generality, that the "initial conditions" $\xi_{jn_\epsilon}^\epsilon$ all take values in some compact set A. Next, the uniform integrability of the set in (4.5b) for each θ implies that we can drop the truncation K in (4.9) without changing the limits. It follows that from this point on, the proof of Theorem 2.2 implies this theorem if

$$\frac{1}{n_\epsilon}\sum_{i=jn_\epsilon}^{jn_\epsilon+n_\epsilon-1} E_{jn_\epsilon}^\epsilon g_i^\epsilon(x_k^\alpha,\xi_i(x_k^\alpha)) \tag{4.10}$$

converges in mean to $\bar{g}(x_k^\alpha)$, when the values of the initial conditions $\{\xi_{jn_\epsilon}^\epsilon\}$ are confined to an arbitrary compact set A. This latter fact is implied by (A4.6). Thus, all we need to do is to show that the approximations of (4.7) by (4.8) and (4.8) by (4.9) are valid. The main problem, which was not encountered in Theorem 2.2, is the dependence of $\{\xi_n^\epsilon\}$ on $\{\theta_n^\epsilon\}$.

In evaluating the conditional expectations below, it will be sufficient to suppose that the values of the initial conditions $\{\xi_{jn_\epsilon}^\epsilon\}$ are confined to some arbitrary compact set A for all j,ϵ. The conditional expectation of a typical summand $E_{jn_\epsilon}^\epsilon g_{jn_\epsilon+i,K}^\epsilon(\theta_{jn_\epsilon+i}^\epsilon,\xi_{jn_\epsilon+i}^\epsilon)$, for $i=2,3,\ldots$, in (4.7) can be written as the iterated conditional expectation

$$E_{jn_\epsilon}^\epsilon E_{jn_\epsilon+1}^\epsilon \cdots E_{jn_\epsilon+i-1}^\epsilon g_{jn_\epsilon+i,K}^\epsilon(\theta_{jn_\epsilon+i}^\epsilon,\xi_{jn_\epsilon+i}^\epsilon). \tag{4.11}$$

It will be shown that (4.11) can be written as

$$\int\cdots\int P(\xi_{jn_\epsilon}^\epsilon,d\widetilde{\xi}_1|\theta_{jn_\epsilon}^\epsilon)\cdots P(\widetilde{\xi}_{i-1},d\widetilde{\xi}_i|\theta_{jn_\epsilon}^\epsilon)g_{jn_\epsilon+i,K}^\epsilon(\theta_{jn_\epsilon}^\epsilon,\widetilde{\xi}_i)+\rho_{jn_\epsilon}(\epsilon,i), \tag{4.12}$$

where if $n_\epsilon = m$, a constant value (i.e., not depending on ϵ), then the error term satisfies

$$\max_{i\le n_\epsilon} E\left|\rho_{jn_\epsilon}(\epsilon, i)\right| \to 0, \tag{4.13}$$

as $\epsilon \to 0$ and $j \to \infty$. It will be seen that there are $n_\epsilon \to \infty$ such that $\epsilon n_\epsilon \to 0$ and (4.13) holds.

Note that the multiple integral term in (4.12) equals

$$E^\epsilon_{jn_\epsilon} g^\epsilon_{jn_\epsilon+i,K}(\theta^\epsilon_{jn_\epsilon}, \xi_i(\theta^\epsilon_{jn_\epsilon})), \tag{4.14}$$

which is the typical summand in (4.8). The representation (4.12) and (4.13), for each constant value $n_\epsilon = m$, will be proved by approximating "backwards" in (4.11), and using the fact that if $\epsilon n_\epsilon \to 0$, then the uniform integrability of the $\{Y^\epsilon_n\}$ implies that there are $\delta_\epsilon \to 0$ (see also (2.14)) such that

$$\limsup_{\epsilon,j} P\left\{\sup_{0\le i\le n_\epsilon}\left|\theta^\epsilon_{jn_\epsilon+i} - \theta^\epsilon_{jn_\epsilon}\right| \ge \delta_\epsilon\right\} = 0. \tag{4.15}$$

All approximations in what follows are in the sense of convergence in the mean and hold as $\epsilon \to 0$ and $j \to \infty$, and they are uniform in $i \le n_\epsilon$. Let $F(\cdot)$ be a bounded and continuous real-valued function on $H \times \Xi$. By (4.15), (A1.7), (A4.7), and the continuity of $F(\cdot)$,

$$E^\epsilon_{jn_\epsilon+i-1} F(\theta^\epsilon_{jn_\epsilon+i}, \xi^\epsilon_{jn_\epsilon+i})$$

can be approximated by

$$E^\epsilon_{jn_\epsilon+i-1} F(\theta^\epsilon_{jn_\epsilon}, \xi^\epsilon_{jn_\epsilon+i}).$$

This last expression can be written as

$$\int P^\epsilon_{jn_\epsilon+i-1}\left(\xi^\epsilon_{jn_\epsilon+i-1}, d\xi\middle|\theta^\epsilon_{jn_\epsilon+i-1}\right) F(\theta^\epsilon_{jn_\epsilon}, \xi).$$

Using (A4.2), (A4.3), (A4.7), (A1.7), and (4.15), the expression above can be approximated by

$$\int P\left(\xi^\epsilon_{jn_\epsilon+i-1}, d\xi\middle|\theta^\epsilon_{jn_\epsilon}\right) F(\theta^\epsilon_{jn_\epsilon}, \xi) = \widehat{F}_{i-1}(\theta^\epsilon_{jn_\epsilon}, \xi^\epsilon_{jn_\epsilon+i-1}),$$

where $\widehat{F}_{i-1}(\cdot)$ is bounded and continuous by (A4.3) and the continuity of $F(\cdot)$.

Next, go back one step in time and use the same procedure to approximate

$$E^\epsilon_{jn_\epsilon+i-2}\widehat{F}_{i-1}(\theta^\epsilon_{jn_\epsilon}, \xi^\epsilon_{jn_\epsilon+i-1})$$

by

$$\int P\left(\xi^\epsilon_{jn_\epsilon+i-2}, d\xi\middle|\theta^\epsilon_{jn_\epsilon}\right) \widehat{F}_{i-1}(\theta^\epsilon_{jn_\epsilon}, \xi) = \widehat{F}_{i-2}(\theta^\epsilon_{jn_\epsilon}, \xi^\epsilon_{jn_\epsilon+i-2}),$$

where $\widehat{F}_{i-2}(\cdot)$ is bounded and continuous.

Continuing this procedure yields (4.12), with the error term satisfying (4.13) if $g^\epsilon_{jn_\epsilon+i}(\cdot)$ does not depend on $(\epsilon, jn_\epsilon + i)$. But (A4.4) says that $\{g^\epsilon_i(\cdot)\}$ is equicontinuous on each compact (θ, ξ)-set. This yields (4.12) and (4.13) for $n_\epsilon = m$, a constant.

Since the only properties used to get (4.13) are the moduli of equicontinuity on each compact set, the convergence (4.3), and the limit result (4.15), it is clear that there are $n_\epsilon \to \infty$ slowly enough so that (4.13) continues to hold. Thus, the approximation (4.8) is obtained. The approximation of (4.8) by (4.9) requires that we replace $\theta^\epsilon_{jn_\epsilon}$ in (4.8) by the random variable (see the proof of Theorem 2.3)

$$\sum_{k=1}^{l_\alpha} x^\alpha_k I_{\{\theta^\epsilon_{jn_\epsilon} \in B^\alpha_k\}}.$$

The proof of this with asymptotically negligible error is identical to the procedure just used to get the approximation (4.8). □

Weakening the continuity conditions. The assumptions can often be weakened in special cases. For simplicity, suppose that the P^ϵ_n in (4.1) does not depend on ϵ or n. Let ξ^ϵ_n consist of two components, $\xi^{\epsilon,s}_n$ and $\xi^{\epsilon,e}_n$, where $\xi^{\epsilon,e}_n$ is exogenous in that

$$P\{\xi^{\epsilon,e}_{n+1} \in \cdot | \mathcal{F}^\epsilon_n\} = P\{\xi^{\epsilon,e}_{n+1} \in \cdot | \xi^{\epsilon,e}_n\}.$$

We can write, in obvious notation,

$$P\left\{\xi^\epsilon_1 \in d\xi^s d\xi^e \middle| \xi^\epsilon_0, \theta\right\} = P\left\{\xi^{\epsilon,s}_1 \in d\xi^s \middle| \xi^{\epsilon,s}_0, \xi^{\epsilon,e}_0, \theta\right\} P\left\{\xi^{\epsilon,e}_1 \in d\xi^e \middle| \xi^{\epsilon,e}_0\right\}.$$

Then we have the following extension of Theorem 4.1. The requirements are weakened further in Subsection 4.4.

Theorem 4.2. *Assume the conditions of Theorem 4.1 for the present case, but with the following changes. The continuity in (A4.4) is only in (θ, ξ^s), uniformly in n, ϵ, and in ξ^e in each compact set, and $P\left\{\xi^{\epsilon,s}_1 \in d\xi^s \middle| \xi^{\epsilon,s}_0, \xi^{\epsilon,e}_0, \theta\right\}$ is weakly continuous only in (θ, ξ^s_0), uniformly in each compact $\xi^{\epsilon,e}_0$-set. Then the conclusions of Theorem 4.1 hold.*

8.4.2 Decreasing Step Size $\epsilon_n \to 0$

The development for the decreasing-step-size case requires only minor modifications of the conditions in Theorem 4.1, analogously to what was done in Theorem 2.3 to adapt the proof of Theorem 2.2. For the sake of completeness and ease of reference, the modified conditions will now be listed.

(A4.8) For each θ and n there is a transition function $P_n(\cdot, \cdot|\theta)$ such that $P_n(\cdot, A|\cdot)$ is measurable for each Borel set A in the range space Ξ of ξ_n and

$$P\{\xi_{n+1} \in \cdot | \mathcal{F}_n\} = P_n(\xi_n, \cdot | \theta_n).$$

(A4.9) For each fixed θ, there is a transition function $P(\xi, \cdot|\theta)$ such that

$$P_n(\xi, \cdot|\theta) \Rightarrow P(\xi, \cdot|\theta), \text{ as } n \to \infty,$$

where the limit is uniform on each compact (θ, ξ) set.

(A4.10) $g_n(\cdot)$ is continuous on each compact (θ, ξ) set uniformly in n.

(A4.11) The set

$$\{g_n(\theta_n, \xi_n)\} \tag{4.16a}$$

is uniformly integrable. For any compact set $A \subset \Xi$, each of the sets

$$\{g_{n+j}(\theta, \xi_{n+j}(\theta)), j \geq 0, \text{ all } \xi_n(\theta) \in A;\ n \geq 0\}, \quad \text{for } \theta \in H \tag{4.16b}$$

is uniformly integrable.

(A4.12) There is a continuous function $\bar{g}(\cdot)$ such that for each $\theta \in H$ and each compact set $A \subset \Xi$,

$$\lim_{n,m} \frac{1}{m} \sum_{i=n}^{n+m-1} E_n \left[g(\theta, \xi_i(\theta)) - \bar{g}(\theta)\right] I_{\{\xi_n \in A\}} = 0,$$

where the limit is in mean.

The $\lim_{n,m}$ means that the limit is taken as $n \to \infty$ and $m \to \infty$ simultaneously in any way at all. The following theorem concerns decreasing step size algorithms with state-dependent noise. Random step sizes can be handled with the approach of Subsection 2.3. The remarks on extensions below Theorem 2.1 hold here also.

Theorem 4.3. *Assume algorithm*

$$\theta_{n+1} = \Pi_H \left[\theta_n + \epsilon_n Y_n\right],$$

conditions (5.1.1), (A2.1), (A2.2), (A2.4), (A2.6), (A2.8), (A4.3), (A4.7)–(A4.12), *and any one of the constraint set conditions* (A4.3.1), (A4.3.2), *or* (A4.3.3). *Then the conclusions of Theorem* 2.3 *hold. If the constraint set H is dropped, the conclusions continue to hold under* (A1.10), *where a bounded invariant set replaces the limit set. The extensions concerning chain recurrence hold if the corresponding conditions are satisfied.*

8.4.3 The Invariant Measure Method: Constant Step Size

An alternative approach to the convergence for the state-dependent noise case uses the "invariant measure" method. This approach was the original general method for treating this problem [127, 142].

The method of Theorem 4.1 is a development of the approach in [145, Appendix 2]. The conclusions of the main result for the state-dependent noise problem in [145, Theorem 3.1] were stated in terms of the invariant measures of the process $\{\xi_n(\theta)\}$, instead of as in our Theorem 4.1. The conditions in the general result [145, Theorem 3.1] are in reality not very different from those we used in Theorem 4.1, even though Theorem 4.1 required only that the averages for the particular sequences in (A4.6) exist, and not that there be a unique invariant measure. Since the phrasing in terms of invariant measures has a useful intuitive content, we will restate that theorem (without the proof) and the required conditions in the terminology of this chapter. The result and proof in [145] are a generalization of the original "invariant measure" work of [142], and the modifications in [127] for the constant-step-size case, where the dynamical term $\bar{g}(\theta)$ in the mean ODE was defined directly as a mean using the invariant measures of the fixed-θ processes $\xi_n(\theta)$. The weak continuity condition was used more heavily in the proof of Theorem 4.1 than in the proofs in [127, 142, 145]. The invariant measure type of proof is an example of a proof based on sample occupation measures; it yields a lot of intuition concerning the averaging procedure. It is a little more complicated than that given in Theorem 4.1, but the sample occupation measure method on which it is based has many other applications.

To use the invariant measure approach, the following conditions will be needed.

(A4.13) Let $\mu(\cdot|\theta)$ denote the invariant measures under the transition function $P(\xi, \cdot|\theta)$. The set $\{\mu(\cdot|\theta), \theta \in H\}$ is tight.

(A4.14) There is a continuous function $g(\cdot)$ such that

$$g_n^\epsilon(\theta_n^\epsilon, \xi_n^\epsilon) = g(\theta_n^\epsilon, \xi_n^\epsilon), \tag{4.17}$$

$\{g(\theta_n^\epsilon, \xi_n^\epsilon); \epsilon, n\}$ is uniformly integrable, and there is a $K_0 < \infty$ such that for all stationary processes $\{\xi_n(\theta)\}$,

$$\sup_{\theta \in H} E|g(\theta, \xi_j(\theta))| < K_0. \tag{4.18}$$

(A4.15) Either (4.19a) or (4.19b) holds:

$$\text{For each } \theta, \quad \mu(\cdot|\theta) \text{ is unique.} \tag{4.19a}$$

There is a measurable function $\bar{g}(\cdot)$ such that for each θ and initial condition $\xi_0(\theta)$,

$$\lim_N \frac{1}{N} \sum_{n=0}^{N-1} Eg(\theta, \xi_n(\theta)) = \bar{g}(\theta). \tag{4.19b}$$

Under (4.19a), define

$$\bar{g}(\theta) = \int g(\theta, \xi) \mu(d\xi|\theta).$$

The following theorem is proved in [145, Theorem 3.1]. The comments below Theorem 2.1 hold here also. There is an obvious analog of the "time-dependent" model and result of Theorem 2.6.

Theorem 4.4. *Assume algorithm* (1.1), (A1.1), (A1.4), (A1.5), (A1.7), (A4.1)–(A4.3), (A4.13)–(A4.15), *and any one of the constraint set conditions* (A4.3.1), (A4.3.2), *or* (A4.3.3). *Then the conclusions of Theorem* 2.1 *hold. Also,* $\bar{g}(\cdot)$ *is a continuous function of* θ. *Now drop the constraint set, assume* (A1.10), *and modify the conditions so that* (A4.13) *and* (4.18) *hold for* H *being an arbitrary compact set. Then the conclusions of Theorem* 2.1 *continue to hold. The extensions concerning chain recurrence hold if the corresponding conditions are satisfied.*

Comments on the proof. The structure of the proof is quite different from that used in Theorem 4.1; only the basic idea will be discussed. Suppose that $g(\cdot)$ is bounded, for simplicity. The key problem in the proof of Theorem 4.1 concerns the approximation of the sum of conditional expectations in (2.22), namely, the approximation of

$$\frac{\epsilon}{\Delta} \sum_{i \in I_j^{\epsilon,\Delta}} E_{j\Delta/\epsilon}^{\epsilon} \left[g(\theta^\epsilon(j\Delta), \xi_i^\epsilon) - \bar{g}(\theta^\epsilon(j\Delta)) \right]$$

by the expectation for the fixed-θ process with parameter $\theta(t)$ if $j\Delta \to t$. The invariant measure method evaluates the sum

$$\frac{\epsilon}{\Delta} \sum_{i \in I_j^{\epsilon,\Delta}} E_{j\Delta/\epsilon}^{\epsilon} g(\theta_{j\Delta/\epsilon}^{\epsilon}, \xi_i^\epsilon) \tag{4.20}$$

somewhat differently than the method of Subsection 4.1.

Let $n_\epsilon = \Delta/\epsilon$ and define the conditional mean sample occupation measure

$$R(j, \epsilon, \theta, \cdot) = \frac{1}{n_\epsilon} \sum_{i=jn_\epsilon}^{jn_\epsilon + n_\epsilon - 1} P\left\{ \xi_i^\epsilon \in \cdot \middle| \theta_{j\Delta/\epsilon}^\epsilon = \theta, \xi_{j\Delta/\epsilon}^\epsilon \right\}. \tag{4.21}$$

Then (4.20) can be written in terms of $R(j, \epsilon, \theta_{j\Delta/\epsilon}^\epsilon, \cdot)$ as

$$\int g(\theta_{j\Delta/\epsilon}^\epsilon, \xi) R(j, \epsilon, \theta_{j\Delta/\epsilon}^\epsilon, d\xi).$$

Thus, to evaluate the limit of (4.20) as $j\Delta \to t$, we need only get the possible limits of the set of random measures $\{R(j, \epsilon, \theta_{j\Delta/\epsilon}^\epsilon, \cdot), j\Delta \to t, \epsilon \to 0\}$. The proofs in [127, 142, 145] show that these limits are invariant measures of the fixed-θ chain under the parameter $\theta(t)$.

An extension: Non-Markov state-dependent noise processes. The central idea of the proof concerns the weak convergence of the occupation measures defined by $R(\cdot)$ in (4.21). One can start with such an assumption, which can often be verified by other means, and weaken the other conditions. It is not necessary to define a fixed-θ process, except in the limit. Loosely speaking, we will suppose that $R(j, \epsilon, \theta, \cdot)$ converges to a stationary measure which is weakly continuous in θ. For an example, consider the form $Y_n^\epsilon = F(\psi_n, \theta_n^\epsilon, \chi_n)$, where $\psi_{n+1} = q(\psi_n, \theta_n^\epsilon, \chi_n)$, for appropriate functions $F(\cdot), q(\cdot)$, and the exogenous noise process $\{\chi_n\}$. The proof of the following theorem is omitted. Related results are given in the next subsection.

Theorem 4.5. *Assume algorithm* (1.1), (A1.1), (A1.4), (A1.5), (A1.7), *and any one of the constraint set conditions* (A4.3.1), (A4.3.2), *or* (A4.3.3). *Suppose that there is a measurable function $g(\cdot)$ such that $g_n^\epsilon(\theta_n^\epsilon, \xi_n^\epsilon) = g(\theta_n^\epsilon, \xi_n^\epsilon)$, where $\{g(\theta_n^\epsilon, \xi_n^\epsilon); \epsilon, n\}$ is uniformly integrable and $g(\cdot, \xi)$ is continuous in θ, uniformly in each compact ξ-set. For each $\theta \in H$, let there be a unique measure $\mu(\cdot|\theta)$ such that the following holds. The set $\{\mu(\cdot|\theta), \theta \in H\}$ is tight. For arbitrary $\theta \in H, \theta_\epsilon \in H$, and $\theta_\epsilon \to \theta$, and any sequence j, suppose that $R(j, \epsilon, \theta_\epsilon, \cdot)$ converges weakly to $\mu(\cdot|\theta)$ and that*

$$\int g(\theta, \xi) R(j, \epsilon, \theta_\epsilon, d\xi) \to \int g(\theta, \xi) \mu(\xi|\theta) = \bar{g}(\theta)$$

as $\epsilon \to 0$. Then $\bar{g}(\cdot)$ is continuous and the conclusions of Theorem 2.1 *hold.*

8.4.4 Non-Markov Noise and General Forms of the Conditions for Convergence

Theorem 4.1 used certain continuity conditions on the functions $g_n^\epsilon(\cdot)$ and on the transition function of the fixed-θ processes to carry out the averaging that led to the mean ODE. An examination of the structure of the proof of that theorem shows that one can accomplish the same thing a little more directly, by working with the functions $g_n^\epsilon(\cdot)$ all the way through. In what follows we replace assumptions (A4.1)–(A4.5) and (A4.7) with weaker assumptions that achieve the same goal. In any particular application, the conditions of one of the Theorems 4.1, 4.3, 4.4, or 4.5 (or the analogous theorems for the decreasing-step-size case) might be the most easily verifiable. In Chapter 9, Theorem 4.1 is applied to the optimization problem for a stochastic differential equation model, and Theorem 4.6 is applied to the optimization of the queueing problem of Section 2.5. Theorem 4.1 could be applied to the queueing problem also, as in [145], but that would require more care in specifying the topology put on the state space Ξ of the Markov chain. Theorem 4.6 has applications in discrete event dynamical systems, despite the seeming generality of the assumptions. Owing to the forms that the functions $g_n^\epsilon(\cdot)$ often take in these applications, one can use the particular continuity properties on the $g_n^\epsilon(\cdot)$ to verify the conditions directly. We will use the following conditions in which the terms $\mu_{\rho,\epsilon}$ allow time for transients to "settle."

Definition: Fixed-θ process. We *redefine* the fixed-θ process. Let $\{\xi_j^\epsilon(\theta), j \geq n\}$ be the process that would result if $\{\theta_j, j \geq n\}$ were held fixed at the value θ and with the past record $\{\xi_j^\epsilon, j \leq n\}$. It need not be Markov. In the following discussion, A is an arbitrary compact set in Ξ.

(A4.16) For each small $\rho > 0$, there are $0 \leq K_\rho < \infty, \epsilon_\rho > 0$, and integers $m_{\rho,\epsilon}$ and $\mu_{\rho,\epsilon}$, where $m_{\rho,\epsilon} \to \infty$, $\mu_{\rho,\epsilon}/m_{\rho,\epsilon} \to 0$, and $\epsilon m_{\rho,\epsilon} \to 0$ as $\epsilon \to 0$, and measurable functions $g_{n,\rho}^\epsilon(\cdot)$ bounded by K_ρ, such that

$$E\left|g_n^\epsilon(\theta_n^\epsilon, \xi_n^\epsilon) - g_{n,\rho}^\epsilon(\theta_n^\epsilon, \xi_n^\epsilon)\right| \leq \rho \tag{4.22}$$

and

$$E\left|E_n^\epsilon g_{i,\rho}^\epsilon(\theta_n^\epsilon, \xi_i(\theta_n^\epsilon)) - E_n^\epsilon g_{i,\rho}^\epsilon(\theta_i^\epsilon, \xi_i^\epsilon)\right| \leq \rho \tag{4.23}$$

for i and n such that $\mu_{\rho,\epsilon} \leq i - n \leq m_{\rho,\epsilon}$ and $\epsilon \leq \epsilon_\rho$. In (4.23), as in Subsection 4.1, the initial condition is $\xi_n(\theta_n^\epsilon) = \xi_n^\epsilon$.

In the following conditions $m_{\rho,\epsilon}$ and $\mu_{\rho,\epsilon}$ are as defined as in (A4.16), and given arbitrary $\rho > 0$, there is an $\epsilon_\rho > 0$ such that the condition holds.

(A4.17) There is $\nu(\rho) > 0$ such that for any $\mathcal{F}_n^\epsilon$-measurable and $I\!R^r$-valued random variables $\widetilde{\theta}$ and $\widehat{\theta}$ taking values in H and satisfying $P\{|\widehat{\theta} - \widetilde{\theta}| > \nu(\rho)\} < \nu(\rho)$ and $\epsilon \leq \epsilon_\rho$, we have

$$\sup_{i,n: m_{\rho,\epsilon} \geq i-n \geq \mu_{\rho,\epsilon}} E\left|E_n^\epsilon g_{i,\rho}^\epsilon(\widehat{\theta}, \xi_i(\widehat{\theta})) - E_n^\epsilon g_{i,\rho}^\epsilon(\widetilde{\theta}, \xi_i(\widetilde{\theta}))\right| I_{\{\xi_n^\epsilon \in A\}} \leq \rho. \tag{4.24}$$

(A4.18) For each $\theta \in H$ and $\epsilon \leq \epsilon_\rho$,

$$\sup_{i,n: m_{\rho,\epsilon} \geq i-n \geq \mu_{\rho,\epsilon}} E\left|E_n^\epsilon g_{i,\rho}^\epsilon(\theta, \xi_i(\theta)) - E_n^\epsilon g_i^\epsilon(\theta, \xi_i(\theta))\right| I_{\{\xi_n^\epsilon \in A\}} \leq \rho. \tag{4.25}$$

Conditions (A4.17) and (A4.18) imply the convergence in (A4.6), uniformly in θ in any compact set. Analogously to what was done with (A1.11)–(A1.13) in replacing (A1.6) and (A1.9), the next set of conditions is an alternative to (A4.16)–(A4.18). It is assumed that the expectations and conditional expectations are well defined.

(A4.19) There is a continuous function $\bar{g}(\cdot)$ such that for small $\delta > 0, \epsilon > 0$, and some sequence $m \to \infty$ (not depending on ϵ, δ)

$$\frac{1}{m} E\left|\sum_{l=n}^{n+m-1} E_n^\epsilon g_l^\epsilon(\theta, \xi_l(\theta)) - \bar{g}(\theta)\right| I_{\{\xi_n^\epsilon \in A\}} I_{\{|\theta_n^\epsilon - \theta| \leq \delta\}} \to 0 \tag{4.26}$$

uniformly in n, and in θ in any compact set.

(A4.20) For each integer m,

$$\frac{1}{m}E\left|\sum_{l=n}^{n+m-1}\left[E_n^\epsilon g_l^\epsilon(\theta_n^\epsilon,\xi_l(\theta_n^\epsilon)) - E_n^\epsilon g_l^\epsilon(\theta_l^\epsilon,\xi_l^\epsilon)\right]\right| I_{\{\xi_n^\epsilon\in A\}} \to 0. \tag{4.27}$$

uniformly in n, as $\epsilon \to 0$.

(A4.21) For small $\delta > 0, \epsilon > 0$ and each integer m, as $|\widehat{\theta} - \widetilde{\theta}| \to 0$,

$$\frac{1}{m}E\left|\sum_{l=n}^{n+m-1} E_n^\epsilon\left[g_l^\epsilon(\widehat{\theta},\xi_i(\widehat{\theta})) - E_n^\epsilon g_l^\epsilon(\widetilde{\theta},\xi_l(\widetilde{\theta}))\right]\right| I_{\{\xi_n^\epsilon\in A\}} \to 0. \tag{4.28}$$

uniformly in n and ϵ, where $\widetilde{\theta}$ and $\widehat{\theta}$ are in any compact set.

Theorem 4.6. *In Theorem* 4.1, *replace* (A4.1)–(A4.7) *with the uniform integrability of the set in* (4.5a) *and either* (A4.6) *and* (A4.16)–(A4.18) *or* (A4.19)–(A4.21). *Then the conclusions of Theorem* 2.1 *hold.*

Discussion of the proof. Condition (4.22) ensures that (4.7) approximates the equation that precedes it. Condition (4.23) guarantees that (4.8) approximates (4.7). Condition (A4.17) ensures that (4.9) approximates (4.8). Condition (A1.7) allows us to suppose that the values of the initial conditions $\{\xi_{jn_\epsilon}^\epsilon\}$ are confined to a compact set when evaluating the limit of (4.9) as $\epsilon \to 0$. Condition (A4.18) permits us to drop the K in (4.9). The conditions (A4.19)–(A4.21) and the uniform integrability of (4.5a) accomplish the same result a little more directly. The rest of the details are as in the proof of Theorem 4.1.

8.4.5 Observations Depending on the Past of the Iterate Sequence or Working Directly with Y_n^ϵ

Most of the book uses averaging conditions on $E_n^\epsilon Y_n^\epsilon$ which we defined in terms of the memory random variable ξ_n^ϵ, as $g_n^\epsilon(\theta_n^\epsilon,\xi_n^\epsilon)$. This procedure was followed since the conditional expectation, given the past, is often a smoothing operation, and the function $g_n^\epsilon(\cdot,\xi)$ is continuous. For some algorithms, the function $g_n^\epsilon(\cdot)$ depends on the past values of the iterates and has the form $g_n^\epsilon(\theta_l^\epsilon, l \le n, \xi_n^\epsilon)$, but which can be approximated by $g_n^\epsilon(\theta_n^\epsilon,\ldots,\theta_n^\epsilon,\xi_n^\epsilon)$ when ϵ is small, and then all of the theorems can be carried over.

There are stochastic algorithms that do not easily fit such formats and where it is most convenient to work directly in terms of the original observation Y_n^ϵ. The past results can be easily extended, requiring only a change in the notation. For each n, define the "fixed-θ" observation process $\{Y_l(\theta), l \geq n)\}$ as the process that would result if the iterate is held fixed at θ for $l > n$ but with θ_l^ϵ used for $l \leq n$, analogously to the definition of the fixed-$\xi_l(\theta)$ process. Suppose that Y_n^ϵ can be written as $Y_n^\epsilon(\theta_n^\epsilon, U_n^\epsilon)$ for some random variable U_n^ϵ, which in turn might be state dependent. Let $\{U_l(\theta), l \geq n\}$ denote the fixed-θ process, defined analogously to $Y_l(\theta)$. Let E_n^ϵ denote the expectation conditioned on $\xi_n^\epsilon = \{U_l^\epsilon, l < n, \theta_0^\epsilon\}$. Depending on the situation (e.g., Markov evolution), the memory might be representable with fewer terms. We have the following variation of Theorem 4.6.

Theorem 4.7 *In Theorem* 4.1, *replace* (A4.1)–(A4.7) *with the uniform integrability of the set in* (4.5a) *and* (A4.19)–(A4.21) *with the following changes. All* $g_l(\theta, \xi_l(\theta))$ *are replaced by* $Y_l(\theta, U_l(\theta))$ *for the corresponding values of* θ, *and* $g_l^\epsilon(\theta_l^\epsilon, \xi_l^\epsilon)$ *is replaced by* Y_l^ϵ. *Then the conclusions of Theorem* 2.1 *hold.*

Example: The adaptive antenna problem: tracking the optimal discount factor. We return to the problem introduced in Section 3.5. In this example, the discount factor α replaces θ. For notational simplicity, we drop the superscript ϵ. Use algorithm (3.5.4) and (3.5.5), but where α is constrained to an interval $[\underline{\alpha}, \overline{\alpha}] \in (0,1)$. The additive noise and test signal are as in Section 3.5. For purposes of the convergence analysis, we will use the formula for the weights

$$w_l = \left[\sum_{i=1}^{l} [\alpha_{i+1} \cdots \alpha_l] X_i X_i'\right]^{-1} \left[\sum_{i=1}^{l} [\alpha_{i+1} \cdots \alpha_l] X_i \bar{s}_i\right]. \tag{4.29}$$

Owing to the discounting in (4.29), the effects of the sample taken at time $(l-v)h$ decreases geometrically as $v \to \infty$. Truncate the components of (4.29) to the interval $[-L, L]$. For large L, truncation will rarely occur. The main issue in the verification of (A4.19) concerns the effect of the Doppler frequency in (3.5.8) and we ignore the effects of the other quantities. Define $Y_n^\pm = [e_n^\pm]^2$ and set $U_n = (\bar{s}_n, X_n, w_n)$.

The $Y_l^\pm$ can be approximated arbitrarily well (in mean and uniformly in l) by ignoring values of X_v for $l - v \geq M$, for large enough M. Fix M to be large, and let $m > M$ be such that $m\epsilon \to 0$ as $\epsilon \to 0$ and consider $E_n Y_l^\pm$ for $m \geq l - n > M$. As $\mu \to 0$, we can replace $\alpha_{n+1}, \ldots, \alpha_l$ in this expression by either α_l or α_n. Then, due to the independence and Gaussian properties of the noise and its independence of the signal, the operation $E_n Y_l^\pm$ is a smoothing operation. There are bounded and continuous functions $F(\cdot)$ such that $E_n[Y_l^\pm | \alpha_l^\pm, \psi_{j,v}^d; j, l \geq v \geq l - M]$ is approximately $F_l^\pm = F(\alpha_l^\pm, \cos\psi_{j,v}^d, \sin\psi_{j,v}^d; j, l \geq v \geq l - M)$.

Let us consider two simple cases. First, let the Doppler frequencies be constant and mutually incommensurate; $\omega_j^d(t) = \omega_j^d$, all j. Then, (modulo some initial phase which is unimportant and that we ignore) we can write

$$F_l^{\pm} = F(\alpha_l^{\pm}, \cos \omega_j^d vh, \alpha^v \sin \omega_j^d vh; l \geq v \geq l - M, j).$$

If $h\omega_j$ is small for all j, then there are many samples per cycle for each j and for large m and small μ the value of

$$\frac{1}{m} \sum_{l=n}^{n+m-1} E_n F_l^{\pm} \tag{4.30}$$

is very close to $\bar{g}(\alpha_n^{\pm})$ (uniformly in n and in the mean) where

$$\bar{g}(\alpha^{\pm}) = EF(\alpha^{\pm}, \cos(\psi_j - \omega_j^d hv), \sin(\psi_j - \omega_j^d hv); 0 \leq v \leq M, j), \tag{4.31}$$

where the ψ_j are independent and uniformly distributed on $[0, 2\pi]$. Then, with $\bar{g}(\alpha) = [\bar{g}(\alpha^+) - \bar{g}(\alpha^-)]/\delta$, (A4.19) will hold approximately, with an error that goes to zero as $h\omega_j \to 0$. Since, for a "typical" l, any finite set of adjacent samples will be highly correlated with each other, the value of $\bar{g}(\alpha^-)$ will be larger than the value of $\bar{g}(\alpha^+)$ and $\bar{g}(\alpha)$ will be positive. Thus, the value of α will tend to its upper bound, as desired.

Next suppose that the Doppler frequencies vary randomly as independent random walks. In particular, let the Doppler phase for mobile j at the lth sample be $\psi_j(0) + b_j lh + W_j(lh)$, where the $W_j(\cdot)$ are mutually independent Wiener processes, perhaps with different variances. Then, for large $l - n$ and m, small μ, and $\alpha_n \approx \alpha$, (4.30) is well approximated by

$$EF(\alpha^{\pm}, \cos(\psi_j - b_j vh + W_j(vh)), \sin(\psi_j - b_j vh + W_j(vh)); 0 \leq v \leq M, j), \tag{4.32}$$

where the ψ_j are mutually independent and are uniformly distributed on $[0, 2\pi]$.

Since the test signals from the different mobiles are mutually independent, it can be seen that the Doppler frequency of the mobile that is being optimized for has a larger effect than those of the interfering mobiles; see [41] for additional detail.

8.5 Unconstrained Algorithms and the ODE-Stability Method

There has been a heavy emphasis on constrained algorithms. But the same weak convergence methods are effective for the unconstrained problems. The only difference is that we will need to either prove or assume tightness (i.e., boundedness in probability) of the iterate sequence. The extension will be given in this section. The essence is the use of a localization method and Theorem 7.3.6 (see also [127, p. 43]).

Comment on soft constraints. The soft constraint idea of Section 5.5 can be used equally well here. As in Section 5.5, one introduces the constraint via a penalty function in the dynamics, which ensures the tightness property (A5.0).

Assumptions. The following conditions will be used for the constant step size and exogenous noise case. The tightness in (A5.0) is normally proved by a stability argument using Liapunov or perturbed Liapunov functions. A sufficient condition, obtained by such a method, is in Theorem 6.7.5.

(A5.0) $\{\theta_n^\epsilon;\ \epsilon, n\}$ is tight.

(A5.1) For each compact set Q,

$$\{Y_n^\epsilon I_{\{\theta_n^\epsilon \in Q\}}; \epsilon, n\} \text{ is uniformly integrable.}$$

(A5.2) There are measurable functions $g_n^\epsilon(\cdot)$ and random variables β_n^ϵ such that for each compact set Q and $\theta_n^\epsilon \in Q$,

$$E_n^\epsilon Y_n^\epsilon = g_n^\epsilon(\theta_n^\epsilon, \xi_n^\epsilon) + \beta_n^\epsilon.$$

(A5.3) For each compact set Q, the sequences

$$\{g_n^\epsilon(\theta_n^\epsilon, \xi_n^\epsilon) I_{\{\theta_n^\epsilon \in Q\}}; \epsilon, n\}, \tag{5.1a}$$

and

$$\{g_n^\epsilon(\theta, \xi_n^\epsilon); \epsilon, n\},\ \theta \in Q, \tag{5.1b}$$

are uniformly integrable.

(A5.4) There are nonempty compact sets D_i that are the closures of their interiors and satisfy $D_0 \subset D_1^0 \subset D_1 \subset D_2^0 \subset D_2$ (where D_i^0 is the interior of D_i) such that all trajectories of

$$\dot{\theta} = \bar{g}(\theta) \tag{5.2}$$

tend to D_0 as $t \to \infty$ and all trajectories starting in D_1 stay in D_2. Equation (5.2) has a unique solution for each initial condition.

(A5.5) For each compact set Q,

$$\lim_{m,n,\epsilon} \frac{1}{m} \sum_{i=n}^{n+m-1} E_n^\epsilon \beta_i^\epsilon I_{\{\theta_n^\epsilon \in Q\}} = 0 \text{ in mean.}$$

The following condition is a "localization" of (A1.11) and can be used in place of (A1.6). The processes $\{\xi_i(\widetilde{\theta}), i \geq n\}$ are defined as above (A4.16). The proof of Theorem 5.1 is essentially that of Theorem 2.2, with the constraint set H dropped.

Theorem 5.1. *Assume* (A5.0)–(A5.5), (A1.7), *and either* (A4.16)–(A4.18) *or* (A4.19)–A4.21). *Then the conclusions of Theorem* 2.1 *hold, with the ODE* (5.2) *replacing* (2.1), *and a bounded invariant set of* (5.2) *replacing the limit or invariant set in* H.

Proof. Let us work with (A1.6). For $\nu > 0$, let R_ν denote the ν-sphere $R_\nu = \{\theta;\ |\theta| \le \nu\}$. Given a sequence of processes $X^\epsilon(\cdot)$, $X^{\epsilon,\nu}(\cdot)$ are said to be the ν-truncations if $X^{\epsilon,\nu}(t) = X^\epsilon(t)$ up until the first exit from R_ν and satisfy

$$\lim_{K\to\infty} \limsup_{\epsilon} P\left\{\sup_{t\le T} |X^{\epsilon,\nu}(t)| \ge K\right\} = 0 \text{ for each } T < \infty.$$

Let the real-valued truncation function $q^\nu(\cdot)$ on $I\!R^r$ have the following properties: $|q^\nu(\theta)| \le 1$, the partial derivatives up to second order are bounded uniformly in ν and θ, and

$$q^\nu(\theta) = \begin{cases} 1 & \text{if } \theta \in R_\nu, \\ 0 & \text{if } \theta \in I\!R^r - R_{\nu+1}. \end{cases}$$

The proof of the theorem is nearly identical to that of Theorem 2.2. As in the proof of that theorem, for simplicity, we will work only with $q_\epsilon = 0$ and show only the tightness and the weak convergence. For the stochastic approximation algorithm

$$\theta^\epsilon_{n+1} = \theta^\epsilon_n + \epsilon Y^\epsilon_n,$$

define the ν-truncation as

$$\theta^{\epsilon,\nu}_{n+1} = \theta^{\epsilon,\nu}_n + \epsilon Y^{\epsilon,\nu}_n q^\nu(\theta^{\epsilon,\nu}_n),$$

where $Y^{\epsilon,\nu}_n$ means that the observation is taken at parameter value $\theta^{\epsilon,\nu}_n$. Thus, the $\{\theta^{\epsilon,\nu}_n\}$ process stops at the first time that the $(\nu+1)$-sphere is exited and equals the original sequence up to the first time that the ν-sphere is exited. Without loss of generality, suppose that θ^ϵ_0 converges weakly. This can always be ensured by the choice of a suitable subsequence. The proof is divided into three steps.

Step 1. Define $\theta^{\epsilon,\nu}(\cdot)$ to be the piecewise constant interpolation of $\{\theta^{\epsilon,\nu}_n\}$:

$$\theta^{\epsilon,\nu}(0) = \theta^\epsilon_0 \ \text{ and } \ \theta^{\epsilon,\nu}(t) = \theta^{\epsilon,\nu}_n \ \text{ for } t \in [n\epsilon, n\epsilon + \epsilon).$$

Then show that for each ν the ν-truncated sequence $\{\theta^{\epsilon,\nu}(\cdot)\}$ is tight in $D^r[0,\infty)$.

Step 2. Extract a weakly convergent subsequence of $\{\theta^{\epsilon,\nu}(\cdot)\}$. Prove that it has a weak sense limit $\theta^\nu(\cdot)$ satisfying the ODE

$$\dot\theta^\nu = \bar g(\theta^\nu) q^\nu(\theta^\nu), \tag{5.3}$$

with initial condition $\theta(0)$ not depending on ν. This latter fact follows from the weak convergence of $\{\theta_0^\epsilon\}$.

The first two steps are carried out as in Theorem 2.2, with the new dynamical term $\bar{g}(\cdot)q^\nu(\cdot)$.

Step 3. The measure of $\theta^\nu(\cdot)$ depends only on the distribution of $\theta(0)$ and the function $\bar{g}(\cdot)q^\nu(\cdot)$. Use the uniqueness of the solution to (5.2) and the stability conditions to prove that the measures of the $\theta^\nu(\cdot)$ converge to that of the solution to (5.2) with initial condition $\theta(0)$ as $\nu \to \infty$.

Step 4. To complete the proof via Theorem 7.3.6, we need only show that for each $T > 0$,

$$\limsup_{\nu} \limsup_{\epsilon \to 0} P\left\{\theta^{\epsilon,\nu}(t) \neq \theta^\epsilon(t), \text{ for some } t \leq T\right\} = 0. \tag{5.4}$$

Equation (5.4) will hold if for each $\rho > 0$ and $T > 0$, there is a $K_\rho < \infty$ such that

$$\limsup_{\epsilon \to 0} P\left\{\sup_{t \leq T} |\theta^{\epsilon,\nu}(t)| \geq 2K_\rho\right\} \leq \rho \tag{5.5}$$

for each large ν. Condition (A5.4) implies that for each compact set S_3 there is a compact set S_4 such that the trajectories of (5.2) starting in S_3 never leave S_4. This and (A5.0) imply that for each $\rho > 0$ there is a compact set S_ρ such that for all ν the trajectories of all the weak sense limit process are confined to S_ρ for all time with probability at least $1 - \rho$. Thus,

$$P\left\{\sup_{t \leq T} |\theta^\nu(t)| \geq K\right\}$$

goes to zero uniformly in the limit process $\theta^\nu(\cdot)$ as $K \to \infty$. By the weak convergence, for each ν we have

$$\limsup_{\epsilon \to 0} P\left\{\sup_{t \leq T} |\theta^{\epsilon,\nu}(t)| \geq 2K\right\} \leq P\left\{\sup_{t \leq T} |\theta^\nu(t)| \geq K\right\}.$$

Since the right side can be made as small as desired uniformly in ν by choosing K large, (5.5) holds. Now, Theorem 7.3.6 implies that $\{\theta^\epsilon(\cdot)\}$ is tight, and all weak sense limit processes satisfy (5.2).

The rest of the details are as in the proofs of Theorems 2.1 and 2.2 and are omitted. □

Extensions to other cases. The changes required for the unconstrained analogs of Theorems 2.3–2.5, 3.1–3.3 and 4.1–4.6 should be obvious: They are (A5.0) and the conditions of those theorems localized, where appropriate, analogously to (A5.1)–(A5.5) and (A4.19)–(4.21).

8.6 Two-Time-Scale Problems

8.6.1 *The Constrained Algorithm*

The model. In multiple-time-scale systems, the components of the iterate are divided into groups with each group having its own step-size sequence and they are of different orders. Consider the two-time-scale system of the following form, where the step sizes are ϵ and μ and ϵ is much smaller than μ:

$$\begin{aligned} \theta_{n+1}^{\epsilon,\mu} &= \Pi_{H_\theta}\left[\theta_n^{\epsilon,\mu} + \epsilon Y_n^{\epsilon,\mu,\theta}\right], \\ v_{n+1}^{\epsilon,\mu} &= \Pi_{H_v}\left[v_n^{\epsilon,\mu} + \mu Y_n^{\epsilon,\mu,v}\right]. \end{aligned} \tag{6.1}$$

The iterate is the pair $(\theta_n^{\epsilon,\mu}, v_n^{\epsilon,\mu})$, where $\epsilon/\mu \to 0$ as $\mu \to 0$. Thus, $\theta_n^{\epsilon,\mu}$ is said to be the "slow" component and $v_n^{\epsilon,\mu}$ the fast component. The component θ is confined to the constraint set H_θ, and v is confined to the constraint set H_v. If μ is small, then the convergence rate can be very slow. The bulk of the notation is analogous to that for the single-time-scale algorithm, and will not always be explicitly defined. Let there be "memory" random variables $\xi_n^{\epsilon,\mu} \in \Xi_\theta$ and $\psi_n^{\epsilon,\mu} \in \Xi_v$ (the Ξ_θ and Ξ_v are complete and separable metric spaces) and functions $g_n^{\epsilon,\mu}(\cdot)$ and $k_n^{\epsilon,\mu}(\cdot)$ such that

$$E_n^{\epsilon,\mu} Y_n^{\epsilon,\mu,\theta} = g_n^{\epsilon,\mu}(\theta_n^{\epsilon,\mu}, v_n^{\epsilon,\mu}, \xi_n^{\epsilon,\mu}), \quad E_n^{\epsilon,\mu} Y_n^{\epsilon,\mu,v} = k_n^{\epsilon,\mu}(\theta_n^{\epsilon,\mu}, v_n^{\epsilon,\mu}, \psi_n^{\epsilon,\mu}).$$

The method of analysis is by now standard. One first analyzes the fast system with the $\theta_n^{\epsilon,\mu}$ fixed at arbitrary θ, and shows that there is a limit point $\bar{v}(\theta)$ to which the solutions converge. Then one shows that it is sufficient to analyze the slow system with $v_n^{\epsilon,\mu}$ replaced by $\bar{v}(\theta_n^{\epsilon,\mu})$. The method is laid out in detail in [131] for quite complex singularly perturbed continuous time systems, where the weak sense limits are jump-diffusion processes. The assumptions usually fall into two sets. The first concerns the "fast" system and characterizes the limit as a function of the "current" value $\theta_n^{\epsilon,\mu}$. Apart from a stability condition, they are analogs of the conditions of Theorem 2.2 (or any set used in Section 4) for the second line of (6.1). The second set of conditions consists of those of Theorem 2.2 (or any set used in Section 4) for the first line of (6.1), with $v_n^{\epsilon,\mu}$ replaced by $\bar{v}(\theta_n^{\epsilon,\mu})$. The analysis can be extended to cover problems with any number of time scales, and with decreasing step sizes as well; see also [23, 24, 28]. The stability methods of [131] can be used to prove tightness. The iterate averaging algorithms of Chapter 11 can also be viewed in terms of two time scales. But it is the rate of convergence that is of most interest there.

Rewrite (6.1) in decomposed form as

$$\begin{aligned} \theta_{n+1}^{\epsilon,\mu} &= \theta_n^{\epsilon,\mu} + \epsilon\left[g_n^{\epsilon,\mu}(\theta_n^{\epsilon,\mu}, v_n^{\epsilon,\mu}, \xi_n^{\epsilon,\mu}) + \delta M_n^{\epsilon,\mu,\theta} + \delta Z_n^{\epsilon,\mu,\theta}\right], \\ v_{n+1}^{\epsilon,\mu} &= v_n^{\epsilon,\mu} + \mu\left[k_n^{\epsilon,\mu}(\theta_n^{\epsilon,\mu}, v_n^{\epsilon,\mu}, \psi_n^{\epsilon,\mu}) + \delta M_n^{\epsilon,\mu,v} + \delta Z_n^{\epsilon,\mu,v}\right], \end{aligned} \tag{6.2}$$

where the δZ are the reflection and the δM the martingale difference terms.

Assumptions. It is always assumed that the integrals are well defined.

(A6.0) H_v and H_θ satisfy any of the constraint set conditions (A4.3.1), (A4.3.2), or (A4.3.3).

(A6.1) The sets $\{Y_n^{\epsilon,\mu,\theta}, Y_n^{\epsilon,\mu,v}\}$ are uniformly integrable, and so are (for each θ and v) $\{g_n^{\epsilon,\mu}(\theta, v, \xi_n^{\epsilon,\mu}), k_n^{\epsilon,\mu}(\theta, v, \psi_n^{\epsilon,\mu})\}$. The sets $\{\xi_n^{\epsilon,\mu}, \psi_n^{\epsilon,\mu}\}$ are tight.

(A6.2) The functions $g_n^{\epsilon,\mu}(\cdot, \xi)$ and $k_n^{\epsilon,\mu}(\cdot, \psi)$ are continuous in θ and v, uniformly in n, ϵ, μ and in the ξ and ψ variables in any compact sets.

(A6.3) There is a continuous function $\bar{k}(\cdot)$ such that for each θ and v,

$$\lim_{(n,m)\to\infty,\, (\mu,\epsilon/\mu)\to 0} \frac{1}{m} \sum_{i=n}^{n+m-1} E_n^{\epsilon,\mu} \left[k_i^{\epsilon,\mu}(\theta, v, \psi_i^{\epsilon,\mu}) - \bar{k}(\theta, v)\right] = 0$$

in probability.

(A6.4) For each fixed θ, the ODE $\dot{v} = \bar{k}(\theta, v)$ has a unique globally asymptotically stable point $\bar{v}(\theta)$, and it is continuous in θ.

(A6.5) Conditions (A1.8) and (A1.9) hold for $g_n^{\epsilon,\mu}(\theta_n^{\epsilon,\mu}, \bar{v}(\theta_n^{\epsilon,\mu}), \xi_n^{\epsilon,\mu})$ and, for each θ, for $g_n^{\epsilon,\mu}(\theta, \bar{v}(\theta), \xi_n^{\epsilon,\mu})$.

Theorem 6.1. *Assume* (A6.0)–(A6.5). *Then the conclusions of Theorem* 2.1 *hold for the sequence* $\{\theta_n^{\epsilon,\mu}\}$ *and the mean ODE is* $\dot{\theta} = \bar{g}(\theta) + z$, *where* $z(\cdot)$ *is the reflection term.*

Proof. The proof is only a minor modification of that of Theorem 2.2 and only a few comments will be made. Define the interpolation $v^{\epsilon,\mu}(\cdot)$ of $\{v_n^{\epsilon,\mu}, n < \infty\}$ in the μ time scale and the interpolation $\theta^{\epsilon,\mu}(\cdot)$ of $\{\theta_n^{\epsilon,\mu}, n < \infty\}$ in the ϵ time scale. Then show that for any $T < \infty$,

$$\limsup_{n,i:(i-n)\mu\le T;\, (\mu,\epsilon/\mu)\to 0} E\left|k_i^{\epsilon,\mu}(\theta_i^{\epsilon,\mu}, v_i^{\epsilon,\mu}, \psi_i^{\epsilon,\mu}) - k_i^{\epsilon,\mu}(\theta_n^{\epsilon,\mu}, v_i^{\epsilon,\mu}, \psi_i^{\epsilon,\mu})\right| = 0.$$

A time interval of length τ in the ϵ-scale involves τ/ϵ iterates, hence it is an interval of length $\tau\mu/\epsilon$ in the μ-scale. We have

$$v^{\epsilon,\mu}(t + \tau\mu/\epsilon) - v^{\epsilon,\mu}(\tau\mu/\epsilon) \approx \int_0^t \bar{k}(\theta^{\epsilon,\mu}(\tau), v^{\epsilon,\mu}(s + \tau\mu/\epsilon))ds$$

in the sense that the $\sup_{t\le T}$ of the difference goes to zero in mean, uniformly in τ as $(\mu, \epsilon/\mu) \to 0$. Thus, for any $\delta > 0$, and after a transient period that is arbitrarily short in the ϵ-scale, by the stability property we can suppose that $E|v_n^{\epsilon,\mu} - \bar{v}(\theta_n^{\epsilon,\mu})| \le \delta$. From this point on, the proof is as in Theorem 2.2.

An analog of (A4.19)–(A4.21). There are analogs to all of the various state-dependent noise formulations of Section 4. We simply state an analog of Theorem 4.6 with slightly simplified conditions and analogous notation. In what follows A is an arbitrary compact set. It is assumed that all expectations and conditional expectations are well defined.

(A6.6) There is a continuous function $\bar{k}(\cdot)$ such that for small positive μ and ϵ/μ, and some sequence $m \to \infty$ (not depending on n, ϵ, μ),

$$\frac{1}{m}E\left|\sum_{l=n}^{n+m-1} E_n^{\epsilon,\mu} k_l^{\epsilon,\mu}(\theta, v, \psi_l(\theta, v)) - \bar{k}(\theta, v)\right| I_{\{\psi_n^{\epsilon,\mu}\in A\}} \to 0$$

uniformly in n, ϵ, μ and in θ and v in any compact set.

(A6.7) For each integer m,

$$\frac{1}{m}E\left|\sum_{l=n}^{n+m-1} \left[E_n^{\epsilon,\mu} k_l^{\epsilon,\mu}(\theta_n^{\epsilon,\mu}, v_n^{\epsilon,\mu}, \psi_i(\theta_n^{\epsilon,\mu}, v_n^{\epsilon,\mu})) - E_n^{\epsilon,\mu} k_l^{\epsilon,\mu}(\theta_l^{\epsilon,\mu}, v_l^{\epsilon,\mu}, \psi_l^{\epsilon,\mu}))\right]\right| \times I_{\{\psi_n^{\epsilon,\mu}\in A\}} \to 0$$

uniformly in n as $\mu, \epsilon/\mu \to 0$.

(A6.8) For each integer m, as $|\widehat{\theta} - \widetilde{\theta}| + |\widehat{v} - \widetilde{v}| \to 0$,

$$\frac{1}{m}E\left|\sum_{l=n}^{n+m-1} \left[E_n^{\epsilon,\mu} k_l^{\epsilon,\mu}(\widehat{\theta}, \widehat{v}, \psi_l(\widehat{\theta}, \widehat{v})) - k_l^{\epsilon,\mu}(\widetilde{\theta}, \widetilde{v}, \psi_l(\widetilde{\theta}, \widetilde{v}))\right]\right| I_{\{\psi_n^{\epsilon,\mu}\in A\}} \to 0$$

uniformly in n and in (small) ϵ and ϵ/μ, where the values of $\widehat{\theta}, \widehat{v}, \widetilde{\theta}, \widetilde{v}$ are in any compact set.

(A6.9) Conditions (A4.19)–(A4.21) hold for $g_n^{\epsilon,\mu}(\theta_n^{\epsilon,\mu}, \bar{v}(\theta_n^{\epsilon,\mu}), \xi_n^{\epsilon,\mu})$, and $g_n^{\epsilon,\mu}(\theta, \bar{v}(\theta), \xi_n(\theta, \bar{v}(\theta))$. Each of these sets is uniformly (in ϵ, μ, n) integrable.

Theorem 6.2. *Assume* (A6.0), (A6.1), (A6.4), and (A6.6)–(A6.9). Then the conclusions of Theorem 2.1 hold.

8.6.2 Unconstrained Algorithms: Stability

If the algorithm is unconstrained, then analogously to what was done in Section 5 and in Theorems 6.7.3–6.7.5, Theorems 6.1 and 6.2 hold if the sequence of iterates is tight. The general approach to the proof of stability follows the methods of [131, Chapter 9] for singularly perturbed problems. It will be illustrated here for the simplest case, where the noise is just a martingale difference sequence. The model is

$$\begin{aligned} \theta_{n+1}^{\epsilon,\mu} &= \theta_n^{\epsilon,\mu} + \epsilon g(\theta_n^{\epsilon,\mu}, v_n^{\epsilon,\mu}) + \epsilon \delta M_n^{\epsilon,\mu,\theta}, \\ v_{n+1}^{\epsilon,\mu} &= v_n^{\epsilon,\mu} + \mu \bar{k}(\theta_n^{\epsilon,\mu}, v_n^{\epsilon,\mu}) + \mu \delta M_n^{\epsilon,\mu,v}. \end{aligned} \tag{6.3}$$

Assume (A6.4) and let $W(\theta, v)$ be a nonnegative Liapunov function for the fixed-θ system $\dot{v} = \bar{k}(\theta, v)$. Let $V(\theta)$ be a nonnegative Liapunov function for $\dot{\theta} = \bar{g}(\theta) = g(\theta, \bar{v}(\theta))$. The mixed second partial derivatives of the Liapunov functions are assumed to be bounded. In a sense the stability analysis parallels the convergence analysis in that one wishes to show that $v_n^{\epsilon,\mu} - \bar{v}(\theta_n^{\epsilon,\mu})$ is small and then use this to show that $\{\theta_n^{\epsilon,\mu}\}$ is tight for small μ and ϵ/μ. However these two facts are connected and need to be shown simultaneously.

Expanding the Liapunov function for the fast system yields

$$\begin{aligned} &E_n^{\epsilon,\mu} W(\theta_{n+1}^{\epsilon,\mu}, v_{n+1}^{\epsilon,\mu}) - W(\theta_n^{\epsilon,\mu}, v_n^{\epsilon,\mu}) = \mu W_v'(\theta_n^{\epsilon,\mu}, v_n^{\epsilon,\mu}) \bar{k}(\theta_n^{\epsilon,\mu}, v_n^{\epsilon,\mu}) \\ &\quad + \epsilon W_\theta'(\theta_n^{\epsilon,\mu}, v_n^{\epsilon,\mu}) g(\theta_n^{\epsilon,\mu}, v_n^{\epsilon,\mu}) + O(1)\mu^2 |\bar{k}(\theta_n^{\epsilon,\mu}, v_n^{\epsilon,\mu})|^2 \\ &\quad + O(1)\epsilon^2 |g(\theta_n^{\epsilon,\mu}, v_n^{\epsilon,\mu})|^2 + O(1)\epsilon\mu |\bar{k}(\theta_n^{\epsilon,\mu}, v_n^{\epsilon,\mu})| |g(\theta_n^{\epsilon,\mu}, v_n^{\epsilon,\mu})| \\ &\quad + O(\epsilon^2) + O(\mu^2). \end{aligned} \tag{6.4}$$

The second and fourth and fifth terms on the right side are due to the fact that $\theta_n^{\epsilon,\mu}$ changes in time. This must be compensated for by using the stability properties of $\dot{\theta} = \bar{g}(\theta)$. We have

$$\begin{aligned} &E_n^{\epsilon,\mu} V(\theta_{n+1}^{\epsilon,\mu}) - V(\theta_n^{\epsilon,\mu}) = \epsilon V_\theta'(\theta_n^{\epsilon,\mu}) g(\theta_n^{\epsilon,\mu}, \bar{v}(\theta_n^{\epsilon,\mu})) \\ &\quad + \epsilon V_\theta'(\theta_n^{\epsilon,\mu}) \left[g(\theta_n^{\epsilon,\mu}, v_n^{\epsilon,\mu}) - g(\theta_n^{\epsilon,\mu}, \bar{v}(\theta_n^{\epsilon,\mu})) \right] + O(1)\epsilon^2 |g(\theta_n^{\epsilon,\mu}, v_n^{\epsilon,\mu})|^2 \\ &\quad + O(\epsilon^2). \end{aligned} \tag{6.5}$$

(6.5) will be used to dominate the "errors" in (6.4), via crude inequalities.

Assumptions. The c_i below are constants.

(A6.6) There is a function $\phi(\cdot)$ such that

$$W_v'(\theta, v)\bar{k}(\theta, v) \le -\phi^2(v - \bar{v}(\theta)) + c_1, \qquad \lim_{|x| \to \infty} \phi(x) = \infty.$$

(A6.7) There is a function $\chi(\cdot)$ such that

$$V_\theta'(\theta)\bar{g}(\theta)) \le -\chi^2(\theta) + c_2, \qquad \lim_{|\theta| \to \infty} \chi(\theta) = \infty.$$

$$
\textbf{(A6.8)} \qquad \begin{gathered}
\sup_{\theta,v} \frac{|W'_\theta(\theta,v)g(\theta,v)| + \left|\bar{k}(\theta,v)\right|^2 + |g(\theta,v)|^2}{\phi^2(\theta-\bar{v}(\theta)) + \chi^2(\theta) + 1} < \infty \\
\sup_\theta \frac{|V_\theta(\theta)|^2}{\chi^2(\theta)+1} < \infty, \quad \sup_{\theta,v} \frac{|g(\theta,v) - g(\theta,\bar{v}(\theta))|^2}{\phi^2(\theta-\bar{v}(\theta))+1} < \infty.
\end{gathered}
$$

Theorem 6.3 *Under the above assumptions,* $\{\theta_n^{\epsilon,\mu}, v_n^{\epsilon,\mu}\}$ *is tight for small* $\mu, \epsilon/\mu > 0$.

Proof. For large C and small $\mu, \epsilon/\mu$, define the composite Liapunov function $\bar{V}(\theta, v) = (C\epsilon/\mu)W(\theta, v) + V(\theta)$, We have

$$
\begin{aligned}
&E_n^{\epsilon,\mu}\bar{V}(\theta_{n+1}^{\epsilon,\mu}, v_{n+1}^{\epsilon,\mu}) - \bar{V}(\theta_n^{\epsilon,\mu}, v_n^{\epsilon,\mu}) \le -C\epsilon\phi^2(v_n^{\epsilon,\mu} - \bar{v}(\theta_n^{\epsilon,\mu})) - \epsilon\chi^2(\theta_n^{\epsilon,\mu}) \\
&+ \epsilon O(1)(C+1) + CO(1)(\epsilon\mu)|\bar{k}(\theta_n^{\epsilon,\mu}, v_n^{\epsilon,\mu})|^2 + CO(1)(\epsilon^3/\mu)|g(\theta_n^{\epsilon,\mu}, v_n^{\epsilon,\mu})|^2 \\
&+ \epsilon^2 O(1)|g(\theta_n^{\epsilon,\mu}, v_n^{\epsilon,\mu})|^2 + C(\epsilon^2/\mu)W'_\theta(\theta_n^{\epsilon,\mu}, v_n)g(\theta_n^{\epsilon,\mu}, v_n^{\epsilon,\mu}) \\
&+ \epsilon V'_\theta(\theta_n^{\epsilon,\mu})[g(\theta_n^{\epsilon,\mu}, v_n^{\epsilon,\mu}) - g(\theta_n^{\epsilon,\mu}, \bar{v}(\theta_n^{\epsilon,\mu}))].
\end{aligned}
\tag{6.6}
$$

To dominate the last line of (6.6) use the inequality $|ab| \le |a|^2/K + |b|^2 K$ for any $K > 0$ and the last line of (A6.8). Using (A6.6)–(A6.8), for large C and small μ and ϵ/μ, yields

$$
\begin{aligned}
&E_n^{\epsilon,\mu}\bar{V}(\theta_{n+1}^{\epsilon,\mu}, v_{n+1}^{\epsilon,\mu}) - \bar{V}(\theta_n^{\epsilon,\mu}, v_n^{\epsilon,\mu}) \\
&\qquad \le -(C\epsilon/2)\phi^2(v_n^{\epsilon,\mu} - \bar{v}(\theta_n^{\epsilon,\mu})) - (\epsilon/2)\chi^2(\theta_n^{\epsilon,\mu}) + \epsilon CO(1)
\end{aligned}
\tag{6.7}
$$

which implies the theorem. □

For the correlated noise case, the Liapunov functions $W(\cdot)$ and $V(\cdot)$ need to be augmented by perturbations as in Theorem 6.7.3.

9

Applications: Proofs of Convergence

9.0 Outline of Chapter

In this chapter, we apply the techniques of Chapter 8 to examples from Chapters 2 and 3, and illustrate the flexibility and usefulness of the weak convergence approach. Sections 1 to 3 are concerned with function minimization, where the function is of the ergodic or average cost per unit time (over the infinite time interval) type. In such cases, one can only get estimates of derivatives over finite time intervals, and it needs to be shown that the averaging implicit in stochastic approximation yields the convergence of reasonable algorithms. In these sections, the results of Section 8.4 are applied to continuous-time and discrete-event dynamical systems that are of interest over a long time period. For example, one might wish to optimize or improve the average cost per unit time performance by sequentially adjusting the parameter θ. The variety of such applications and the literature connected with them are rapidly increasing. The examples will demonstrate the power of the basic ideas as well as the convenience of using them in difficult problems. Section 1 is a general introduction to some of the key issues. The step sizes will generally be constant, but virtually the same conditions can be used for the case $\epsilon_n \to 0$.

Section 4 deals with some of the signal processing applications in Chapter 3 for both constrained and unconstrained algorithms. For the unconstrained case, stability or stability–ODE methods are used. The basic algorithm is the one for the classical parameter identification or interference cancellation problem. We also treat the adaptive equalizer problem of Section 3.4.2 when there is a training sequence and show how the stability and strict positive real assumptions on the underlying linear systems models are used to prove the ultimate convergence of the algorithm. In Section 5, convergence is proved for the Proportional Fair Sharing algorithm of Section 3.6.

9.1 Optimization of the Average Cost per Unit Time: Introduction

9.1.1 General Comments

Let θ be an adjustable parameter of a dynamical system and let $x(\cdot, \theta)$ denote the system state process, when the parameter is constant at value θ. For example, $x(\cdot, \theta)$ might be the path of a stochastic dynamical system defined by a stochastic difference or differential equation, or it might be a stochastic discrete event dynamical system. For a given bounded cost rate $c(\theta, x)$, define

$$L_T(\theta, x(0)) = \frac{1}{T} E \int_0^T c(\theta, x(t, \theta)) dt \tag{1.1}$$

and define

$$L(\theta, x(0)) = \lim_{T\to\infty} L_T(\theta, x(0)), \tag{1.2}$$

assuming that the limit exists. In well-posed problems, the noise that "drives" the system generally has a "mixing" effect, and one expects that $L(\theta, x(0))$ will not depend on the initial condition $x(0)$. In this heuristic discussion, we drop the $x(0)$ in $L(\theta, x(0))$, and suppose that

$$\lim_{T_2 - T_1 \to\infty} \frac{1}{T_2 - T_1} \frac{\partial}{\partial\theta} E \int_{T_1}^{T_2} c(\theta, x(t, \theta)) dt = L_\theta(\theta)$$

does not depend on $x(0)$.

Suppose that we can control and change at will the values taken on by the parameter θ and that a sample path of the dynamical system can be observed for any chosen parameter value. We wish to minimize $L(\theta)$ by the sequential adjustment of θ using estimates of the derivatives made from measurements of the actual sample path. Much of the recent growing interest in stochastic approximation methods is due to the development of estimators of the derivatives of functions such as $L_T(\theta, x(0))$ of the IPA or related types [84, 85, 101, 186, 243, 261] or of the pathwise mean square derivative type [44]. In applications, the estimators are taken over a finite time interval, and the parameter value at which the estimator is taken is changed at each update time. However, one actually wishes to use these estimators for the infinite-time problem of minimizing $L(\theta)$. This conflict between the finite time estimation intervals and the ergodic cost criterion has led to various ad hoc approaches. A frequently used method requires that the successive estimation intervals go to infinity. For the types of applications of concern here, this is not necessary and it can be harmful to the rate of convergence.

The problems in Sections 1 to 3 have noise of the Markov-state-dependent type. It will be seen that the type of averaging used in the theorems of Chapter 8 (say, in condition (A8.1.9) or (A8.4.6) to get the ODE that characterizes the asymptotic behavior) gives us great freedom in the construction of the algorithm. The individual observations Y_n^ϵ need not be (even asymptotically) unbiased estimators of the gradient at the current parameter values. One basically does what is more natural: Keep the successive updating time intervals uniformly bounded, and continue to update the estimator of the derivative as if there were no change in the parameter value. Such results illustrate the essence of the "local averaging" idea in the ODE method. The meaning of these comments will be illustrated in the examples to follow.

Since the variety of possible algorithms is potentially endless, no set of conditions will cover all possible cases. Nevertheless, it will be seen that the proof techniques used in Chapter 8 require conditions that are quite weak. Let us recapitulate the basic ideas of the proofs: The first one is to repose the problem as a *martingale problem*, which allows us to replace the noise terms with appropriate conditional expectations given the past and greatly facilitates the averaging. Then, for the Markov state-dependent-noise case of Section 8.4, we are confronted by the fact that the noise at step n can depend on the values of the state at that time as well as at previous times. This is handled in a convenient way by the use of a Markov model for the joint (noise, state) process and by imposing appropriate weak continuity conditions on the transition function. In doing the local averaging to get the appropriate mean ODE, simple weak continuity assumptions and the slow change of the parameter value for small ϵ allow us to average as if the state did not change. These few basic ideas underlie the bulk of the results of Section 8.4 and the conditions that need to be verified in the applications.

For the queueing example in Section 2.5 (Section 3), there is a regenerative structure. However, updating at regenerative intervals is not generally needed and might not even be a good idea; it is certainly inadvisable when the regenerative periods are very long. The gradient estimators used in Sections 2 and 3 have an "additive structure." For the sake of simplicity, we use periodic updating times in the examples, but the additive structure allows considerably greater flexibility in the choices, as discussed at length in [145]. Roughly, any intuitively reasonable way of selecting the updating intervals works.

In Section 2 we apply the ideas to a stochastic differential equation (SDE) model, where the sample derivative is obtained from the differential equation for the pathwise mean square sense derivative [82]. This is the SDE analog of the IPA estimator; it has been in use for a long time. In such examples one must prove stability of the solution to the derivative estimator equation (that can be a nontrivial job), and there is no regenerative structure to help. The right side of the mean ODE is the negative of the gradient of the stationary cost with respect to the parameter. When stability of the mean square sense pathwise derivative process cannot be proved, one can use various forms of finite differences.

As they are stated, the theorems in Chapters 6 to 8 do not *explicitly* deal with the optimization of an average cost over an infinite interval. It is the form of the averaging conditions (A8.1.9) or (A8.4.6) and the "additive" nature of the gradient estimators that yield the average cost per unit time, or ergodic, results.

9.1.2 A Simple Illustrative SDE Example

We will now illustrate the general comments of the previous subsection with a simple continuous-time model. This will be done in an informal way; no conditions will be given, because we only wish to explain the basic ideas. The complete details for a more general model will be given in the next section. The ideas and issues raised in this subsection also hold for discrete-time models, as will be seen in Section 3. We wish to minimize the cost (1.2), where $x(\cdot,\theta)$ is the solution to the parameterized stochastic differential equation

$$dx(t,\theta) = b(x(t,\theta),\theta)dt + dw, \tag{1.3}$$

where $w(\cdot)$ is a Wiener process.

For notational simplicity, suppose that both x and θ are real valued, and let the function $b(\cdot)$ be smooth enough so that the following calculations make sense. The allowed (bounded) range for θ is $H = [a,b]$, $b > a$. For initial condition $x(0,\theta) = x(0)$, a fixed parameter value θ and the cost rate $c(\theta, x(t,\theta))$, define the average cost per unit time on $[0,T]$ by (1.1). Suppose that $L_T(\theta, x(0))$ is continuously differentiable with respect to θ, and let the derivative $L_{T,\theta}(\theta, x(0))$ be bounded uniformly in θ and T. Suppose that for each θ the limit $L(\theta) = \lim_T L_T(\theta, x(0))$ exists, does not depend on $x(0)$, and that the limit of $L_{T,\theta}(\theta, x(0))$ exists (denoted by $L_\theta(\theta)$). Then $L_\theta(\theta)$ is the derivative of $L(\theta)$ with respect to θ.

A common stochastic approximation procedure for updating θ via a gradient type search is based on the fact that $L_{T,\theta}(\theta, x(0))$ is a good estimator of $L_\theta(\theta)$ if T is large and $x(\cdot,\theta)$ is "well behaved." Pursuing this idea, for any $T > 0$, let us update the parameter at times $nT, n = 1, 2, \ldots,$ as follows. Letting θ_n^ϵ denote the nth choice of the parameter, use it on the time interval $[nT, nT+T)$ to get an unbiased (or asymptotically unbiased) estimator $\widehat{Y}_n^\epsilon$ of the derivative $-L_{T,\theta}(\theta_n^\epsilon, x(nT, \theta_n^\epsilon))$. Note that the initial condition used at the start of the new estimation interval is $x(nT, \theta_n^\epsilon)$, which itself depends on the past $\{\theta_i^\epsilon, i < n\}$ iterates. Let $x^\epsilon(\cdot)$ (with $x^\epsilon(0) = x(0)$)

denote the actual physical state process with the time varying θ_n^ϵ used, that is, on $[nT, nT+T)$, $x^\epsilon(t) = x(t, \theta_n^\epsilon)$, satisfying (1.3), with the "initial condition" at time nT being

$$x(nT, \theta_n^\epsilon) = x^\epsilon(nT). \tag{1.4}$$

Then update according to

$$\theta_{n+1}^\epsilon = \Pi_H \left[\theta_n^\epsilon + \epsilon \widehat{Y}_n^\epsilon\right]. \tag{1.5}$$

The use of $\widehat{Y}_n^\epsilon$ is equivalent to "restarting" the estimation procedure at each nT with new initial condition $x^\epsilon(nT)$.

To formally derive the mean ODE for algorithm (1.5), define the state dependent "noise process" $\xi_n^\epsilon = x^\epsilon(nT)$ and suppose that the limit

$$\bar{g}(\theta) = -\lim_{n,m} \frac{1}{m} \sum_{i=n}^{n+m-1} E_n^\epsilon L_{T,\theta}(\theta, x(iT, \theta))$$

of the conditional expectations exists (in the mean as n and m go to infinity). Suppose that the process $\{x(nT, \theta), n = 0, 1, \ldots\}$ has a unique invariant measure denoted by $\mu(\cdot|\theta)$. Then a formal application of Theorem 8.4.1 yields the mean ODE

$$\dot{\theta} = -\int L_{T,\theta}(\theta, \xi)\mu(d\xi|\theta) + z, \tag{1.6}$$

where $z(\cdot)$ serves only to keep the solution in the interval $[a, b]$.

The integral in (1.6) is not $L_\theta(\theta)$, the desired value. For the right side of (1.6) to be close to $-L_\theta(\theta)$, we would generally need at least T to be large, which reduces the dependence of $L_{T,\theta}(\theta, \xi)$ on ξ. For this reason, it is often suggested that T depend on ϵ or n or both and that T go to infinity as one or both of these quantities go to their limits. This would not be desirable in a practical algorithm. [In such algorithms, ϵ is often fixed at some small value.] Letting $T \to \infty$ as $n \to \infty$ slows down the rate at which new information is used in the updates and lowers the rate of convergence. It also raises the additional question of the rate of increase of T with n that would be used.

Comments and review of the derivation of (1.6). To obtain the true gradient descent ODE, it is not necessary to increase the lengths of the estimation intervals. Indeed, it can be obtained from a simpler and more natural procedure. To see how to get it, we need to first specify the method used to get the gradient estimator when θ is held fixed. One generally introduces an auxiliary "derivative" process $y(\cdot, \theta)$. For IPA estimators [84, 101, 161], this would be the pathwise derivative of $x(\cdot, \theta)$ with respect to θ; for likelihood ratio estimators [161, 200, 210] this would be the score function. Other methods, such as RPA (rare perturbation analysis) [35], use auxiliary information that represents the difference between the path $x(\cdot, \theta)$ and a perturbed one.

For the case of this subsection with θ fixed, an appropriate auxiliary process is the pathwise mean square sense derivative process $y(\cdot, \theta)$. It is the process defined by

$$\lim_{\delta\theta \to 0} E \left| \frac{x(t, \theta + \delta\theta) - x(t, \theta)}{\delta\theta} - y(\theta, t) \right|^2 = 0.$$

In this mean square sense, we can write $y(t, \theta) = x_\theta(t, \theta)$ and

$$\dot{y}(t, \theta) = b_x(x(t, \theta), \theta) y(t, \theta) + b_\theta(x(t, \theta), \theta). \tag{1.7}$$

Since the initial condition $x(0)$ does not depend on θ, $y(0, \theta) = 0$. Define $u(\cdot, \theta) = (x(\cdot, \theta), y(\cdot, \theta))$. A suitable estimator of $L_{T,\theta}(\theta, x(0))$ is

$$\frac{1}{T} \int_0^T \lambda(\theta, u(s, \theta)) ds, \tag{1.8a}$$

where

$$\lambda(\theta, u(s, \theta)) = \frac{\partial c(\theta, x(s, \theta))}{\partial \theta} = c_x(\theta, x(s, \theta)) y(s, \theta) + c_\theta(\theta, x(s, \theta)) \tag{1.8b}$$

and $c(\cdot)$ is as in (1.1).

Using the definition of $x^\epsilon(\cdot)$ given above (1.4) as the actual physical process with the actual time-varying parameter used, the complete procedure that led to (1.6) can be described as follows. Define the process $y_0^\epsilon(\cdot)$ by

$$\dot{y}_0^\epsilon(t) = b_x(x^\epsilon(t), \theta_n^\epsilon) y_0^\epsilon(t) + b_\theta(x^\epsilon(t), \theta_n^\epsilon), \quad t \in [nT, nT + T),$$

with $y_0^\epsilon(nT) = 0$. Define $u_0^\epsilon(\cdot) = (x^\epsilon(\cdot), y_0^\epsilon(\cdot))$. The quantity

$$\widehat{Y}_n^\epsilon = -\frac{1}{T} \int_{nT}^{nT+T} \lambda(\theta_n^\epsilon, u_0^\epsilon(s)) ds \tag{1.9}$$

is an unbiased estimator of $-L_{T,\theta}(\theta, x^\epsilon(nT))$ at $\theta = \theta_n^\epsilon$. The use of this estimator in (1.5) leads to the incorrect mean ODE (1.6). The use of $y_0^\epsilon(nT) = 0$ resets the gradient estimation procedure so that on the interval $[nT, nT+T)$ we are actually estimating the θ-gradient of the cost on $[0, T]$ for the process starting at state value $x^\epsilon(nT)$ and parameter value θ_n^ϵ.

The correct limit equation. Now, let us see how the true gradient descent ODE (1.14) can be obtained, without increasing the intervals between updates. Let $x^\epsilon(\cdot)$ be as above but replace the $y_0^\epsilon(\cdot)$ process by the process defined by

$$\dot{y}^\epsilon(t) = b_x(x^\epsilon(t), \theta_n^\epsilon) y^\epsilon(t) + b_\theta(x^\epsilon(t), \theta_n^\epsilon), \quad t \in [nT, nT + T), \tag{1.10}$$

with $y^\epsilon(0) = 0$. Define $u^\epsilon(\cdot) = (x^\epsilon(\cdot), y^\epsilon(\cdot))$. Then use the estimator

$$\widehat{Y}_n^\epsilon = -\frac{1}{T}\int_{nT}^{nT+T} \lambda(\theta_n^\epsilon, u^\epsilon(s))ds \tag{1.11}$$

in (1.5). Thus, $y^\epsilon(\cdot)$ is never reset. Only the parameter value is adjusted at times nT. In general, (1.11) would not be an unbiased estimator of $-L_{T,\theta}(\theta_n^\epsilon, x^\epsilon(nT))$.

Now redefine the noise process to be $\xi_n^\epsilon = u^\epsilon(nT)$, define $g(\theta, \xi) = E[Y_n^\epsilon | \xi_n^\epsilon = \xi, \theta_n^\epsilon = \theta]$, and let E_n^ϵ denote the expectation conditioned on $u^\epsilon(nT)$. Let $\xi_n(\theta)$ denote the fixed-θ process, as used in Section 8.4, and $u(\cdot, \theta)$ the process $u^\epsilon(\cdot)$ with θ fixed. Suppose that the limit

$$\begin{aligned}\bar{g}(\theta) &= \lim_{n,m} \frac{1}{m} \sum_{i=n}^{n+m-1} E_n^\epsilon g(\theta, \xi_i(\theta)) \\ &= -\lim_{n,m} \frac{1}{mT} \int_{nT}^{nT+mT} E_n^\epsilon \lambda(\theta, u(s,\theta))ds\end{aligned} \tag{1.12}$$

exists in mean and is a constant for each $\theta \in [a, b]$, where (as in Section 8.4) the "initial" condition for $\{\xi_i(\theta), i \geq n\}$ and $\{u(s, \theta), s \geq nT\}$ in (1.12) is $\xi_n(\theta) = \xi_n^\epsilon$. Due to the assumed convergence in mean in (1.12),

$$\bar{g}(\theta) = -\lim_{n,m} \frac{1}{mT} E \int_{nT}^{nT+mT} \lambda(\theta, u(s, \theta))ds = -L_\theta(\theta). \tag{1.13}$$

This is what we want, because the mean ODE has the gradient descent form

$$\dot{\theta} = -L_\theta(\theta) + z \tag{1.14}$$

for the desired cost function, in lieu of the "biased" (1.6). The term $z(\cdot)$ in (1.14) serves only to keep the solution in $[a, b]$. In the parlance of the literature (e.g., [161]), (1.14) results when we do not "reset the accumulator"; that is, the $y^\epsilon(nT)$ are not set to zero. Verifying the conditions of Theorem 8.4.1 involves proving the stability of $y^\epsilon(\cdot)$, and this might be the hardest part of the proof of convergence. If the sequence $\{\xi_n^\epsilon\}$ cannot be proved to be tight, then one might need to reset the $y^\epsilon(\cdot)$ process from time to time, but then the convergence assertion would need to be modified.

We choose to work with update intervals of fixed length T. In fact, these intervals can depend on n in a more or less arbitrary way (as noted in [145]), because we work with the integral of $\lambda(\cdot)$ and θ_n^ϵ varies slowly for small ϵ. The averaging principles used in Section 8.4 are fundamental to the success of the applications.

Comment on the derivative process. If the optimization is to be done via simulation, then we choose the desired form of the process. However, a numerical method will generally be used to approximate both the $x^\epsilon(\cdot)$

and the $y^\epsilon(\cdot)$ processes. Thus, there is an additional question concerning the relations between the estimators for the approximations and those of the original model. Some results on this problem are in [148]. The problem is more serious if the optimization is to be done on-line. Then we would not know the exact model, and we could not compute the exact pathwise derivative process. It appears that there is some robustness in this approximation, but the problem must be kept in mind in any on-line application. The same issues arise in the "discrete event" cases, such as in Section 3.

9.2 A Continuous-Time Stochastic Differential Equation Example

The system. The discussion of the SDE model introduced in Section 1 will be continued with more detail under a more general setup. We first work with the mean square sense pathwise derivatives to get appropriate estimates of the derivatives of the cost function, then we discuss finite difference methods. Let θ be real valued (for notational simplicity only) and $x \in I\!R^k$. Let $b(\cdot)$ be an $I\!R^k$-valued and continuously differentiable function of (x, θ) with bounded x and θ first partial derivatives. Let $\sigma(\cdot)$ be a continuously differentiable matrix-valued function of x with bounded first partial derivatives. The fixed-θ state process is defined by

$$dx(t,\theta) = b(x(t,\theta),\theta)dt + \sigma(x(t,\theta))dw(t), \quad \theta \in [a,b], \tag{2.1}$$

where $w(t)$ is a standard vector-valued Wiener process. Define the auxiliary process $y(t,\theta)$ by

$$dy(t,\theta) = b_x(x(t,\theta),\theta)y(t,\theta)dt + b_\theta(x(t,\theta),\theta)dt + (\sigma,y)(t,\theta)dw(t), \tag{2.2}$$

where the vector $(\sigma, y)(t,\theta)dw(t)$ is defined by its components

$$\sum_{j,p} \frac{\partial \sigma_{ij}(x(t,\theta))}{\partial x_p} y_p(t,\theta)dw_j(t), \quad i = 1,\ldots,k.$$

The process $y(t,\theta)$ is the pathwise mean square sense derivative of $x(t,\theta)$ with respect to θ. Under our smoothness conditions on $b(\cdot)$ and $\sigma(\cdot)$, both (2.1) and (2.2) have unique strong (and weak) sense solutions for each initial condition. Define $u(\cdot,\theta) = (x(\cdot,\theta), y(\cdot,\theta))$. Let $c(\cdot,\cdot)$ be a bounded, real-valued and continuously differentiable function of (θ, x), with bounded x and θ first derivatives, and define $L_T(\theta, x(0)) = \int_0^T c(\theta, x(s,\theta))ds/T$ as in Section 1.

The SA procedure. Use the method of Section 1, where we update at intervals nT, for $n = 1, 2, \ldots$, with θ_n^ϵ being the parameter value used on

$[nT, nT+T)$. Let $x^\epsilon(0) = x(0), y^\epsilon(0) = 0$, and on the interval $[nT, nT+T)$ use

$$dx^\epsilon(t) = b(x^\epsilon(t, \theta_n^\epsilon), \theta_n^\epsilon)dt + \sigma(x^\epsilon(t, \theta_n^\epsilon))dw(t), \tag{2.3}$$

$$dy^\epsilon(t) = b_x(x^\epsilon(t), \theta_n^\epsilon)y^\epsilon(t)dt + b_\theta(x^\epsilon(t), \theta_n^\epsilon)dt + (\sigma, y)^\epsilon(t, \theta_n^\epsilon)dw(t), \tag{2.4}$$

where the components of the vector $(\sigma, y)^\epsilon(t, \theta_n^\epsilon)dw(t)$ are defined as in the expression below (2.2) with $x(\cdot)$ and $y(\cdot)$ replaced by $x^\epsilon(\cdot)$ and $y^\epsilon(\cdot)$, and θ by θ_n^ϵ, respectively.

The $y^\epsilon(\cdot)$ is not a pathwise derivative process because the θ-argument depends on time. Nevertheless, the rate of change of θ_n^ϵ is slow for small ϵ, so we expect that $y^\epsilon(t)$ will be a good approximation to the derivative at time t at the parameter value used at that time. Set $u^\epsilon(\cdot) = (x^\epsilon(\cdot), y^\epsilon(\cdot))$ and define

$$Y_n^\epsilon = -\frac{1}{T}\int_{nT}^{nT+T}\left[\sum_j c_{x_j}(\theta_n^\epsilon, x^\epsilon(s))y_j^\epsilon(s) + c_\theta(\theta_n^\epsilon, x^\epsilon(s))\right] ds. \tag{2.5}$$

Define $g(\cdot)$ by

$$g(\theta, x, y) = E\left[Y_n^\epsilon | \theta_n^\epsilon = \theta, x^\epsilon(nT) = x, y^\epsilon(nT) = y\right],$$

and note that it does not depend on either n or ϵ.

We assume the following additional conditions. For each $\theta \in [a, b]$, $L_t(\theta, x(0))$ converges as $t \to \infty$. For each compact set $A \in \mathbb{R}^{2k}$ and $\theta \in [a, b]$,

$$\{u(t, \theta), t < \infty, u(0, \theta) \in A\} \text{ is tight,} \tag{2.6}$$

$$\{u^\epsilon(t), t < \infty\} \text{ is tight,} \tag{2.7}$$

$$\{y(t, \theta);\ t < \infty, u(0, \theta) \in A\} \text{ is uniformly integrable for } \theta \in [a, b], \tag{2.8}$$

and

$$\{y^\epsilon(t), t < \infty\} \text{ is uniformly integrable.} \tag{2.9}$$

Also suppose that there is a continuous function (of θ) $\bar{g}(\cdot)$, such that

$$\lim_n \frac{1}{n}\sum_{i=0}^{n-1} E_0^\epsilon g(\theta, x(iT, \theta), y(iT, \theta)) = \bar{g}(\theta), \tag{2.10}$$

where the limit is in the mean and is uniform in the initial condition $(x(0, \theta), y(0, \theta))$ in any compact set.

We now verify the conditions of Theorem 8.4.1. Conditions (A8.4.1) and (A8.4.3) hold by the weak sense uniqueness of the solution to (2.1) and the smoothness of the coefficient functions. (A8.4.2) holds because the transition function for $\xi_n^\epsilon = u^\epsilon(nT)$ does not depend on n or ϵ. Condition (8.4.5a) holds by (2.9), and (8.4.5b) holds by (2.8). Condition (A8.4.6) holds by (2.10) and (A8.4.7) holds by (2.6). (A8.1.1) holds by (2.9), and

(A8.1.4) holds since $\beta_n^\epsilon = 0$. Condition (A8.4.4) holds by the continuity and boundedness of the derivatives of the dynamical and cost rate functions and (A4.1.7) holds by (2.7). Thus, Theorem 8.4.1 is applicable. Analogously to the situation in Section 1, we have

$$\bar{g}(\theta) = -\lim_n \frac{1}{nT} \int_0^{nT} E\lambda(\theta, y(s,\theta))ds = -L_\theta(\theta), \tag{2.11}$$

and the mean ODE has the gradient descent form

$$\dot{\theta} = \bar{g}(\theta) + z = -L_\theta(\theta) + z, \tag{2.12}$$

where $z(\cdot)$ serves only to keep the solution in $[a, b]$. The fact that the limit in (2.11) is $-L_\theta(\theta)$ follows from the convergence in (2.10) and the convergence of $L_t(\theta, x(0))$. Usually, the most difficult conditions to verify are (2.6)–(2.9) for the $y^\epsilon(\cdot)$-process.

Finite difference methods. The main difficulties in applications are usually verifying the tightness and uniform integrability conditions on the $y^\epsilon(t)$ and $y(t,\theta)$. These difficulties can be alleviated by using finite differences rather than the mean square derivative process $y^\epsilon(\cdot)$. Two forms will be discussed, the first being the more traditional, using separate runs for the different components of the difference. The second uses a single run and provides a possible alternative since it can be used on line, although the noise effects might be large because we cannot have the advantages of the use of common random numbers. There is an obvious analog for discrete event systems when common random numbers cannot be used.

The assumptions are weaker than those required for the pathwise mean square derivative method. Both methods to be discussed will use the following conditions. Suppose that there is a unique weak sense solution to (2.1) for each θ and $x(0)$. Let $c(\cdot)$ be bounded, and let

$$E\left[\int_0^T c(\theta, x(s,\theta))ds \middle| x(0) = x\right]$$

be continuous in (x, θ). Suppose that the transition function

$$P\left\{x(t,\theta) \in \cdot | x(0) = x, \theta\right\}$$

is weakly continuous in (x,θ), and let there be a unique invariant measure $\mu(\cdot|\theta)$ for each fixed θ.

A finite difference alternative: Simultaneous runs. Given the finite difference value $\delta\theta > 0$, replace the integrand in (2.5) by

$$\frac{c(\theta_n^\epsilon + \delta\theta, x(s, \theta_n^\epsilon + \delta\theta)) - c(\theta_n^\epsilon - \delta\theta, x(s, \theta_n^\epsilon - \delta\theta))}{2(\delta\theta)}. \tag{2.13}$$

Two separate simulations are used, one for $\{\theta_n^\epsilon + \delta\theta\}$ and one for $\{\theta_n^\epsilon - \delta\theta\}$. We thus run two processes $x^{\epsilon,\pm}(\cdot)$ defined by $x^{\epsilon,\pm}(0) = x(0)$, and on $[nT, nT+T)$ set $x^{\epsilon,\pm}(\cdot) = x(\cdot, \theta_n^\epsilon \pm \delta\theta)$ with initial condition at nT defined recursively by $x(nT, \theta_n^\epsilon \pm \delta\theta) = x^{\epsilon,\pm}(nT)$. Generally, one would want to use the same Wiener process to drive the two processes. This (common random variables) form often yields essentially the same path properties as the use of the derivative process. Finally, suppose that $\{x^{\epsilon,\pm}(t); \epsilon, t\}$ and $\{\mu(\cdot|\theta), \theta \in [a,b]\}$ are tight.

The discussion will be in terms of the invariant measures of Theorem 8.4.4, but the analog under the conditions of Theorem 8.4.1 should be obvious. The ξ_n^ϵ is now just $\xi_n^\epsilon = x^\epsilon(nT)$. Under the given conditions, Theorem 8.4.4 yields that the mean ODE is

$$\dot{\theta} = -\frac{1}{2(\delta\theta)} \int \left[c(\theta+\delta\theta,\xi)\mu(d\xi|\theta+\delta\theta) - c(\theta-\delta\theta,\xi)\mu(d\xi|\theta-\delta\theta)\right] + z, \quad (2.14)$$

where $z(\cdot)$ serves only to keep the solution in $[a,b]$. Due to the additive way that the two terms appear in (2.13), we do not need to have a unique invariant measure of the pair $\{x(nT, \theta+\delta\theta), x(nT, \theta-\delta\theta)\}$ for each θ, only of $\{x(nT,\theta)\}$ for each θ.

Comparison with the pathwise derivative method. The finite difference approach can be either easier or harder than the pathwise derivative approach. The dimension of the SDEs to be solved in each case is the same. If $\sigma(x)$ actually depends on x, then the pathwise derivative procedure cannot be conducted on-line, because we need to know the Wiener process to get $y(\cdot,\theta)$. If $\sigma(x)$ does not depend on x, then the equation for $y(\cdot,\theta)$ or $y^\epsilon(\cdot)$ is linear in the y-variable (but with time-varying coefficients), and it is simpler to solve. The pathwise derivative can then be obtained on-line, at least in principle. The finite difference method can be used for cases where $c(\cdot)$ and/or $b(\cdot)$ are not smooth, such as where $c(\cdot)$ is an indicator function of an event of interest.

Finite differences with only one run. As an alternative to the simultaneous run method just discussed, a single run can be used; this might be considered when the optimization must be done on-line where simultaneous runs might not be possible, although there is little experience with such a method. Given $T > 0$, use $\theta_n^\epsilon + \delta\theta$ on the interval $[2nT, 2nT+T)$ and then $\theta_n^\epsilon - \delta\theta$ on $[2nT+T, 2nT+2T)$. Let $x^\epsilon(\cdot)$ denote the actual process with the $\theta_n^\epsilon \pm \delta\theta$ being used on the appropriate alternating time intervals. The appropriate fixed-θ process, which we call $\widehat{x}(\cdot,\theta)$, uses parameter value $\theta+\delta\theta$ on $[0,T)$ and then alternates between $\theta-\delta\theta$ and $\theta+\delta\theta$ on successive intervals of length T. We use

$$\widehat{Y}_n^\epsilon = -\frac{1}{2T(\delta\theta)} \int_0^T \left[c(\theta_n^\epsilon + \delta\theta, x^\epsilon(nT+s)) - c(\theta_n^\epsilon - \delta\theta, x^\epsilon(nT+T+s))\right] ds.$$

The analysis follows the lines of Theorem 8.4.4 (alternatively, Theorem 8.4.1 can be used if desired). The main additional problem is due to the fact that the transition function for the fixed-θ process depends periodically on time.

Let $\xi_n^+(\theta) = \widehat{x}(2nT, \theta)$ and $\xi_n^-(\theta) = \widehat{x}(2nT + T, \theta)$. Suppose that the stationary processes $\xi_n^+(\theta)$ and $\xi_n^-(\theta)$ exist and have unique invariant measures denoted by $\mu^+(\cdot|\theta)$ and $\mu^-(\cdot|\theta)$, respectively. Let $\{\mu^\pm(\cdot|\theta), \theta \in [a,b]\}$ and $\{x^\epsilon(t); t, \epsilon\}$ be tight. Define

$$g^+(\theta, \xi) = \frac{1}{2T(\delta\theta)} \int_0^T E\left[c(\theta + \delta\theta, x(s, \theta + \delta\theta)) | x(0) = \xi\right] ds,$$

$$g^-(\theta, \xi) = \frac{1}{2T(\delta\theta)} \int_0^T E\left[c(\theta - \delta\theta, x(s, \theta - \delta\theta)) | x(0) = \xi\right] ds.$$

The function $\bar{g}(\theta)$ in the mean ODE is now

$$\bar{g}(\theta) = -\int \left[g^+(\theta, \xi)\mu^+(d\xi|\theta) - g^-(\theta, \xi)\mu^-(d\xi|\theta)\right]. \tag{2.15}$$

Define $P_T(\widetilde{\xi}, \cdot|\theta + \delta\theta) = P\{x(T, \theta + \delta\theta) \in \cdot | x(0) = \widetilde{\xi}\}$. Then

$$\mu^-(d\xi|\theta) = \int \mu^+(d\widetilde{\xi}|\theta) P_T(\widetilde{\xi}, d\xi|\theta + \delta\theta). \tag{2.16}$$

Now compare this one-run procedure to an alternative one-run procedure, where the process is restarted each T units of time at the same fixed initial value $x(0)$, and using $\theta \pm \delta\theta$ for the alternate restarts (assuming that such restarts were possible in the application). This would yield a right side of the form (2.15), but the $\mu^\pm$ are replaced by the measure that is concentrated on the fixed initial condition $x(0)$. We expect that this "restarting method" would be inferior to the original "one continuous run" procedure, since $\mu^\pm(\cdot|\theta)$ defined earlier would be much closer to the desired values $\mu(\cdot|\theta \pm \delta\theta)$, particularly for large T. The situation would be a little more complicated if θ were vector valued, but the general idea is the same. The methods using one run would seem to be inferior to the simultaneous run method if the same Wiener process were used in all of the simultaneous runs.

9.3 A Discrete Example: A GI/G/1 Queue

We will now prove convergence of the algorithm (2.5.10) for the controlled queueing problem, where the cost to be minimized is (2.5.1). Recall the following definitions and assumptions from Section 2.5, which will be retained. The queue has a renewal arrival process where the distribution function of the interarrival times is continuous, a single server, and a general service time distribution $F(\cdot|\theta)$ parameterized by θ. The interarrival intervals σ_n have bounded mean square value and the property that $\sup_v E[\sigma - v|\sigma \geq v] < \infty$. Letting ζ denote the sample service time, define the random variable $\chi(\theta) = F(\zeta|\theta)$. The function $F_\theta^{-1}(\chi|\theta)$ is assumed to exist and be continuous in $\theta \in [a, b]$, uniformly in χ. We suppose that the supremum over $\theta \in [a, b]$ of the mean service times is less than the mean interarrival time. Thus, the mean length of a busy period is bounded, uniformly in $\theta \in [a, b]$.

Recall the definition $Z(\theta) = F_\theta^{-1}(\chi(\theta)|\theta)$. In the physical system with a small step size $\epsilon > 0$, θ_0^ϵ is used up to the time of departure of the Nth customer, then θ_1^ϵ is used up to the time of departure of the $2N$th customer, and so on. For the actual physical system with the time-varying parameter, ζ_i^ϵ denotes the actual service time of the ith customer, and Z_i^ϵ the θ-derivative of the inverse function at ζ_i^ϵ and whatever the parameter value is at the time of that service. Let v_i^ϵ be the index of the first arrival in the busy period in which customer i arrives. Recall that the stochastic approximation algorithm (2.5.10) is

$$\theta_{n+1}^\epsilon = \Pi_{[a,b]}\left[\theta_n^\epsilon + \epsilon \widehat{Y}_n^\epsilon - \epsilon K_\theta(\theta_n^\epsilon)\right],$$

where $\widehat{Y}_n^\epsilon$ in (2.5.9) is

$$\widehat{Y}_n^\epsilon = -\frac{1}{N} \sum_{i=nN+1}^{nN+N} \sum_{j=v_i^\epsilon}^{i} Z_j^\epsilon. \tag{3.1}$$

A key assumption in the use of IPA is (2.5.6); namely,

$$\widehat{Z}_m(\theta) \to \widehat{L}_\theta(\theta) \tag{3.2}$$

with probability one and in mean as $m \to \infty$, where $\widehat{L}(\theta)$ is defined by (2.5.1) and $\widehat{Z}_m(\theta)$ is defined by (2.5.5). The proof of this fact is one of the major results in IPA. Proofs under various conditions, and further references are in [84, 161, 231]. In fact, the convergence of (3.2) is used only in evaluating (3.9), where we will actually need the form of (3.2) given by

$$E\widehat{Z}_m(\theta) \to \widehat{L}_\theta(\theta), \tag{3.3}$$

uniformly in θ in the desired interval, when the queue starts empty. Without the uniformity requirement, (3.3) is weaker than (3.2). The uniformity is important if the convergence is to be physically meaningful, since otherwise there would be some $\widetilde{\theta}$ and $\theta_\alpha \to \widetilde{\theta}$ as $\alpha \to \infty$, such that the time needed for the iterates to be in a small neighborhood of the limit $\widehat{L}_\theta(\theta_a)$ goes to infinity as $\alpha \to \infty$. We will simply assume (3.3) rather than give additional conditions that guarantee it. This simplifies the discussion and allows the proof to remain valid as more general conditions guaranteeing the asymptotic unbiasedness (3.3) are found.

Recall that, for notational simplicity, in Section 2.5 we supposed that the queue starts empty, but we drop that assumption now. The following additional assumptions will be used. Recall that $u \vee v = \max\{u, v\}$, and use the usual convention that a sum over a set of indices starting at p and ending at q is zero if $q < p$. Suppose that, if $\widetilde{\theta}_n$ is any random variable with values in $[a, b]$ and measurable on the data available up to and including departure $n-1$, then

$$E\left|\sum_{j=v_i}^{v_i \vee (i-k)-1} Z_j(\widetilde{\theta}_j)\right| \to 0, \tag{3.4}$$

as $k \to \infty$, uniformly in i and $\{\widetilde{\theta}_j\}$, and that

$$\{Z_j(\widetilde{\theta}_j); j, \text{ all possible such } \widetilde{\theta}_j\} \text{ is uniformly integrable.} \tag{3.5}$$

The sum in (3.4) is over the set of indices starting at v_i, the beginning of the busy period in which the customer who is the ith departure actually arrives, up to at most departure $(i-k)$ if that is bigger than v_i. Otherwise, the sum is zero. For large k the sum will be zero with a high probability.

Since $\chi(\theta)$ is uniformly distributed on the unit interval for each θ,

$$EZ(\theta) = \int_0^1 F_\theta^{-1}(x|\theta)dx,$$

which is continuous in θ. It follows from (3.5) and the uniform θ-continuity assumption on F_θ^{-1} that (for a sequence $\gamma \to 0$) if $\widetilde{\theta}_n^\gamma$ and $\widehat{\theta}_n^\gamma$ are any random variables with values in $[a, b]$ and that are measurable on the data up to and including the $(n-1)$st departure. Then

$$E|Z_n(\widetilde{\theta}_n^\gamma) - Z_n(\widehat{\theta}_n^\gamma)| \to 0 \tag{3.6}$$

uniformly in n if $|\widetilde{\theta}_n^\gamma - \widehat{\theta}_n^\gamma| \to 0$ in probability uniformly in n as $\gamma \to 0$.

The mean ODE will be

$$\dot{\theta} = -L_\theta(\theta) + z = -\widehat{L}_\theta(\theta) - K_\theta(\theta) + z, \tag{3.7}$$

where $z(\cdot)$ serves only to keep $\theta(t) \in [a, b]$. Let $\epsilon \to 0$ and $n \to \infty$ such that $n\epsilon \to \infty$. Then the possible limit points of θ_n^ϵ (as $n \to \infty$ and $\epsilon \to 0$) are the stationary points of $L_\theta(\cdot)$ in the interval $[a, b]$ and the end points a (resp., b) if $L_\theta(\cdot)$ is positive at $\theta = a$ (negative at $\theta = b$, resp.).

In [145] the "invariant measure" Theorem 8.4.4 was used, but Theorem 8.4.6 is somewhat simpler to use on this problem and will be used here. To prepare for the verification of the conditions of that theorem, let us write out the negative of the sum (3.1) as follows (a sum is defined to be zero if the upper index of summation is less than the lower index of summation):

$$\begin{aligned}
&\left\{Z^\epsilon_{v^\epsilon_{(nN+1)}} + \ldots + Z^\epsilon_{nN}\right\} + \left[Z^\epsilon_{nN+1}\right] \\
&\quad + \left\{Z^\epsilon_{v^\epsilon_{(nN+2)}} + \ldots + Z^\epsilon_{nN}\right\} + \left[Z^\epsilon_{(nN+1)\vee v^\epsilon_{(nN+2)}} + Z^\epsilon_{nN+2}\right] \\
&\quad \vdots \\
&\quad + \left\{Z^\epsilon_{v^\epsilon_{(nN+N)}} + \ldots + Z^\epsilon_{nN}\right\} \\
&\quad + \left[Z^\epsilon_{(nN+1)\vee v^\epsilon_{(nN+N)}} + \ldots + Z^\epsilon_{(nN+N-1)\vee v^\epsilon_{(nN+N)}} + Z^\epsilon_{nN+N}\right].
\end{aligned} \tag{3.8}$$

The ith line in (3.8) is the ith inner sum in (3.1).

In (3.8), we have separated out (by curly brackets) the terms up to departure nN (the "past") from the "future" terms, which are in the square brackets. However, the indices $v^\epsilon_{(nN+i)}$ depend on the "future." If the queue is emptied at the (nN)th departure, then the terms in the curly brackets are all zero. If the queue is not empty at the (nN)th departure, then all the terms in the curly brackets are equal up until the first departure that empties the queue (or $nN+N$ if the queue is never empty on $[nN+1, nN+N]$). The curly bracketed terms associated with departures after that time are zero.

Let Q_n denote the number of customers in the queue and τ_n the time elapsed since the last arrival of a customer, both measured just after the departure of the nth customer. Let ψ_{nN} denote the first curly bracketed term in (3.8). This depends on the data up to and including the time of the (nN)th departure. Define $\xi^\epsilon_n = (Q_{nN}, \tau_{nN}, \psi_{nN})$. This is the "effective memory" in the system at the start of the update interval $[nN+1, nN+N]$. The sequence $\{(\theta^\epsilon_n, \xi^\epsilon_n)\}$ is a Markov chain. Let E^ϵ_n denote the expectation conditioned on ξ^ϵ_n and all the other data up to the time of the departure of the (nN)th customer. The parameter to be used on the next set of services $[nN+1, \ldots, nN+N]$ is $\theta = \theta^\epsilon_n$. Following the usage in Section 8.4, the fixed θ-chain $\{\xi_i(\theta), i \geq n\}$ starting at time n would use initial condition $\xi_n(\theta) = \xi^\epsilon_n$ and then evolve with the parameter being fixed at the given value θ.

We now discuss the remaining conditions of Theorem 8.4.6. The properties of the interarrival times and conditions (3.4) and (3.5) imply the uniform integrability of $\{\widehat{Y}^\epsilon_n; \epsilon, n\}$. The tightness of $\{\xi^\epsilon_n\}$ is a consequence of (3.4), the uniform integrability (3.5), and the stability assumptions on the queue. By the tightness and uniform integrability properties, without loss of generality, for the purposes of the proof we can suppose that the ξ^ϵ_n are bounded. Then, for each large but fixed integer m, the continuity of the service times in θ implies that (A8.4.20) and (A8.4.21) hold, since the mean fraction of the busy periods (over the interval of departures $[nN+1, nN+mN]$) whose shape is affected by a small change in θ is small (uniformly in n) for small changes in the values of the θ.

It remains to verify (A8.4.19) with the appropriate $\bar{g}(\cdot)$. For $l \geq n$, let $\widehat{Y}_l(\theta)$ denote the value of $\widehat{Y}_l^\epsilon$ if the parameter were fixed at the value θ starting at iterate $n+1$, with the initial condition $\widehat{Y}_n(\theta) = \widehat{Y}_n^\epsilon$. We have

$$\begin{aligned} &\frac{1}{m}\sum_{l=n}^{n+m-1} E_n^\epsilon \widehat{Y}_l(\theta) \\ &= -\frac{1}{Nm} E_n^\epsilon \sum_{l=n}^{n+m-1} \sum_{i=lN+1}^{lN+N} \sum_{j=v_i(\theta)}^{i} Z_j(\theta). \end{aligned} \tag{3.9}$$

Now, the properties of the arrival process, (3.4), (3.5), and the stability properties of the queue, imply that the limit will be the same if we suppose that the queueing process restarts with an empty queue after the (nN)th departure. Thus we only need that the expectation of the right side of (3.9) goes (uniformly in θ in the desired set as $m \to \infty$) to the continuous limit $-\widehat{L}_\theta(\theta)$. This is equivalent to (3.3). Thus, (A8.4.19) holds for $\bar{g}(\theta) = -\widehat{L}_\theta(\theta) - K_\theta(\theta)$.

9.4 Signal Processing Problems

A sampling of results will be given to illustrate some of the techniques. The reader is referred to the large amount of literature cited in Chapter 3 for further discussion. Consider the constrained form of the problem in Section 3.4, with the algorithm

$$\theta_{n+1}^\epsilon = \Pi_H \left[\theta_n^\epsilon + \epsilon\phi_n \left(y_n - \phi_n'\theta_n^\epsilon\right)\right], \tag{4.1}$$

with H being the hyperrectangle $\prod_{i=1}^r [a_i, b_i]$, $-\infty < a_i < b_i < \infty$. Represent H by inequalities $q_i(\theta) \leq 0, i \leq 2r$. The algorithm (4.1) models the classical parameter identification and noise cancellation problems and the adaptive equalizer problem when there is a training sequence.

Suppose that

$$\{|\phi_n|^2, |y_n|^2\} \text{ is uniformly integrable.} \tag{4.2}$$

Let $\mathcal{F}_n^\epsilon$ be the minimal σ-algebra that measures $\{\theta_0^\epsilon, \phi_i, y_i, i < n\}$, and let E_n^ϵ be the expectation conditioned on $\mathcal{F}_n^\epsilon$. Suppose that there is a positive definite symmetric matrix Q and a vector S such that

$$\lim_{n,m} \frac{1}{m} \sum_{i=n}^{n+m-1} E_n^\epsilon \left[\phi_i\phi_i' - Q\right] = 0, \tag{4.3}$$

$$\lim_{n,m} \frac{1}{m} \sum_{i=n}^{n+m-1} E_n^\epsilon \left[\phi_i y_i - S\right] = 0, \tag{4.4}$$

where the limit (as n and m go to infinity) is in the sense of convergence in probability. Let q_ϵ be a sequence of non-negative integers. Then, by Theorem 8.2.2 $\{\theta^\epsilon(\epsilon q_\epsilon + \cdot)\}$ is tight and as $\epsilon \to 0$ any weak sense limit process satisfies

$$\dot{\theta} = S - Q\theta + z, \ z(t) \in -C(\theta(t)). \tag{4.5}$$

In other words, as $\epsilon \to 0$ through a convergent subsequence, the support of the process $\theta^\epsilon(\epsilon q_\epsilon + \cdot)$ is eventually concentrated on the solutions of (4.5) whose initial condition is the limit in distribution of $\{\theta^\epsilon_{q_\epsilon}\}$ along the convergent subsequence.

As usual, $z(\cdot)$ serves only to keep the solution in H. If $\epsilon q_\epsilon \to \infty$, then by the definitions of $C(\theta)$ and $z(\cdot)$, the limit process has the constant value θ defined by

$$S - Q\theta = \sum_{i \in A(\theta)} \lambda_i q_{i,\theta}(\theta), \tag{4.6}$$

where $A(\theta)$ is the set of active constraints at θ and $\lambda_i \geq 0$. Equation (4.6) implies that if $a_i < 0$ and $b_i > 0$ are large enough, then the limit point satisfies $\theta \equiv \bar{\theta} = Q^{-1}S$.

Remark on the assumptions (4.3) and (4.4). These conditions imply a type of "local regularity" on the ϕ_n, y_n processes. In the classical parameter identification problem of Subsection 3.2.1 where $y_n = \phi_n'\bar{\theta} + \nu_n$, if the products $\phi_n\nu_n$ average to zero, we have $S = Q\bar{\theta}$ (with the definitions used in this section) and the ODE then reduces to $\dot{\theta} = -Q(\theta - \bar{\theta}) + z$. For this problem, under a weaker local averaging condition (see Subsections 6.8.1 and 8.2.5), the ODE becomes the differential inclusion $\dot{\theta} \in -G(\theta - \bar{\theta}) + z$ where G is a compact and convex set of positive definite matrices. If the constraint set covers a large enough region around $\bar{\theta}$ and $\epsilon q_\epsilon \to \infty$, then $\{\theta^\epsilon(\epsilon q_\epsilon + \cdot)\}$ converges weakly to the process with constant value $\bar{\theta}$.

The conditions on the noise processes are viewed somewhat differently in the control and communications theory literature. In the adaptive noise cancellation problem, one chooses the form of the output of the adaptive filter to be $\phi_n'\theta$ and then tries to find the best value of θ. It is commonly supposed that the limits in (4.3) and (4.4) do exist, at least locally in time. In the parameter identification problem, it is supposed that there is a linear system whose measured output is $y_n = \phi_n\bar{\theta} + \nu_n$, where ν_n is the observation noise. Then, one would like to know the weakest conditions under which the algorithm will asymptotically yield the value of the parameter $\bar{\theta}$. For example, there might be one type of input for a short period, then none whatsoever for a very long time, then another for a short time, than none for a long time, and so on. As a practical matter, the identification problem loses much of its meaning unless a good quality of identification occurs within a reasonable time, with a high probability. Because of this, conditions that ensure that the mean ODE is the differential inclusion with

G being a compact convex set of positive definite matrices are of sufficient generality for the practical problem if $\bar{\theta}$ is to be identified.

A stability argument for an unconstrained algorithm. Drop the constraint set H in (4.1), let $Q > 0$, and let $\{\phi_n, y_n\}$ be bounded. Recall the definition $\bar{\theta} = Q^{-1}S$. Define $\widetilde{\theta} = \theta - \bar{\theta}$ and $\widetilde{\theta}_n = \theta_n - \bar{\theta}$, and rewrite the algorithm as

$$\begin{aligned}\widetilde{\theta}_{n+1}^{\epsilon} &= \widetilde{\theta}_n^{\epsilon} + \epsilon \left[\phi_n y_n - \phi_n \phi_n' \widetilde{\theta}_n^{\epsilon} - \phi_n \phi_n' \bar{\theta}\right] \\ &= \widetilde{\theta}_n^{\epsilon} + \epsilon \left[(\phi_n y_n - S) - (\phi_n \phi_n' - Q)\, \widetilde{\theta}_n^{\epsilon} - (\phi_n \phi_n' - Q)\, \bar{\theta} - Q\widetilde{\theta}_n^{\epsilon}\right].\end{aligned}$$

Suppose that the two components of the perturbation

$$\begin{aligned}\delta v_n^d(\widetilde{\theta}) &= -\epsilon \sum_{i=n}^{\infty} \Pi(n+1,i) E_n^{\epsilon} \left[\phi_i \phi_i' - Q\right] \left(\bar{\theta} + \widetilde{\theta}\right) \\ &\quad + \epsilon \sum_{i=n}^{\infty} \Pi(n+1,i) E_n^{\epsilon} \left[\phi_i y_i - S\right]\end{aligned} \tag{4.7}$$

are of order $O(\epsilon)$ for each $\widetilde{\theta} + \bar{\theta}$. Then by Theorem 10.5.1, there is a positive K such that for small ϵ, $\sup_n E|\theta_n^{\epsilon}|^2 \leq K$. Then Theorem 8.5.1 yields that the mean ODE is just $\dot{\widetilde{\theta}} = -Q\widetilde{\theta}$ and that if $\epsilon q_{\epsilon} \to \infty$, then $\{\theta^{\epsilon}(\epsilon q_{\epsilon} + \cdot)\}$ converges weakly to the process with constant value $\bar{\theta}$.

The least squares algorithm. Now suppose that the algorithm is the constrained form of the least squares estimator (1.1.12):

$$\theta_{n+1} = \Pi_H \left[\theta_n + \left[\Phi_{n+1}\right]^{-1} \phi_n \left(y_n - \phi_n' \theta_n\right)\right], \tag{4.8}$$

where $\Phi_n = \sum_{i=0}^{n-1} \phi_i \phi_i'$. Let $\Phi_n / n \to Q > 0$ with probability one. Then, under the conditions (4.2)–(4.4), the mean ODE is

$$\dot{\theta} = Q^{-1}S - \theta + z, \; z(t) \in -C(\theta(t)),$$

and similarly for the form (1.1.14) and (1.1.15) or (1.1.16) and (1.1.17). The differential inclusion form of Theorem 8.2.5 can also be used. The constraints that define H might appear to be "fixed," but if they are used for algorithmic purposes only and the iterate hovers near the boundary, then one would enlarge the constraint set until it appeared that there was convergence to an interior point. Practical algorithms should be flexible.

The adaptive equalizer: Stability, convergence and the strict positive real condition. Let us return to the problem of Subsection 3.4.2 with the constant step size $\epsilon_n = \epsilon$.

The algorithm (3.4.8) is

$$\theta_{n+1}^{\epsilon} = \theta_n^{\epsilon} + \epsilon\phi_n^{\epsilon}\left[\psi_n - \widehat{\psi}_n^{\epsilon}\right], \tag{4.9}$$

where ϕ_n^{ϵ} and ϕ_n are defined above (3.4.7). Recall that $\widehat{\psi}_n^{\epsilon} = (\phi_n^{\epsilon})'\theta_n^{\epsilon}$ and $\psi_n = \phi_n'\bar{\theta}$. It was shown that we have the representation $\psi_n - \widehat{\psi}_n^{\epsilon} = -A^{-1}(q^{-1})v_n^{\epsilon}$, where the transfer function $A(z)$ is defined below (3.4.8) and $v_n^{\epsilon} = (\phi_n^{\epsilon})'\widetilde{\theta}_n^{\epsilon}$, $\widetilde{\theta} = \theta - \bar{\theta}$. In (3.4.12), the algorithm was rewritten as

$$\widetilde{\theta}_{n+1}^{\epsilon} = \widetilde{\theta}_n^{\epsilon} - \epsilon\phi_n^{\epsilon}A^{-1}(q^{-1})v_n^{\epsilon}. \tag{4.10}$$

Suppose that $A(z)$ has its roots strictly outside the unit circle in the complex plane (i.e., the inverse transfer function $A^{-1}(q^{-1})$ is asymptotically stable). We also suppose that there is an $\alpha > 0$ such that the real part of $A(z)$ is at least α on the unit circle in the complex plane. This latter condition is known as the *strict positive real* condition in the literature on adaptive systems. It is widely used, although it does not cover some important cases. Since the roots of $A(z)$ are strictly outside the unit circle, there are $\{\gamma_n\}$ and real $C_0 > 0$ and $0 \le C_1 < 1$ such that $|\gamma_n| \le C_0 C_1^n$ and

$$A^{-1}(q^{-1}) = \sum_{i=0}^{\infty} \gamma_i q^{-i}.$$

Owing to this, it can be supposed without loss of generality that $\phi_n = 0$ for $n < 0$. It is supposed, as is common in applications, that the transfer function $B(q^{-1})$ defined above (3.4.9) also has its roots strictly outside the unit circle and that $\{\psi_n\}$ is bounded. Thus the y_n in (3.4.5) are also uniformly bounded.

The algorithm (4.10) is not too easy to analyze rigorously as it is written. Because of this, we first go through a formal analysis to show the basic role of the strict positive real condition, and then we make some practical modifications that greatly simplify the rigorous analysis. It will be seen that the analysis is essentially deterministic.

A formal analysis of (4.10). First, consider the stability of $\{\widetilde{\theta}_n^{\epsilon}\}$ with algorithm (4.10). Using the Liapunov function $V(\widetilde{\theta}) = |\widetilde{\theta}|^2$, we can write

$$\begin{aligned} &V(\widetilde{\theta}_{n+1}^{\epsilon}) - V(\widetilde{\theta}_n^{\epsilon}) \\ &= -2\epsilon v_n^{\epsilon} A^{-1}(q^{-1})(v_n^{\epsilon}) + O(\epsilon^2)|\phi_n^{\epsilon}|^2 \left[A^{-1}(q^{-1})(v_n^{\epsilon})\right]^2. \end{aligned} \tag{4.11}$$

The transfer function $A^{-1}(z)$ is also strictly positive real and there is an $\alpha_0 > 0$ such that for all n [165, 225],

$$\sum_{i=0}^{n} v_i^{\epsilon} A^{-1}(q^{-1})v_i^{\epsilon} \ge \alpha_0 \sum_{i=0}^{n} |v_i^{\epsilon}|^2. \tag{4.12}$$

If $\phi_n^\epsilon \neq 0$ for $n < 0$, then a negative constant independent of n is added to the right side.

The problem in the analysis stems from the second-order term in (4.11). [For example, in the analysis in [225], the θ_n^ϵ in $\widehat{\psi}_n^\epsilon = (\phi_n^\epsilon)'\theta_n^\epsilon$ was changed to θ_{n+1}^ϵ, which is actually not available until after the iteration. With this change, the corresponding second-order term could be properly bounded.] In this preliminary and formal discussion, we will neglect the second-order term. Then we have

$$V(\widetilde{\theta}_n^\epsilon) - V(\widetilde{\theta}_0^\epsilon) \leq -2\epsilon\alpha_0 \sum_{i=0}^{n-1} |v_i^\epsilon|^2. \tag{4.13}$$

Equation (4.13) implies that $\{|\widetilde{\theta}_n^\epsilon|\}$ is nonincreasing and $v_n^\epsilon \to 0$ as $n \to \infty$. Hence, $\psi_n - \widehat{\psi}_n^\epsilon = -A^{-1}(q^{-1})(v_n^\epsilon) \to 0$.

Now, assuming that $|\widetilde{\theta}_n^\epsilon|$ is nonincreasing, we will establish the convergence of θ_n^ϵ. Suppose that there is a positive definite matrix P such that

$$\liminf_{n,m} \frac{1}{m} \sum_{i=n}^{n+m-1} E_n^\epsilon \left[\phi_i \phi_i' - P\right] \geq 0, \tag{4.14}$$

in probability, in the sense of positive definite matrices, as n and m go to infinity. Since $\widehat{\psi}_n^\epsilon - \psi_n \to 0$, there is a sequence $m_n \to \infty$ slowly enough such that

$$\lim_n \sup_{n \leq i \leq n+m_n} |\theta_i^\epsilon - \theta_n^\epsilon| = 0$$

and

$$\lim_n \sum_{j=n}^{n+m_n} \left|(\phi_j^\epsilon)'\phi_j^\epsilon - \phi_j'\phi_j\right| = 0.$$

Then we can write

$$\lim_n \left[V(\widetilde{\theta}_{n+m_n}^\epsilon) - V(\widetilde{\theta}_n^\epsilon)\right] \leq -2\epsilon\alpha_0 \liminf_n (\widetilde{\theta}_n^\epsilon)' \left[\sum_{i=n}^{n+m_n-1} \phi_i\phi_i'\right] \widetilde{\theta}_n^\epsilon. \tag{4.15}$$

If $\widetilde{\theta}_n^\epsilon \not\to 0$, then (4.14) implies that the right side of (4.15) goes to minus infinity, which is impossible. Note that it was not required that $\epsilon \to 0$. The fact that we did not require a small ϵ is due to the neglect of the second-order term; this will be changed below.

A practical truncated algorithm. The preceding development was formal only because we neglected the second-order term in (4.11), but it illustrated the fundamental role of the strict positive real condition in the proof of stability, a fact that was first explicitly recognized by Ljung [165] and is widely used in adaptive control problems.

We will make two reasonable modifications to the algorithm. First, the constraint H defined below (4.1) will be used for large enough a_i, b_i such that $\bar{\theta} \in H^0$. The actual values of ψ_n were assumed to be uniformly bounded; thus, there is a $K < \infty$ such that $|\psi_n| \le K$ for all n and ω. It then makes sense to bound the estimates $\widehat{\psi}_n^\epsilon$ as well, and we will bound them by $K_1 > K$. Thus the revised algorithm is

$$\theta_{n+1}^\epsilon = \Pi_H \left[\theta_n^\epsilon + \epsilon \phi_n^\epsilon \left(\psi_n - \widehat{\psi}_n^\epsilon\right)\right], \tag{4.16}$$

where

$$\widehat{\psi}_n^\epsilon = \Pi_{[-K_1, K_1]} \left[(\phi_n^\epsilon)'\theta_n^\epsilon\right]. \tag{4.17}$$

This modified algorithm was analyzed in [138], where it was shown that the strict positive real condition plays the same role as in the formal analysis and that there are $\alpha_i > 0$ such that (this is (4.12))

$$\sum_{i=0}^{n} (\phi_i^\epsilon)'\widetilde{\theta}_i^\epsilon(\widehat{\psi}_i^\epsilon - \psi_i) = \sum_{i=0}^{n} v_i^\epsilon(\widehat{\psi}_i^\epsilon - \psi_i) \ge \alpha_1 \sum_{i=0}^{n} |v_i^\epsilon|^2 \tag{4.18}$$

and

$$\sum_{i=0}^{n} v_i^\epsilon(\widehat{\psi}_i^\epsilon - \psi_i) \ge \alpha_2 \sum_{i=0}^{n} |\widehat{\psi}_i^\epsilon - \psi_i|^2. \tag{4.19}$$

If $\phi_n^\epsilon \ne 0, n < 0$, then some negative number that does not depend on n is to be added to the right-hand side.

For the truncated algorithm (4.16) and (4.17), we have

$$V(\widetilde{\theta}_{n+1}^\epsilon) - V(\widetilde{\theta}_n^\epsilon) \le -2\epsilon v_n^\epsilon[\widehat{\psi}_n^\epsilon - \psi_n] + O(\epsilon^2)|\phi_n^\epsilon|^2[\widehat{\psi}_n^\epsilon - \psi_n]^2. \tag{4.20}$$

The inequality $\le$ appears since the truncation in (4.16) reduces the error, since $\bar{\theta} \in H^0$. Using (4.19) yields that, for some $\alpha_2 > 0$,

$$V(\widetilde{\theta}_n^\epsilon) - V(\widetilde{\theta}_0^\epsilon) \le -2\epsilon\alpha_2 \sum_{i=0}^{n-1} |\widehat{\psi}_i^\epsilon - \psi_i|^2 + O(\epsilon^2) \sum_{i=0}^{n-1} |\phi_i^\epsilon|^2 |\widehat{\psi}_i^\epsilon - \psi_i|^2. \tag{4.21}$$

Since the ϕ_n^ϵ are bounded and $V(\theta) \ge 0$, (4.21) implies that, for small enough ϵ, $\lim_n |\widehat{\psi}_n^\epsilon - \psi_n| = 0$ with probability one and $|\widetilde{\theta}_n^\epsilon|^2$ is nonincreasing. Thus, for small ϵ and large enough n, there is no truncation of the $\widehat{\psi}_n^\epsilon$. Furthermore, the convergence of $\widehat{\psi}_n^\epsilon - \psi_n$ to zero is equivalent to $v_n^\epsilon \to 0$ since

$$v_n^\epsilon = (\phi_n^\epsilon)'\widetilde{\theta}_n^\epsilon = \left[(\phi_n^\epsilon)'\theta_n^\epsilon - \phi_n'\bar{\theta}\right] + \left[\phi_n - \phi_n^\epsilon\right]'\bar{\theta},$$

where the first term on the right side equals $\widehat{\psi}_n^\epsilon - \psi_n$ and the second is proportional to it (see the definitions below (3.4.6)). The proof that $\lim_n \theta_n^\epsilon = \bar{\theta}$ for each small ϵ is completed under (4.14), as was done for the untruncated algorithm.

9.5 Proportional Fair Sharing

We now give the proof of convergence of the PFS algorithm in Section 3.6 under one set of assumptions. The notation of that section will be used. Weaker assumptions and other forms of the problem are in [146, 147]. The last part of (A5.1a) below is unrestrictive and is used to assure that when a component $\theta_{n,i}$ is very small there is a nonzero chance that user i will be chosen, no matter what the values of the other components of θ_n. It guarantees that the mean rate function $\bar{h}_i(\theta)$ defined in (3.1) is positive when θ^i is small. All of the assumptions hold under Rayleigh fading if the channels are independent. The density condition is used only to show that the limit point is unique.

(A5.0) Let ξ_n denote the past data $\{R_l : l \leq n\}$. For each i, n, ξ_n,

$$h_{n,i}(\theta, \xi_n) = E_n r_{n+1,i} I_{\{r_{n+1,i}/(d_i+\theta^i) \geq r_{n+1,j}/(d_j+\theta^j), j \neq i\}}$$

is continuous in $\theta \in I\!R^N_+$. Let $\delta > 0$ be arbitrary. Then in the set $\{\theta : \theta^i \geq \delta, i \leq N\}$, the continuity is uniform in n and in ξ_n.

(A5.1a) $\{R_n, n < \infty\}$ is stationary. Define $\bar{h}_i(\cdot)$ by the stationary expectation:

$$\bar{h}_i(\theta) = E r_i I_{\{r_i/(d_i+\theta^i) \geq r_j/(d_j+\theta^j), j \neq i\}}, \quad i \leq N. \tag{5.1}$$

Also,

$$\lim_{m,n\to\infty} \frac{1}{m} \sum_{l=n}^{n+m-1} \left[E_n r_{l+1,i} I_{\{r_{l+1,i}/(d_i+\theta^i) \geq r_{l+1,j}/(d_j+\theta^j), j \neq i\}} - \bar{h}_i(\theta) \right] = 0 \tag{5.2}$$

in the sense of probability. There are small positive δ and δ_1 such that

$$P\left\{ r_{n,i}/d_i \geq r_{n,j}/(d_j - \delta) + \delta_1, j \neq i \right\} > 0, \quad i \leq N. \tag{5.3}$$

(A5.1b) R_n is defined on some bounded set and has a bounded density.

On the smoothness of $\bar{h}(\cdot)$. By (A5.1b), $\bar{h}(\cdot)$ is Lipschitz continuous. To see this in the two dimensional case, let $p(\cdot)$ denote the density of R_n, and write $w = (d_1 + \theta^1)/(d_2 + \theta^2)$. Then

$$\bar{h}_1(\theta) = \int r_1 I_{\{r_1/r_2 \geq w\}} p(r_1, r_2) dr_1 dr_2,$$

which is Lipschitz continuous with respect to w, since the area of the region where the indicator is not zero is a differentiable function of w. Under (A5.0) and (A5.1a), Theorem 8.2.2 implies that the mean ODE is (3.6.7):

$$\dot{\theta}^i = \bar{h}_i(\theta) - \theta^i, \quad i \leq N. \tag{5.4}$$

The proof of the next theorem depends on the monotonicity property given in Section 4.4, which is fundamental to the analysis of a large class of stochastic algorithms which model competitive or cooperative behavior. The solution to (5.4) with initial condition $\theta(0)$ is denoted by $\theta(t|\theta(0))$.

Theorem 5.1. *Assume* (A5.0) *and* (A5.1). *Then the limit point* $\bar{\theta}$ *of* (5.4) *is unique.*

Proof. All paths tend to some compact set as $t \to \infty$. Without loss of generality (say, by shifting the time origin), we can suppose where needed that there is a $\delta > 0$ such that $\theta^i(t|\theta(0)) \geq \delta$ for all t. Since the path is initially monotonically increasing in each coordinate when started near the origin (since $\bar{h}(\theta) - \theta \gg 0$ for θ small), Theorem 4.4.1 implies that it will be monotonically nondecreasing in each coordinate for all t, for any initial condition close to the origin. Thus, there is a unique limit point for $\theta(t|\theta(0))$ for each $\theta(0)$ near the origin. Let $\bar{\theta}$ and $\widetilde{\theta}$ be two such limit points for paths corresponding, resp., to initial conditions $\bar{\theta}(0)$ and $\widetilde{\theta}(0)$ close to the origin. Since both paths are monotonically increasing near the origin, without loss of generality, we can suppose that for some small $t_0 > 0$, $\theta(t_0|\widetilde{\theta}(0)) \gg \bar{\theta}(0)$. Then, by letting $\theta(t_0|\widetilde{\theta}(0))$ be the initial condition for a new path, we can suppose that $\theta(t|\widetilde{\theta}(0) \geq \theta(t|\bar{\theta}(0))$ for all t. Thus, $\widetilde{\theta} \geq \bar{\theta}$. An analogous argument yields that $\bar{\theta} \geq \widetilde{\theta}$. we can conclude that there is a unique limit point, say $\bar{\theta}$, for all paths starting sufficiently close to the origin. Hence $\bar{\theta}$ is an equilibrium point for (5.4); i.e., $h(\bar{\theta}) = \bar{\theta}$. Furthermore, $\theta(t|\theta(0)) \leq \bar{\theta}$ for all $\theta(0)$ close to the origin.

Now, consider the path starting at an arbitrary initial condition $\widehat{\theta} \leq \bar{\theta}$. After some small time $t_0 > 0$, all components of the path will be positive and $\theta(t_0|\widehat{\theta}) \geq \theta(0)$ for some $\theta(0)$ arbitrarily close to the origin. Then, the monotonicity argument (and slightly shifting the time origin of one of the paths) of the above paragraph yields that $\theta(t|\widehat{\theta}) \geq \theta(t|\theta(0))$ for all t. Hence any limit point of $\theta(t|\widehat{\theta})$ must be no smaller than $\bar{\theta}$. But, by the monotonicity again, $\theta(t|\widehat{\theta}) \leq \theta(t|\bar{\theta}) = \bar{\theta}$ for all t. We can conclude that $\theta(t|\widehat{\theta}) \to \bar{\theta}$ as $t \to \infty$.

Define the set $Q(\theta) = \{x : x \geq \theta\}$. Now, consider an arbitrary initial condition $\theta(0)$. The monotonicity argument can be used again to show that all limit points of the path are in $Q(\bar{\theta})$. It only remains to show that any path starting in $Q(\bar{\theta})$ must ultimately go to $\bar{\theta}$ also. So far, we have used only the monotonicity property and not any other aspect of the original stochastic approximation process that led to (5.4). The rest of the details involve the properties of the argmax rule and an "stochastic approximation" argument. Suppose that there is a point $\widetilde{\theta} \in Q(\bar{\theta})$, $\widetilde{\theta} \neq \bar{\theta}$, such that

$$\dot{U}(\widetilde{\theta}) = \sum_i (\bar{h}_i(\widetilde{\theta}) - \widetilde{\theta}^i)/(d_i + \widetilde{\theta}^i) \geq 0. \tag{5.5}$$

Since $\widetilde{\theta} \geq \bar{\theta}$ and $\widetilde{\theta} \neq \bar{\theta}$, (5.5) implies that

$$\sum_i (\bar{h}_i(\widetilde{\theta}) - \bar{\theta}^i)/(d_i + \bar{\theta}^i) > 0. \tag{5.6}$$

Consider the algorithm (3.6.4) started at $\bar{\theta}$, but with the slot allocation rule

$$\arg\max_{i \leq N} \left\{ r_{n+1,i}/(d_i + \widetilde{\theta}^i) \right\}$$

used at time n. Let $\widetilde{I}^\epsilon_{n+1,i}$ denote the indicator function of the event that user i is chosen at time n with this rule. Modulo a second order error of value $O(\epsilon)t$, the expansion (3.6.10) and the maximizing property of I^ϵ_{n+1} yield

$$U(\theta^\epsilon(t)) - U(\bar{\theta}) = \epsilon \sum_i \sum_{l=0}^{t/\epsilon - 1} \frac{r_{l+1,i} I^\epsilon_{l+1,i} - \theta^\epsilon_{l,i}}{d_i + \theta^\epsilon_{l,i}} \geq \epsilon \sum_i \sum_{l=0}^{t/\epsilon - 1} \frac{r_{l+1,i} \widetilde{I}^\epsilon_{l+1,i} - \theta^\epsilon_{l,i}}{d_i + \theta^\epsilon_{l,i}} \tag{5.7}$$

where θ^ϵ_l (and $\theta^\epsilon(\cdot)$) is the solution to (3.6.4).

The arguments of Theorem 8.2.2 together with (5.7) imply that as $\epsilon \to 0$ the limit $\theta(\cdot)$ satisfies

$$U(\theta(t)) - U(\bar{\theta}) = \int_0^t \sum_i \frac{\bar{h}_i(\theta(s)) - \theta^i(s)}{d_i + \theta^i(s)} ds \geq \int_0^t \sum_i \frac{\bar{h}_i(\widetilde{\theta}) - \theta^i(s)}{d_i + \theta^i(s)} ds.$$

This, together with the inequality in (5.6) implies that

$$\dot{U}(\theta(t))\Big|_{t=0} = \sum_i \frac{\bar{h}_i(\bar{\theta}) - \bar{\theta}^i}{d_i + \bar{\theta}^i} \geq \sum_i \frac{\bar{h}_i(\widetilde{\theta}) - \bar{\theta}^i}{d_i + \bar{\theta}^i} > 0.$$

But the first sum is zero since $\bar{h}(\bar{\theta}) = \bar{\theta}$. Thus, we have a contradiction to (5.5) and can conclude that $\dot{U}(\theta) < 0$ for all $\theta \in Q(\bar{\theta}) - \{\bar{\theta}\}$. This implies that $\dot{U}(\theta(\cdot|\widetilde{\theta}))$ is strictly decreasing when the path is in $Q(\bar{\theta}) - \{\bar{\theta}\}$, which implies that any path starting at some $\theta(0) \in Q(\bar{\theta})$ must end up at $\bar{\theta}$. Thus, $\bar{\theta}$ is the unique limit point of (5.5), irrespective of the initial condition. Hence it is asymptotically stable. □

Recall the utility function $U(\cdot)$ in (3.6.9). We can see intuitively from the argument below (3.6.9) that (3.6.4) and (3.6.5) is, in the limit, a type of "steepest ascent" algorithm for a strictly concave function. However, since the allowed directions of ascent at each value of θ depend on θ, there is no a priori guarantee of any type of maximization. The following theorem is one way of quantifying this idea. The same idea works for an algorithm based on any smooth strictly concave utility function. A short proof is in [146]; see [147] for additional details.

Theorem 5.2. *Assume* (A5.0) *and* (A5.1). *Then there is no assignment policy which yields a limit throughput* $\widetilde{\theta} \neq \bar{\theta}$ *such that* $U(\widetilde{\theta}) \geq U(\bar{\theta})$.

10

Rate of Convergence

10.0 Outline of Chapter

The traditional definition of *rate of convergence* refers to the asymptotic properties of normalized errors about the limit point $\bar{\theta}$. For the Robbins–Monro algorithm with $\epsilon_n = \epsilon$, it is concerned with the asymptotic properties of $U_n^\epsilon = (\theta_n^\epsilon - \bar{\theta})/\sqrt{\epsilon}$ for large n and small ϵ. If $\epsilon_n \to 0$, then it is concerned with the asymptotic properties of $U_n = (\theta_n - \bar{\theta})/\sqrt{\epsilon_n}$. Define the processes $U^\epsilon(\cdot)$ by $U^\epsilon(t) = U_i^\epsilon$, for $t \in [i\epsilon, i\epsilon + \epsilon)$, and $U^n(\cdot)$ by $U^n(t) = U_{n+i}$, for $t \in [t_{n+i} - t_n, t_{n+i+1} - t_n)$. Then, under broad conditions, $U^\epsilon(T + \cdot)$ and $U^n(\cdot)$ converges to a stationary Gauss–Markov process as $\epsilon \to 0, T \to \infty$, or $n \to \infty$. Let R denote the stationary covariance of this process. Then loosely speaking, asymptotically, $\theta_n^\epsilon - \bar{\theta}$ is normally distributed with mean zero and covariance ϵR and $\theta_n - \bar{\theta}$ is normally distributed with mean zero and covariance $\epsilon_n R$ (there might be a nonzero mean for the Kiefer–Wolfowitz algorithm, due to the finite difference bias). These variances, together with the scaling, are used as a measure of the rate of convergence. This chapter is concerned with this definition of rate of convergence. Until Section 9, if the algorithm is constrained, then it is supposed that $\bar{\theta}$ is interior to the constraint set. Reference [71] proves the asymptotic normality of the normalized iterates for the classical algorithms.

An alternative notion of rate of convergence comes from the theory of large deviations [61, 63, 114]. With this alternative definition, one fixes a small region G containing the limit point $\bar{\theta}$ and then shows (under appropriate conditions and probability one convergence) that the probability that

$\{\theta_i, i \geq n\}$ will ever leave this neighborhood after it next enters is bounded above by e^{-c/ϵ_n}, where $c > 0$ depends on the noise and stabilizing effects in the algorithm. The analog for the constant-step-size algorithm is that the probability of escape on any finite time interval (resp., the mean escape time) is bounded above by $e^{-c/\epsilon}$ (resp., bounded below by $e^{c/\epsilon}$); see Section 6.10 for further details on this alternative approach.

Section 1 deals with the constant $\epsilon_n = \epsilon$ case. The first result, Theorem 1.1, contains the basic ideas of all the cases. The proof is simplest because the noise is a "martingale difference" sequence. The methods of proving tightness by the use of truncated processes and the method of characterizing the limit and showing its stationarity are basic to all the subsequent results. To simplify the proof and divide the work of the chapter into natural parts, it is assumed that $\{(\theta_n^\epsilon - \bar{\theta})/\sqrt{\epsilon}, n \geq n_\epsilon\}$ is tight for some sequence n_ϵ. This tightness is usually proved by a stability method and is treated in Section 4 for the case where $\epsilon_n \to 0$ and $\theta_n \to \bar{\theta}$ with probability one. The case $\epsilon_n = \epsilon$ is treated in Section 5.

The extension of Theorem 1.1 to the correlated noise case is in Theorem 1.3. In this theorem, we make an assumption that a certain process converges weakly to a Wiener process. This is done because there is much literature on the convergence of such processes to a Wiener process, and it yields a more general theorem. Section 6 shows that the assumption holds under quite reasonable "mixing" type conditions on the noise.

The minor alterations to Section 1 needed for the decreasing-step-size algorithm are in Section 2. The Kiefer–Wolfowitz algorithm is treated in Section 3, where we use the normalization $U_n = n^\beta(\theta_n - \bar{\theta})$ for appropriate $\beta > 0$. Section 4 is devoted to proving tightness of the sequence of normalized iterates for the various processes under probability one convergence. Owing to the probability one convergence, the proof involves only a local stability analysis about $\bar{\theta}$. Section 5 treats the same problem, when there is only weak convergence. Section 6 contains some results concerning the weak convergence of a sequence of processes to a Wiener process. The random directions Kiefer–Wolfowitz algorithm is treated in Section 7. It is shown that there can be considerable advantages to it, but it must be used with care, especially if there are serious concerns for bias in the finite difference estimators or if the number of observations is not large. The state-dependent-noise case is treated in Section 8. A brief outline of the problem where the algorithm is constrained and the limit point $\bar{\theta}$ is on the boundary of the constraint set is in Section 9. The so-called strong diffusion approximation aims at estimating the asymptotic difference between the normalized error process and a Wiener process in terms of an "iterated logarithm;" this is not dealt with in the book, but see [105, 157, 159, 188, 191].

10.1 Exogenous Noise: Constant Step Size

This chapter is concerned with the rate of convergence, assuming that convergence occurs. The algorithms can be constrained or not. In the constrained cases in Sections 1–8, it will always be supposed that the path is concentrated about some point $\bar{\theta}$ *strictly inside* (relaxed in Section 9) the constraint set for large time and small ϵ. Let $\{q_\epsilon\}$ be a sequence of non-negative integers. Recall the definition of the process $\theta^\epsilon(\cdot)$: For $n \geq 0$, $\theta^\epsilon(t) = \theta^\epsilon_n$ on the interval $[n\epsilon, n\epsilon+\epsilon)$, and $\theta^\epsilon(t) = \theta^\epsilon_0$ for $t < 0$. The processes $\theta^\epsilon(\epsilon q_\epsilon + \cdot)$ introduced below will be used, where generally $\epsilon q_\epsilon \to \infty$ because we are concerned with the asymptotic behavior of θ^ϵ_n for small ϵ and large n; this shifting of the time origin is a good way of bringing the asymptotic behavior to the foreground.

10.1.1 Martingale Difference Noise

In this subsection, $E^\epsilon_n Y^\epsilon_n = g^\epsilon_n(\theta^\epsilon_n)$ and the algorithm of interest is

$$\begin{aligned} \theta^\epsilon_{n+1} &= \Pi_H\left[\theta^\epsilon_n + \epsilon Y^\epsilon_n\right] = \theta^\epsilon_n + \epsilon Y^\epsilon_n + \epsilon Z^\epsilon_n \\ &= \theta^\epsilon_n + \epsilon g^\epsilon_n(\theta^\epsilon_n) + \epsilon \delta M^\epsilon_n + \epsilon Z^\epsilon_n. \end{aligned} \tag{1.1}$$

Assumptions. The $\lim_{n,m}$ and $\lim_{n,m,\epsilon}$ in (A1.5) and (A1.6) are as $n \to \infty, m \to \infty$, and $\epsilon \to 0$ in any way at all. Criteria for the crucial stability Assumption (A1.3) will be given in Sections 4 and 5. The "delay" N_ϵ is to account for a transient period, because for an arbitrary initial condition θ^ϵ_0, some time will be required for the iterates to settle near $\bar{\theta}$. In fact, the p_ϵ in (A1.3) is generally at most of the order of $|\log \epsilon|/\epsilon$ for small ϵ. Assumption (A1.5) holds if (the classical condition) $g^\epsilon_n(\theta) = \bar{g}(\theta)$ for all n and ϵ, since $\bar{g}(\bar{\theta}) = 0$. In addition, it allows the possibility that the $g^\epsilon_n(\bar{\theta})$ will only "locally average" to $\bar{g}(\theta)$, as might occur if only a few of the components of θ were changed at each step. For example, consider the two-dimensional gradient descent case defined by

$$\theta^\epsilon_{n+1} = \begin{cases} \theta^\epsilon_n - \epsilon e_1\left[f_{\theta^1}(\theta^\epsilon_n) + \psi^\epsilon_n\right], & n \text{ odd}, \\ \theta^\epsilon_n - \epsilon e_2\left[f_{\theta^2}(\theta^\epsilon_n) + \psi^\epsilon_n\right], & n \text{ even}, \end{cases}$$

where e_i is the unit vector in the ith coordinate direction and ψ^ϵ_n is the observation noise.

(A1.1) $\{Y^\epsilon_n I_{\{|\theta^\epsilon_n - \bar{\theta}| \leq \rho\}}; \epsilon, n\}$ is uniformly integrable for small $\rho > 0$.

(A1.2) There is a sequence of non-negative and nondecreasing integers N_ϵ such that $\{\theta^\epsilon(\epsilon N_\epsilon + \cdot)\}$ converges weakly to the process with constant value $\bar{\theta}$ strictly inside the constraint set H.

(A1.3) There are nondecreasing and non-negative integers p_ϵ (that can be taken to be greater than N_ϵ) such that

$$\{(\theta^\epsilon_{p_\epsilon+n} - \bar{\theta})/\sqrt{\epsilon};\ \epsilon > 0, n \geq 0\} \quad \text{is tight.} \tag{1.2}$$

(A1.4) $E^\epsilon_n Y^\epsilon_n = g^\epsilon_n(\theta^\epsilon_n)$, where $g^\epsilon_n(\cdot)$ is continuously differentiable for each n and ϵ, and can be expanded as

$$g^\epsilon_n(\theta) = g^\epsilon_n(\bar{\theta}) + [g^\epsilon_{n,\theta}(\bar{\theta})]'(\theta - \bar{\theta}) + o(|\theta - \bar{\theta}|),$$

where $o(\cdot)$ is uniform in n and ϵ, and $\{g^\epsilon_{n,\theta}(\bar{\theta}); \epsilon, n\}$ is bounded.

(A1.5) The following holds:

$$\lim_{n,m,\epsilon} \frac{1}{\sqrt{m}} \sum_{i=n}^{n+tm-1} g^\epsilon_i(\bar{\theta}) = 0$$

uniformly in $t \geq 0$ in some bounded interval.

(A1.6) There is a Hurwitz matrix A (i.e., the real parts of the eigenvalues of A are negative) such that

$$\lim_{n,m,\epsilon} \frac{1}{m} \sum_{i=n}^{n+m-1} \left[g^{\epsilon,\prime}_{i,\theta}(\bar{\theta}) - A\right] = 0. \tag{1.3}$$

(A1.7) For some $p > 0$ and small $\rho > 0$,

$$\sup_{\epsilon,n} E|\delta M^\epsilon_n|^{2+p} I_{\{|\theta^\epsilon_n - \bar{\theta}| \leq \rho\}} < \infty, \tag{1.4}$$

and there is a non-negative definite matrix Σ_1 such that for small $\rho > 0$,

$$E^\epsilon_n \delta M^\epsilon_n \left[\delta M^\epsilon_n\right]' I_{\{|\theta^\epsilon_n - \bar{\theta}| \leq \rho\}} \to \Sigma_1 \tag{1.5}$$

in probability as $n \to \infty$ and $\epsilon \to 0$. [Convergence in mean in (1.5) is then implied by (1.4).]

Definition of the normalized error process. Let $\{q_\epsilon\}$ be a sequence of non-negative integers going to infinity such that $(q_\epsilon - p_\epsilon)\epsilon \to \infty$. Define $U^\epsilon_n = (\theta^\epsilon_{q_\epsilon+n} - \bar{\theta})/\sqrt{\epsilon}$, and let $U^\epsilon(\cdot)$ denote the piecewise constant right continuous interpolation (always with interpolation intervals ϵ) of the sequence $\{U^\epsilon_n\}$ on $[0, \infty)$. Define $W^\epsilon(\cdot)$ on $(-\infty, \infty)$ by

$$\begin{aligned} W^\epsilon(t) &= \sqrt{\epsilon} \sum_{i=q_\epsilon}^{q_\epsilon+t/\epsilon-1} \delta M^\epsilon_i,\ t \geq 0 \\ &= -\sqrt{\epsilon} \sum_{i=q_\epsilon+t/\epsilon}^{q_\epsilon-1} \delta M^\epsilon_i,\ t < 0. \end{aligned} \tag{1.6}$$

The proof of the following theorem contains the basic ideas that will be used for the correlated noise and decreasing-step-size cases.

Theorem 1.1. *Assume algorithm* (1.1) *and* (A1.1)–(A1.7). *Then the sequence* $\{U^\epsilon(\cdot), W^\epsilon(\cdot)\}$ *converges weakly in* $D^r[0,\infty) \times D^r(-\infty,\infty)$ *to a limit denoted by* $(U(\cdot), W(\cdot))$, *and*

$$U(t) = U(0) + \int_0^t AU(s)ds + W(t), \tag{1.7a}$$

equivalently written in differential form as

$$dU = AUdt + dW, \tag{1.7b}$$

where $W(\cdot)$ *is a Wiener processes with covariance matrix* Σ_1 *and* $U(\cdot)$ *is stationary.*

Remark. The path space $D^r(-\infty,\infty)$ can also be used for $U^\epsilon(\cdot)$, but the proof is a little easier as done here. If q_ϵ is simply no smaller than p_ϵ, there will still be weak convergence to a solution of (1.7), but that solution need not be stationary. Alternatively, we could fix the initial condition $\theta_0^\epsilon = \theta_0$ and show that the piecewise constant interpolations (with interpolation interval ϵ) of the ratios $(\theta_n^\epsilon - \theta_0(\epsilon n))/\sqrt{\epsilon}$ converge to a solution to (1.7), where $\theta_0(\cdot)$ is the solution to the mean ODE with initial condition θ_0. But the asymptotics of the θ_n^ϵ for large ϵn are of greater interest.

Proof. The criterion of Theorem 7.3.3 will be used to prove tightness. Then the martingale method of Section 7.4, which was used in the proof of Theorem 8.2.1, will be used to characterize the limit process. It is always assumed (without loss of generality) that if the algorithm is constrained, then the distance between $\bar{\theta}$ and the boundary of the constraint set H is greater than μ for the (arbitrarily small) values of $\mu > 0$ that will be used. By Theorem 7.3.6, it is sufficient to prove the tightness and to characterize the weak sense limit processes on each finite interval. For any $\mu > 0$ and $T > 0$, the weak convergence (A1.2) implies that

$$\limsup_\epsilon P\left\{ \sup_{-T/\epsilon \le i \le T/\epsilon} \left|\theta_{q_\epsilon+i}^\epsilon - \bar{\theta}\right| \ge \mu/2 \right\} = 0. \tag{1.8}$$

Hence, by modifying $\{Y_{q_\epsilon+n}^\epsilon, |n| \le T/\epsilon\}$ on a set of arbitrarily small measure, without altering the assumptions, it can be supposed (as justified by Theorem 7.3.6) that

$$\sup_{-T/\epsilon \le i \le T/\epsilon} \left|\theta_{q_\epsilon+i}^\epsilon - \bar{\theta}\right| < \mu \tag{1.9}$$

for small ϵ on each interval $[-T,T]$ of concern. Thus, without loss of generality, Z_n^ϵ in (1.1) can be dropped, and it can be supposed that (1.9) holds for any given T and μ.

Dropping the Z_n^ϵ terms, rewrite the algorithm in the expanded form:

$$\theta_{n+1}^\epsilon = \theta_n^\epsilon + \epsilon \left[g_n^\epsilon(\bar{\theta}) + [g_{n,\theta}^\epsilon(\bar{\theta})]'(\theta_n^\epsilon - \bar{\theta}) + [y_n^\epsilon(\theta_n^\epsilon)]'(\theta_n^\epsilon - \bar{\theta}) + \delta M_n^\epsilon \right], \tag{1.10}$$

where the term $y_n^\epsilon(\theta)$ satisfies $y_n^\epsilon(\theta) = o(\theta - \bar{\theta})/|\theta - \bar{\theta}|$ uniformly in n and ϵ, and the order of $o(\cdot)$ is uniform in n and ϵ by (A1.4). Now shift the time origin to the right by q_ϵ and divide all terms in (1.10) by $\sqrt{\epsilon}$. For simplicity, we write n in lieu of $n + q_\epsilon$ for the subscripts of $(g, \theta, \delta M)$. Then, recalling the definition of U_n^ϵ, (1.10) yields

$$\begin{aligned} U_{n+1}^\epsilon &= U_n^\epsilon + \epsilon A U_n^\epsilon + \sqrt{\epsilon} \left[g_n^\epsilon(\bar{\theta}) + \delta M_n^\epsilon \right] \\ &\quad + \epsilon \left[g_{n,\theta}^{\epsilon,\prime}(\bar{\theta}) - A \right] U_n^\epsilon + \epsilon [y_n^\epsilon(\theta_n^\epsilon)]' U_n^\epsilon. \end{aligned} \tag{1.11}$$

Owing to (A1.3) and the fact that the noise δM_n^ϵ is a martingale difference, it would not be difficult to show that the sequence $\{U^\epsilon(\cdot)\}$ is tight in $D^r[0, \infty)$. For more general noise processes, the main difficulty in proving tightness stems from the possible unboundedness of $\{U^\epsilon(\cdot)\}$ on some time interval $[0, T]$. Then, the possible unboundedness is handled by a standard truncation procedure [127, pp. 42–45]. For simplicity of the overall presentation, that truncation technique will also be used in this proof. We will work with truncated $\{U^\epsilon(\cdot)\}$, prove its tightness, and characterize the limits of its weakly convergent subsequences. It will then be shown that the truncation is not necessary by proving that the sequence of truncated processes is uniformly (in the truncation level) bounded in probability on each finite interval. The uniform boundedness in probability will be shown by using the weak convergence of the sequence of truncated processes, for each fixed truncation level, and characterizing the limit as the solution to a truncated form of (1.7).

The truncated processes. For each integer M, let $q_M(\cdot)$ be a continuous real-valued function on $I\!R^r$ satisfying: $0 \le q_M(x) \le 1$, $q_M(x) = 1$ for $|x| \le M$, and $q_M(x) = 0$ for $|x| \ge M + 1$. Define $U_n^{\epsilon,M}$ by $U_0^{\epsilon,M} = U_0^\epsilon$ and for $n \ge 0$ set

$$\begin{aligned} U_{n+1}^{\epsilon,M} &= U_n^{\epsilon,M} + \epsilon A U_n^{\epsilon,M} q_M(U_n^{\epsilon,M}) + \sqrt{\epsilon} \left[g_n^\epsilon(\bar{\theta}) + \delta M_n^\epsilon \right] \\ &\quad + \epsilon \left[g_{n,\theta}^{\epsilon,\prime}(\bar{\theta}) - A \right] U_n^{\epsilon,M} q_M(U_n^{\epsilon,M}) + \epsilon [y_n^\epsilon(\theta_n^\epsilon)]' U_n^{\epsilon,M} q_M(U_n^{\epsilon,M}). \end{aligned} \tag{1.12}$$

Note that $U_n^\epsilon = U_n^{\epsilon,M}$ until the first time (denoted by $N^{\epsilon,M}$) that U_n^ϵ exceeds M in norm and that the term with coefficient $\sqrt{\epsilon}$ is not truncated.

Fix M and T until further notice. Let $U^{\epsilon,M}(\cdot)$ denote the piecewise constant right continuous interpolation of the $\{U_n^{\epsilon,M}\}$. By (A1.5), the sequence defined by

$$\sqrt{\epsilon} \sum_{i=0}^{t/\epsilon - 1} g_i^\epsilon(\bar{\theta})$$

goes to zero uniformly on $[0, T]$ as $\epsilon \to 0$. It will be shown at the end of the proof that $\{W^\epsilon(\cdot)\}$ defined in (1.6) is tight in $D^r(-\infty, \infty)$ and converges weakly to the asserted Wiener process. Assuming this convergence result now, the sequence of processes defined by the piecewise constant right continuous interpolations of the term in (1.12) with coefficient $\sqrt{\epsilon}$ is tight in $D^r[0, \infty)$. The sequence of processes defined by the piecewise constant right continuous interpolation of the sums of the second and the fourth terms on the right-hand side of (1.12) are tight in $D^r[0, \infty)$ by the boundedness of $U_n^{\epsilon,M} q_M(U_n^{\epsilon,M})$ and $\{g_{n,\theta}^\epsilon(\bar{\theta}); n, \epsilon\}$. The sequence of interpolated processes defined by the sums of the last term on the right side of (1.12) is tight by the boundedness of $U_n^{\epsilon,M} q_M(U_n^{\epsilon,M})$ and the fact that we can suppose that (1.9) holds for small $\mu > 0$. Furthermore, its weak sense limit is identically zero.

Putting the pieces together and noting that $\{U_0^\epsilon\}$ is tight by (A1.3), it is seen that $\{U^{\epsilon,M}(\cdot), \epsilon > 0\}$ is tight in $D^r[0, \infty)$. Extract a weakly convergent subsequence (as $\epsilon \to 0$) of $\{U^{\epsilon,M}(\cdot), W^\epsilon(\cdot)\}$ in $D^r[0, \infty) \times D^r(-\infty, \infty)$ with limit denoted by $(U^M(\cdot), W(\cdot))$. Strictly speaking, the Wiener process limit $W(\cdot)$ should be indexed by M, because the distribution of the pair $(U^{\epsilon,M}(\cdot), W^\epsilon(\cdot))$ depends on M. Since the distribution of the limit of $W^\epsilon(\cdot)$ does not depend on M, the index M will be omitted for notational simplicity.

It will be shown that the sequence of processes $J^{\epsilon,M}(\cdot)$ defined by

$$J^{\epsilon,M}(t) = \sum_{i=0}^{t/\epsilon - 1} \epsilon \left[g_{i,\theta}^{\epsilon,\prime}(\bar{\theta}) - A \right] U_i^{\epsilon,M} q_M(U_i^{\epsilon,M}) \tag{1.13}$$

has the "zero process" as a limit. This fact and the weak convergence of $(U^{\epsilon,M}(\cdot), W^\epsilon(\cdot))$ imply that $(U^M(\cdot), W(\cdot))$ satisfies

$$U^M(t) = U^M(0) + \int_0^t AU^M(s) q_M(U^M(s)) ds + W(t). \tag{1.14}$$

Bounds on the truncated processes. It will next be shown that for each positive T,

$$\lim_{K \to \infty} \sup_M P \left\{ \sup_{t \le T} |U^M(t)| \ge K \right\} = 0. \tag{1.15}$$

Owing to the tightness assumption (A1.3), for the purpose of showing the tightness of $\{U_n^\epsilon(\cdot)\}$ in this paragraph, we can assume without loss of generality that the set of initial conditions $\{U_0^\epsilon\}$ is bounded, and (1.11) holds as it is written for $n \ge 0$. By this assumption, $U^M(0)$ is bounded uniformly in (ω, M) and is independent of the selected convergent subsequence.

By virtue of (1.14), there is a real number C not depending on M such that for any real $\tau \in [0, T]$,

$$E \sup_{t \le \tau} \left|U^M(t)\right|^2 \le CE \left|U_0^M\right|^2 + CE \left[\int_0^\tau \left|U^M(s) q_M(U^M(s))\right| ds \right]^2 + CE \sup_{t \le \tau} \left|W(t)\right|^2.$$

Define $V^M(\tau) = E \sup_{t \le \tau} |U^M(t)|^2$. Now, using the assumed bound on the initial condition, Schwarz inequality and the martingale inequality (see (4.1.5)),

$$E \sup_{s \le t} |W(s)|^2 \le 4E|W(t)|^2 \tag{1.16}$$

yield that there is a real number C_1 not depending on M or T such that

$$V^M(t) \le C_1 + C_1 T \int_0^t V^M(s) ds + C_1 t, \quad t \le T. \tag{1.17}$$

The Bellman–Gronwall inequality then implies that

$$V^M(t) \le C_1(1+T) e^{C_1 T^2}, \quad t \le T.$$

This implies that for any small $\rho > 0$ and any $T > 0$, there is a $K_{\rho,T} < \infty$ such that

$$\sup_M P\left\{ \sup_{t \le T} |U^M(t)| \ge K_{\rho,T} \right\} \le \rho, \tag{1.18}$$

which is equivalent to (1.15). The fact that any subsequence has a further subsequence that converges weakly to a solution of (1.7) implies that

$$\limsup_\epsilon P\left\{ \sup_{t \le T} |U^{\epsilon,M}(t)| \ge 2K \right\} \le P\left\{ \sup_{t \le T} |U^M(t)| \ge K \right\}. \tag{1.19}$$

Since we need only verify the tightness criterion for each bounded time interval, the results (1.15), (1.19), and the fact that for $N > M$, $U^{\epsilon,M}(t) = U^{\epsilon,N}(t) = U^\epsilon(t)$ until the first time that $|U^{\epsilon,M}(t)|$ is larger than M, imply that the M-truncation is not needed in the tightness proof and that $\{U^\epsilon(\cdot)\}$ is also tight in $D^r[0,\infty)$. It is clear that any weak limit of $\{U^\epsilon(\cdot), W^\epsilon(\cdot)\}$ satisfies (1.7).

Stationary limit process. Now we work with the original untruncated processes $U^\epsilon(\cdot)$, and show that the weak sense limit process $U(\cdot)$ is stationary. The technique is a useful "shifting" method using the fact that for any positive T, the limit of the shifted sequence $\{U^\epsilon(-T+\cdot)\}$ also satisfies (1.7) with starting time $-T$.

For each positive T, define the process $U_T^\epsilon(t) = U^\epsilon(-T+t), t \ge 0$, with initial condition $U_T^\epsilon(0) = U^\epsilon(-T) = (\theta^\epsilon_{q_\epsilon - T/\epsilon} - \bar\theta)/\sqrt{\epsilon}$. Then $U_T^\epsilon(T) = U^\epsilon(0)$. By (A1.3) and the fact that $\epsilon(q_\epsilon - p_\epsilon) \to \infty$, the set $\{U^\epsilon(-T) \colon \epsilon\}$ is tight and the set (indexed by T) of all possible weak sense limits of $U^\epsilon(-T)$ (as $\epsilon \to 0$) is tight. By the arguments in the previous parts of the proof, for each $T > 0$ the sequence of untruncated processes $\{U^\epsilon(\cdot), U_T^\epsilon(\cdot)\}$ is tight in $D^{2r}[0,\infty)$. For each T, choose a weakly convergent subsequence of

$$\{U^\epsilon(\cdot), U_T^\epsilon(\cdot), W^\epsilon(\cdot)\}$$

in $D^{2r}[0,\infty) \times D^r(-\infty,\infty)$ with limit denoted by $(U(\cdot), U_T(\cdot), W(\cdot))$. Strictly speaking, $W(\cdot)$ should also be indexed by T because the joint distribution of the triple $(U^\epsilon(\cdot), U_T^\epsilon(\cdot), W^\epsilon(\cdot))$ depends on T. However the distribution of the weak sense limits of $\{W^\epsilon(\cdot)\}$ does not depend on T, so we omit the index T on $W(\cdot)$. Since $U_T(\cdot)$ satisfies (1.7) on $[0,\infty)$, with Wiener process $W(\cdot)$, we can write

$$U_T(T) = e^{AT} U_T(0) + \int_{-T}^{0} e^{-As} dW(s).$$

By (1.7),

$$U(t) = e^{At} U(0) + \int_0^t e^{A(t-s)} dW(s).$$

Since $U_T(T) = U(0)$ with probability one, for $t \geq 0$ we have

$$U(t) = e^{A(t+T)} U_T(0) + \int_{-T}^{t} e^{A(t-s)} dW(s).$$

The random vector $U_T(0)$ depends on T and on the selected weakly convergent subsequence, but the set of all possible initial conditions $\{U_T(0), T > 0\}$ is tight, as noted above. Hence, whatever the sequence of initial conditions $\{U_T(0), T > 0\}$ is, as $T \to \infty$ we have

$$e^{AT} U_T(0) \to 0$$

in probability since A is Hurwitz. Thus, we can write (in the sense of equivalence of probability law on $D^r[0,\infty) \times D^r(-\infty,\infty)$)

$$U(t) = \int_{-\infty}^{t} e^{A(t-s)} dW(s). \tag{1.20}$$

The process defined by (1.20) is the stationary solution to (1.7).

The reason for using $D^r(-\infty,\infty)$ for the path space of $W^\epsilon(\cdot)$ is simply so that $U(\cdot)$ could be written as (1.20), an integral of the entire past of the Wiener process that is the stationary solution of (1.7).

Limits of $J^{\epsilon,M}(\cdot)$. It will next be shown that $J^{\epsilon,M}(\cdot)$, defined in (1.13), converges weakly to the "zero process" for each M, as $\epsilon \to 0$. Let $n_\epsilon \to \infty$

as $\epsilon \to 0$, such that $\Delta_\epsilon = \epsilon n_\epsilon \to 0$. Suppose without loss of generality that t is an integral multiple of Δ_ϵ, and define the index set

$$I_j^{\epsilon,\Delta_\epsilon} = \{i : i\epsilon \in [j\Delta_\epsilon, j\Delta_\epsilon + \Delta_\epsilon)\}.$$

We can now write

$$\begin{aligned} J^{\epsilon,M}(t) &= \sum_{j=0}^{t/\Delta_\epsilon - 1} \Delta_\epsilon \frac{1}{n_\epsilon} \sum_{i \in I_j^{\epsilon,\Delta_\epsilon}} \left[g_{i,\theta}^{\epsilon,\prime}(\bar{\theta}) - A \right] \\ &\quad \times \left[\left(U_i^{\epsilon,M} q_M(U_i^{\epsilon,M}) - U_{jn_\epsilon}^{\epsilon,M} q_M(U_{jn_\epsilon}^{\epsilon,M}) \right) + U_{jn_\epsilon}^{\epsilon,M} q_M(U_{jn_\epsilon}^{\epsilon,M}) \right]. \end{aligned} \tag{1.21}$$

By the tightness of $\{U^{\epsilon,M}(\cdot)\}$ and the continuity of the limit process $U^M(\cdot)$, for each positive T,

$$\sup_{jn_\epsilon \le T} \sup_{i \in I_j^{\epsilon,\Delta_\epsilon}} \left| U_i^{\epsilon,M} q_M(U_i^{\epsilon,M}) - U_{jn_\epsilon}^{\epsilon,M} q_M(U_{jn_\epsilon}^{\epsilon,M}) \right| \tag{1.22}$$

goes to zero in probability (and in mean) as $\epsilon \to 0$. It is also bounded, uniformly in ϵ. This implies that the $\sup_{t \le T}$ of the term in (1.21) containing the difference of the two U-terms goes to zero in mean as $\epsilon \to 0$. Condition (A1.6) implies that the remaining term goes to zero in mean as $\epsilon \to 0$.

Proof of the properties of $W(\cdot)$. For any bounded stopping time τ, the martingale difference property and (1.4) imply that for $s \ge 0$,

$$E_{\tau/\epsilon}^\epsilon \left| W^\epsilon(\tau + s) - W^\epsilon(\tau) \right|^2 = O(s),$$

where $O(s)$ is uniform in τ. Thus, tightness in $D^r(-\infty, \infty)$ follows from the criterion of Theorem 7.3.3. In the rest of this part of the proof, $t \in (-\infty, \infty)$ and τ is a positive real number. To characterize the limit $W(\cdot)$, it is sufficient to fix $T > 0$ and focus our attention on $[-T, T]$. As noted in connection with (1.9), for purposes of characterizing $W(\cdot)$, without loss of generality, for any $\mu > 0$ we can suppose that $|\theta^\epsilon(t) - \bar{\theta}| \le \mu$ for $|t| \le T$ and all small ϵ. This fact, together with (A1.7) and Theorem 7.4.2, will be used to get the desired result. For notational simplicity, the case of real-valued $W^\epsilon(\cdot)$ will be dealt with, but except for the notation (scalar vs. vector), the proof is the same in general.

Let $F(\cdot)$, $h(\cdot)$ and $s_j \le t, j \le p$, be as defined in Theorem 7.4.2. Then

$$\begin{aligned} F(W^\epsilon(t+\tau)) - F(W^\epsilon(t)) &= \sum_{i=t/\epsilon}^{(t+\tau)/\epsilon - 1} \sqrt{\epsilon} F_w(W^\epsilon(\epsilon i)) \delta M_i^\epsilon \\ &+ \frac{1}{2} \sum_{i=t/\epsilon}^{(t+\tau)/\epsilon - 1} \epsilon F_{ww}(W^\epsilon(\epsilon i)) \left| \delta M_i^\epsilon \right|^2 + \text{“error term,”} \end{aligned}$$

where the error term is

$$\sum_{i=t/\epsilon}^{(t+\tau)/\epsilon-1} \epsilon \int_0^1 \left[F_{ww}(W^\epsilon(\epsilon i) + \sqrt{\epsilon} s \delta M_i^\epsilon) - F_{ww}(W^\epsilon(\epsilon i))\right] ds \, |\delta M_i^\epsilon|^2 .$$

By (1.4) and the boundedness of $F(\cdot)$, the sequence of coefficients of ϵ in the last sum is uniformly (in ϵ, i) integrable. This, the smoothness and boundedness properties of $F(\cdot)$, and the fact that $\sqrt{\epsilon}\sup_{t/\epsilon \le i \le (t+\tau)/\epsilon} |\delta M_i^\epsilon|$ goes to zero in probability as $\epsilon \to 0$ imply that the error term goes to zero as $\epsilon \to 0$. Now (1.5) and the martingale difference property yield

$$\begin{aligned}&\lim_\epsilon Eh(W^\epsilon(s_j), j \le p) \\ &\quad \times E_{t/\epsilon}^\epsilon \left[F(W^\epsilon(t+\tau)) - F(W^\epsilon(t)) - \frac{\epsilon}{2}\Sigma_1 \sum_{i=t/\epsilon}^{(t+\tau)/\epsilon-1} F_{ww}(W^\epsilon(\epsilon i)) \right] = 0.\end{aligned}$$

Letting $\epsilon \to 0$ and using Theorem 7.4.2 yield that the weak sense limit $W(\cdot)$ is a Wiener process with covariance Σ_1. □

Criterion for convergence to a Wiener process. Theorem 1.2 follows from the last part of the proof and is a useful way of showing that a sequence of martingales converges to a Wiener process.

Theorem 1.2. *Let $\{V_n^\epsilon\}$ and $\{\delta N_n^\epsilon\}$ be $\mathbb{R}^k$-valued random variables and $\mathcal{F}_n^\epsilon$ the minimal σ-algebra that measures $\{V_i^\epsilon, \delta N_i^\epsilon, i < n\}$ (with E_n^ϵ being the associated conditional expectation operator). For $t \ge 0$, define $N^\epsilon(t) = \sum_{i=0}^{t/\epsilon-1} \sqrt{\epsilon}\delta N_i^\epsilon$. Suppose that $E_n^\epsilon \delta N_n^\epsilon = 0$ with probability one for all n and ϵ and that there are $p > 0$ and a matrix Σ such that*

$$\sup_{n,\epsilon} E\,|\delta N_n^\epsilon|^{2+p} < \infty$$

and

$$E_n^\epsilon \delta N_n^\epsilon (\delta N_n^\epsilon)' \to \Sigma$$

in probability as $n \to \infty$ and $\epsilon \to 0$. Then $\{N^\epsilon(\cdot)\}$ converges weakly to a Wiener process with covariance matrix Σ. Let $V^\epsilon(\cdot)$ denote the right continuous piecewise constant interpolation of $\{V_n^\epsilon\}$ on $[0,\infty)$ with interpolation interval ϵ. Suppose that $(V^\epsilon(\cdot), N^\epsilon(\cdot)) \Rightarrow (V(\cdot), N(\cdot))$ in $D^{2r}[0,\infty)$ with $\mathcal{F}_t^V$ being the minimal σ-algebra that measures $\{V(s), N(s), s \le t\}$. Then $N(\cdot)$ is an $\mathcal{F}_t^V$-Wiener process. The analogous results hold on $(-\infty,\infty)$.

10.1.2 Correlated Noise

The algorithm of interest can be written as

$$\begin{aligned} \theta_{n+1}^{\epsilon} &= \Pi_H(\theta_n^{\epsilon} + \epsilon Y_n^{\epsilon}) \\ &= \theta_n^{\epsilon} + \epsilon Y_n^{\epsilon} + \epsilon Z_n^{\epsilon} \\ &= \theta_n^{\epsilon} + \epsilon g_n^{\epsilon}(\theta_n^{\epsilon}, \xi_n^{\epsilon}) + \epsilon \delta M_n^{\epsilon} + \epsilon Z_n^{\epsilon}. \end{aligned} \tag{1.23}$$

Assumptions. Assumptions (A1.1)–(A1.3) will continue to be used. Let $Y_n^{\epsilon}(\bar{\theta})$ denote the observation at iterate n if parameter value $\theta = \bar{\theta}$ is used at that time. (A1.9) is not a strong condition in view of the fact that $\theta_n^{\epsilon} \to \bar{\theta}$. The use of assumptions (A1.3) and (A1.10) allows a "separation of labor" and leads to a stronger result, because these assumptions can be shown to hold under a wide variety of conditions. Condition (A1.10) will be dealt with in Section 6.

(A1.8) There are random variables ξ_n^{ϵ} taking values in a complete separable metric space Ξ (as above (A8.1.5)) and measurable functions $g_n^{\epsilon}(\cdot)$ such that

$$E_n^{\epsilon} Y_n^{\epsilon} = g_n^{\epsilon}(\theta_n^{\epsilon}, \xi_n^{\epsilon}),$$

where E_n^{ϵ} is the expectation conditioned on the σ-algebra $\mathcal{F}_n^{\epsilon}$ that measures $\{\theta_0^{\epsilon}, Y_{i-1}^{\epsilon}, Y_{i-1}^{\epsilon}(\bar{\theta}), \xi_i^{\epsilon}, i \le n\}$.

(A1.9) Define $\delta M_n^{\epsilon}(\bar{\theta}) = Y_n^{\epsilon}(\bar{\theta}) - E_n^{\epsilon} Y_n^{\epsilon}(\bar{\theta})$. Then for small $\rho > 0$, and any sequence $\epsilon \to 0$ and $n \to \infty$ such that $\theta_n^{\epsilon} \to \bar{\theta}$ in probability,

$$E \left| \delta M_n^{\epsilon} - \delta M_n^{\epsilon}(\bar{\theta}) \right|^2 I_{\{|\theta_n^{\epsilon} - \bar{\theta}| \le \rho\}} \to 0.$$

(A1.10) The sequence of processes $W^{\epsilon}(\cdot)$ defined on $(-\infty, \infty)$ by (1.6) with δM_n^{ϵ} replaced by $Y_n^{\epsilon}(\bar{\theta})$ converges weakly in $D^r(-\infty, \infty)$ to a Wiener process $W(\cdot)$, with covariance matrix Σ_2.

(A1.11) $g_n^{\epsilon}(\cdot, \xi)$ is continuously differentiable for each n, ϵ, and ξ, and can be expanded as

$$g_n^{\epsilon}(\theta, \xi) = g_n^{\epsilon}(\bar{\theta}, \xi) + [g_{n,\theta}^{\epsilon}(\bar{\theta}, \xi)]'(\theta - \bar{\theta}) + [y_n^{\epsilon}(\theta, \xi)]'(\theta - \bar{\theta}), \tag{1.24a}$$

where

$$y_n^{\epsilon}(\theta, \xi) = \int_0^1 \left[g_{n,\theta}^{\epsilon}(\bar{\theta} + s(\theta - \bar{\theta}), \xi) - g_{n,\theta}^{\epsilon}(\bar{\theta}, \xi) \right] ds, \tag{1.24b}$$

and if $\delta_n^{\epsilon} \to 0$ as $\epsilon \to 0$ and $n \to \infty$, then

$$E \left| y_n^{\epsilon}(\theta_n^{\epsilon}, \xi_n^{\epsilon}) \right| I_{\{|\theta_n^{\epsilon} - \bar{\theta}| \le \delta_n^{\epsilon}\}} \to 0 \tag{1.24c}$$

as $\epsilon \to 0$ and $n \to \infty$.

(A1.12) The set

$$\{g^\epsilon_{n,\theta}(\bar\theta, \xi^\epsilon_n);\ \epsilon, n \geq q_\epsilon\} \quad \text{is uniformly integrable,} \tag{1.25}$$

where q_ϵ satisfies the condition above (1.6).

(A1.13) There is a Hurwitz matrix A such that

$$\frac{1}{m}\sum_{i=n}^{n+m-1}\left[E^\epsilon_n g^{\epsilon,\prime}_{i,\theta}(\bar\theta, \xi^\epsilon_i) - A\right] \to 0 \tag{1.26}$$

in probability as $\epsilon \to 0$ and $n, m \to \infty$.

Theorem 1.3. *Assume algorithm* (1.23), (A1.1)–(A1.3), *and* (A1.8)–(A1.13). *Then the sequence* $\{U^\epsilon(\cdot), W^\epsilon(\cdot)\}$ *converges weakly in* $D^r[0,\infty) \times D^r(-\infty,\infty)$ *to a limit denoted by* $(U(\cdot), W(\cdot))$, *which satisfies* (1.7) *where* $W(\cdot)$ *is a Wiener processes with covariance matrix* Σ_2, *and* $U(\cdot)$ *is stationary.*

Proof. The proof is the same as that of Theorem 1.1, except for the new definition of $W^\epsilon(\cdot)$ (owing to the use of $Y^\epsilon_n(\bar\theta)$) and the replacement of $g^{\epsilon,\prime}_{n,\theta}(\bar\theta)$ with $g^{\epsilon,\prime}_{n,\theta}(\bar\theta, \xi)$ in the process $J^{\epsilon,M}(\cdot)$. Equation (1.8) still holds, and (1.9) can be assumed for any $T > 0$ without loss of generality. Hence, the Z_n terms can still be omitted in the development.

Dropping the Z^ϵ_n terms, rewrite the algorithm in the expanded form:

$$\begin{aligned}\theta^\epsilon_{n+1} &= \theta^\epsilon_n + \epsilon\left[g^\epsilon_n(\theta^\epsilon_n, \xi^\epsilon_n) + \delta M^\epsilon_n\right] \\ &= \theta^\epsilon_n + \epsilon\Big[g^\epsilon_n(\bar\theta, \xi^\epsilon_n) + [g^\epsilon_{n,\theta}(\bar\theta, \xi^\epsilon_n)]'(\theta^\epsilon_n - \bar\theta) \\ &\qquad + \delta M^\epsilon_n + [y^\epsilon_n(\theta^\epsilon_n, \xi^\epsilon_n)]'(\theta^\epsilon_n - \bar\theta)\Big].\end{aligned} \tag{1.27}$$

Define $\delta N^\epsilon_n = \delta M^\epsilon_n - \delta M^\epsilon_n(\bar\theta)$, and recall that $Y^\epsilon_n(\bar\theta) = g^\epsilon_n(\bar\theta, \xi^\epsilon_n) + \delta M^\epsilon_n(\bar\theta)$ by the definition of $\delta M^\epsilon_n(\bar\theta)$. Now, repeating what was done in the proof of Theorem 1.1, we shift the time origin to the right by q_ϵ and divide all terms in (1.27) by $\sqrt{\epsilon}$. Again, for simplicity, we write n in lieu of $n + q_\epsilon$ for the subscripts of $\xi^\epsilon, g^\epsilon, \theta^\epsilon, \delta M^\epsilon$, and so on. Then

$$\begin{aligned}U^\epsilon_{n+1} &= U^\epsilon_n + \epsilon A U^\epsilon_n + \sqrt{\epsilon}Y^\epsilon_n(\bar\theta) + \sqrt{\epsilon}\delta N^\epsilon_n \\ &\quad + \epsilon\left[g^{\epsilon,\prime}_{n,\theta}(\bar\theta, \xi^\epsilon_n) - A\right]U^\epsilon_n + \epsilon[y^\epsilon_n(\theta^\epsilon_n, \xi^\epsilon_n)]'U^\epsilon_n.\end{aligned} \tag{1.28}$$

Repeating the truncation procedure used in the proof of Theorem 1.1, we have

$$\begin{aligned}U^{\epsilon,M}_{n+1} &= U^{\epsilon,M}_n + \epsilon A U^{\epsilon,M}_n q_M(U^{\epsilon,M}_n) + \sqrt{\epsilon}Y^\epsilon_n(\bar\theta) + \sqrt{\epsilon}\delta N^\epsilon_n \\ &\quad + \epsilon[g^{\epsilon,\prime}_{n,\theta}(\bar\theta, \xi^\epsilon_n) - A]U^{\epsilon,M}_n q_M(U^{\epsilon,M}_n) \\ &\quad + \epsilon[y^\epsilon_n(\theta^\epsilon_n, \xi^\epsilon_n)]'U^{\epsilon,M}_n q_M(U^{\epsilon,M}_n).\end{aligned} \tag{1.29}$$

By (A1.9), the sequence of processes defined by

$$\sqrt{\epsilon}\sum_{i=0}^{t/\epsilon-1}\delta N_i^\epsilon$$

converges weakly to the zero process. The sequence $\{W^\epsilon(\cdot)\}$ is tight by (A1.10) and the weak sense limit is a Wiener process. The sequence of processes $J^{\epsilon,M}(\cdot)$ defined by

$$J^{\epsilon,M}(t)=\sum_{i=0}^{t/\epsilon-1}\epsilon\left[g_{i,\theta}^{\epsilon,\prime}(\bar{\theta},\xi_i^\epsilon)-A\right]U_i^{\epsilon,M}q_M(U_i^{\epsilon,M})$$

is tight by the uniform integrability condition (A1.12). It will be shown that it has the "zero process" as its limit. By these facts and the arguments concerning tightness in the proof of Theorem 1.1, $\{U^{\epsilon,M}(\cdot)\}$ is tight for each M. The rest of the proof is as for Theorem 1.1.

We need only show that $\{J^{\epsilon,M}(\cdot)\}$ converges weakly to the "zero process"; this will also follow the method and terminology of the proof of Theorem 1.1. We can write

$$\begin{aligned}J^{\epsilon,M}(t) &= \sum_{j=0}^{t/\Delta_\epsilon-1}\Delta_\epsilon\frac{1}{n_\epsilon}\sum_{i\in I_j^{\epsilon,\Delta_\epsilon}}\left[g_{i,\theta}^{\epsilon,\prime}(\bar{\theta},\xi_i^\epsilon)-A\right]\\ &\quad\times\left[\left(U_i^{\epsilon,M}q_M(U_i^{\epsilon,M})-U_{jn_\epsilon}^{\epsilon,M}q_M(U_{jn_\epsilon}^{\epsilon,M})\right)+U_{jn_\epsilon}^{\epsilon,M}q_M(U_{jn_\epsilon}^{\epsilon,M})\right].\end{aligned}\tag{1.30}$$

By the tightness of $\{U^{\epsilon,M}(\cdot)\}$ and the continuity of the limit processes $U^M(\cdot)$, for each positive T,

$$\sup_{jn_\epsilon\le T}\sup_{i\in I_j^{\epsilon,\Delta_\epsilon}}\left|U_i^{\epsilon,M}q_M(U_i^{\epsilon,M})-U_{jn_\epsilon}^{\epsilon,M}q_M(U_{jn_\epsilon}^{\epsilon,M})\right|\tag{1.31}$$

is bounded and goes to zero in probability as $\epsilon\to 0$. This fact and the uniform integrability condition (A1.12) imply that the $\sup_{t\le T}$ of the terms in (1.30) that contains the difference of the two U-terms goes to zero in mean as $\epsilon\to 0$. By (A1.12) and (A1.13), the remaining term goes to zero in mean as $\epsilon\to 0$. □

10.2 Exogenous Noise: Decreasing Step Size

We continue to follow the approach taken in the previous section, by assuming the key tightness condition (A2.3), which will be dealt with further in Sections 4 and 5.

10.2.1 Martingale Difference Noise

We keep the basic structure of Subsection 1.1, with only the modifications required due to the fact that $\epsilon_n \to 0$. The algorithm is

$$\theta_{n+1} = \Pi_H(\theta_n + \epsilon_n Y_n)$$

and can be written in the expanded form

$$\theta_{n+1} = \theta_n + \epsilon_n Y_n + \epsilon_n Z_n = \theta_n + \epsilon_n g_n(\theta_n) + \epsilon_n \delta M_n + \epsilon_n Z_n. \tag{2.1}$$

Except (A2.0), the following assumptions are essentially those of Section 1, with ϵ dropped.

Assumptions.

(A2.0) Write $(\epsilon_n/\epsilon_{n+1})^{1/2} = 1 + \mu_n$. Then either (a) or (b) holds:

(a) $\epsilon_n = 1/n$; then $\mu_n = 1/(2n) + o(\epsilon_n)$.

(b) $\mu_n = o(\epsilon_n)$.

(A2.1) $\{Y_n I_{\{|\theta_n - \bar\theta| \le \rho\}}\}$ is uniformly integrable for small $\rho > 0$.

(A2.2) For $\bar\theta$ an isolated stable point of the ODE $\dot\theta = \bar g(\theta)$ in the interior of H, $\{\theta^n(\cdot)\}$ converges weakly to the process with constant value $\bar\theta$.

(A2.3) $\{(\theta_n - \bar\theta)/\sqrt{\epsilon_n}\}$ is tight.

(A2.4) $E_n Y_n = g_n(\theta_n)$, where $g_n(\cdot)$ is continuously differentiable for each n, and can be expanded as

$$g_n(\theta) = g_n(\bar\theta) + g'_{n,\theta}(\bar\theta)(\theta - \bar\theta) + o(|\theta - \bar\theta|),$$

where $o(\cdot)$ is uniform in n.

(A2.5) $\displaystyle\lim_{n,m} \frac{1}{\sqrt{m}} \sum_{i=n}^{n+mt-1} g_i(\bar\theta) = 0$, where the limit is uniform on some bounded t-interval.

(A2.6) There is a Hurwitz matrix A such that

$$\lim_{n,m} \frac{1}{m} \sum_{i=n}^{n+m-1} \left[g'_{i,\theta}(\bar\theta) - A\right] = 0,$$

and if (A2.0a) holds, then $(A + I/2)$ is also a Hurwitz matrix.

(A2.7) For some $p > 0$ and small $\rho > 0$,

$$\sup_n E|\delta M_n|^{2+p} I_{\{|\theta_n - \bar\theta| \le \rho\}} < \infty. \tag{2.2}$$

There is a non-negative definite matrix Σ_1 such that for small $\rho > 0$,

$$E_n \delta M_n [\delta M_n]' I_{\{|\theta_n - \bar{\theta}| \le \rho\}} \to \Sigma_1 \tag{2.3}$$

in probability as $n \to \infty$.

Definitions. Define $U_n = (\theta_n - \bar{\theta})/\sqrt{\epsilon_n}$, and let $U^n(\cdot)$ denote the piecewise constant right continuous interpolation (with interpolation intervals $\{\epsilon_i\}$) of the sequence $\{U_i, i \ge n\}$ on $[0, \infty)$. Define $W^n(\cdot)$ on $(-\infty, \infty)$ by

$$\begin{aligned} W^n(t) &= \sum_{i=n}^{m(t_n+t)-1} \sqrt{\epsilon_i} \delta M_i, \quad t \ge 0 \\ &= -\sum_{i=m(t_n+t)}^{n-1} \sqrt{\epsilon_i} \delta M_i, \quad t < 0. \end{aligned} \tag{2.4}$$

Theorem 2.1. *Assume algorithm* (2.1) *and* (A2.0)–(A2.7). *Then the sequence* $\{U^n(\cdot), W^n(\cdot)\}$ *converges weakly in* $D^r[0, \infty) \times D^r(-\infty, \infty)$ *to a limit denoted by* $(U(\cdot), W(\cdot))$, *where* $W(\cdot)$ *is a Wiener process with covariance matrix* Σ_1 *and* $U(\cdot)$ *is stationary. Under* (A2.0b), (1.7) *holds. Under* (A2.0a), (1.7) *is replaced by*

$$\begin{aligned} U(t) &= U(0) + \int_0^t (A + I/2) U(s) ds + W(t) \\ &= \int_{-\infty}^t e^{(A+I/2)(t-s)} dW(s). \end{aligned} \tag{2.5}$$

Proof. The proof is essentially that of Theorem 1.1, with minor modifications in the notation. As in Subsection 1.1 we can take $Z_n = 0$ in the analysis without loss of generality. It follows from the definitions of θ_n, U_n, and $y_n(\cdot)$ that

$$\begin{aligned} U_{n+1} = &\left(\frac{\epsilon_n}{\epsilon_{n+1}}\right)^{1/2} U_n + \left(\frac{\epsilon_n}{\epsilon_{n+1}}\right)^{1/2} \Big[\epsilon_n A U_n + \sqrt{\epsilon_n} \left(g_n(\bar{\theta}) + \delta M_n \right) \\ &+ \epsilon_n \left(g'_{n,\theta}(\bar{\theta}) - A \right) U_n + \epsilon_n y'_n(\theta_n) U_n \Big], \end{aligned} \tag{2.6}$$

where $y_n(\theta_n)$ satisfies $y_n(\theta_n) = o(\theta_n - \bar{\theta})/|\theta_n - \bar{\theta}|$, where the order of $o(\cdot)$ is uniform in n. For each n, define the truncated process $\{U_i^{n,M}, i \ge n\}$ analogously to what was done in Subsection 1.1; namely, $U_n^{n,M} = U_n$ and for $i \geq n$,

$$
\begin{aligned}
U_{i+1}^{n,M} &= U_i^{n,M} + \left(\left(\frac{\epsilon_n}{\epsilon_{n+1}} \right)^{1/2} - 1 \right) U_i^{n,M} q_M(U_i^{n,M}) \\
&+ \left(\frac{\epsilon_n}{\epsilon_{n+1}} \right)^{1/2} \Big[\epsilon_n A U_i^{n,M} q_M(U_i^{n,M}) + \sqrt{\epsilon_n} \left(g_n(\bar{\theta}) + \delta M_n \right) \\
&+ \epsilon_n \left(g'_{n,\theta}(\bar{\theta}) - A \right) U_i^{n,M} q_M(U_i^{n,M}) + \epsilon_n y'_n(\theta_n) U_i^{n,M} q_M(U_i^{n,M}) \Big].
\end{aligned}
\tag{2.7}
$$

Let $U^{n,M}(\cdot)$ denote the continuous parameter interpolation of $\{U_i^{n,M}, i \geq n\}$ on $[0, \infty)$ with interpolation intervals $\{\epsilon_i, i \geq n\}$. Under our conditions, the tightness of $\{U^{n,M}(\cdot)\}$ is evident for each fixed M by the arguments in the proof of Theorem 1.1. Now write $(\epsilon_n/\epsilon_{n+1})^{1/2} = 1 + \mu_n$. The contribution to the limit (as $n \to \infty$) of the term involving the μ_n times the square bracketed term in (2.7) is the "zero process." Suppose that (A2.0b) holds. Then, the second term on the right of the equality sign of (2.7) is $U_i^{n,M} o(\epsilon_n) q_M(U_i^{n,M})$. Due to the $o(\epsilon_n)$ component, its contribution to the weak sense limit of $\{U^{n,M}(\cdot)\}$ is the "zero process." Now suppose (A2.0a). Then the second term on the right of the equal sign of (2.7) is essentially $U_i^{n,M} q_M(U_i^{n,M}) \epsilon_i/2$. The rest of the proof closely parallels the arguments used for Theorem 1.1 and is omitted. □

10.2.2 Optimal Step Size Sequence

From a formal point of view, for large n the mean square error $E[\theta_n - \bar{\theta}][\theta_n - \bar{\theta}]'$ is ϵ_n times the stationary covariance of $U(t)$ and is minimized by using $\epsilon_n = 1/n$, among the choices in (A2.0). This is an asymptotic result. It is a useful guide to practice, but letting the step size decrease too fast can lead to problems with nonrobustness and slower convergence in practice, because the effects of large noise values early in the procedure might be hard to overcome in a reasonable period of time. It will be seen in the next chapter that the effective rate $O(1/n)$ for the squared error can be achieved by iterate averaging, while letting the step size decrease more slowly than $1/n$ in the sense that $n\epsilon_n \to \infty$. (See also the comments on observation averaging in Subsection 1.3.4.)

Let Σ denote the covariance matrix of the driving Wiener process. Using Theorem 2.1 as a guide, let us choose $\epsilon_n = K/n$ where K is a *nonsingular matrix,* and determine the best K. [The interpolation intervals will always be $\{1/n\}$.] The matrix K can be absorbed into the function $g_n(\cdot)$ and the matrix Σ, so that the A and Σ are replaced with KA and $K\Sigma K'$, respectively. Then the limit $U(\cdot)$ satisfies

$$dU = \left(KA + \frac{I}{2} \right) U dt + K \Sigma^{\frac{1}{2}} dw,$$

where $w(\cdot)$ is a standard Wiener process (i.e., the covariance matrix is the identity). The stationary covariance is

$$\int_0^\infty e^{(KA+I/2)t} K\Sigma K' e^{(A'K'+I/2)t} dt.$$

The trace of this matrix is minimized by choosing $K = -A^{-1}$, which yields the asymptotically optimal covariance

$$A^{-1}\Sigma(A^{-1})'. \tag{2.7a}$$

10.2.3 Correlated Noise

The algorithm is (1.23) with $\epsilon_n \to 0$ and can be written as

$$\begin{aligned} \theta_{n+1} &= \Pi_H(\theta_n + \epsilon_n Y_n) \\ &= \theta_n + \epsilon_n Y_n + \epsilon_n Z_n \\ &= \theta_n + \epsilon_n g_n(\theta_n, \xi_n) + \epsilon_n \delta M_n + \epsilon_n Z_n. \end{aligned} \tag{2.8}$$

Assumptions. Assumptions (A2.0)–(A2.3) will continue to be used. Following the approach of Subsection 1.2, let $Y_n(\bar{\theta})$ denote the observation at iterate n if $\theta = \bar{\theta}$ is the parameter value used at that time.

(A2.8) There are measurable functions $g_n(\cdot)$ such that

$$E_n Y_n = g_n(\theta_n, \xi_n),$$

where E_n is the expectation conditioned on the σ-algebra $\mathcal{F}_n$ that measures $\{\theta_0, Y_{i-1}, Y_{i-1}(\bar{\theta}), \xi_i, i \le n\}$.

(A2.9) Define $\delta M_n(\bar{\theta}) = Y_n(\bar{\theta}) - E_n Y_n(\bar{\theta})$. Then for small $\rho > 0$, as $n \to \infty$,

$$E\left|\delta M_n - \delta M_n(\bar{\theta})\right|^2 I_{\{|\theta_n - \bar{\theta}| \le \rho\}} \to 0.$$

(A2.10) The sequence of processes $W^n(\cdot)$ defined on $(-\infty, \infty)$ by (2.4) with δM_n replaced by $Y_n(\bar{\theta})$ converges weakly in $D^r(-\infty, \infty)$ to a Wiener process $W(\cdot)$, with covariance matrix Σ_2.

(A2.11) $g_n(\cdot, \xi)$ is continuously differentiable for each n and ξ, can be expanded as

$$g_n(\theta, \xi) = g_n(\bar{\theta}, \xi) + g'_{n,\theta}(\bar{\theta}, \xi)(\theta - \bar{\theta}) + y'_n(\theta, \xi)(\theta - \bar{\theta}), \tag{2.9a}$$

where

$$y_n(\theta, \xi) = \int_0^1 \left[g_{n,\theta}(\bar{\theta} + s(\theta - \bar{\theta}), \xi) - g_{n,\theta}(\bar{\theta}, \xi)\right] ds, \tag{2.9b}$$

and if $\delta_n \to 0$ and $\rho > 0$ is small, then

$$\lim_n E\left|y_n(\theta_n, \xi_n)\right| I_{\{|\theta_n - \bar{\theta}| \le \delta_n\}} = 0. \tag{2.9c}$$

(A2.12) $\{g_{n,\theta}(\bar{\theta}, \xi_n)\}$ is uniformly integrable.

(A2.13) There is a Hurwitz matrix A such that

$$\frac{1}{m} \sum_{i=n}^{n+m-1} E_n \left[g'_{i,\theta}(\bar{\theta}, \xi_i) - A\right] \to 0 \tag{2.10}$$

in probability as $n, m \to \infty$. Under (A2.0a), $(A + I/2)$ is also a Hurwitz matrix.

The proof of the following theorem simply requires the analogs of the changes in the proof of Theorem 2.1 that were required to go from Theorem 1.1 to Theorem 2.1.

Theorem 2.2. *Assume algorithm* (2.8), *conditions* (A2.0)–(A2.3), *and* (A2.8)–(A2.13). *Then the conclusions of Theorem* 2.1 *hold, where* $W(\cdot)$ *is a Wiener processes with covariance matrix* Σ_2.

10.3 Kiefer–Wolfowitz Algorithm

10.3.1 Martingale Difference Noise

Let $0 < c_n \to 0$ be the finite difference intervals, and let e_i denote the unit vector in the ith coordinate direction. The basic Kiefer–Wolfowitz algorithm of (8.3.1) will be used; namely,

$$\theta_{n+1} = \Pi_H\left(\theta_n + \epsilon_n \frac{Y_n^- - Y_n^+}{2c_n}\right) = \theta_n + \epsilon_n \frac{Y_n^- - Y_n^+}{2c_n} + \epsilon_n Z_n, \tag{3.1}$$

where $Y_n^\pm = (Y_{n,1}^\pm, \ldots, Y_{n,r}^\pm)$. The $Y_{n,i}^\pm$ are the observations at time n at the parameter values $\theta_n \pm e_i c_n, i = 1, \ldots, r$. It is supposed that $\epsilon_n \to 0$, because otherwise it cannot be assumed that $c_n \to 0$. The set $\{Y_n^\pm\}$ will be assumed to be uniformly integrable. We first consider the case of two-sided differences, since it has a better convergence rate than the one-sided case. The result for the one-sided difference case will be discussed after the theorem is presented.

Define $\delta M_{n,i}^\pm = Y_{n,i}^\pm - E_n Y_{n,i}^\pm$ and $\delta M_n = (\delta M_{n,1}^- - \delta M_{n,1}^+, \ldots, \delta M_{n,r}^- - M_{n,r}^+)$. We consider the classical cases $\epsilon_n = 1/n^\gamma$, $c_n = c/n^\alpha$, and $0 < \alpha < \gamma \le 1$. Define $U_n = (\theta_n - \bar{\theta})n^\beta$, where β satisfies $\beta = 2\alpha$, and $\beta + \alpha - \gamma/2 = 0$. Thus, $\alpha = \gamma/6$. Other choices of (α, γ) result in a lower rate of convergence. Let $U^n(\cdot)$ denote the right continuous piecewise constant interpolation of

$\{U_i, i \geq n\}$ with interpolation intervals $\{\epsilon_i, i \geq n\}$, and define $W^n(\cdot)$ by (2.4) but with the current definition of δM_n used.

Assumptions. Criteria for (A3.2) will be given in Sections 4 and 5.

(A3.1) $\{Y_n^{\pm} I_{\{|\theta_n - \bar{\theta}| \leq \rho\}}\}$ is uniformly integrable for small $\rho > 0$.

(A3.2) $\{U_n\}$ is tight.

(A3.3) There are measurable functions $g_{n,i}(\cdot)$ such that $E_n Y_{n,i}^{\pm} = g_{n,i}(\theta_n \pm e_i c_n)$ where the partial derivatives of $g_{n,i}(\cdot)$ up to order three are continuous. Define $g_n(\cdot) = (g_{n,1}(\cdot), \ldots, g_{n,r}(\cdot))$.

(A3.4) Define $B_n(\theta) = (B_{n,1}(\theta), \ldots, B_{n,r}(\theta))$, where $B_{n,i}(\theta) = g_{n,i,\theta^i\theta^i\theta^i}(\theta)/(3!)$. There is a $B(\bar{\theta})$ such that

$$\lim_{n,m} \frac{1}{m} \sum_{i=n}^{n+m-1} \left[B_i(\bar{\theta}) - B(\bar{\theta})\right] = 0.$$

(A3.5) Define $\gamma_{n,i}(\cdot) = g_{n,i,\theta^i}(\cdot)$ and $\gamma_n(\cdot) = (\gamma_{n,1}(\cdot), \ldots, \gamma_{n,r}(\cdot))$. Let

$$\lim_{n,m} \epsilon_n^{2/3} \sum_{i=n}^{n+mt-1} \gamma_i(\bar{\theta}) = 0,$$

where the limit is uniform on some bounded t-interval, and

$$\gamma_n(\theta) = \gamma_n(\bar{\theta}) + \gamma'_{n,\theta}(\bar{\theta})(\theta - \bar{\theta}) + o(\theta - \bar{\theta}),$$

where the order of $o(\cdot)$ is uniform in n and $\{\gamma_i(\bar{\theta}), \gamma_{i,\theta}(\bar{\theta})\}$ is bounded. There is a Hurwitz matrix A such that

$$\lim_{n,m} \frac{1}{m} \sum_{i=n}^{n+m-1} \left[\gamma'_{i,\theta}(\bar{\theta}) + A\right] = 0. \tag{3.2}$$

If $\epsilon_n = 1/n$, then $A + \beta I$ is also a Hurwitz matrix, where $\beta = 2\alpha = \gamma/3$.

Remark on (A3.5). In many applications, $g_n(\cdot)$ does not depend on n, and then (A3.4) and (A3.5) always hold. In the classical case when the Kiefer–Wolfowitz procedure is used for the minimization of a real-valued function $f(\cdot)$ and the observations take the form $f(\theta_n \pm e_i c_n) + \delta M_{n,i}^{\pm}$, $i = 1, \ldots, r$, we have $g_{n,i}(\cdot) = f(\cdot)$ for all i and n, $\gamma_n(\cdot) = f_\theta(\cdot)$, and the matrix A is $-f_{\theta\theta}(\bar{\theta})$.

Theorem 3.1. *Assume algorithm* (3.1), *and conditions* (A2.2), (A2.7), *and* (A3.1)–(A3.5). *If* $\gamma < 1$, *then* $(U^n(\cdot), W^n(\cdot))$ *converges weakly in*

$D^r[0,\infty) \times D^r(-\infty,\infty)$ to $(U(\cdot), W(\cdot))$, where $U(\cdot)$ is the stationary process defined by

$$U(t) = \int_{-\infty}^{t} e^{A(t-s)}[dW(s)/(2c) - c^2 B(\bar{\theta})ds] \tag{3.3a}$$

or

$$dU = AUdt - c^2 B(\bar{\theta})dt + dW/(2c), \tag{3.3b}$$

and $W(\cdot)$ is a Wiener process with covariance matrix Σ_1. If $\gamma = 1$, replace A with $A + \beta I$ in (3.3).

Proof. As in Theorem 1.1, we can drop the term Z_n and suppose without loss of generality that $|\theta_n - \bar{\theta}| \leq \mu$ for any small $\mu > 0$. Now, write (3.1) in terms of its components as

$$\theta_{n+1,i} = \theta_{n,i} + \epsilon_n \left[\frac{g_{n,i}(\theta_n - e_i c_n) - g_{n,i}(\theta_n + e_i c_n)}{2c_n} \right] + \epsilon_n \frac{\delta M_{n,i}}{2c_n},$$

$i = 1, \ldots, r$. Expanding the bracketed finite difference ratio yields

$$-g_{n,i,\theta^i}(\theta_n) - \frac{g_{n,i,\theta^i\theta^i\theta^i}(\theta_n)c_n^2}{3!} + o(c_n^2).$$

Note that the second-order terms do not show up owing to the use of central finite differences. Now use the definition of the function $\gamma_n(\cdot)$ and its expansion in (A3.5) to rewrite the algorithm as

$$\begin{aligned} \theta_{n+1} = \theta_n &+ \epsilon_n \left[-\gamma'_{n,\theta}(\bar{\theta})(\theta_n - \bar{\theta}) - B_n(\theta_n) c_n^2 \right] \\ &+ \epsilon_n \left[-\gamma_n(\bar{\theta}) + o(\theta_n - \bar{\theta}) + o(c_n^2) + \frac{\delta M_n}{2c_n} \right]. \end{aligned} \tag{3.4}$$

Subtract $\bar{\theta}$ from both sides of (3.4), multiply both sides by $(n+1)^\beta$, and use the definition of U_n and the fact that $(n+1)^\beta / n^\beta = 1 + \beta/n + o(\epsilon_n)$ to get

$$\begin{aligned} U_{n+1} &= \left(1 + \frac{\beta}{n} + o(\epsilon_n)\right) U_n - \epsilon_n \frac{(n+1)^\beta}{n^\beta} \gamma'_{n,\theta}(\bar{\theta}) U_n \\ &- \epsilon_n \frac{(n+1)^\beta}{n^{2\alpha}} c^2 B_n(\theta_n) - \epsilon_n (n+1)^\beta \gamma_n(\bar{\theta}) \\ &+ \epsilon_n \frac{o(|\theta_n - \bar{\theta}|)}{|\theta_n - \bar{\theta}|} |U_n| + \epsilon_n (n+1)^\beta o(c_n^2) \\ &+ \sqrt{\epsilon_n} \frac{n^\alpha (n+1)^\beta}{2cn^{\gamma/2}} \delta M_n. \end{aligned} \tag{3.5}$$

It is clear from this equation that if there is weak convergence, we need $\beta - 2\alpha \leq 0$ and $\beta + \alpha - \gamma/2 \leq 0$. The largest value of β is obtained, with equality replacing the inequality, and such a β will be used in the sequel. Thus, we can write

$$U_{n+1} = \left(1 + \frac{\beta}{n}\right) U_n - \epsilon_n \gamma'_{n,\theta}(\bar{\theta}) U_n - \epsilon_n c^2 B_n(\bar{\theta}) + \sqrt{\epsilon_n} \frac{\delta M_n}{2c} + \epsilon_n \rho_n, \tag{3.6}$$

where ρ_n represents the "other" terms in (3.5). The contribution of the ρ_n terms to the limit is zero. The expression (3.6) is of the same form as used in Theorem 2.1, and the proof of that theorem yields this one. □

One-sided finite differences. For the one-sided finite difference case, at the nth iteration an observation is taken at θ_n and at the points $\theta_n + c_n e_i, i = 1, \ldots, r$. Write the algorithm as

$$\theta_{n+1} = \Pi_H \left(\theta_n - \epsilon_n \frac{Y_n}{c_n} \right), \tag{3.7}$$

where $Y_n = (Y_{n,1} - Y_{n,0}, \ldots, Y_{n,r} - Y_{n,0})$. The $Y_{n,i}$ are the observations at $\theta_n + c_n e_i$ and $Y_{n,0}$ is the observation at θ_n. For the sake of simplicity, we consider the classical case where, for a smooth enough real-valued function $f(\cdot)$, $E_n Y_{n,i} = f(\theta_n + c_n e_i)$, $E_n Y_{n,0} = f(\theta_n)$, and $f_\theta(\bar{\theta}) = 0$. Define $\delta M_n = Y_n - E_n Y_n$. The algorithm can be written as

$$\theta_{n+1,i} = \theta_{n,i} - \epsilon_n \left[\frac{f(\theta_n + c_n e_i) - f(\theta_n)}{c_n} \right] + \epsilon_n \frac{\delta M_{n,i}}{c_n} + \epsilon_n Z_{n,i}. \tag{3.8}$$

Note that the martingale difference δM_n in (3.8) is not the same as that for the algorithm (3.1) and will generally have a different covariance, because there is one observation common to all the iterates.

The following assumptions will be used. The partial derivatives of $f(\cdot)$ up to order two are continuous. Let $\epsilon_n = 1/n^\gamma$, $c_n = c/n^\alpha$, where α, β, and γ satisfy $\beta + \alpha = \gamma/2$ and $\beta = \alpha$. Let $A = -f_{\theta\theta}(\bar{\theta})$ be negative definite, and if $\gamma = 1$, let $A + \beta I$ also be negative definite. Suppose that, for small $\rho > 0$ the set $\{Y_{n,i} I_{\{|\theta_n - \bar{\theta}| \le \rho\}}; i = 0, \ldots r, n > 0\}$ is uniformly integrable and let (A2.2) and (A2.7) hold. Finally, suppose that $\{U_n\}$ is tight, where $U_n = n^\beta(\theta_n - \bar{\theta})$. Define $\widehat{B}(\bar{\theta}) = (\widehat{B}_1(\bar{\theta}), \ldots, \widehat{B}_r(\bar{\theta}))$, where $\widehat{B}_i(\theta) = f_{\theta^i\theta^i}(\theta)/2$. The best values are $\beta = \gamma/4$ for the one-sided case vs. $\beta = \gamma/3$ for the two-sided case. Thus, the two-sided case has a faster convergence rate.

Theorem 3.2. *Assume the conditions listed above and define $U^n(\cdot)$ as in Theorem 3.1 and $W^n(\cdot)$ as in (2.4), but with the current definitions of $\{U_n, \delta M_n\}$ used. Then $(U^n(\cdot), W^n(\cdot))$ converges weakly in $D^r[0,\infty) \times D^r(-\infty,\infty)$ to $(U(\cdot), W(\cdot))$ satisfying*

$$dU = AUdt - c\widehat{B}(\bar{\theta})dt + dW/c, \tag{3.9}$$

where $W(\cdot)$ is a Wiener process with covariance Σ_1. The process $U(\cdot)$ is stationary. If $\gamma = 1$, then $A + \beta I$ replaces A in (3.9).

Remark on the proof. The proof is the same as that of Theorem 3.1. Dropping the Z_n term, the form (3.8) can be written as

$$\theta_{n+1,i} = \theta_{n,i} + \epsilon_n \left[-f_{\theta^i}(\theta_n) - \frac{1}{2} f_{\theta^i\theta^i}(\theta_n) c_n + o(c_n) + \frac{\delta M_{n,i}}{c_n} \right].$$

Subtracting $\bar{\theta}$ from both sides, writing

$$f_\theta(\theta) = f_{\theta\theta}(\bar{\theta})(\theta - \bar{\theta}) + o(\theta - \bar{\theta}),$$

and multiplying all terms by $(n+1)^\beta$ yields

$$\begin{aligned} U_{n+1} &= \left(I + \frac{\beta}{n} + o(\epsilon_n)\right) U_n + \epsilon_n \frac{(n+1)^\beta}{n^\beta} \left[-f_{\theta\theta}(\bar{\theta}) U_n - \widehat{B}(\bar{\theta}) c \frac{n^\beta}{n^\alpha} \right] \\ &+ \sqrt{\epsilon_n} \frac{(n+1)^\beta n^\alpha}{n^{\gamma/2}} \frac{\delta M_n}{c} + \epsilon_n \rho_n, \end{aligned}$$

where the asymptotic contribution of the ρ_n is zero. The best value of β satisfies $\beta = \alpha$ and $\beta + \alpha = \gamma/2$. The rest of the proof is the same as that for the Robbins–Monro case.

10.3.2 Correlated Noise

Occasionally, Kiefer–Wolfowitz algorithms arise where the noise is correlated from iterate to iterate. The analysis depends on how the noise enters the algorithm. The form assumed in (8.3.2) is reasonably broad and illustrates a general approach to the proof. In fact, the proof is quite similar to what was done for the Robbins–Monro procedure when the noise is correlated. As for the previous cases, in the interest of simplicity of development and breadth of the result, it is convenient to divide and specialize the work, by making the general assumptions concerning the weak convergence of the appropriate form of $W^n(\cdot)$ and the tightness of $\{U^n(\cdot)\}$. These will be returned to later in this chapter.

We use the algorithm in the expanded form coming from (8.3.2) and use two-sided difference intervals. Thus, we suppose that the algorithm can be represented in the form

$$\begin{aligned} \theta_{n+1} &= \theta_n + \epsilon_n g_n(\theta_n, \xi_n) - \epsilon_n B(\theta_n) c_n^2 + \epsilon_n o(c_n^2) \\ &+ \frac{\epsilon_n}{2c_n} \delta M_n + \frac{\epsilon_n}{2c_n} \widetilde{g}_n(\theta_n, \xi_n) + \epsilon_n Z_n. \end{aligned} \tag{3.10}$$

Assumptions. Note that the $g_n(\cdot)$ used here is not the same as that used in Subsection 3.1. Suppose that $B(\cdot)$ is continuous and that $\widetilde{g}_n(\cdot)$ is twice continuously differentiable in θ for each ξ and can be written in the form

$$\widetilde{g}_n(\theta, \xi_n) = \widetilde{g}_n(\bar{\theta}, \xi_n) + \widetilde{g}'_{n,\theta}(\bar{\theta}, \xi_n)(\theta - \bar{\theta}) + \widetilde{y}_n |\theta - \bar{\theta}|^2, \tag{3.11}$$

where for small $\rho > 0$,

$$\sup_n E|\widetilde{y}_n|^2 I_{\{|\theta_n - \bar{\theta}| \le \rho\}} < \infty. \tag{3.12}$$

Suppose that the sequence of processes defined by

$$W^n(t) = \sum_{i=n}^{m(t_n+t)-1} \sqrt{\epsilon_i}\left[\widetilde{g}_i(\bar{\theta},\xi_i) + \delta M_n(\bar{\theta})\right] \tag{3.13}$$

converges weakly to a Wiener process $W(\cdot)$, where $\delta M_n(\theta)$ is the martingale difference at iterate n if the parameter value θ is used at that time.

For $T > 0$ and $n \le i \le m(t_n + T)$, define the sums

$$\Gamma_i^n = \sum_{j=i}^{m(t_n+T)-1} E_i g_j(\bar{\theta},\xi_j), \qquad \Lambda_i^n = \sum_{j=i}^{m(t_n+T)-1} E_i \widetilde{g}'_{j,\theta}(\bar{\theta},\xi_j), \tag{3.14}$$

and suppose that for each $T > 0$,

$$\begin{aligned} &\sup_n \sup_{n\le i\le m(t_n+T)} E\left[|\Lambda_i^n|^2 + |\Gamma_i^n|^2\right] < \infty, \\ &\sup_n \sup_{n\le i\le m(t_n+T)} E\left[|g_i(\bar{\theta},\xi_i)|^2 + |\widetilde{g}_{i,\theta}(\bar{\theta},\xi_i)|^2 + |\delta M_i(\bar{\theta})|^2\right] < \infty. \end{aligned} \tag{3.15}$$

Theorem 3.3. *Assume the conditions and the algorithm form listed above and let ϵ_n, c_n, α, β, and γ be as in Theorem 3.1. Set $U_n = n^\beta(\theta_n - \bar{\theta})$. Assume* (A2.2), (A2.9), (A2.11), (A2.13), *and* (A3.2). *Then the conclusions of Theorem* 3.1 *hold for* $\{U^n(\cdot), W^n(\cdot)\}$.

Outline of proof. Drop the term $\epsilon_n Z_n$ as in Theorems 2.1 and 3.1, and rewrite the algorithm in expanded form as

$$\begin{aligned} \theta_{n+1} &= \theta_n + \epsilon_n\left[g_n(\bar{\theta},\xi_n) + g'_{n,\theta}(\bar{\theta},\xi_n)(\theta_n - \bar{\theta})\right] \\ &\quad + \epsilon_n y'_n(\theta_n,\xi_n)(\theta_n - \bar{\theta}) - \epsilon_n B(\theta_n)c_n^2 + \epsilon_n o(c_n^2) + \frac{\epsilon_n}{2c_n}\delta M_n \\ &\quad + \frac{\epsilon_n}{2c_n}\left[\widetilde{g}_n(\bar{\theta},\xi_n) + \widetilde{g}'_{n,\theta}(\bar{\theta},\xi_n)(\theta_n - \bar{\theta}) + \widetilde{y}_n|\theta_n - \bar{\theta}|^2\right], \end{aligned} \tag{3.16}$$

where $y_n(\cdot)$ is defined in (A2.11). Subtracting $\bar{\theta}$ from both sides of (3.16), multiplying both sides by $(n+1)^\beta = n^\beta(1 + \beta/n + o(1/n))$, and dropping the clearly negligible terms (while retaining the less obviously negligible terms) yields

$$\begin{aligned} U_{n+1} &= \left(1 + \frac{\beta}{n} + o(\epsilon_n)\right) U_n + \epsilon_n A U_n - \epsilon_n c^2 B(\bar{\theta}) \\ &\quad + \frac{\sqrt{\epsilon_n}}{2c}\widetilde{g}_n(\bar{\theta},\xi_n) + \frac{\sqrt{\epsilon_n}}{2c}\delta M_n(\bar{\theta}) \\ &\quad + \frac{\sqrt{\epsilon_n}}{2c}\left[\delta M_n - \delta M_n(\bar{\theta})\right] + \epsilon_n\left[g'_{n,\theta}(\bar{\theta},\xi_n) - A\right] U_n \\ &\quad + \frac{O(\epsilon_n)}{n^{\gamma/6}}\widetilde{y}_n|U_n|^2 + \epsilon_n y'_n(\theta_n,\xi_n)U_n \\ &\quad + \frac{\sqrt{\epsilon_n}}{n^{\gamma/6}} g_n(\bar{\theta},\xi_n) + \frac{\sqrt{\epsilon_n}}{2cn^{\gamma/3}}\widetilde{g}'_{n,\theta}(\bar{\theta},\xi_n)U_n. \end{aligned} \tag{3.17}$$

In proving tightness and weak convergence, one uses the M-truncated iterate U_n^M as in Theorems 1.1 and 2.1. It can then be shown that the terms in the third and fourth lines contribute nothing to the limit, and are not an impediment to proving tightness. The main difficulty comes from the terms on the last line. One needs to show that they contribute nothing to the limit. In other words, one needs to show that the processes

$$\begin{aligned} G^n(t) &= \sum_{i=n}^{m(t_n+t)-1} \frac{\sqrt{\epsilon_i}}{i^{\gamma/6}} g_i(\bar{\theta}, \xi_i), \\ H^n(t) &= \sum_{i=n}^{m(t_n+t)-1} \frac{\sqrt{\epsilon_i}}{i^{\gamma/3}} \widetilde{g}'_{i,\theta}(\bar{\theta}, \xi_i) U_i \end{aligned}$$

converge weakly to the "zero process," as $n \to \infty$.

A perturbed state argument can be used for this, analogously to the arguments used in Theorems 8.3.2 and 6.1 in Section 6, although other methods (such as laws of large numbers or rate of growth conditions) might also be used. Define the sums

$$\begin{aligned} G_i^n &= \sum_{j=n}^{i-1} \frac{\sqrt{\epsilon_j}}{j^{\gamma/6}} g_j(\bar{\theta}, \xi_j), \\ H_i^n &= \sum_{j=n}^{i-1} \frac{\sqrt{\epsilon_j}}{j^{\gamma/3}} \widetilde{g}'_{j,\theta}(\bar{\theta}, \xi_j) U_j. \end{aligned}$$

Define the perturbations

$$\delta G_i^n = \frac{\sqrt{\epsilon_i}}{i^{\gamma/6}} \Gamma_i^n, \qquad \delta H_i^n = \frac{\sqrt{\epsilon_i}}{i^{\gamma/3}} \Lambda_i^n U_i,$$

where Γ_i^n and Λ_i^n are given in (3.14), and the perturbed variables

$$\widetilde{G}_i^n = G_i^n + \delta G_i^n, \qquad \widetilde{H}_i^n = H_i^n + \delta H_i^n.$$

The condition (3.15) implies that

$$\lim_n P\left\{ \sup_{n \le i \le m(t_n+T)} |\delta G_i^n| \ge \mu \right\} = 0$$

for any $\mu > 0$ and $T > 0$, and similarly for δH_i^n. Thus, it is enough to show the desired result for the perturbed variables. The quantities

$$\sup_{n \le i \le m(t_n+T)} \left[|\widetilde{G}_i^n| + |\widetilde{H}_i^n| \right]$$

are not *a priori* guaranteed to be bounded in probability uniformly in n for each T. Nevertheless, a truncation argument (as in Theorem 1.1) can be used to circumvent this difficulty, if necessary. Similarly we can truncate U_n where needed.

Now assuming the boundedness of $\widetilde{G}_i^n, \widetilde{H}_i^n$ and U_n, a direct computation using (3.15) yields that, for $n \le i \le m(t_n + T)$,

$$E_i \left[\widetilde{G}_{i+1}^n - \widetilde{G}_i^n \right] = o(\epsilon_i) q_i^n,$$

where $\{|q_i^n|; i, n\}$ is uniformly integrable, with a similar result for $\widetilde{H}_i^n$. Then an application of Theorem 7.4.3 and the arbitrariness of T yields that the piecewise constant interpolations of $\widetilde{G}_i^n$ and $\widetilde{H}_i^n$, with intervals $\{\epsilon_i, i \ge n\}$, converge to the zero process as $n \to \infty$, for each $T > 0$. Hence, $G^n(\cdot)$ and $H^n(\cdot)$ converge to the zero process. With this given, the rest of the proof is the same as that of Theorem 2.2. □

Random bias terms. In Theorem 3.3, we assumed that the functions $B_n(\cdot)$ are not random. This can easily be weakened. Replace $B_n(\theta_n)$ with $B_n(\theta_n, \xi_n)$, and suppose that the set

$$\{B_n(\theta_n, \xi_n), B_n(\bar{\theta}, \xi_n)\}$$

is uniformly integrable. Let

$$\lim_n E|B_n(\theta_n, \xi_n) - B_n(\bar{\theta}, \xi_n)| = 0,$$

and suppose that there is a $B(\bar{\theta})$ such that, as $n, m \to \infty$,

$$\frac{1}{m} \sum_{i=n}^{n+m-1} E_n \left[B_i(\bar{\theta}, \xi_i) - B(\bar{\theta}) \right] \to 0,$$

in probability. Then Theorem 3.3 continues to hold.

10.4 Tightness of the Normalized Iterates: Decreasing Step Size, W.P.1 Convergence

In this section, the tightness (A2.3) will be proved for algorithm (2.1) under the assumption that $\theta_n \to \bar{\theta}$ with probability one. The proofs are of the stability type and use various forms of "local" Liapunov functions. To facilitate the understanding of the basic ideas, we start with the simple case where the noise is just a martingale difference.

10.4.1 Martingale Difference Noise: Robbins–Monro Algorithm

The following assumptions will be used.

(A4.1) There is a $\bar{\theta}$ in the interior of the constraint set H such that $\theta_n \to \bar{\theta}$ with probability one.

(A4.2) For small $\rho > 0$, $\{Y_n I_{\{|\theta_n - \bar{\theta}| \leq \rho\}}\}$ is uniformly integrable, and there is a function $\bar{g}(\cdot)$ such that for $|\theta_n - \bar{\theta}| \leq \rho$

$$E_n Y_n = \bar{g}(\theta_n). \tag{4.1}$$

(A4.3) There is a $0 < K < \infty$ such that for small $\rho > 0$,

$$\sup_n E_n |Y_n|^2 I_{\{|\theta_n - \bar{\theta}| \leq \rho\}} \leq K \text{ w.p.1.} \tag{4.2}$$

(A4.4) There is a Hurwitz matrix A such that

$$\bar{g}(\theta) = A(\theta - \bar{\theta}) + o(|\theta - \bar{\theta}|). \tag{4.3}$$

Remark. By the stability theory of ordinary differential equations, for any symmetric, positive definite matrix C and Hurwitz matrix A, the Liapunov equation

$$A'P + PA = -C \tag{4.4}$$

has a unique symmetric and positive definite solution P. Let the positive definite symmetric pair (P, C) satisfy (4.4), and let λ denote the largest positive number such that $C \geq \lambda P$. The proof of the next theorem uses the Liapunov function $V(\theta) = (\theta - \bar{\theta})' P (\theta - \bar{\theta})$. Finally, a condition on the step size sequence is needed.

(A4.5) Either

$$\epsilon_n = 1/n \text{ and } \lambda > 1, \tag{4.5a}$$

or $\epsilon_n \to 0$, and for each $T > 0$

$$\liminf_n \min_{n \geq i \geq m(t_n - T)} \frac{\epsilon_n}{\epsilon_i} = 1. \tag{4.5b}$$

Remarks on the conditions. (A4.2) will be relaxed in the next subsection. Suppose that there is a real-valued twice continuously differentiable function $f(\cdot)$ such that $\bar{g}(\theta) = -f_\theta(\theta)$ and that the Hessian matrix $A = -f_{\theta\theta}(\bar{\theta})$ is negative definite. Letting $P = I$, the identity matrix, we have $C = -(A' + A) = -2A$. Then the condition on $\lambda > 1$ in (4.5a) is just that the smallest eigenvalue of $-A$ is larger than $1/2$. If $\epsilon_n = K/n$ for a positive definite matrix K (that is absorbed into $\bar{g}(\cdot)$), then $\lambda > 1$ for large enough K. The requirement that $\lambda > 1$ in (4.5a) is consistent with the requirement in (A2.6) that $(A + I/2)$ be a Hurwitz matrix, when $\epsilon_n = 1/n$. Condition (4.5b) holds if $\epsilon_n = 1/n^\gamma$, for $\gamma \in (0, 1)$.

Theorem 4.1. *The tightness condition* (A2.3) *holds under* (A4.1)–(A4.5).

Proof. Since $\theta_n \to \bar{\theta}$, and $\bar{\theta}$ is in the interior of H, we can suppose that $Z_n = 0$ without loss of generality. Let $\bar{\theta} = 0$ for notational simplicity. To prove the tightness we need to verify (7.3.3b) for $\theta_n/\sqrt{\epsilon_n}$ replacing A_n; namely, that for each small $\mu > 0$, there are finite M_μ and K_μ such that

$$P\left\{\frac{|\theta_n|}{\sqrt{\epsilon_n}} \geq K_\mu\right\} \leq \mu, \quad \text{for } n \geq M_\mu.$$

Let $\rho > 0$ be small enough. Since $\theta_n \to \bar{\theta} = 0$ with probability one, given any small $\nu > 0$, there is an $N_{\nu,\rho}$ such that $|\theta_n| \leq \rho$ for $n \geq N_{\nu,\rho}$ with probability $\geq 1 - \nu$. By modifying the processes on a set of probability at most ν, we can suppose that $|\theta_n| \leq \rho$ for $n \geq N_{\nu,\rho}$ and that all the assumptions continue to hold. Denote the modified process by $\{\theta_n^\nu\}$. If $\{\theta_n^\nu/\sqrt{\epsilon_n}\}$ is shown to be tight for each $\rho > 0, \mu > 0$, then the original sequence $\{U_n\}$ is tight. Thus for the purposes of the tightness proof, and by shifting the time origin if needed, it can be supposed without loss of generality that $|\theta_n| \leq \rho$ for all n for the *original process,* where $\rho > 0$ is arbitrarily small.

Owing to the quadratic form of $V(\theta) = \theta' P\theta$ and (A4.2),

$$E_n V(\theta_{n+1}) - V(\theta_n) = 2\epsilon_n \theta_n' P\bar{g}(\theta_n) + O(\epsilon_n^2) E_n |Y_n|^2. \tag{4.6}$$

Using the expansion (4.3), (A4.3), and the fact that we can suppose that $|\theta_n| \leq \rho$ for arbitrarily small $\rho > 0$, it is seen that the right side of (4.6) is

$$2\epsilon_n \theta_n' PA\theta_n + O(\epsilon_n)|\theta_n| o(|\theta_n|) + O(\epsilon_n^2).$$

Noting that $\theta' PA\theta = \theta' A' P\theta$ and $2\theta' PA\theta = \theta'(PA + A'P)\theta = -\theta' C\theta \leq -\lambda\theta' P\theta$, we have

$$E_n V(\theta_{n+1}) - V(\theta_n) \leq -\lambda_1 \epsilon_n V(\theta_n) + O(\epsilon_n^2),$$

where $0 < \lambda_1 < \lambda$. By taking ρ small enough, it can be supposed that λ_1 is arbitrarily close to λ. Thus, there is a real number K_1 such that for all $m \geq 0$,

$$EV(\theta_{m+1}) \leq \prod_{i=0}^{m} (1 - \lambda_1 \epsilon_i) EV(\theta_0) + K_1 \sum_{i=0}^{m} \prod_{j=i+1}^{m} (1 - \lambda_1 \epsilon_j) \epsilon_i^2. \tag{4.7}$$

We need to show that the right side of (4.7) is of the order of ϵ_m for $m \geq 0$. It will be shown that each of the two terms on the right side of (4.7) is of the correct order. To simplify the notation in the development, suppose (without loss of generality, since we can shift the time origin) that $\lambda_1 \epsilon_i < 1$ for all i. For the first term on the right of (4.7), it is sufficient to work with the approximation

$$\exp\left[-\lambda_1 \sum_{i=0}^{m-1} \epsilon_i\right] = e^{-\lambda_1 t_m}$$

and to show that it is $O(\epsilon_m)$. Equivalently, we can let m in (4.7) be $m = m(\tau) - 1$, work with

$$\prod_{i=0}^{m(\tau)-1} (1 - \lambda_1 \epsilon_i),$$

and show that its exponential approximation satisfies

$$e^{-\lambda_1 t_{m(\tau)}} = e^{-\lambda_1 \tau} = O(\epsilon_{m(\tau)}). \tag{4.8}$$

By the order relation (4.8), we mean that the ratio of the left side to $\epsilon_{m(\tau)}$ is bounded in τ.

For $m = m(\tau) - 1$, the second term on the right side of (4.7) can be approximated by the left side of (4.9), and it is sufficient to show that

$$\int_0^\tau e^{-\lambda_1(\tau - s)} \epsilon_{m(s)} ds = O(\epsilon_{m(\tau)}). \tag{4.9}$$

Consider case (4.5a), under which $\epsilon_{m(\tau)}$ is of the order of $e^{-\tau}$, and we can suppose that $\lambda_1 > 1$. Then, using $\lambda_1 > 1$, the left side of (4.8) is $o(\epsilon_{m(\tau)})$ and the left side of (4.9) is $O(\epsilon_{m(\tau)})$.

Now, consider case (4.5b), fix $T > 0$, and (without loss of generality) work with τ that are integral multiples of T. Write the left side of (4.9) as

$$\sum_{k=0}^{\tau/T-1} e^{-k\lambda_1 T} \int_{\tau-kT-T}^{\tau-kT} e^{-\lambda_1(\tau-kT-s)} \epsilon_{m(s)} ds. \tag{4.10}$$

Given small $\kappa > 0$ satisfying $(1 + \kappa) < e^{\lambda_1 T}$, suppose that the time origin is shifted enough so that we can suppose that (A4.5b) implies that

$$\min_{n \geq i \geq m(t_n - T)} \frac{\epsilon_n}{\epsilon_i} \geq \frac{1}{1 + \kappa}$$

for all n. Then

$$\epsilon_i \leq \epsilon_{m(\tau)}(1 + \kappa)^k, \text{ for } \tau - t_i \in [kT - T, kT]. \tag{4.11}$$

Using (4.11) and successively approximating the right-hand side of (4.10), we obtain the following upper bound:

$$\frac{1}{\lambda_1} \sum_{k=0}^{\tau/T} e^{-\lambda_1 kT} (1 + \kappa)^{k+1} \epsilon_{m(\tau)}.$$

This is of the order of $\epsilon_{m(\tau)}$, since $(1 + \kappa) < e^{\lambda_1 T}$.

The demonstration that the left side of (4.8) is $O(\epsilon_{m(\tau)})$ uses the fact that $\epsilon_{m(\tau)} \geq O(1)/(1 + \kappa)^{\tau/T} \geq O(1)e^{-\lambda_1 \tau}$. □

10.4.2 Correlated Noise

In all the correlated noise cases, E_n, $\mathcal{F}_n$, and ξ_n are as defined in Subsection 8.2.3 (equivalently, in (A2.8)). We will use the following assumptions.

(A4.6) For small $\rho > 0$, $\{Y_i I_{\{|\theta_n - \bar{\theta}_n| \leq \rho\}}\}$ is uniformly integrable, and there are measurable functions $g_n(\cdot)$ such that, for $|\theta_n - \bar{\theta}_n| \leq \rho$,

$$E_n Y_n = g_n(\theta_n, \xi_n). \tag{4.12}$$

(A4.7) $g_n(\theta, \xi)$ can be expanded as

$$g_n(\theta, \xi) = g_n(\bar{\theta}, \xi) + g'_{n,\theta}(\bar{\theta}, \xi)(\theta - \bar{\theta}) + y'_n(\theta, \xi)(\theta - \bar{\theta}), \tag{4.13}$$

where

$$y_n(\theta, \xi) = \int_0^1 \left[g_{n,\theta}(\bar{\theta} + s(\theta - \bar{\theta}), \xi) - g_{n,\theta}(\bar{\theta}, \xi)\right] ds, \tag{4.14}$$

and

$$y_n(\theta, \xi) \to 0 \tag{4.15}$$

uniformly in (n, ξ) as $\theta - \bar{\theta} \to 0$.

(A4.8) There is a $0 < K < \infty$ such that for small $\rho > 0$,

$$E|Y_n|^2 I_{\{|\theta_n - \bar{\theta}| \leq \rho\}} \leq K < \infty. \tag{4.16}$$

Note that (4.16) uses E instead of E_n as in (A4.3). Recall the definition $\Pi(n, i) = \prod_{j=n}^{i}(1 - \epsilon_j)$. Define the "discounted" sequences

$$\Lambda_n^d = \sum_{i=n}^{\infty} \epsilon_i \Pi(n+1, i) E_n g_i(\bar{\theta}, \xi_i), \tag{4.17}$$

$$\Gamma_n^d = \sum_{i=n}^{\infty} \epsilon_i \Pi(n+1, i) E_n \left[g'_{i,\theta}(\bar{\theta}, \xi_i) - A\right], \tag{4.18}$$

where A is a Hurwitz matrix, which plays the same role as the matrix A in Theorem 4.1.

(A4.9) For small $\rho > 0$,

$$\sup_n E|\Lambda_{n+1}^d| \left(1 + |Y_n|\right) I_{\{|\theta_n - \bar{\theta}| \leq \rho\}} / \epsilon_n < \infty, \tag{4.19}$$

$$\sup_n E|\Gamma_{n+1}^d| \left(1 + |Y_n| + |Y_n|^2\right) I_{\{|\theta_n - \bar{\theta}| \leq \rho\}} / \epsilon_n < \infty. \tag{4.20}$$

Remark on (4.15). Condition (4.15) will hold for the "linear" or "bilinear" algorithms of Chapter 3, where $y_n(\theta, \xi) = 0$. It will also hold if the functions $g_n(\cdot)$ have uniformly (in n, ξ) bounded second-order θ-derivatives in a neighborhood of $\bar{\theta}$, or if the derivatives $g_{n,\theta}(\cdot,\xi)$ are continuous in θ in a small neighborhood of $\bar{\theta}$, uniformly in ξ. The condition also holds if the noise is only a martingale difference, since then ξ does not appear in (4.12) and $g_{n,\theta}(\cdot)$ is continuous in a neighborhood of $\bar{\theta}$. Thus, Theorem 4.2 generalizes Theorem 4.1 for the martingale difference noise case.

Theorem 4.2. *Let P, C be symmetric, positive definite matrices satisfying* (4.4) *for the matrix A of* (4.18), *and let λ be the largest positive number such that $C \geq \lambda P$. Assume* (A4.1) *and* (A4.5)–(A4.9). *Then* (A2.3) *holds.*

Proof. Let $\bar{\theta} = 0$ for notational simplicity. Again, without loss of generality, we can suppose that $Z_n = 0$. A perturbed Liapunov function method will be used. Using the Liapunov function $V(\theta) = \theta' P \theta$ yields

$$E_n V(\theta_{n+1}) - V(\theta_n) = 2\epsilon_n \theta_n' P g_n(\theta_n, \xi_n) + O(\epsilon_n^2) E_n |Y_n|^2.$$

As in the previous proof, for any small $\rho > 0$ we can suppose without loss of generality that $|\theta_n| \leq \rho$ for all n, and shift the origin such that $\epsilon_n \ll 1$ for all n. Rewrite the right side as

$$\begin{aligned} 2\epsilon_n \theta_n' P A \theta_n &+ O(\epsilon_n^2) E_n |Y_n|^2 + O(\epsilon_n) |\theta_n|^2 |y_n(\theta_n, \xi_n)| \\ &+ 2\epsilon_n \theta_n' P \left[g_{n,\theta}'(0, \xi_n) - A \right] \theta_n + 2\epsilon_n \theta_n' P g_n(0, \xi_n). \end{aligned} \tag{4.21}$$

Given $\rho > 0$, define ρ_1 by

$$\sup_{n,\omega,|\theta| \leq \rho} |y_n(\theta, \xi_n)| = \rho_1.$$

By (4.15), $\rho_1 \to 0$ as $\rho \to 0$. Thus, the term $O(\epsilon_n)|\theta_n|^2 |y_n(\theta_n, \xi_n)|$ in (4.21) is bounded above by $O(\epsilon_n)\rho_1 |\theta_n|^2$, where ρ_1 can be assumed to be as small as needed.

The second term of (4.21) will be easy to handle. The last two terms are more difficult to treat due to the correlation in the noise. They will be treated with the perturbed Liapunov function defined by

$$V_n(\theta_n) = V(\theta_n) + 2\theta_n' P \Gamma_n^d \theta_n + 2\theta_n' P \Lambda_n^d = V(\theta_n) + \delta V_n(\theta_n).$$

The perturbation is small in the sense of (4.19) and (4.20). The perturbation component $2\theta_n' P \Gamma_n^d \theta_n$ is used to cancel the term

$$2\epsilon_n \theta_n' P \left[g_{n,\theta}'(0, \xi_n) - A \right] \theta_n$$

in (4.21) and the component $2\theta_n' P \Lambda_n^d$ is used to cancel the term $2\epsilon_n \theta_n' P g_n(0, \xi_n)$.

By expanding $E_n \delta V_{n+1}(\theta_{n+1}) - \delta V_n(\theta_n)$, the negative of the last two terms of (4.21) appear. Thus, by expanding $E_n V_{n+1}(\theta_{n+1}) - V_n(\theta_n)$ and canceling terms where possible, we can write

$$E_n V_{n+1}(\theta_{n+1}) - V_n(\theta_n) = 2\epsilon_n \theta_n' P A \theta_n + \text{ error term}, \tag{4.22}$$

where the error term can be written as

$$
\begin{aligned}
&O(\epsilon_n^2)E_n|Y_n|^2 + \rho_1 O(\epsilon_n)|\theta_n|^2 \\
&\quad + O(\epsilon_{n+1})|\theta_n|^2|E_n\Gamma_{n+1}^d| + E_n\left|\Gamma_{n+1}^d\right|\left[O(\epsilon_n)|\theta_n||Y_n| + O(\epsilon_n^2)|Y_n|^2\right] \\
&\quad + O(\epsilon_{n+1})|E_n\Lambda_{n+1}^d||\theta_n| + E_nO(\epsilon_n)|Y_n||\Lambda_{n+1}^d|.
\end{aligned}
$$

Now taking expectations, bound the first component of the error term by using (4.16); bound the second component by using the fact that ρ_1 is arbitrarily small; bound the last two lines by using (4.19) and (4.20) and the assumed small bounds on the θ_n. The bounds and (4.4) yield

$$
EV_{n+1}(\theta_{n+1}) - EV_n(\theta_n) \le -\epsilon_n E\theta_n' C\theta_n + \lambda_p\epsilon_n E|\theta_n|^2 + O(\epsilon_n^2), \tag{4.23}
$$

where $\lambda_p \to 0$ as $\rho \to 0$. Thus, for $\lambda_1 < \lambda$ but arbitrarily close to it, we can write

$$
EV_{n+1}(\theta_{n+1}) \le (1 - \epsilon_n\lambda_1)EV_n(\theta_n) + O(\epsilon_n^2). \tag{4.24}
$$

With (4.24) in hand, the proof in Theorem 4.1 implies that $EV_n(\theta_n) \le O(\epsilon_n)$ for large n. Thus, for large n

$$
EV(\theta_n) + 2E\left[\theta_n' P\Gamma_n^d\theta_n + \theta_n' P\Lambda_n^d\right] \le O(\epsilon_n). \tag{4.25}
$$

Using the fact that $|\theta_n| \le \rho$, where ρ is small, (4.19) and (4.20) imply that $E|\theta_n' P\Gamma_n^d\theta_n + \theta_n' P\Lambda_n^d| = O(\epsilon_n)$. Thus, (4.25) yields that $EV(\theta_n) = E\theta_n' P\theta_n = O(\epsilon_n)$, which implies (A2.3). □

10.4.3 Kiefer–Wolfowitz Algorithm

Martingale difference noise. The algorithm (3.1) (equivalently, (3.4)) will be used. The result for the one-sided difference procedure of Theorem 3.2 is proved in a similar way. The following assumptions will be needed. If $\gamma_n(\bar\theta)$ does not depend on n and $\gamma_\theta(\bar\theta)$ is a Hurwitz matrix, then Γ_n^d and Λ_n^d are not needed.

(A4.10) $\epsilon_n = 1/n^\gamma$, $c_n = c/n^\alpha$, $0 < \gamma \le 1$. Let α, β, and γ satisfy $\beta = 2\alpha$, $\beta + \alpha = \gamma/2$.

(A4.11) For small $\rho > 0$,

$$
\sup_n E\left[|B_n(\theta_n)| + |Y_n^\pm|^2\right] I_{\{|\theta_n - \bar\theta| \le \rho\}} < \infty.
$$

(A4.12) There is a Hurwitz matrix A such that

$$
\Gamma_n^d = -\sum_{i=n}^{\infty} \epsilon_i \Pi(n+1, i)\left[\gamma_{i,\theta}'(\bar\theta) + A\right] = O(\epsilon_n).
$$

If $\gamma = 1$, then $A + \beta I$ is also a Hurwitz matrix. In addition,

$$\Lambda_n^d = -\sum_{i=n}^{\infty} \epsilon_i^{2/3} \Pi(n+1, i) \gamma_i(\bar{\theta}) = O(\epsilon_n^{2/3}).$$

Theorem 4.3. *Assume* (A3.3), *the expansion of* $\gamma_n(\cdot)$ *and the boundedness of the terms in* (A3.5), (A4.1), *and* (A4.10)–(A4.12). *Then* $\{U_n\}$ *is tight, where* $U_n = n^{\beta}(\theta_n - \bar{\theta})$.

Proof. We work with $\gamma < 1$; the proof for $\gamma = 1$ is similar. The proof is similar to that of Theorem 4.2, except that it is more convenient to work with U_n instead of $\theta_n - \bar{\theta}$. As in Theorem 4.2, we can suppose that $|\theta_n - \bar{\theta}|$ is as small as desired. Dropping the Z_n term as in Theorems 4.1 and 4.2, rewrite (3.5) as

$$\begin{aligned} U_{n+1} &= \left(1 + \frac{\beta}{n} + o(\epsilon_n)\right) U_n + \epsilon_n A U_n \\ &\quad - \epsilon_n \left[\gamma'_{n,\theta}(\bar{\theta}) + A\right] U_n + O(\epsilon_n) + \epsilon_n \nu_n(|U_n|) \\ &\quad + o(\epsilon_n) + \sqrt{\epsilon_n} q_n \delta M_n - \epsilon_n n^{\beta} \gamma_n(\bar{\theta}), \end{aligned} \tag{4.26}$$

where $\{q_n\}$ is a bounded sequence of real numbers and $\nu_n = o(|\theta_n - \bar{\theta}|)/|\theta_n - \bar{\theta}|$ and can be assumed to be as small as desired.

If $\gamma_{n,\theta}(\bar{\theta})$ does not depend on n, then it must equal $-A$ due to the fact that $\gamma_{n,\theta}(\bar{\theta}) = \gamma_{\theta}(\bar{\theta})$ is assumed to be Hurwitz. If in addition $\gamma(\bar{\theta}) = 0$ a straightforward estimation using the Liapunov function $V(U) = U'PU$ could be used. Owing to the allowed n-dependence, the perturbed Liapunov function $V_n(U_n) = V(U_n) + 2U_n' P \Gamma_n^d U_n + 2U_n P \Lambda_n^d$ will be used to the same end. Working with the form of U_n given by (4.26) directly, one shows that

$$EV_{n+1}(U_{n+1}) \le (1 - \epsilon_n \lambda_1) EV_n(U_n) + O(\epsilon_n),$$

where $0 < \lambda_1 < \lambda$ but is arbitrarily close to it for large n. This yields that $EV_n(U_n) \le O(1)$, analogously to (but simpler than) what was done in the proof of Theorem 4.2. Then the facts that $U_n' P \Gamma_n^d U_n = O(\epsilon_n)|U_n|^2$ and that $|U_n' P \Lambda_n^d| = O(\epsilon_n^{2/3})(1 + |U_n|^2)$ yield that $EV(U_n) = O(1)$. The details are left to the reader. □

10.5 Tightness of the Normalized Iterates: Weak Convergence

10.5.1 Unconstrained Algorithm

In Section 4 it was supposed that $\theta_n \to \bar{\theta}$ with probability one. Since this assumption (and a shifting of the time origin) allowed us to suppose that

$|\theta_n - \bar{\theta}|$ is small for all n, a "local" analysis was possible. When we do not have probability one convergence, the local Liapunov function arguments of Section 4 cannot be used and a more global stability argument is required. This is simplest when there is no constraint set H, and the algorithm is

$$\theta_{n+1} = \theta_n + \epsilon_n Y_n. \tag{5.1}$$

The cases $\epsilon_n \to 0$ and $\epsilon_n \equiv \epsilon$ will be dealt with simultaneously. The following assumptions will be needed. The sequence ξ_n and the notation E_n are as defined in Section 6.1 or Subsection 8.3.2. The following conditions and (A4.5) will be used, where K_i are arbitrary positive numbers whose values can change from usage to usage. The conditions simplify for the martingale difference noise case, since then $g_n(\cdot)$ depends only on θ. In this case, if $g_n(\theta) = \bar{g}(\theta)$, then the $\Gamma_n^d(\theta)$ in (5.2) is zero.

(A5.1) $\bar{\theta}$ is a globally asymptotically stable (in the sense of Liapunov) point of the ODE $\dot{\theta} = \bar{g}(\theta)$.

(A5.2) The non-negative and continuously differentiable function $V(\cdot)$ is a Liapunov function for the ODE. The mixed second-order partial derivatives are bounded and $|V_\theta(\theta)|^2 \le K_1(V(\theta)+1)$.

(A5.3) There is a $\lambda > 0$ such that

$$V_\theta'(\theta)\bar{g}(\theta) \le -\lambda V(\theta).$$

If $\epsilon_n = 1/n$, then $\lambda > 1$.

(A5.4) For each $K > 0$,

$$\sup_n E|Y_n|^2 I_{\{|\theta_n - \bar{\theta}| \le K\}} \le K_1 E(V(\theta_n)+1),$$

where K_1 does not depend on K.

(A5.5) There are measurable functions $g_n(\cdot)$ such that $E_n Y_n = g_n(\theta_n, \xi_n)$.

(A5.6) The sum

$$\Gamma_n^d(\theta) = \sum_{i=n}^{\infty} \epsilon_i \Pi(n+1,i) E_n \left[g_i(\theta,\xi_i) - \bar{g}(\theta)\right] \tag{5.2}$$

is well defined in that the sum of the norms of the summands is integrable for each θ, and

$$E|\Gamma_n^d(\theta_n)|^2 = O(\epsilon_n^2)\,(EV(\theta_n)+1). \tag{5.3}$$

(A5.7) $E\left|\Gamma_{n+1}^d(\theta_{n+1}) - \Gamma_{n+1}^d(\theta_n)\right|^2 = O(\epsilon_n^4)\,(EV(\theta_n)+1).$

Theorem 5.1. *Assume algorithm* (5.1), *and conditions* (A4.5) *and* (A5.1)–(A5.7). *Let there be a symmetric and positive definite matrix P such that $V(\cdot)$ satisfies*

$$V(\theta) = (\theta - \bar{\theta})' P (\theta - \bar{\theta}) + o(|\theta - \bar{\theta}|^2) \tag{5.4}$$

for small $|\theta - \bar{\theta}|$. *Then* $\{(\theta_n - \bar{\theta})/\sqrt{\epsilon_n}\}$ *is tight. If there is a symmetric and positive definite* P_1 *such that for all* θ,

$$V(\theta) \geq (\theta - \bar{\theta})' P_1 (\theta - \bar{\theta}), \tag{5.5}$$

then

$$\sup_n E \frac{|\theta_n - \bar{\theta}|^2}{\epsilon_n} < \infty. \tag{5.6}$$

Now let $\epsilon_n = \epsilon$, *redefine* $\Gamma_n^d(\theta)$ *using* $\epsilon_n = \epsilon$, *and let* θ_n^ϵ *and* ξ_n^ϵ *be used in the assumptions. Suppose that* $\{\theta^\epsilon(\cdot)\}$ *converges weakly (perhaps after shifting the time origins) to the process with constant value* $\bar{\theta}$. *Then, under* (5.4), *there are* $n_\epsilon < \infty$ *such that* $\{(\theta_n^\epsilon - \bar{\theta})/\sqrt{\epsilon}; n \geq n_\epsilon, \epsilon\}$ *is tight, and under* (5.5),

$$\limsup_{\epsilon \to 0} \sup_{n \geq n_\epsilon} E \frac{|\theta_n^\epsilon - \bar{\theta}|^2}{\epsilon} < \infty. \tag{5.7}$$

In fact, n_ϵ *satisfies* $e^{-\lambda_1 \epsilon n_\epsilon} |\theta_0 - \bar{\theta}| = O(\sqrt{\epsilon})$, *where* $0 < \lambda_1 < \lambda$ *but is arbitrarily close to it.*

Comment. In the proof, (5.3) and (A5.7) are used to bound $EV_\theta'(\theta_n)\Gamma_{n+1}^d(\theta_n)$ and $EV_\theta'(\theta_n)[\Gamma_{n+1}^d(\theta_{n+1}) - \Gamma_{n+1}^d(\theta_n)]$ via the Schwarz inequality. But other conditions can be used for this, depending on the form of $\Gamma_n^d(\theta)$.

Proof. [See the proof of Theorems 6.7.3 and 4.2 for related calculations.] The proof for $\epsilon_n \to 0$ will be given. The proof for $\epsilon_n = \epsilon$ follows along the same lines and is slightly simpler. We can write

$$E_n V(\theta_{n+1}) - V(\theta_n) = \epsilon_n V_\theta'(\theta_n) g_n(\theta_n, \xi_n) + O(\epsilon_n^2) E_n |Y_n|^2. \tag{5.8}$$

A perturbed Liapunov function method will be used to help average out the first term on the right. Define the perturbed Liapunov function

$$V_n(\theta_n) = V(\theta_n) + V_\theta'(\theta_n)\Gamma_n^d(\theta_n).$$

It is easy to see that

$$\begin{aligned} &E_n V_\theta'(\theta_{n+1})\Gamma_{n+1}^d(\theta_{n+1}) - V_\theta'(\theta_n)\Gamma_n^d(\theta_n) \\ &\quad = -\epsilon_n V_\theta'(\theta_n)\left[g_n(\theta_n, \xi_n) - \bar{g}(\theta_n)\right] \\ &\qquad + E_n V_\theta'(\theta_{n+1})\Gamma_{n+1}^d(\theta_{n+1}) - E_n V_\theta'(\theta_n)\Gamma_{n+1}^d(\theta_n) \\ &\qquad + \epsilon_{n+1} V_\theta'(\theta_n)\Gamma_{n+1}^d(\theta_n). \end{aligned} \tag{5.9}$$

Using (5.8), (5.9), (A5.2), (A5.4), and (A5.6)–(A5.7) yields

$$EV_{n+1}(\theta_{n+1}) - EV_n(\theta_n) \leq -\epsilon_n \lambda_1 EV(\theta_n) + O(\epsilon_n^2) \tag{5.10}$$

for large n, where $\lambda_1 < \lambda$ but is arbitrarily close to it.

The bound on $\Gamma_n^d(\theta)$ of (5.3) and the bound in (A5.2) yield that $EV_\theta'(\theta_n)\Gamma_n^d(\theta_n) = O(\epsilon_n)E(V(\theta_n)+1)$. Thus,

$$EV_{n+1}(\theta_{n+1}) - EV_n(\theta_n) \le -\epsilon_n\lambda_1 EV_n(\theta_n) + O(\epsilon_n^2)$$

for large n and a new $\lambda_1 > 0$ arbitrarily close to the old one. Equivalently,

$$EV_{n+1}(\theta_{n+1}) \le (1-\epsilon_n\lambda_1)EV_n(\theta_n) + O(\epsilon_n^2). \tag{5.11}$$

By (5.11), (A4.5), and the proof of Theorem 4.1,

$$EV_n(\theta_n) = O(\epsilon_n).$$

Using the bound on $\Gamma_n^d(\theta)$ in terms of $V(\theta_n)$ in (5.3) yields

$$EV(\theta_n) = O(\epsilon_n). \tag{5.12}$$

Thus, there is a $K_2 > 0$ such that for any $K > 0$ and all n,

$$P\left\{\frac{V(\theta_n)}{\epsilon_n} \ge K\right\} \le \frac{K_2}{K},$$

which implies that $\{V(\theta_n)/\epsilon_n\}$ is tight. The tightness of the normalized iterate sequence follows from the local quadratic structure (5.4). The result (5.6) follows from (5.5) and (5.12). □

The constrained algorithm. The method of proof of Theorem 5.1 extends directly to the constrained algorithm if the projection reduces the error measured by the value of the Liapunov function (see Theorem 5.2 in what follows). The proof of Theorem 5.2 follows from the proof of Theorem 5.1. The various growth conditions (for large $|\theta|$) that were used in Theorem 5.1 are not needed because H is bounded. Analogs of Theorems 4.1 to 4.3 can also be proved, whether or not the algorithm is constrained. When the projection does not reduce the error, then the analysis is much more difficult since one must construct a Liapunov function that takes the boundary into account; see Section 9 and [39].

Theorem 5.2. *Assume the constrained algorithm with constraint set H, where $\bar{\theta} \in H^0$ and H satisfies condition* (A4.3.2). *Assume that conditions of Theorem* 5.1 *hold in H, except that* (5.3) *and* (A5.7) *are changed to*

$$\begin{aligned} &E|\Gamma_n^d(\theta_n)|^2 = O(\epsilon_n^2), \\ &E|\Gamma_{n+1}^d(\theta_{n+1}) - \Gamma_{n+1}^d(\theta_n)| = O(\epsilon_n^2). \end{aligned}$$

Let the projection onto H reduce the value of $V(\cdot)$ in the sense that $V(\Pi_H(\theta)) \le V(\theta)$ for all θ. Then the conclusions of Theorem 5.1 *hold.*

10.5.2 Local Methods for Proving Tightness

The proof of Theorem 5.1 is based on a "global" stability argument, and does not use any other information on the asymptotic properties of the θ_n or θ_n^ϵ sequences. At the other extreme, the method of proof of Theorem 4.1 is "purely" local. It depends on the fact that $\theta_i - \bar{\theta}$ is arbitrarily small (with an arbitrarily high probability) for $i \in [n, \infty)$ for large enough n, which is guaranteed by the probability one convergence assumed in that theorem. Probability one convergence is not actually needed to do a local analysis; it is enough that $|\theta_n - \bar{\theta}|$ (or $|\theta_n^\epsilon - \bar{\theta}|$) be small for a long enough time interval with a high probability. Let us examine the proof of Theorem 4.1. The inequality (4.7) contains two terms. The second is due to the noise and other "perturbations," and the first is due to the initial condition. It was seen that it is easy to show that the second term is of the correct order. The first term was more difficult to treat, since the initial time in the analysis is arbitrary.

With an eye to getting analogs of the local methods of Theorems 4.1–4.3 for the weak convergence case, consider the following way of dealing with the first term for arbitrary initial time. In what follows $0 < \mu < \mu_1$ are arbitrarily small numbers. Suppose that

$$\lim_n P\left\{|\theta_n - \bar{\theta}| \geq \mu\right\} = 0. \tag{5.13}$$

Suppose for the moment that for each large n there is $\nu_n > 0$ such that $n - \nu_n \to \infty$ and

$$\prod_{i=n-\nu_n}^{n} (1 - \lambda_1 \epsilon_i) = O(\epsilon_n), \tag{5.14}$$

$$\lim_n P\left\{\sup_{\nu_n - n \leq i \leq n} |\theta_i - \bar{\theta}| \geq \mu_1\right\} = 0. \tag{5.15}$$

We know that (5.13) holds if there is weak convergence to a limit process that is concentrated at $\bar{\theta}$. Thus if one could verify (5.14) and (5.15), the proof of Theorem 4.1 implies that $\{(\theta_n - \bar{\theta})/\sqrt{\epsilon_n}\}$ is tight. A similar idea would work to get an analog of the tightness conclusions of Theorems 4.2 and 4.3, and for the constant $\epsilon_n = \epsilon$ case. The next theorem gives one result of this type, and requires the following assumptions.

Assumptions.

(A5.8) There is a $\bar{\theta}$ in the interior of the constraint set H such that $\theta^n(\cdot)$ converges weakly to the process with constant value $\bar{\theta}$.

(A5.9) The perturbations Γ_n^d and Λ_n^d defined in (4.17) and (4.18) are $O(\epsilon_n)$.

(A5.10) There is a $0 < K < \infty$ such that for small $\rho > 0$ and all n,

$$E_n |Y_n|^2 I_{\{|\theta_n - \bar{\theta}| \leq \rho\}} \leq K < \infty.$$

Let P and C be symmetric, positive definite matrices satisfying (4.4) for the Hurwitz matrix A of (4.18), and let λ be the largest positive number such that $C \geq \lambda P$. Define $V(\theta) = (\theta - \bar{\theta})' P (\theta - \bar{\theta})$.

(A5.11) There are integers $\nu_n \to \infty$ such that $0 \leq n - \nu_n$ for large n, and

$$\exp(-\lambda_1(t_n - t_{n-\nu_n})) = O(\epsilon_n), \tag{5.16}$$

$$\lim_n \sum_{j=n-\nu_n}^{n} \epsilon_i^2 = 0, \tag{5.17}$$

where $\lambda_1 < \lambda$ but is arbitrarily close to it.

Comment on (A5.11). This is the only condition that needs comment. First, let $\epsilon_n = \epsilon$. Then t_n is replaced by ϵn and ν_n by ν_ϵ. Equation (5.16) requires that $\nu_\epsilon = O(|\log \epsilon|/\epsilon)$. With this, (5.17) is of the order of $\epsilon |\log \epsilon|$ which goes to zero as $\epsilon \to 0$. Now, let $\epsilon_n = 1/n^\gamma, \gamma \in (0, 1]$. Condition (5.17) always holds if $\gamma \in (0.5, 1]$ and $n - \nu_n \to \infty$. We can always find $\nu_n \to \infty$ such that (5.16) holds. In particular, for $\gamma = 1$, we can use $\nu_n = n - n^{1-1/\lambda_1}$. For $\gamma \in (0, 0.5)$, we can use $\nu_n = O(n^\gamma \log n)$, for which (5.17) is bounded by the order of $\log n / n^\gamma$.

Theorem 5.3. *Assume* (A4.5) (*if* $\epsilon_n \to 0$), (A4.6), (A4.7), *and* (A5.8)–(A5.11). *Then*

$$\lim_n P\left\{ \sup_{0 \leq n - \nu_n \leq i \leq n} V(\theta_i) \geq \mu_1 \right\} = 0 \tag{5.18}$$

and $\{(\theta_n - \bar{\theta})/\sqrt{\epsilon_n}\}$ *is tight. For the constant* $\epsilon_n = \epsilon$ *case, assume the natural analogs of the conditions. Then, there are* $n_\epsilon < \infty$ *such that* $\{(\theta_n^\epsilon - \bar{\theta})/\sqrt{\epsilon}; n \geq n_\epsilon, \epsilon\}$ *is tight. The analogous result holds for the Kiefer–Wolfowitz procedure.*

Proof. Let $\bar{\theta} = 0$. Let $\mu_2 > \mu_1 > \mu > 0$ be arbitrarily small. By the weak convergence and shifting the time origin, for any small $\delta > 0$, without loss of generality we can suppose that

$$P\{V(\theta_i) \geq \mu\} \leq \delta, \quad \text{all } i. \tag{5.19}$$

We work in the intervals $[n - \nu_n, n]$ for each n. For the purposes of proving the tightness, in view of (5.19) we can suppose that $V(\theta_{n-\nu_n}) \leq \mu$. We wish to prove that the probability that some $V(\theta_i)$ exceeds μ_1 on the interval is small. Thus without loss of generality, we can suppose that $V(\theta_i) \leq \mu_2$ on the interval. Keep in mind that all the μ, μ_i are small.

Now return to the proof of Theorem 4.2. Adding (4.22) and the error term below it, using (A5.9) and (A5.10), we have (4.23), with the expectation replaced by a conditional expectation:

$$E_i V_{i+1}(\theta_{i+1}) - V_i(\theta_i) \leq -\epsilon_i \theta_i' C \theta_i + \lambda_0 \epsilon_i |\theta_i|^2 + O(\epsilon_i^2). \tag{5.20}$$

where we can suppose that λ_0 is as small as needed (by letting μ_2 be small). Henceforth, for arbitrary n, let $i \in [n - \nu_n, n]$.

For suitably large $K > 0$ define

$$V_i^1(\theta_i) = V_i(\theta_i) + K \sum_{j=i}^{n} \epsilon_j^2.$$

Then, (5.20) yields

$$E_i V_{i+1}^1(\theta_{i+1}) - E_i V_i^1(\theta_i) \leq 0,$$

and $\{V_i^1(\theta_i)\}$ is a supermartingale on the interval $i \in [n-\nu_n, n]$. By (A5.9), $V_i^1(\theta_i) \geq -K_1\epsilon_i$ for some $K_1 > 0$. Thus by the supermartingale inequality (4.1.6),

$$\begin{aligned} &P_{n-\nu_n}\left\{ \sup_{n-n_\nu \leq i \leq n} V_i^1(\theta_i) \geq \mu_1 \right\} \\ &\quad \leq \frac{1}{\mu_1}\left[O(\epsilon_{n-\nu_n}) + V(\theta_{n-\nu_n}) + K \sum_{j=n-\nu_n}^{n} \epsilon_j^2 \right]. \end{aligned} \tag{5.21}$$

By (5.17), (5.19) and (5.21), the probability that $V(\theta_i) \leq \mu_1$ for all $i \in [n - \nu_n, n]$ goes to one as $n \to \infty$ for each $\mu_1 > 0$. To prove the tightness, we need only show that for each small $\delta > 0$ and large $c > 0$ there is an $N < \infty$ such that

$$P\{V(\theta_n) \geq \epsilon_n c\} \leq \delta, \text{ for } n \geq N. \tag{5.22}$$

We have shown that to prove (5.22) for each large n, we can suppose that for any small $\mu_1 > 0$, $V(\theta_i)$ is less than μ_1 on the interval $[n - \nu_n, n]$ with an arbitrarily high probability. The tightness proof then follows from the proofs of Theorems 4.1 and 4.2.

The proof with constant $\epsilon_n = \epsilon$ is essentially the same. □

10.6 Weak Convergence to a Wiener Process

The assumptions (A1.10) and (A2.10) and related conditions were stated as they were because there is a vast literature concerning the weak convergence of a sequence of processes to a Wiener process, when each of the processes is composed of the sums of "small effects." Various mixing and ergodic type conditions can be used [25, 68, 104]. To illustrate the possibilities, in this section we give one useful set of sufficient conditions, which is based on a weak ergodic theorem and is a consequence of Theorems 7.4.3 and 7.4.4. The theorems for the rate of convergence of stochastic approximation algorithms depend heavily on limit theorems for sequences of random processes. The perturbed test function methods of Section 7.4 are powerful tools for getting such limit theorems. While these have been used previously in this book, they play a more important role here. The reader is referred to [127, 129] for further reading. The method used here is called the perturbed test function–direct averaging method in [127], and is further developed in [129]. Although our interest is mainly in the discrete-time problem in this book, it is equally applicable to the continuous-time problem; see, for example, the results in [132].

For a sequence of $I\!R^r$-valued random variables $\{\psi_n\}$ and $i \geq n$, define

$$X_i^n = \sum_{j=n}^{i-1} \sqrt{\epsilon_j}\psi_j, \quad X^n(t) = X_i^n, \quad \text{for } t \in [t_i - t_n, t_{i+1} - t_n). \tag{6.1}$$

Let $\mathcal{F}_i$ denote the minimum σ-algebra that measures $\{\psi_j, j < i\}$, and write E_i for the associated conditional expectation. Define

$$\Gamma_n = \sum_{j=n}^{\infty} E_n \psi_j.$$

The following conditions will be used, where it is assumed that the sums are well defined.

(A6.0) The following equations hold:

$$\lim_{N\to\infty} \sup_n E \left| \sum_{j=n+N}^{\infty} E_n \psi_j \right| = 0,$$

$$\lim_{N\to\infty} \sup_n \left| \sum_{i=n+N}^{\infty} E\psi_n \psi_i' \right| = 0.$$

(A6.1) The sets $\{|\psi_i|^2\}$ and $\{|\sum_{j=i}^{\infty} E_i \psi_j|^2\}$ are uniformly integrable.

(A6.2) There is a sequence $m_n \to \infty$ such that $\epsilon_n m_n = \delta_n \to 0$ and

$$\lim_n \sup_{m_n \geq j \geq 0} \left| \frac{\epsilon_{n+j}}{\epsilon_n} - 1 \right| = 0. \tag{6.2a}$$

Also,

$$\limsup_n \sup_{i \geq n} \frac{\epsilon_i}{\epsilon_n} < \infty, \tag{6.2b}$$

and $\sum_n \epsilon_n = \infty$.

(A6.3) There is a matrix Σ_0 such that

$$\frac{1}{m} \sum_{j=n}^{n+m-1} E_n \psi_j \psi_j' - \Sigma_0 \to 0$$

in probability as $n, m \to \infty$.

(A6.4) There is a matrix Σ_1 such that

$$\frac{1}{m} \sum_{j=n}^{n+m-1} \sum_{k=j+1}^{\infty} E_n \psi_j \psi_k' - \Sigma_1 \to 0$$

in probability as $n, m \to \infty$.

(A6.5) Either $\sqrt{\epsilon_{n+1}}/\sqrt{\epsilon_n} = 1 + o(\epsilon_n)$ and $\epsilon_n \to 0$, or $\epsilon_n = 1/n$.

Comments on the conditions. The uniform integrability in (A6.1) and the convergence in probability in (A6.3) and (A6.4) imply the convergence in mean. Condition (A6.0) can be weakened to allow various groupings of the terms before taking the absolute values. The perturbed test function method of Theorem 7.4.3 will be used and Γ_i will be the perturbation. A discounted perturbation could be used instead. Alternatively, we could work on each interval $[0, T]$ and use the perturbation $\sum_{j=i}^{m(t_n+T)-1} \sqrt{\epsilon_j} E_i \psi_j$.

Condition (A6.4) is implied by somewhat simpler looking formulas. Suppose that

$$E \left| \sum_{k=j+N}^{j+N+M} E_j \psi_k \right|^2 \to 0 \tag{6.3a}$$

uniformly in j as N and M go to infinity and that there are matrices $\Sigma^{(k)}$ such that for each k

$$\frac{1}{m} \sum_{j=n}^{n+m-1} E_n \psi_j \psi_{j+k}' - \Sigma^{(k)} \to 0 \tag{6.3b}$$

in mean as n and m go to infinity, where

$$\sum_k |\Sigma^{(k)}| < \infty. \tag{6.3c}$$

Then (A6.4) holds with $\Sigma_1 = \sum_{k=1}^{\infty} \Sigma^{(k)}$. To prove this, write the sum in (A6.4) as

$$\frac{1}{m} \sum_{j=n}^{n+m-1} E_n \psi_j \sum_{k=j+1}^{\infty} E_{j+1} \psi_k'.$$

By (A6.1) and (6.3a), this can be approximated (uniformly in n and m) by

$$\frac{1}{m} \sum_{j=n}^{n+m-1} E_n \psi_j \sum_{k=j+1}^{j+N} E_{j+1} \psi_k'$$

for large N. Rewrite the last expression as

$$\sum_{k=1}^{N} \frac{1}{m} \sum_{j=n}^{n+m-1} E_n \psi_j \psi'_{j+k}.$$

Now (6.3b) and (6.3c) yield the assertion.

Further remarks on the conditions. If $\{\psi_n\}$ is second-order stationary, with mean zero and covariance $R(l) = E\psi_0\psi'_l, l \in (-\infty, \infty)$, then the Σ in Theorem 6.1 equals $\sum_{l=-\infty}^{\infty} R(l)$. The conditions on the $\{\psi_n\}$ hold if they are stationary Gauss–Markov and zero mean with a summable correlation function.

Define ϕ_k by

$$\sup_i \sup_{A \in \mathcal{F}_{i+k}, B \in \mathcal{F}_i} |P\{A|B\} - P\{A\}| = \phi_k.$$

If $\lim_k \phi_k = 0$, then $\{\psi_n\}$ is said to be a ϕ-mixing process [25] (or uniform mixing [68]). If the ψ_n are bounded with $E\psi_n = 0$ and $\sum_n \phi_n^{1/2} < \infty$, then (A6.0) and (A6.1) hold, and (A6.3) and (A6.4) hold if the expressions therein go to zero when the expectation replaces the conditional expectation [127]. Other results for weak limits of such mixing processes are in [25, 68].

Theorem 6.1. *Under* (A6.0)–(A6.5), $\{X^n(\cdot)\}$ *converges weakly to a Wiener process with covariance matrix* $\Sigma = \Sigma_0 + \Sigma_1 + \Sigma'_1$.

Proof. Tightness of $\{X^n(\cdot)\}$ is proved via Theorem 7.4.3. One need only check the conditions for each fixed $T > 0$. Let $F(\cdot)$ be a real-valued function on $\mathbb{R}^r$ with compact support that has continuous partial derivatives up to order two. For $i \geq n$, define

$$\delta F_i^n = \sqrt{\epsilon_i} F'_x(X_i^n)\Gamma_i$$

and the perturbed test function

$$F^n(t) = F(X_i^n) + \delta F_i^n, \; t \in [t_i - t_n, t_{i+1} - t_n).$$

If Condition (7.4.9) does not hold *a priori*, work with a truncated process and then take limits as the truncation level goes to infinity as in Theorem 1.1. To avoid the introduction of extra notation, we simply assume that (7.4.9) holds. To check (7.4.8), note that for any $\rho > 0$ and $T > 0$,

$$P\left\{ \sup_{m(t_n+T) \geq i \geq n} \sqrt{\epsilon_i}|\Gamma_i| \geq \rho \right\} \leq \sum_{i=n}^{m(t_n+T)} P\left\{ |\Gamma_i|^2 \geq \rho^2/\epsilon_i \right\}, \tag{6.4}$$

$$P\left\{|\Gamma_i|^2 \geq \rho^2/\epsilon_i\right\} \leq \frac{\epsilon_i}{\rho^2} E|\Gamma_i|^2 I_{\{|\Gamma_i| \geq \rho^2/\epsilon_i\}}. \tag{6.5}$$

By the uniform integrability of the sets in (A6.1), $E|\Gamma_i|^2 I_{\{|\Gamma_i| \geq \rho^2/\epsilon_i\}} \to 0$ as $i \to \infty$. Using this, (6.5), and the fact that $\sum_{i=n}^{m(t_n+T)} \epsilon_i \approx T$, we see that the right side of (6.4) goes to zero as $n \to \infty$.

To complete the proof of tightness of $\{X^n(\cdot)\}$ we need only verify the uniform integrability condition below (7.4.10) in Theorem 7.4.3. Let $t_{i+1} - t_n \leq T$. Then

$$\begin{aligned} E_i F(X_{i+1}^n) - F(X_i^n) &= \sqrt{\epsilon_i} F_x'(X_i^n) E_i \psi_i + \frac{\epsilon_i}{2} E_i \psi_i' F_{xx}(X_i^n) \psi_i \\ &+ \frac{1}{2} E_i \psi_i' \int_0^1 \left[F_{xx}(X_i^n + s(X_{i+1}^n - X_i^n)) - F_{xx}(X_i^n)\right] ds\, \psi_i \end{aligned}$$

and

$$\begin{aligned} E_i \delta F_{i+1}^n - \delta F_i^n &= -\sqrt{\epsilon_i} F_x'(X_i^n) E_i \psi_i \\ &+ \left[\sqrt{\epsilon_{i+1}} E_i F_x'(X_{i+1}^n) - \sqrt{\epsilon_i} E_i F_x'(X_i^n)\right] \sum_{j=i+1}^{\infty} E_{i+1} \psi_j. \end{aligned}$$

Using the first condition of (A6.5) and expanding $F_x(\cdot)$ and $\sqrt{\epsilon_{i+1}}$ and combining the terms yields

$$\begin{aligned} &\left[E_i F^n(t_{i+1} - t_n) - F^n(t_i - t_n)\right]/\epsilon_i \\ &= \left[\frac{1}{2} E_i \psi_i' F_{xx}(X_i^n) \psi_i + E_i \psi_i' F_{xx}(X_i^n) \sum_{j=i+1}^{\infty} E_{i+1} \psi_j\right] + \rho_i^n, \end{aligned} \tag{6.6}$$

where $\lim_n \sup_{m(t_n+T) \geq i \geq n} E|\rho_i^n| = 0$. Since the bracketed term is uniformly (in n, i) integrable, the uniform integrability condition of Theorem 7.4.3 holds and we have the desired tightness. The limit processes are continuous since $\sqrt{\epsilon_n}\psi_n \to 0$ in the sense that (6.4) and (6.5) hold for ψ_i replacing Γ_i.

If $\epsilon_n = 1/n$, then the expansions alluded to yield another non-negligible term in (6.6), namely,

$$E_i F_x'(X_i^n) \Gamma_{i+1}/2. \tag{6.7}$$

These are also uniformly (in n, i) integrable and tightness again follows from Theorem 7.4.3.

Assume the first situation of (A6.5). By the definition of $\widehat{A}^n$ given in (7.4.7), the right side of (6.6) can be written as $\widehat{A}^n F^n(t_i - t_n)$. Now, Theorem 7.4.4 will be used to show that the weak sense limit is the asserted Wiener process. Since (7.4.12) and (7.4.13) are implied by the calculations done in the above proof of tightness, only the equivalent of (7.4.14) need be shown. Let m_n satisfy (A6.2), and let $\Sigma_{m,kl}$ denote the klth element of

Σ_m, for $m = 0, 1$. Then, in particular, we must show that

$$\frac{1}{m_i} \sum_{j=i}^{i+m_i-1} E_i \left[\psi_j' F_{xx}(X_j^n)\psi_j - \sum_{k,l} F_{x_k x_l}(X_j^n)\Sigma_{0,kl} \right] \to 0 \tag{6.8}$$

and

$$\frac{1}{m_i} \sum_{j=i}^{i+m_i-1} E_i \left[\psi_j' F_{xx}(X_j^n) \sum_{k=j+1}^{\infty} E_{j+1}\psi_k - \sum_{k,l} F_{x_k x_l}(X_j^n)\Sigma_{1,kl} \right] \to 0 \tag{6.9}$$

in mean as n and i go to infinity.

By (A6.1), the tightness of $\{X^n(\cdot)\}$, and the fact that any weak limit must be continuous with probability one, $F_{xx}(X_j^n)$ can be approximated by $F_{xx}(X_i^n)$ in (6.8) and (6.9) without changing the limit. Now, (A6.3) and (A6.4) yield (7.4.14) with $\Sigma = \Sigma_0 + \Sigma_1 + \Sigma_1'$.

If $\epsilon_n = 1/n$, then we have to deal with (6.7) and show that if $F_x'(X_j^n)\Gamma_{j+1}$ replaces the summand in (6.8), then the limit is zero. By the tightness and continuity of any weak sense limit process of $\{X^n(\cdot)\}$, X_j^n can be replaced by X_i^n without changing the limit. Then the "zero limit" will follow from the first parts of (A6.0) and (A6.1). □

If $\epsilon_n = \epsilon$, a constant, Theorem 6.1 takes the following form.

Theorem 6.2. *Assume the conditions of Theorem 6.1, except those concerning the step size, with the following changes: ϵ replaces ϵ_n, and $m_\epsilon \to \infty$ replaces m_n, where $\epsilon m_\epsilon \to 0$. Replace the superscript n by ϵ and E_i by E_i^ϵ. Then $\{X^\epsilon(\cdot)\}$ converges weakly to a Wiener process with covariance matrix $\Sigma = \Sigma_0 + \Sigma_1 + \Sigma_1'$.*

10.7 Random Directions: Martingale Difference Noise

Recall the random directions Kiefer–Wolfowitz algorithm (5.6.1):

$$\theta_{n+1} = \theta_n + \epsilon_n d_n \frac{Y_n^- - Y_n^+}{2c_n} + \epsilon_n Z_n, \tag{7.1}$$

where d_n is the random direction used in the nth iteration; see also [213, 226, 227, 228, 229]. Recall that the observations are taken at parameter values $\theta_n \pm c_n d_n$. Let $\mathcal{F}_n$ denote the minimal σ-algebra, which measures $\{Y_{i-1}^\pm, d_{i-1}, \theta_i, i \le n\}$, and let $\mathcal{F}_n^d$ also measure d_n. Let E_n^d denote the expectation conditioned on $\mathcal{F}_n^d$.

Comment on the random directions method for the Robbins–Monro procedure. The discussion here and the general literature on the random directions methods concern applications to the Kiefer–Wolfowitz procedure, where noisy estimates of finite differences are used. In many applications of the Robbins–Monro procedure, the value of the vector Y_n is no harder to compute than the real-valued directional value $d_n' Y_n$, no matter what the dimension is. This is usually because Y_n would have to be computed in any case to get the directional value in these applications.

If it were much simpler to get the directional value for any given direction, then the obvious random directions form of the Robbins–Monro algorithm would be advantageous. We can view Y_n as a "noisy" derivative, as it is in many of the applications in Chapters 1 to 3. Whether a random directions form will be advantageous will depend on the details of the computation of the general "noisy" directional derivative.

Assumptions. In Section 5.6, it was seen that the proof of the convergence of random directions algorithm is essentially the same as that of the classical Kiefer–Wolfowitz algorithm. The rate of convergence results in Section 3 for the general Kiefer-Wolfowitz algorithm can be readily applied to the random directions algorithm. Because of its great current interest, and to allow us to focus on the essential elements in the comparison of the basic features of various random directions algorithms, we will spell out the conditions in the standard case. It will be supposed that the noise terms are martingale differences and that there is a function $f(\cdot)$ such that

$$E_{\mathcal{F}_n^d} Y_n^{\pm} = f(\theta_n \pm c_n d_n).$$

Define $\delta M_n^{\pm} = Y_n^{\pm} - f(\theta_n \pm c_n d_n)$. Then we can write

$$\begin{aligned}\theta_{n+1} &= \theta_n - \epsilon_n \frac{d_n}{2c_n} \left[f(\theta_n + c_n d_n) - f(\theta_n - c_n d_n) \right] \\ &+ \epsilon_n \frac{d_n(\delta M_n^- - \delta M_n^+)}{2c_n} + \epsilon_n Z_n.\end{aligned}$$

Define $\delta M_n = (\delta M_n^- - \delta M_n^+)$, and suppose that $f(\cdot)$ and its mixed partial derivatives up to order three are continuous. Let $f_\theta(\bar{\theta}) = 0$, and let the Hessian matrix $-A = f_{\theta\theta}(\bar{\theta})$ be positive definite.

Suppose that $\theta^n(\cdot)$ converges weakly to the process with constant value $\bar{\theta}$, where $\bar{\theta}$ is strictly inside the constraint set H if the algorithm is constrained. Let $\bar{\theta}$ be an asymptotically stable (in the sense of Liapunov) point of the ODE $\dot{\theta} = -f_\theta(\theta)$. Thus, without loss of generality, we ignore the Z_n term henceforth. The conditions on (ϵ_n, c_n) of Theorem 3.1 are retained. Thus, $\epsilon_n = 1/n^\gamma$, $c_n = c/n^\alpha$, $U_n = (\theta_n - \bar{\theta})n^\beta$, $\alpha = \gamma/6$, and $\beta = 2\gamma$. The random variables d_n are assumed to be mutually independent, identically distributed, with a distribution that is symmetric with respect to reflection about each coordinate axis, $E_{\mathcal{F}_n} d_n d_n' = I$ and $|d_n|^2 = r$. These conditions can readily be generalized to allow correlations.

The $\delta M_n^{\pm}$ term might depend on the product $c_n d_n$, because the associated observations are taken at the points $\theta_n \pm c_n d_n$, and the location would in general affect the noise. However, since $c_n \to 0$, it is often the case that the dependence vanishes as $n \to \infty$. We will use (A2.7) with E_n replaced by E_n^d in (2.3). Using this and the fact that $E_{\mathcal{F}_n}(d_n d_n') = I$, we have

$$\lim_n E_{\mathcal{F}_n} d_n d_n' E_{\mathcal{F}_n^d} (\delta M_n)^2 I_{\{|\theta_n - \bar{\theta}| \leq \mu\}} = \Sigma_1 I \text{ in probability}$$

(and in mean) for any $\mu > 0$. Note that δM_n is real valued, and so is Σ_1. Furthermore, the limit covariance does not depend on the dimension r, except that the dimension of the identity matrix I is r.

We will also require the tightness of $\{U_n\}$. If there is probability one convergence, then the proof of tightness is just that for the classical Kiefer–Wolfowitz procedure in Theorem 4.3. If there is weak but not probability one convergence, then the proof of tightness would be similar to that in Section 5. In the current case, the $(\Lambda_n^d, \Gamma_n^d)$ used in Theorem 5.1 are zero and $f(\cdot)$ is the Liapunov function. Again, the random directions property would not play an important role in the proof.

In preparation for the rate of convergence theorem, expand the finite difference as

$$f(\theta_n + c_n d_n) - f(\theta_n - c_n d_n) = 2d_n' f_\theta(\theta_n) c_n + 2\widetilde{B}(\theta_n, d_n) c_n^3 + o(c_n^3),$$

where

$$\widetilde{B}(\theta_n, d_n) = \frac{1}{3!} \sum_{i,j,k} f_{\theta^i \theta^j \theta^k}(\theta_n) d_{n,i} d_{n,j} d_{n,k}.$$

The $o(c_n^3)$ term will not affect the result and is dropped henceforth for simplicity. Define $B(\theta_n) = E_{\mathcal{F}_n} d_n \widetilde{B}(\theta_n, d_n)$. Then the algorithm can be rewritten as

$$\theta_{n+1} = \theta_n - \epsilon_n f_\theta(\theta_n) - \epsilon_n B(\theta_n) c_n^2 + \epsilon_n d_n \frac{\delta M_n}{2c_n} + \epsilon_n \rho_n,$$

where

$$\rho_n = (I - d_n d_n') f_\theta(\theta_n) - \left[d_n \widetilde{B}(\theta_n, d_n) - B(\theta_n) \right] c_n^2.$$

The term ρ_n is a martingale difference. The covariance matrix of $\rho_n I_{\{|\theta_n - \bar{\theta}| \leq \mu\}}$ goes to zero as $n \to \infty$ for any $\mu > 0$. Thus, the term ρ_n plays no role in the limit. The proof of Theorem 3.1 yields the following result.

Theorem 7.1. *Under the preceding assumptions, with $\Sigma_1 I$ (with real-valued Σ_1) replacing the matrix Σ_1 in Theorem 3.1, the conclusions of Theorem 3.1 hold for the random directions algorithm.*

10.7.1 Comparison of Algorithms

In reference [135], $\{d_n\}$ is chosen to be mutually independent with each d_n being uniformly distributed on the unit sphere in $\mathbb{R}^r$. Modify that choice by replacing the unit sphere by the sphere of radius $\sqrt{r}$. Thus, $|d_n|^2 = r$. This will be referred to as the *spherical method.* In [226], the directions $d_n = (d_{n,1}, \ldots, d_{n,r})$ were selected so that the set of components $\{d_{n,i}; n, i\}$ were mutually independent, with $d_{n,i}$ taking values ± 1 each with probability $1/2$; that is, the direction vectors are the corners of the unit cube in $\mathbb{R}^r$; this will be called the *Bernoulli method.* It will be seen that the performance of the two methods is essentially the same. This similarity is also borne out by a comparison via simulation. Indeed, the essential facts are that $|d_n|^2 = r$, $E_{\mathcal{F}_n} d_n d_n' = I$, and the assumed form of the symmetry of the distribution of d_n. Hence any other method with these properties should perform similarly (asymptotically).

The asymptotic rate formulas. Note that the value of Σ_1 does not depend on the method of choosing the random directions. The bias term equals

$$B(\bar{\theta}) = \frac{1}{3!} E d_n \sum_{i,j,k} f_{\theta^i \theta^j \theta^k}(\bar{\theta}) d_{n,i} d_{n,j} d_{n,k}.$$

The symmetry of the distribution of d_n implies that $E d_{n,l} d_{n,i} d_{n,j} d_{n,k} = 0$ unless the indices i, j, k, l occur in pairs. Thus, using the fact that the order of taking derivatives is irrelevant (i.e., $f_{\theta^l \theta^i \theta^i}(\bar{\theta}) = f_{\theta^i \theta^l \theta^i}(\bar{\theta}) = f_{\theta^i \theta^i \theta^l}(\bar{\theta})$), the lth component of $B(\bar{\theta})$ equals

$$B_l(\bar{\theta}) = E d_{n,l}^2 \sum_i f_{\theta^l \theta^i \theta^i}(\bar{\theta}) d_{n,i}^2(\bar{\theta})/2.$$

For the Bernoulli method, $|d_{n,i}| = 1$, so

$$B_l(\bar{\theta}) = \sum_i f_{\theta^l \theta^i \theta^i}(\bar{\theta})/2.$$

It is a little harder to calculate the direction vectors for the spherical method, because the components $d_{n,i}, i = 1, \ldots, r$, of d_n are not mutually independent. However the fact that given $d_{n,i}$, $\{d_{n,j}, j \neq i\}$ are uniformly distributed on the sphere in $\mathbb{R}^{r-1}$ whose squared radius is $r^2 - |d_{n,i}|^2$ implies that for large r the bias terms of the two methods are very close. The bias term can conceivably grow as $O(r)$. This chance of a large bias is not surprising because the size of the difference interval is $c_n|d_n|$, which grows as $\sqrt{r}$. Note that for the standard Kiefer–Wolfowitz procedure $B_l(\bar{\theta}) = f_{\theta^l \theta^l \theta^l}(\bar{\theta})/3!$, which will be much smaller than that for any of the random directions methods unless either many of the mixed third partial derivatives are very small or they essentially cancel each other.

General comments on random directions methods. For small bias and large dimension r, the rate of convergence result suggests that the random directions algorithms can be very useful, since Σ_1 does not depend on r. To get a reasonable understanding of the general usefulness, one needs to examine the nonasymptotic behavior, in which phase both described methods of getting d_n behave similarly and are strongly affected by the $O(r)$ dependence of the variance of ρ_n. It will be seen that the random directions methods can behave poorly unless n is sufficiently large, but there can be significant advantages. The performance of the algorithms will now be examined for two interesting classes of problems for small or moderate values of n. Only the spherical and Bernoulli cases will be discussed.

A no-noise case. Let $f(\theta) = \theta'\theta/2$, a quadratic form, and suppose that there is no observation noise. Owing to the quadratic form, there is no bias in a finite difference estimate of the derivative. The Kiefer–Wolfowitz algorithm then takes the form

$$\theta_{n+1} = \theta_n - \epsilon_n f_\theta(\theta_n) = \theta_n - \epsilon_n \theta_n$$

and

$$f(\theta_n) = \prod_{i=0}^{n-1} (1-\epsilon_i)^2 f(\theta_0).$$

Using the approximation $\log(1-\epsilon) \approx -\epsilon - \epsilon^2/2$, up to the second order in the step sizes ϵ_i the above expression is approximated by

$$\exp\left[-2\sum_{i=0}^{n-1}\epsilon_i - \sum_{i=0}^{n-1}\epsilon_i^2\right] f(\theta_0) = \exp\left[-2t_n - \sum_{i=0}^{n-1}\epsilon_i^2\right] f(\theta_0).$$

To account for the fact that $2r$ observations are used per step in the Kiefer–Wolfowitz algorithm, we compare $Ef(\theta_n)$ for the random directions method with $Ef(\theta_{n/r})$ for the Kiefer–Wolfowitz procedure, for which we use

$$Ef(\theta_{n/r}) \approx \exp\left[-2t_{n/r} - \sum_{i=0}^{n/r-1}\epsilon_i^2\right] f(\theta_0). \tag{7.2}$$

Let the random directions satisfy the conditions of Theorem 7.1. The random directions algorithm takes the form

$$\theta_{n+1} = \theta_n - \epsilon_n d_n d_n' \theta_n = (1-\epsilon_n)\theta_n + \epsilon_n \left(I - d_n d_n'\right)\theta_n.$$

Then

$$Ef(\theta_{n+1}) = (1-\epsilon_n)^2 Ef(\theta_n) + \epsilon_n^2 E\theta_n' \left(I - d_n d_n'\right)\left(I - d_n d_n'\right)\theta_n/2.$$

Using the facts that $d_n' d_n = r$ and $E_n(I - d_n d_n')(I - d_n d_n') = (r-1)I$, the right-hand side equals

$$\left[(1-\epsilon_n)^2 + \epsilon_n^2(r-1)\right] Ef(\theta_n).$$

Up to the second order in ϵ_i, $Ef(\theta_n)$ is approximated by

$$\exp\left[-2t_n + \sum_{i=0}^{n-1} \epsilon_i^2 (r-2)\right] f(\theta_0). \tag{7.3}$$

Of course, if $r\epsilon_n^2$ is large then higher powers of $r\epsilon_n^2$ will be needed in the exponent.

Let $\epsilon_n < 1$ for all n. The case of greatest interest for the random directions method is where the dimension r is large, and one effect of large r can be seen by comparing (7.2) and (7.3). For large values of r, the performance of the random directions method can actually deteriorate for small and moderate values of n, unless ϵ_n is selected appropriately small. The best values of the step sizes ϵ_n are not the same for the two procedures, and any comparison should be based on the use of step-size sequences appropriate to the particular procedure. The random directions method requires smaller ϵ_n, at least for the initial stages, but this reduces the advantage gained by using only two observations per step. When r is large, the Kiefer–Wolfowitz procedure uses many observations per step, but the function value is guaranteed to decrease monotonically (for this example and if $\epsilon_n < 1$). The function will also decrease monotonically for the random directions algorithm if $\epsilon_n^2 r$ is small enough, but this requires smaller values of ϵ_n.

Let the step sizes be small constants. Then minimizing the exponent of (7.3) with respect to this constant step size yields that its optimal value is scaled inversely proportional to the dimension r. This scaling reduces any advantage of the random directions idea, even apart from the problem of possibly large bias for nonquadratic $f(\cdot)$.

If $\epsilon_n \to 0$, then t_n grows faster than $\sum_{i=0}^{n-1} \epsilon_i^2$. Thus the deleterious role of r in (7.3) eventually diminishes as n increases, and the random directions method gains the advantage. The comparison makes it clear that the short- or moderate-term behavior of the random directions algorithm might be worse than that of the classical Kiefer–Wolfowitz method, and that in any case, care must be used in choosing the step sizes.

This discussion has merely opened up the question of what to do if the dimension is large, but relatively few observations can be taken. One might consider letting the norm of the direction vector grow from 1 to $\sqrt{r}$ in some way as n increases.

Noisy observation with few significant components. In the preceding comparison, each of the components of θ were of equal importance.

In many applications where the dimension is large, relatively few of the components of θ are significant. One would expect that the random directions methods would have a significant advantage when the number of important components of θ is much fewer than r. Suppose that there are q components of θ that are significant. Without loss of generality, assume that the first q components $(\theta^1, \ldots, \theta^q)$ are the important ones. Let $q \ll r$, and define $f(\theta) = \sum_{i=1}^q (\theta^i)^2/2$. Write $\bar{\theta}_n = \{\theta_{n,i}, i \leq q\}$. Let the finite difference interval be fixed at $c > 0$.

For the classical Kiefer–Wolfowitz procedure, suppose that one observes the finite difference estimates

$$[f(\theta_n + ce_i) - f(\theta_n - ce_i) - \delta M_{n,i}]/(2c), \quad i \leq r,$$

and let the random vectors $\delta \bar{M}_n = (\delta M_{n,i}, i \leq q)$ be mutually independent with mean zero and covariance $\sigma^2 I_q$, where I_q is the $q \times q$ identity matrix. Then

$$\begin{aligned} \bar{\theta}_{n+1} &= (1 - \epsilon_n)\bar{\theta}_n + \frac{1}{2c}\epsilon_n \delta \bar{M}_n, \\ Ef(\theta_{n+1}) &= (1 - \epsilon_n)^2 Ef(\theta_n) + \frac{1}{8c^2}\epsilon_n^2 q\sigma^2. \end{aligned} \tag{7.4}$$

For the random directions method, let the finite difference estimates be

$$\frac{\left[f(\theta_n + cd_n) - f(\theta_n - cd_n) - \delta \widehat{M}_n\right]}{2c}.$$

and suppose that the random variables $\delta \widehat{M}_n$ are mutually independent, are independent of the sequence of random directions, and have mean zero and variance σ^2. Let $\bar{d}_n$ denote the vector of the first q components of d_n. Then,

$$\bar{\theta}_{n+1} = \frac{(1 - \epsilon_n)\bar{\theta}_n + \epsilon_n \bar{d}_n \delta \widehat{M}_n}{2c} + \epsilon_n \left(I_q - \bar{d}_n \bar{d}_n'\right) \bar{\theta}_n$$

and

$$\begin{aligned} Ef(\theta_{n+1}) &= (1 - \epsilon_n)^2 Ef(\theta_n) + \frac{1}{8c^2}\epsilon_n^2 q\sigma^2 \\ &\quad + \frac{1}{2} E\bar{\theta}_n' \left(\bar{d}_n \bar{d}_n' - I_q\right) \left(\bar{d}_n \bar{d}_n' - I_q\right) \bar{\theta}_n. \end{aligned} \tag{7.5}$$

For the Bernoulli case, $\bar{d}_n' \bar{d}_n = q$, and this equals

$$Ef(\theta_{n+1}) = \left[(1 - \epsilon_n)^2 + \epsilon_n^2 (q - 1)\right] Ef(\theta_n) + \frac{1}{8c^2}\epsilon_n^2 q\sigma^2. \tag{7.6}$$

For the spherical case, (7.6) is a good approximation for r much larger than q. Thus, ignoring bias considerations, a comparison of (7.4) (with n replaced by n/r) and (7.6) suggests that the random directions method might be superior when the number of significant parameters is much smaller than the dimension, even for moderate values of n if $\epsilon_n^2 q$ is small.

10.8 State-Dependent Noise

All the results in Sections 1 to 5 can be extended to the Markov state-dependent noise case dealt with in Section 8.4 and in the examples of Sections 2.4 and 2.5. We will confine our attention to the constant-step-size algorithm

$$\theta_{n+1}^{\epsilon} = \Pi_H\left(\theta_n^{\epsilon} + \epsilon Y_n^{\epsilon}\right), \tag{8.1}$$

where we can write $Y_n^{\epsilon} = g_n^{\epsilon}(\theta_n^{\epsilon}, \xi_n^{\epsilon}) + \delta M_n^{\epsilon}$, where the terms have the meanings given in Section 8.4. Neglecting the constraint, the normalized process $U_n^{\epsilon} = (\theta_n^{\epsilon} - \bar{\theta})/\sqrt{\epsilon}$ satisfies

$$U_{n+1}^{\epsilon} = U_n^{\epsilon} + \sqrt{\epsilon}\left[g_n^{\epsilon}(\theta_n^{\epsilon}, \xi_n^{\epsilon}) + \delta M_n^{\epsilon}\right]. \tag{8.2}$$

In Theorem 1.3, we supposed that the sequence of processes $W^{\epsilon}(\cdot)$ defined by

$$W^{\epsilon}(t) = \sqrt{\epsilon} \sum_{i=q_\epsilon}^{q_\epsilon + t/\epsilon - 1} Y_i^{\epsilon}(\bar{\theta})$$

(with the appropriate definition for $t < 0$) converges to a Wiener process. Owing to the state-dependent property, such an assumption is not adequate in the current case, and we need to work directly with the perturbed test function method of the last part of Theorem 7.4.4.

Assumptions. The assumptions of Theorem 8.4.1 will be used, as well as (A1.2), (A1.3), and those listed below. One could also base the proof on the assumptions of the other theorems of Section 8.4, but it would then be more complicated. As in Section 1, $\bar{\theta}$ is an asymptotically stable point of the mean ODE $\dot{\theta} = \bar{g}(\theta)$ strictly interior to H, $\epsilon(q_\epsilon - p_\epsilon) \to \infty$, where p_ϵ is defined in (A1.3), and the interpolated process $U^{\epsilon}(\cdot)$ is defined in the same way. In the following assumptions, $\rho > 0$ is an arbitrarily small number. Also, recall the definition of the fixed-θ process $\xi_i(\theta)$ of Subsection 8.4.1. Owing to the Markov structure, if g_i^{ϵ} does not depend on ϵ or n, then $\Gamma^{\epsilon,d}(\cdot)$ defined in (8.3) is a solution to the discounted Poisson equation.

(A8.1) For small $\rho > 0$, $\{|Y_n^{\epsilon}|^2 I_{\{|\theta_n^{\epsilon} - \bar{\theta}| \le \rho\}}\}$ is uniformly integrable.

Define

$$\Gamma_n^{\epsilon,d}(\theta) = \sum_{i=n}^{\infty} (1-\epsilon)^{i-n} E_n^{\epsilon}\left[g_i^{\epsilon}(\theta, \xi_i(\theta)) - \bar{g}(\theta)\right], \tag{8.3}$$

where we use the convention of Subsection 8.4.1 that when E_n^{ϵ} is used, the initial condition is $\xi_n(\theta) = \xi_n^{\epsilon}$.

(A8.2) For the initial conditions ξ_n^{ϵ} confined to any compact set,

$$\{|\Gamma_n^{\epsilon,d}(\theta_n^{\epsilon})|^2 I_{\{|\theta_n^{\epsilon} - \bar{\theta}| \le \rho\}}, |\Gamma_n^{\epsilon,d}(\bar{\theta})|^2; n, \epsilon\} \text{ is uniformly integrable,}$$

and

$$E\left|E_n^\epsilon\Gamma_{n+1}^{\epsilon,d}(\theta_{n+1}^\epsilon) - \Gamma_{n+1}^{\epsilon,d}(\theta_n^\epsilon)\right|^2 I_{\{|\theta_n^\epsilon-\bar{\theta}|\leq\rho\}} = O(\epsilon^2).$$

(A8.3) There is a Hurwitz matrix A such that

$$\bar{g}(\theta) = A(\theta - \bar{\theta}) + o(|\theta - \bar{\theta}|).$$

(A8.4) There is a matrix $\Sigma_0 = \{\sigma_{0,ij}; i,j = i,\ldots,r\}$ such that, as $n, m \to \infty$,

$$\frac{1}{m}\sum_{i=n}^{n+m-1} E_n^\epsilon\left[\delta M_i^\epsilon(\delta M_i^\epsilon)' - \Sigma_0\right] I_{\{|\theta_n^\epsilon-\bar{\theta}|\leq\rho\}} \to 0$$

in probability.

(A8.5) There is a matrix $\bar{\Sigma}_0 = \{\bar{\sigma}_{0,ij}; i,j = i,\ldots,r\}$ such that, as $n, m \to \infty$,

$$\frac{1}{m}\sum_{i=n}^{n+m-1}\left[E_n^\epsilon g_i^\epsilon(\bar{\theta},\xi_i(\bar{\theta}))(g_i^\epsilon(\bar{\theta},\xi_i(\bar{\theta})))' - \bar{\Sigma}_0\right] \to 0$$

in probability.

(A8.6) The matrix-valued function $G_n^\epsilon(\cdot)$ defined by

$$G_n^\epsilon(\theta,\xi_n^\epsilon)I_{\{|\theta-\bar{\theta}|\leq\rho\}} = E_n^\epsilon\left[\Gamma_{n+1}^{\epsilon,d}(\theta_n^\epsilon)[Y_n^\epsilon]' I_{\{|\theta_n^\epsilon-\bar{\theta}|\leq\rho\}}\big|\theta_n^\epsilon = \theta\right]$$

is continuous in (θ, ξ_n^ϵ), uniformly in n, ϵ, when $|\theta - \bar{\theta}| < \rho$.

(A8.7) There is a matrix $\Sigma_1 = \{\sigma_{1,ij}, i,j = i,\ldots,r\}$ such that, as $n, m \to \infty$,

$$\frac{1}{m}\sum_{i=n}^{n+m-1}\left[E_n^\epsilon G_i^\epsilon(\bar{\theta},\xi_i(\bar{\theta})) - \Sigma_1\right] \to 0$$

in probability.

Comment on the tightness condition (A1.3). All of the results of Section 5 carry over to the state-dependent-noise case. One simply uses the perturbation that is appropriate to the state dependent noise case; namely, the $\Gamma_n^d(\theta)$ in (5.2) is replaced by the $\epsilon\Gamma_n^{\epsilon,d}(\theta)$ defined by (8.3).

Theorem 8.1. *Assume the conditions listed above. Then there is a Wiener process $W(\cdot)$ with covariance matrix $\Sigma = \{\sigma_{ij}; i,j = i,\ldots,r\} = \Sigma_0 + \bar{\Sigma}_0 + \Sigma_1 + \Sigma_1'$ such that $\{U^\epsilon(\cdot)\}$ converges weakly to a stationary solution of*

$$dU = AUdt + dW. \tag{8.4}$$

Proof. As in Section 1, the criterion of Theorem 7.4.3 will be used to prove tightness. But, owing to complications stemming from the state dependence, the direct method of characterizing the limit process in Theorem 1.3 cannot be used here. Instead, we will apply the powerful perturbed test function method of the last part of Theorem 7.4.4.

As in Section 1, by shifting the time origin to q_ϵ (i.e., by starting at the q_ϵth iterate), we can suppose that $q_\epsilon = 0$. Similarly, we can suppose that $|\theta_n^\epsilon - \bar{\theta}|$ is as small as desired, and drop the Z_n^ϵ terms. Again, as in Theorem 1.1, the process $U^\epsilon(\cdot)$ needs to be truncated: One proves the result for the truncated processes and then shows that the truncation is not needed. However, to minimize terminology in this proof, and to avoid repetition of the steps in Theorem 1.1 that concern the truncation, we will simply suppose that the $U^\epsilon(\cdot)$ are bounded.

Let $F(\cdot)$ be a real-valued continuous function on $\mathbb{R}^r$, with compact support, whose partial derivatives up to order three are continuous. We can write

$$\begin{aligned} E_n^\epsilon F(U_{n+1}^\epsilon) - F(U_n^\epsilon) &= \sqrt{\epsilon} F_u'(U_n^\epsilon) g_n^\epsilon(\theta_n^\epsilon, \xi_n^\epsilon) \\ &+ \frac{\epsilon}{2} (g_n^\epsilon(\theta_n^\epsilon, \xi_n^\epsilon))' F_{uu}(U_n^\epsilon) g_n^\epsilon(\theta_n^\epsilon, \xi_n^\epsilon) \\ &+ \frac{\epsilon}{2} E_n^\epsilon (\delta M_n^\epsilon)' F_{uu}(U_n^\epsilon) \delta M_n^\epsilon \\ &+ \epsilon [\text{error term}](\epsilon, n), \end{aligned} \tag{8.5}$$

where the last "error" term satisfies $\lim_\epsilon \sup_n E|[\text{error term}](\epsilon, n)| = 0$.

We will use the perturbed test function

$$F_n^\epsilon(U_n^\epsilon) = F(U_n^\epsilon) + \delta F_n^\epsilon(\theta_n^\epsilon),$$

where

$$\delta F_n^\epsilon(\theta_n^\epsilon) = \sqrt{\epsilon} F_u'(U_n^\epsilon) \Gamma_n^{\epsilon,d}(\theta_n^\epsilon).$$

Expanding the perturbation term yields

$$\begin{aligned} E_n^\epsilon \delta F_{n+1}^\epsilon(\theta_{n+1}^\epsilon) - \delta F_n^\epsilon(\theta_n^\epsilon) &= -\sqrt{\epsilon} F_u'(U_n^\epsilon) \left[g_n^\epsilon(\theta_n^\epsilon, \xi_n^\epsilon) - \bar{g}(\theta_n^\epsilon) \right] \\ &+ \sqrt{\epsilon} E_n^\epsilon \left[F_u'(U_{n+1}^\epsilon) - F_u'(U_n^\epsilon) \right] \Gamma_{n+1}^{\epsilon,d}(\theta_{n+1}^\epsilon) \\ &+ \sqrt{\epsilon} F_u'(U_n^\epsilon) E_n^\epsilon \left[\Gamma_{n+1}^{\epsilon,d}(\theta_{n+1}^\epsilon) - \Gamma_{n+1}^{\epsilon,d}(\theta_n^\epsilon) \right]. \end{aligned}$$

Now, combining terms, making cancellations where possible, and using the expansion

$$\sqrt{\epsilon} \bar{g}(\theta_n^\epsilon) = \epsilon A U_n^\epsilon + \epsilon |U_n^\epsilon| o(|\theta_n^\epsilon - \bar{\theta}|)/(|\theta_n^\epsilon - \bar{\theta}|)$$

given by (A8.2), (A8.3) and the fact that $|\theta_n^\epsilon - \bar{\theta}|$ can be assumed to be as small as desired (since the U_n^ϵ are truncated) yields

$$\begin{aligned} E_n^\epsilon F_{n+1}^\epsilon(U_{n+1}^\epsilon) - F_n^\epsilon(U_n^\epsilon) &= \epsilon F_u'(U_n^\epsilon)AU_n^\epsilon \\ &+\frac{\epsilon}{2}(g_n^\epsilon(\theta_n^\epsilon,\xi_n^\epsilon))'F_{uu}(U_n^\epsilon)g_n^\epsilon(\theta_n^\epsilon,\xi_n^\epsilon) \\ &+\frac{\epsilon}{2}E_n^\epsilon(\delta M_n^\epsilon)'F_{uu}(U_n^\epsilon)\delta M_n^\epsilon \\ &+\epsilon E_n^\epsilon(Y_n^\epsilon)'F_{uu}(U_n^\epsilon)\Gamma_{n+1}^\epsilon(\theta_n^\epsilon) + \epsilon\rho_n^\epsilon, \end{aligned} \tag{8.6}$$

where $\lim_\epsilon \sup_n E|\rho_n^\epsilon| = 0$.

Now we use Theorems 7.4.3 and 7.4.4, following the pattern of the proof of Theorem 6.1. To prove tightness by Theorem 7.4.3, we need to show that for each $T > 0$ and $\mu > 0$,

$$\lim_\epsilon P\left\{\sup_{0\le i\le T/\epsilon} |\delta F_i^\epsilon(\theta_i^\epsilon)| \ge \mu\right\} = 0.$$

This follows from the uniform integrability in (A8.2) in the same manner that the analogous result was proved in (6.4) and (6.5). Similarly, the set (indexed by (ϵ, n)) of coefficients of ϵ on the right side of (8.6) can be shown to be uniformly integrable. Then Theorem 7.4.3 implies the tightness of $\{U^\epsilon(\cdot)\}$. The limit processes are continuous since $\sqrt{\epsilon}Y_n^\epsilon \to 0$ in the sense that, for each $T > 0, \mu > 0$,

$$\lim_\epsilon \sup_n P\left\{\sup_{n\le i\le n+T/\epsilon} \sqrt{\epsilon}|Y_i^\epsilon| \ge \mu\right\} = 0.$$

The limit process will be identified via the identification of its operator as in the last part of Theorem 7.4.4. The $\bar{g}(X)$ in Theorem 7.4.4 is AX here. Let $m_\epsilon \to \infty$ with $\epsilon m_\epsilon \to 0$. We need to evaluate the following limits, in the sense of convergence in probability:

$$\frac{1}{2}\lim_{n,\epsilon} \frac{1}{m_\epsilon} \sum_{i=n}^{n+m_\epsilon-1} E_n^\epsilon(\delta M_i^\epsilon)'F_{uu}(U_i^\epsilon)\delta M_i^\epsilon, \tag{8.7}$$

$$\frac{1}{2}\lim_{n,\epsilon} \frac{1}{m_\epsilon} \sum_{i=n}^{n+m_\epsilon-1} E_n^\epsilon(g_i^\epsilon(\theta_i^\epsilon,\xi_i^\epsilon))'F_{uu}(U_i^\epsilon)g_i^\epsilon(\theta_i^\epsilon,\xi_i^\epsilon), \tag{8.8}$$

and

$$\lim_{n,\epsilon} \frac{1}{m_\epsilon} \sum_{i=n}^{n+m_\epsilon-1} E_n^\epsilon(Y_i^\epsilon)'F_{uu}(U_i^\epsilon)\Gamma_{i+1}^\epsilon(\theta_i^\epsilon). \tag{8.9}$$

Owing to the tightness of $\{U^\epsilon(\cdot)\}$ and the fact that the weak sense limit processes are continuous, the U_i^ϵ in these expressions can be changed to U_n^ϵ

without changing the limits. Then use the method of Theorem 8.4.1 and the conditions (A8.4)–(A8.7) to show that (8.7) has the same limit as

$$\frac{1}{2}\sum_{j,k} F_{u_j u_k}(U_n^\epsilon)\sigma_{0,jk},$$

(8.8) has the same limit as

$$\frac{1}{2}\sum_{j,k} F_{u_j u_k}(U_n^\epsilon)\bar{\sigma}_{0,jk},$$

and (8.9) has the same limit as

$$\sum_{j,k} F_{u_j u_k}(U_n^\epsilon)\sigma_{1,jk}.$$

Thus, the operator of the weak sense limit process is defined by

$$F_u'(U)AU + \frac{1}{2}\sum_{j,k} F_{u_j u_k}(U)\sigma_{jk}.$$

Then Theorem 7.4.4 yields the asserted Wiener process such that the weak sense limit processes satisfy (8.4). The stationarity is proved by a "time shifting" argument similar to the type used in the proof of Theorem 1.1 and is left to the reader. □

10.9 Limit Point on the Boundary

The previous sections dealt with the case where either the algorithm is unconstrained or the limit point $\bar{\theta}$ is interior to the constraint set H. It was proved that the normalized iterate processes $U^n(\cdot)$ or $U^\epsilon(\cdot)$ converges to a stationary Gaussian diffusion process, whose covariance (together with the scale factor) is taken to be a measure of the rate of convergence. When the algorithm is constrained and the limit point is on the boundary, the analysis is much more complicated. The stability methods that were used to prove tightness of the normalized iterates do not carry over, and getting the desired representation of the limit and proving its stationarity are more difficult. The full details for the case where H is a hyperrectangle are in [37, 39] and only a brief outline will be given; see also [134] for a synopsis of some simple cases. Note that the theory of large deviations can also be used [62, 63] to get estimates of the probability that some small region will be exited after some time, which is an alternative approach to the rate of convergence.

Let $H = \prod_{i=1}^r [a_i, b_i]$, where $-\infty < a_i < b_i < \infty$. Suppose that for some $i \le r$ the limit point $\bar{\theta}$ is on the part of the boundary where $\theta^i = b_i$, and

$\bar{g}_i(\bar{\theta}) > 0$, so that the dynamics in the limit would push the (unconstrained) iterate out of H. Then it can be shown that the asymptotic values of the error components $U_{n,i}$ (or $U^\epsilon_{n,i}$) are arbitrarily small, and this component can be eliminated from the analysis. The analogous situation holds if $\bar{\theta}^i = a_i$ and $\bar{g}_i(\bar{\theta}) < 0$. So, one need only work with the components i for which $\bar{g}_i(\bar{\theta}) = 0$.

Let us consider the case where $\theta_n \to \bar{\theta}$ with probability one and (A2.0) holds. (The weak convergence case is treated in [37].) Transform the coordinates so that $\bar{\theta} = 0$. Define the matrix $A = \bar{g}_\theta(\bar{\theta})$ as in (A1.6) or (A1.13) and define the deterministic *Skorohod or reflected ODE problems*

$$
\begin{aligned}
\dot{x} &= Ax + \dot{z} && \text{under (A2.0b)},\\
\dot{x} &= [A + I/2]x + \dot{z} && \text{under (A2.1a)}.
\end{aligned}
\tag{9.1}
$$

Here, $x(0) \in H$, A (resp., $A + I/2$) is assumed to be a Hurwitz matrix, and $\dot{z}$ is the smallest force that will keep the solution in H if it tries to leave there. The tightness proofs in Section 4 used a quadratic Liapunov function for these systems with the $\dot{z}$ dropped. Suppose that the solution converges to the origin for any initial condition. Then one can construct a Liapunov function that can be used to prove the tightness of $\{U_n\}$ and $U^\epsilon(\cdot)$; see [39]. The construction is based on results in [64] and is quite complicated since it must take the effects of the boundary into account. The main difficulties are due to the requirements of smoothness and boundedness of the derivatives, and that the value of the Liapunov function must decrease, not only along paths in H, but also under projections back to H from points not in H but close to $\bar{\theta}$.

Now drop the coordinates i for which $\bar{g}_i(\bar{\theta}) \neq 0$. Thus, we can suppose that $\bar{g}(\theta) = 0$. For notational simplicity and without loss of generality, suppose that $\bar{\theta}^i = a_i = 0$, for $i \leq k \leq r$, and $\bar{\theta}^i \in (a_i, b_i)$ for $k < i \leq r$ (in the latter case, $a_i < 0 < b_i$). Then the appropriate state space for the U_n and the limit processes is the hyperplane

$$
K = \{x : x_i \geq 0 \text{ for } i \leq k, x_i \text{ unconstrained for } k < i \leq r\}.
$$

Then, under the additional conditions used in the previous sections, the weak sense limits are the stationary processes with values in K defined by

$$
\begin{aligned}
dU &= AU dt + dW + dZ && \text{under (A2.0b)},\\
dU &= [A + I/2]U dt + dW + dZ && \text{under (A2.1a)},
\end{aligned}
\tag{9.2}
$$

where $W(\cdot)$ is a Wiener process as in the previous theorems, and $Z(\cdot)$ is the minimal force needed to keep the solution in K. The components $Z_i(\cdot)$ are identically zero for $k < i \leq r$. Full details are in the cited references.

The stationary covariance of processes such as (1.7) is easy to compute, but that for (9.2) is very hard. With constraints, there can be a great improvement in the trace of the covariance due to the dropping of the components i with $\bar{g}_i(\bar{\theta}) \neq 0$. Otherwise, evaluations by simulation indicate that there is usually some (but not necessarily great) improvement.

11
Averaging of the Iterates

11.0 Outline of Chapter

The difficulty of selecting a good step-size sequence $\{\epsilon_n\}$ has been a serious handicap in applications. In a fundamental paper, Polyak and Juditsky [195] showed that (loosely speaking) if ϵ_n goes to zero slower than $O(1/n)$, the averaged sequence $\sum_{i=1}^{n} \theta_i/n$ converges to its limit at an optimal rate. This result implies that we should use larger than usual gains and let the off-line averaging take care of the increased noise effects that are due to the larger step size, with substantial overall improvement. In addition, the basic stochastic approximation algorithm tends to be more robust with a larger step size in that it is less likely to get stuck at an early stage and more likely to have a faster initial convergence.

In Section 1, it is shown that the iterate averaging improves the asymptotic properties whenever there is a "classical" rate of convergence theorem of the types derived in Chapter 10 for both the decreasing-step-size and the constant-step-size algorithms. In other words, improvement in the rate of convergence due to iterate averaging is generic in stochastic approximation. Intuitive insight is provided by relating the behavior to that of a two-time-scale discrete algorithm in Section 2, where it is seen that the key property is the separation of the time scales of the averaged and original processes, which requires that $\epsilon_n \to 0$ more slowly than $1/n$, the "time scale" of the averaged process. The practical value of the averaging method and the predictions of the general theory have been supported by many simulations.

The usual idea is to select the step sizes so that an appropriate measure of the rate of convergence is maximized. Chapter 10 provides a convenient measure via the asymptotic covariance of the normalized processes $U^n(\cdot)$. Suppose that $\theta_n \to \bar{\theta}$ either with probability one or weakly. Then, as in Chapter 10, under appropriate conditions $(\theta_n - \bar{\theta})/\sqrt{\epsilon_n}$ converges in distribution to a normally distributed random variable with mean zero and some positive definite covariance matrix V_0. The matrix V_0 is often considered to be a measure of the "rate of convergence," taken together with the scale factors or step sizes ϵ_n.

Suppose that $\epsilon_n \to 0$ "slower" than $O(1/n)$. In particular, suppose that

$$\epsilon_n/\epsilon_{n+1} = 1 + o(\epsilon_n). \tag{0.1}$$

Define

$$\Theta_n = \frac{1}{n}\sum_{1}^{n} \theta_i. \tag{0.2}$$

Then [194] and [195] (and subsequently generalized in [47, 55, 149, 151, 152, 158, 265, 267], where the last reference corrected an error in [149]) showed that $\sqrt{n}(\Theta_n - \bar{\theta})$ converges in distribution to a normally distributed random variable with mean zero and covariance $\bar{V}$, where $\bar{V}$ is the smallest possible in some sense (to be defined in Section 2). If (0.1) holds and the sequence $\{\epsilon_n\}$ has some mild regularity properties, then the value of $\bar{V}$ does not otherwise depend on the particular sequence $\{\epsilon_n\}$. This weakening of the requirements on $\{\epsilon_n\}$ has important implications for applications. It says that the step sizes should be "relatively large." (Recall that the form $\epsilon_n = O(1/n)$ is the "classical" one recommended for the "tail" of the procedure, although it was well known that such a sequence leads to poor "finite time" behavior.) Indeed, in an asymptotic sense, one cannot do better than using (0.1) and (0.2) with real-valued ϵ_n, even in comparison with an algorithm where ϵ_n are allowed to be matrix-valued. The results of Ruppert [211] for a one-dimensional case contain similar conclusions. The development here is relatively simple, and we obtain much insight into the underlying reasons for the success of the iterate averaging method. The structure of the matrix $\bar{V}$ implies that the asymptotic errors are smaller in the more important directions.

Section 1 obtains a useful averaging result that follows directly from "classical" rate of convergence results using a "minimal window" of averaging, smaller than that used in (0.2). The result also holds for the Kiefer–Wolfowitz and constant-step-size algorithms. The iterate averaging does not affect the bias term in the Kiefer–Wolfowitz algorithm, but it does reduce the effects of the noise around the bias just as for the Robbins–Monro procedure. If we robustify the algorithm by adding periodic perturbations, similarly to what was discussed in Section 5.8, then a window of averaging of the order of the "minimum" might be appropriate. Then the iterate averaging would partially compensate for the "rate of convergence reducing" effects of the robustifying perturbations. The size of the "window of averaging" is extended to the maximal one in Section 3.

When there is only weak rather than probability one convergence, the result in Section 3 requires that there be no constraint set. This is due to the difficulties associated with treating the effects of the reflection terms Z_n, even if the limit point is inside the constraint set. These difficulties can be overcome by using the fact that the iterate spends a "very long time" near the limit point (Sections 6.10 and 10.5.2), even if there is weak but not probability one convergence. Results such as those in Section 3 can be extended to the decentralized/asynchronous algorithms of Chapter 12 and to the state-dependent-noise case of Section 8.4, but more work needs to be done on these more complicated classes of problems to obtain a reasonably simple development. Of course, results of the type in Section 1 for the smaller window of averaging do hold for these cases whenever there is a classical rate of convergence result, and the time scale separation idea of Section 2 still makes sense.

It was noted in [149] that the success of the averaging idea is due to the fact that the "time scales" of the original sequence $\{\theta_n\}$ and the averaged sequence $\{\Theta_n\}$ are "separated," with the time scale of the former sequence being the "faster" one. Of course, time scale separation arguments have been used in the analysis of stochastic approximation from the very beginning. But, the separation of the scales of the iterate $\{\theta_n\}$ and the noise $\{\xi_n\}$ sequences was the issue. Now there is an additional time scale to be considered, namely, that of the Θ_n-sequence. This idea will be discussed further in Section 2.

The averaging method is an off-line procedure in the sense that the stochastic approximation procedure is not influenced by Θ_n. In many cases, θ_n is of primary interest since it is the "operating" or physical parameter. We cannot simply substitute Θ_n for θ_n in the stochastic approximation algorithm without losing the advantages of iterate averaging. Thus, there is still the question of how we can exploit the averaging to improve the behavior of the $\{\theta_n\}$. This issue is dealt with in [152] for "linear" algorithms of the type appearing in parameter estimators and adaptive noise cancellers.

Section 4 contains an impressive application of iterate averaging to the parameter identification problem of Subsection 3.1.1. It is shown that the stochastic approximation algorithm with iterate averaging has the same asymptotic properties as the optimal least squares algorithm of Subsection 1.1.3 and is computationally much simpler. Some data for a constant step size form of the algorithm is given, and this supports the advantages of iterate averaging. The observation-averaging approach discussed in Subsection 1.3.4 has similar advantages.

11.1 Rate of Convergence of the Averaged Iterates: Minimal Window of Averaging

11.1.1 Robbins–Monro Algorithm: Decreasing Step Size

In this subsection, we work with the algorithm $\theta_{n+1} = \Pi_H(\theta_n + \epsilon_n Y_n)$, where $\epsilon_n \to 0$ and the limit point is denoted by $\bar{\theta}$. Recall the "interpolated time" $t_n = \sum_{i=0}^{n-1} \epsilon_i$, the definitions of the interpolated process, namely, $\theta^n(\cdot)$, and the interpolated normalized process $U^n(\cdot)$

$$\left.\begin{aligned} \theta^n(t) &= \theta_{n+i} \\ U^n(t) &= (\theta_{n+i} - \bar{\theta})/\sqrt{\epsilon_{n+i}} \end{aligned}\right\} \text{ for } t \in [t_{n+i} - t_n, t_{n+i+1} - t_n),\ i \geq 0.$$

These interpolations will be referred to as being on the ϵ_n-scale, to distinguish them from the $1/n$-scale introduced later. By (0.1), $n\epsilon_n \to \infty$.

In this section, we redefine the iterate average to be

$$\Theta_n = \frac{\epsilon_n}{t} \sum_{i=n}^{n+t/\epsilon_n - 1} \theta_i \tag{1.1a}$$

and define the associated normalized error

$$\widehat{U}^n(t) = \frac{\sqrt{\epsilon_n}}{t} \sum_{i=n}^{n+t/\epsilon_n - 1} (\theta_i - \bar{\theta}). \tag{1.1b}$$

In (1.1), the *window of averaging* is $t/\epsilon_n = o(n)$ for arbitrary real $t > 0$. Theorem 1.1 will show that it is the minimal window for which iterate averaging will improve the performance. The value of t can be made as large as desired in (1.1) and can go to infinity slowly as $n \to \infty$. Two-sided averages can also be used instead of the one-sided average in (1.1).

To reduce the essential requirements for studying the rate of convergence for the averaged iterates and to connect it with that of the classical algorithms, the following condition will be used. Recall that Chapter 10 was devoted to its verification.

(A1.1) There is a Hurwitz matrix A and a symmetric and positive-definite matrix Σ such that $\{U^n(\cdot)\}$ converges weakly to $U(\cdot)$, where $U(\cdot)$ is the stationary solution to

$$dU = AUdt + \Sigma^{\frac{1}{2}} dw, \tag{1.2}$$

where $w(\cdot)$ is a standard Wiener process (i.e., its covariance matrix is the identity matrix).

The "minimal window" of averaging convergence theorem. We establish the following theorem. It shows that the iterate averaging yields substantial improvements even with the minimal window of averaging, under essentially any conditions that guarantee the classical asymptotic normality of the $U^n(\cdot)$, such as was developed in Chapter 10.

Theorem 1.1. *Assume* (0.1) *and* (A1.1). *Define* $\bar{V} = A^{-1\prime}\Sigma(A')^{-1}$. *For each* t, $\widehat{U}^n(t)$ *converges in distribution to a normally distributed random variable with mean zero and covariance* $V_t = \bar{V}/t + O(1/t^2)$.

Proof. Define the covariance matrix $R(s) = EU(t+s)U'(t)$, where $U(\cdot)$ is the stationary solution to (1.2), and define

$$\bar{R} = EU(0)U'(0) = \int_0^\infty e^{As}\Sigma e^{A's} ds. \tag{1.3}$$

Then

$$R(s) = e^{As}\bar{R}, \ R(-s) = R'(s), \ s \geq 0, \tag{1.4}$$

and

$$\int_{-\infty}^\infty R(s)ds = -A^{-1}\bar{R} - \bar{R}(A^{-1})' = A^{-1}\Sigma(A^{-1})' = \bar{V}. \tag{1.5}$$

The first equality of (1.5) is obvious by evaluating the integral on the left side. The second equality of (1.5) follows by noting that $\bar{R}$ defined in (1.3) solves the algebraic Ricatti equation

$$-\bar{R}A' - A\bar{R} = \Sigma$$

and multiplying this on the right by $(A^{-1})'$ and on the left by A^{-1}.

Define the processes

$$\widetilde{U}^n(t) = \frac{1}{\sqrt{t}} \int_0^t U^n(s)ds, \ \ \widetilde{U}(t) = \frac{1}{\sqrt{t}} \int_0^t U(s)ds.$$

Since $R(s) \to 0$ exponentially as $s \to \infty$, we can write

$$\begin{aligned} \text{Cov } \widetilde{U}(t) &= \frac{1}{t} \int_0^t \int_0^t R(s-\tau)dsd\tau \\ &= \int_{-\infty}^\infty R(s)ds + O(1/t) = \bar{V} + O(1/t). \end{aligned} \tag{1.6}$$

The basic result on the averaged iterates is obtained by relating $U^n(t)$ to $\widetilde{U}^n(t)$. Equation (0.1) implies that for any $t < \infty$,

$$\max\{i - n : 0 \leq t_i - t_n \leq t\} \cdot \epsilon_n / t \overset{n}{\to} 1. \tag{1.7}$$

Equivalently, asymptotically, there are t/ϵ_n iterates in the interpolated time interval $[0, t]$ for $U^n(\cdot)$. The relation (1.7) would not hold if $\epsilon_n = O(1/n)$.

By the definition of $\widetilde{U}^n(\cdot)$, for $i \geq n$

$$\begin{aligned} \sqrt{t}\widetilde{U}^n(t) &= \sum_{i:t_i - t_n \leq t} \left((\theta_i - \bar{\theta}) \, \epsilon_i^{-1/2} \right) \epsilon_i \\ &= \sum_{i:t_i - t_n \leq t} (\theta_i - \bar{\theta}) \left(\epsilon_i^{1/2} - \epsilon_n^{1/2} \right) + \epsilon_n^{1/2} \sum_{i:t_i - t_n \leq t} (\theta_i - \bar{\theta}). \end{aligned} \tag{1.8}$$

By the weak convergence of $U^n(\cdot)$ in (A1.1), $\widetilde{U}^n(\cdot) \Rightarrow \widetilde{U}(\cdot)$ in $D^r[0, \infty)$. Then (0.1) and (A1.1) imply that the first sum in (1.8) goes to zero in probability as $n \to \infty$. By (1.7) and the weak convergence in (A1.1), the difference between $\sqrt{t}\widehat{U}^n(t)$ and the second sum in (1.8) goes to zero in probability as $n \to \infty$, which yields the desired conclusion. □

The optimality of the "rate of convergence" of the sequence of averages. Let us suppose that the normalized error defined by $U_n = (\theta_n - \bar{\theta})/\sqrt{\epsilon_n}$ converges in distribution to a normally distributed random variable $\widehat{U}$ with mean zero, as implied by (A1.1). Repeating some of the discussion in Chapter 10, the covariance of $\widehat{U}$, taken together with the step size sequence ϵ_n, is a traditional measure of the rate of convergence of θ_n to the limit $\bar{\theta}$. In this sense, the best (asymptotic) value of ϵ_n is $O(1/n)$. Pursuing this value, let $\epsilon_n = K/n$, where K is a nonsingular matrix. To get the best asymptotic rate, one needs to get the best K. Under appropriate conditions, it is a classical result (see Theorem 10.2.1, the discussion below it on the optimal step size sequence and [16, 135, 250]) that $U^n(\cdot) \Rightarrow U_0(\cdot)$, where $U_0(\cdot)$ is the stationary solution to

$$dU_0 = \left(KA + \frac{I}{2} \right) U_0 dt + K\Sigma^{\frac{1}{2}} dw, \tag{1.9}$$

where $w(\cdot)$ is a standard Wiener process. It is obviously required that $(KA + I/2)$ be a Hurwitz matrix, a condition that is not needed for the iterate averaging result to hold.

The stationary covariance matrix of the solution of (1.9) is given by (1.3) with $KA + I/2$ replacing A and $K\Sigma K'$ replacing Σ. Using $\Sigma = A\bar{V}A'$, it can be written as

$$\int_0^\infty e^{(B+I/2)t} B\bar{V}B' e^{(B'+I/2)t} dt, \tag{1.10}$$

where $B = KA$. By minimizing either the trace of (1.10) over B or (1.10) over B in the sense of positive definite matrices, the best value of the matrix K is given by $B = -I$, namely, $K = -A^{-1}$. With this asymptotically optimal value of K, the stationary covariance of $U_0(t)$ is just $\bar{V}$.

Loosely speaking, for the asymptotically optimal matrix-valued step size $\epsilon_n = K/n, K = -A^{-1}$, we have $(\theta_n - \bar{\theta}) \sim N(0, \bar{V}/n)$. For the centered iterate average,

$$(\Theta_n - \bar{\theta}) = \frac{\epsilon_n}{t} \sum_{i=n}^{n+t/\epsilon_n - 1} (\theta_i - \bar{\theta}),$$

Theorem 1.1 yields (modulo an error of $O(\epsilon_n/t^2)$ in the variance)

$$(\Theta_n - \bar{\theta}) \sim N(0, \epsilon_n \bar{V}/t) = N\left(0, \frac{\bar{V}}{\text{\# of iterates in window}}\right). \tag{1.11}$$

To get this result, we do not need to know the matrix A, which would generally be very difficult to obtain. We also use step sizes that go to zero more slowly than $O(1/n)$ yielding better performance in the transient period.

11.1.2 *Constant Step Size*

Now, turn to the constant-step-size algorithm $\theta_{n+1}^\epsilon = \Pi_H(\theta_n^\epsilon + \epsilon Y_n^\epsilon)$. Let q_ϵ be a nondecreasing sequence of positive integers, and (as in Chapter 10) define the interpolation of the normalized iterates by $U^\epsilon(t) = (\theta_{q_\epsilon+n}^\epsilon - \bar{\theta})/\sqrt{\epsilon}$ for t on the interval $[\epsilon n, \epsilon n + \epsilon)$. Write $t_\epsilon = \epsilon q_\epsilon$ and suppose that $t_\epsilon \to \infty$. Letting $t_\epsilon \to \infty$ accounts for the transient period. Suppose that $\{U^\epsilon(\cdot)\}$ converges weakly to the stationary solution to (1.2), where A is a Hurwitz matrix (see Chapter 10). Let $t_1 + t_2 = t > 0$, where $t_i \geq 0$ and define the average of the iterates

$$\Theta_n^\epsilon = \frac{\epsilon}{t} \sum_{(t_\epsilon - t_2)/\epsilon}^{(t_\epsilon + t_1)/\epsilon - 1} \theta_i^\epsilon. \tag{1.12}$$

Define

$$\widehat{U}^\epsilon(t) = \frac{\sqrt{\epsilon}}{t} \sum_{(t_\epsilon - t_2)/\epsilon}^{(t_\epsilon + t_1 - 1)/\epsilon} (\theta_i^\epsilon - \bar{\theta}).$$

A proof very close to that of Theorem 1.1 gives the analogous conclusion, namely, that for each t, $\widehat{U}^\epsilon(t)$ converges in distribution to a normally distributed random variable with mean zero and covariance $\bar{V}/t + O(1/t^2)$. Formally, (modulo an error of $O(\epsilon/t^2)$ in the variance)

$$\Theta_n^\epsilon \sim N(\bar{\theta}, \epsilon \bar{V}/t) = N\left(\bar{\theta}, \frac{\bar{V}}{\text{\#of iterates in window}}\right). \tag{1.13}$$

A discounted form defined by

$$\Theta_n^{\epsilon,\alpha} = \alpha\epsilon \sum_{i=-\infty}^{n} (1 - \alpha\epsilon)^{n-i} \theta_i^\epsilon \tag{1.14}$$

can also be used. Then $1/\alpha$ replaces t, and for small positive α and ϵ, the middle term in (1.13) is replaced by $N(0, \alpha\epsilon\bar{V})$ and $1/(\alpha\epsilon)$ becomes the "effective" number of iterates in the "window." The discounted form can be used, in principle, for the tracking of time varying systems, provided that the variation over $1/(\alpha\epsilon)$ iterates is not too big so that there is a good approximation to averaging. This issue is pursued in [167, 168].

11.1.3 Averaging with Feedback and Constant Step Size

Up to now, the averaging was "off line" in that the averaging had no effect on the algorithm for computing θ_n^ϵ. Recall the discussion of averaging with feedback in Section 3.3 and, for $k > 0$, consider the algorithm

$$\theta_{n+1}^\epsilon = \theta_n^\epsilon + \epsilon Y_n^\epsilon + \epsilon k \left(\Theta_n^{\epsilon,\alpha} - \theta_n^\epsilon\right),$$

where $\Theta_n^{\epsilon,\alpha}$ is defined by the discounted form (1.14) and define $\bar{U}_n^{\epsilon,\alpha} = \Theta_n^{\epsilon,\alpha}/\sqrt{\epsilon}$. A formal analysis of the dependence on k and α will indicate the advantages to be gained by feedback. Using $Q = -A$, the limit equations for the normalized variables are

$$\begin{aligned} dU &= -(Q + kI)U dt + k\bar{U} dt + \Sigma^{1/2} dw, \\ d\bar{U} &= \alpha U dt - \alpha \bar{U} dt. \end{aligned} \tag{1.15}$$

The stationary variance is

$$\bar{R} = \int_0^\infty e^{Mt} \bar{\Sigma} e^{M't} dt = \begin{bmatrix} \bar{R}_{11} & \bar{R}_{12} \\ \bar{R}_{21} & \bar{R}_{22} \end{bmatrix},$$

where

$$\bar{\Sigma} = \begin{bmatrix} \Sigma & 0 \\ 0 & 0 \end{bmatrix}, \qquad M = \begin{bmatrix} -(Q + kI) & kI \\ \alpha I & -\alpha I \end{bmatrix}.$$

We have $-\bar{R}M' - M\bar{R} = \bar{\Sigma}$, yielding ($R_{12} = R_{21}'$)

$$\begin{aligned} \Sigma &= \left[\bar{R}_{11}(Q' + kI) - \bar{R}_{12}k\right] + \left[(Q + kI)\bar{R}_{11} - k\bar{R}_{21}\right], \\ 0 &= \left[-\alpha\bar{R}_{11} + \alpha\bar{R}_{12}\right] + \left[(Q + kI)\bar{R}_{12} - k\bar{R}_{22}\right], \\ 0 &= [-\bar{R}_{21} + \bar{R}_{22}] + [-\bar{R}_{12} + \bar{R}_{22}]. \end{aligned}$$

This yields $\bar{R}_{22} = [\bar{R}_{12} + \bar{R}_{21}]/2$ and $[H + O(\alpha)]\bar{R}_{12} = \alpha\bar{R}_{11}$, for some matrix H and $\bar{R}_{11}(Q' + kI) + (Q + kI)\bar{R}_{11} = \Sigma + O(\alpha)$. Hence

$$\bar{R}_{11} = O(1/k), \quad \bar{R}_{22} = O(\alpha/k), \tag{1.16}$$

which is consistent with [151, 152].

11.1.4 Kiefer–Wolfowitz Algorithm

There is a result analogous to Theorem 1.1 for the Kiefer–Wolfowitz procedure with either one- or two-sided finite differences used, as well as for the random directions method, but the statement is a little more complicated due to the bias terms $-c^2B(\bar{\theta})$ in (10.3.3) or $-c\widehat{B}$ in (10.3.9). Define U_n, ϵ_n, c_n, and β as in Theorem 10.3.1. Thus, $\epsilon_n = 1/n^\gamma$ and (0.1) requires that $\gamma < 1$. The next condition replaces (A1.1).

(A1.2) There is a Hurwitz matrix A, a vector $\bar{B}$, and a symmetric and positive-definite matrix Σ, such that $\{U^n(\cdot)\}$ converges weakly to $U(\cdot)$ which is the stationary solution to

$$dU = AUdt - \bar{B}dt + \Sigma^{\frac{1}{2}}dw, \tag{1.17}$$

and $w(\cdot)$ is a standard Wiener process.

Note that (1.17) implies that the stationary mean bias is $A^{-1}\bar{B}$. In Theorem 10.3.1, $\bar{B} = c^2B(\bar{\theta})$. The iterate average is now defined by (1.1a), but the normalized average is redefined to be

$$\widehat{U}^n(t) = n^\beta \frac{\epsilon_n}{t} \sum_{i=n}^{n+t/\epsilon_n - 1} (\theta_i - \bar{\theta}). \tag{1.18}$$

The expression (1.18) is n^β times the iterate average, in lieu of $1/\sqrt{\epsilon_n}$ times the iterate average as used in (1.1b), analogous to the differences in normalization used in Chapter 10 for the Robbins–Monro and the Kiefer–Wolfowitz procedures. Theorem 1.2 covers the one- and two-sided difference interval algorithms, as well as the random directions algorithm.

Theorem 1.2 says that as the size of the window of averaging t grows, the bias does not decrease, but the fluctuations about the bias will have reduced covariance, analogously to the case of Theorem 1.1. Even with substantial bias, the averaging can be very beneficial, because the reduced fluctuations around the bias yields a more robust and reliable procedure. Without the averaging, the asymptotic covariance of $n^\beta(\theta_n - \bar{\theta})$ is the $\bar{R}$ defined by (1.3). The reader should keep in mind that in practice one does not usually let the difference interval go to zero.

Theorem 1.2. *Assume* (A1.2). *Then $\widehat{U}^n(t)$ is asymptotically normal with mean $A^{-1}\bar{B}$ and covariance matrix $\bar{V}/t + O(1/t^2)$.*

Comment on the proof. The proof follows the lines of that of Theorem 1.1, and only a few comments will be made. Under (0.1) and (A1.2), (1.18) can be approximated by

$$\frac{1}{t} \sum_{i=n}^{n+t/\epsilon_n - 1} \left(\theta_i - \bar{\theta}\right) i^\beta \epsilon_i,$$

which equals

$$\frac{1}{t}\int_0^t U^n(s)ds.$$

In turn, this expression converges weakly to

$$\frac{1}{t}\int_0^t U(s)ds, \tag{1.19}$$

where $U(\cdot)$ is the stationary solution to (1.17). The mean of (1.19) is $A^{-1}\bar{B}$, and it does not depend on t due to the stationarity of $U(\cdot)$. The covariance of the integral in (1.19) is $1/t$ times (1.6).

11.2 A Two-Time-Scale Interpretation

A key to understanding why the averaging works when (0.1) holds but does not work when $\epsilon_n = O(1/n)$ can be seen by rewriting the recursive formulas for θ_n and Θ_n in the same time scale. The discussion will be only for the purposes of insight and motivation. Thus, in order not to encumber the notation, we work with a linear one-dimensional model. Let $\epsilon_n = 1/n^\gamma$, where $\gamma \in (0,1)$, and let the algorithm be

$$\theta_{n+1} = \left(1 + \frac{A}{n^\gamma}\right)\theta_n + \frac{\xi_n}{n^\gamma},$$

where $A < 0$. Rewrite the algorithm as

$$\frac{1}{n^{1-\gamma}}(\theta_{n+1} - \theta_n) = \frac{A\theta_n}{n} + \frac{\xi_n}{n}, \tag{2.1}$$

and write the Θ_n of (0.2) recursively as

$$\Theta_{n+1} = \Theta_n - \frac{1}{n+1}\Theta_n + \frac{\theta_{n+1}}{n+1}. \tag{2.2}$$

The purpose of (2.1) is the estimation of the mean $\bar{\theta} = 0$. Due to the presence of the $1/n^{1-\gamma}$ factor, the algorithm (2.1) and (2.2) can be viewed as a two time scale or singularly perturbed stochastic approximation, although the time scale of the first component is "time varying." If $\gamma = 1$, then the scales of $\{\theta_n\}$ and $\{\Theta_n\}$ are the same, and the driving term θ_{n+1} in the Θ_n-equation has a correlation time of the order of that of Θ_n itself. If $\gamma < 1$, then the correlation time of the driving θ_n process in (2.2) gets shorter and shorter (as seen from the point of view of the $1/n$ scale used in (2.1) and (2.2)) as $n \to \infty$, and θ_n behaves more and more like a "white noise." With the "approximately white noise" property in the $1/n$ time scale, the averaging can produce a substantial reduction in the variance, whereas if the noise were highly correlated in this $1/n$ scale, the reduction would be minor.

The scheme of (2.1) and (2.2) loosely resembles the continuous-time, two-time-scale system

$$\begin{aligned} \epsilon\, dz^\epsilon &= a_{11} z^\epsilon dt + dw_1 \\ dx^\epsilon &= a_{22} x^\epsilon dt + a_{21} z^\epsilon dt + dw_2, \end{aligned} \tag{2.3}$$

for a small parameter ϵ. It is shown in [132] that (under suitable stability conditions) the sequence of processes defined by $\{\int_0^t z^\epsilon(s)ds, \epsilon > 0\}$ converges weakly to a Wiener process as $\epsilon \to 0$. This result and the resemblance of (2.3) to the form (2.1) and (2.2) suggest that the sequence of functions of t defined by

$$\frac{1}{\sqrt{n}} \sum_{i=0}^{nt} (\theta_i - \bar{\theta})$$

might converge weakly to a Wiener process with covariance matrix $\bar{V}$ as $n \to \infty$. This is one type of extension of Theorem 1.1 to the maximal window case and is dealt with in [149].

11.3 Maximal Window of Averaging

In the discussion in Sections 1 and 2, the size of the window of averaging was t/ϵ_n (or t/ϵ for the constant-step-size case). Equivalently, $\epsilon_n \times$(size of window) does not go to infinity as $n \to \infty$. We now consider the case where $\epsilon_n \times$(size of window)$\to \infty$.

The algorithm is (6.1.1), where $\mathcal{F}_n$, E_n, and ξ_n are as defined above (6.1.2). Under (A3.2), the algorithm can be written as

$$\theta_{n+1} = \theta_n + \epsilon_n g_n(\theta_n, \xi_n) + \epsilon_n \delta M_n + \epsilon_n Z_n.$$

For a sequence $\{q_n\}$ of real numbers, we will be concerned with either of the averages

$$\Theta_n = \frac{1}{q_n} \sum_{i=n}^{n+q_n-1} \theta_i, \quad \text{or} \quad \Theta_n^0 = \frac{1}{q_n} \sum_{i=0}^{q_n-1} \theta_i. \tag{3.1}$$

Define the corresponding centered and normalized averages,

$$\widehat{U}_n = \frac{1}{\sqrt{q_n}} \sum_{i=n}^{n+q_n-1} (\theta_i - \bar{\theta}), \quad \widehat{U}_n^0 = \frac{1}{\sqrt{q_n}} \sum_{i=0}^{q_n-1} (\theta_i - \bar{\theta}), \tag{3.2}$$

respectively. The following assumptions will be used; the weak convergence case will also be treated. Section 10.4 gives conditions for (3.3).

(A3.1) $\theta_n \to \bar{\theta}$ with probability one, where $\bar{\theta}$ is strictly inside the constraint set H. For small $\rho > 0$ and with perhaps changing the algorithm on a set of arbitrarily small probability,

$$\limsup_n E|\theta_n - \bar{\theta}|^2 I_{\{|\theta_n - \bar{\theta}| \le \rho\}} / \epsilon_n < \infty. \tag{3.3}$$

(A3.2) For small $\rho > 0$, $\sup_n E|Y_n|^2 I_{\{|\theta_n - \bar{\theta}| \le \rho\}} < \infty$. There are measurable functions $g_n(\cdot)$ such that $E_n Y_n = g_n(\theta_n, \xi_n)$.

(A3.3) $n\epsilon_n / \log n \to \infty$; $q_n \to \infty$ such that $\epsilon_n^2 q_n \to 0$ and $\epsilon_n q_n \to \infty$.

(A3.4) The following holds:

$$\lim_{\alpha \uparrow 1} \limsup_n \sup_{m(\alpha t_n) \le i \le n} \left| \frac{\epsilon_i}{\epsilon_n} - 1 \right| = \lim_{\alpha \uparrow 1} \limsup_t \sup_{\alpha t \le s \le t} \left| \frac{\epsilon_{m(s)}}{\epsilon_{m(t)}} - 1 \right| = 0.$$

(A3.5) $\limsup_n \sup_{i \ge n} \epsilon_i / \epsilon_n < \infty$.

(A3.6) $g_n(\cdot, \xi)$ is continuously differentiable for each ξ and can be written as

$$g_n(\theta, \xi) = g_n(\bar{\theta}, \xi) + g'_{n,\theta}(\bar{\theta}, \xi)(\theta - \bar{\theta}) + O(|\theta - \bar{\theta}|^2), \tag{3.4}$$

where $O(\cdot)$ is uniform in ξ and n.

(A3.7) Recall the definition of $Y_n(\bar{\theta})$ given below (10.2.8) and of $\delta M_n(\bar{\theta})$ in (A10.2.9), which is the observation (resp., the martingale difference) at time n when the parameter is set at $\bar{\theta}$ at that time. For small $\rho > 0$,

$$\lim_n E|\delta M_n - \delta M_n(\bar{\theta})|^2 I_{\{|\theta_n - \bar{\theta}| \le \rho\}} = 0. \tag{3.5}$$

There are matrices $R(l)$ and non-negative numbers $\phi(l)$ such that

$$\sum_{l=-\infty}^{\infty} |R(l)| < \infty, \quad \sum_{l=-\infty}^{\infty} \phi(l) < \infty \tag{3.6a}$$

and

$$\frac{1}{m} E \sum_{i=n}^{n+m-1} Y_i(\bar{\theta}) Y'_{i+l}(\bar{\theta}) \to R(l) \tag{3.6b}$$

uniformly in l and n as $m \to \infty$ and

$$\left| E Y_i(\bar{\theta}) Y'_{i+l}(\bar{\theta}) \right| \le \phi(l).$$

(A3.8) Recall the definition of Γ_n^d of (10.4.18); namely,

$$\Gamma_n^d = \sum_{i=n}^{\infty} \epsilon_i \Pi(n+1, i) E_n \left[g'_{i,\theta}(\bar{\theta}, \xi_i) - A \right],$$

where A is a Hurwitz matrix. Then, for small $\rho > 0$,

$$E|\Gamma^d_{n+1}|\,(1+|Y_n|)\,I_{\{|\theta_n-\bar{\theta}|\le\rho\}} = O(\epsilon_n), \tag{3.7a}$$

and

$$E|\Gamma^d_{n+1} - E_n\Gamma^d_{n+1}|^2 = O(\epsilon_n^2). \tag{3.7b}$$

For the second case in (3.1), we will also need the following (which holds if $\epsilon_n = 1/n^\gamma, q_n = n, \gamma > 0.5$, because then t_n is asymptotically proportional to $n^{1-\gamma}$).

(A3.9) $t_n/\sqrt{q_n} \to 0.$

Comments on the conditions. (3.5) is not a strong condition since $\theta_n \to \bar{\theta}$. Note that for $\epsilon_n = 1/n$, $\epsilon_{m(t)} \approx e^{-t}$ and (A3.4) fails. If $\epsilon_n = 1/n^\gamma$ for $0 < \gamma < 1$, then $\epsilon_{m(t)} \approx \text{constant}/t^{\gamma/(1-\gamma)}$ and the condition holds. Note also that (A3.4) implies that $\epsilon_{m(t)}$ goes to zero slower that any exponential function of t. Condition (A3.4) implies that

$$\int_0^t e^{A(t-s)}\left[\frac{\epsilon_{m(s)}}{\epsilon_{m(t)}}\right] ds \to -A^{-1} \tag{3.8}$$

as $t \to \infty$; this fact will be used in the proof. Equation (3.8) follows from the expansion, where $\alpha < 1$ and is close to unity,

$$\begin{aligned}\int_0^t & e^{A(t-s)}\frac{\epsilon_{m(s)}}{\epsilon_{m(t)}}ds \\ &= \int_{t-\alpha t}^t e^{A(t-s)}\frac{\epsilon_{m(s)}}{\epsilon_{m(t)}}ds + \frac{e^{A\alpha t}}{\epsilon_{m(t)}}\int_0^{t-\alpha t} e^{A(t-\alpha t-s)}\epsilon_{m(s)}ds.\end{aligned}$$

By (A3.4), the first term on the right tends to $-A^{-1}$ as $t \to \infty$. The integral in the second term on the right goes to zero as $t \to \infty$, and the factor multiplying the integral goes to zero because (A3.4) implies that $\epsilon_{m(t)} \to 0$ slower than any exponential.

Theorem 3.1. *Assume* (A3.1)–(A3.8). *For the second case in* (3.1), *assume* (A3.9) *as well. Then*

$$\widehat{U}_n = \widetilde{U}_n + Q^1_n + Q^2_n,$$

where $E|Q^1_n| \to 0$, $P\{Q^2_n \neq 0\} \to 0$, *and* $E\widetilde{U}_n\widetilde{U}'_n \to \bar{V}$, *as* $n \to \infty$, *where* $\bar{V}$ *is defined by* (1.5) *with* $\Sigma = \sum_{j=-\infty}^{\infty} R(j)$. *If the w.p.1 convergence condition is replaced by weak convergence and the algorithm is unconstrained, then the assertion in the last sentence continues to hold, provided that*

$$|\Gamma^d_n|/\epsilon_n \tag{3.9}$$

is bounded uniformly in n *with probability one, and the indicator function is dropped in* (A3.2), (3.3), (3.5), *and* (3.7a). *The analog of the last assertion for the constant-step-size case holds. For the discounted algorithm* (1.14), *as* $\epsilon \to 0$, $[\Theta_n^\epsilon - \bar{\theta}]/\sqrt{\epsilon} \sim N(0, \alpha\bar{V} + O(\alpha^2))$.

Comment on the theorem. Under stronger conditions, one can actually get weak convergence results of the type in [149] (modulo a minor correction [151] in the proofs there) for the processes

$$\widehat{U}_n(t) = \frac{1}{\sqrt{q_n}} \sum_{i=n}^{n+q_n t-1} (\theta_i - \bar{\theta}).$$

The proof of such a weak convergence result involves many details, and we wish to continue with a relatively simple development to show the general validity of the iterate averaging idea. The current proof involves essentially standard algebraic manipulations. To see the optimality of the iterate averaging procedure, let $q_n = n$. Ignoring a set of arbitrarily small probability, loosely speaking the theorem asserts that asymptotically

$$\Theta_n - \bar{\theta} = \frac{1}{n} \sum_{n}^{2n-1} (\theta_i - \bar{\theta}) \sim N(0, \bar{V}/n).$$

This is what we get if $\epsilon_n = K/n$, where $K = -A^{-1}$, its asymptotically optimal matrix value. Estimates such as used in Section 10.5.2 can be employed to extend the theorem in the weak convergence case.

Proof. Assume probability one convergence until further notice. Since $\theta_n \to \bar{\theta}$ with probability one, for any $\mu > 0$ the process $\{\theta_i, i \geq N\}$ can be modified on a set whose probability goes to zero as $N \to \infty$ such that the assumptions continue to hold but $|\theta_n - \bar{\theta}| \leq \mu$ for $n \geq N$. By shifting the origin to N, we can suppose that $|\theta_n - \bar{\theta}| \leq \mu$ for all n. In fact, we can suppose that there are real $\mu_n \to 0$ such that $|\theta_n| \leq \mu_n$. This modification and the modification alluded to in (A3.1) give rise to the Q_n^2 term in the statement of the theorem. Let $\mu > 0$ be small enough so that the θ_n are interior to the constraint set. Thus we can set $Z_n = 0$ for all n without loss of generality.

The basic difficulty in getting the averaging result is due to the nonlinearity of $g_n(\cdot, \xi)$. The long-range correlations among the iterates that are due to the nonlinearities need to be appropriately bounded. Keep in mind that the effective time interval over which we must work is larger than that at which the weak convergence arguments of Chapters 8 and 10 hold. This follows from the fact that $\epsilon_n q_n \to \infty$. Furthermore, the averaged process is defined on the time scale with intervals $1/n$ and an interval of time $[a_n, a_n + b]$ on this scale uses the iterates on an interval in the ϵ_n-scale whose length goes to infinity as $a_n \to \infty$.

A linearized algorithm. Henceforth, let $\bar{\theta} = 0$ for notational simplicity. Dropping the Z_n-term, expand the algorithm and introduce $Y_n(\bar{\theta}) = Y_n(0) = g_n(0, \xi_n) + \delta M_n(0)$ as

$$\begin{aligned}\theta_{n+1} &= (I + \epsilon_n A)\theta_n \\ &\quad + \epsilon_n \left[g_n(0,\xi_n) + \left(g'_{n,\theta}(0,\xi_n) - A\right)\theta_n + \delta M_n\right] + \epsilon_n O(|\theta_n|^2) \\ &= (I + \epsilon_n A)\theta_n + \epsilon_n Y_n(0) \\ &\quad + \epsilon_n \left[\left(g'_{n,\theta}(0,\xi_n) - A\right)\theta_n + (\delta M_n - \delta M_n(0)) + O(|\theta_n|^2)\right].\end{aligned}$$

Define the perturbed iterate $\widehat{\theta}_n = \theta_n + \Gamma^d_n \theta_n$, and define the martingale difference $\delta\widetilde{M}_n$ by $\epsilon_n \delta\widetilde{M}_n = (\Gamma^d_{n+1} - E_n\Gamma^d_{n+1})\theta_n$. Then we can write

$$\begin{aligned}\widehat{\theta}_{n+1} &= (I + \epsilon_n A)\widehat{\theta}_n + \epsilon_n Y_n(0) + \epsilon_n \left[\delta M_n - \delta M_n(0)\right] \\ &\quad + \epsilon_n O(|\theta_n|^2) - \epsilon_n A\Gamma^d_n\theta_n + \epsilon_n \Gamma^d_{n+1} Y_n + \epsilon_{n+1} E_n \Gamma^d_{n+1}\theta_n + \epsilon_n \delta\widetilde{M}_n.\end{aligned}$$

Rewrite the iteration in the form

$$\widehat{\theta}_{n+1} = (I+\epsilon_n A)\widehat{\theta}_n + \epsilon_n Y_n(0) + \epsilon_n v_n + \epsilon_n \left[\delta M_n - \delta M_n(0) + \delta\widetilde{M}_n\right], \quad (3.10)$$

where v_n is defined in the obvious way. We work with $\widehat{U}_n$ in (3.2), assuming that $|\theta_n| \le \mu$ for all n, where μ is arbitrarily small. The method for $\widehat{U}^0_n$ is almost identical to that for $\widehat{U}_n$, and will be commented on as needed. If $\widehat{\theta}_i$ replaces θ_i in $\widehat{U}_n$, then the contribution of the perturbation $\Gamma^d_n\theta_n$ is

$$\frac{1}{\sqrt{q_n}} \sum_{i=n}^{n+q_n-1} \Gamma^d_i \theta_i.$$

By (A3.8), this is bounded in mean by

$$O(1)\frac{1}{\sqrt{q_n}} \sum_{i=n}^{n+q_n-1} \epsilon_i \mu \le O(1)\mu\epsilon_n q_n^{1/2},$$

which goes to zero by the middle part of (A3.3). For $\widehat{U}^0_n$, the left side is $O(1)\mu t_n/\sqrt{q_n}$, which goes to zero by (A3.9). Thus, it is sufficient to prove the result with $\widehat{\theta}_n$ replacing θ_n and work with the new definition of $\widehat{U}_n$ given by

$$\widehat{U}_n = \frac{1}{\sqrt{q_n}} \sum_{i=n}^{n+q_n-1} \widehat{\theta}_i.$$

An analogous calculation shows that we can suppose that $\widehat{\theta}_0 = 0$ in (3.10) without loss of generality. To see this, define $\Pi_A(i,n) = \prod_{j=i}^n (I + \epsilon_j A)$ and note that

$$\widehat{\theta}_{n+1} = \Pi_A(0,n)\widehat{\theta}_0 + \sum_{l=0}^{n} \epsilon_l \Pi_A(l+1,n)\kappa_l, \quad (3.11)$$

where κ_n is the sum of all the terms after the $(I+\epsilon_n A)\widehat{\theta}_n$ in (3.10). The contribution of the initial condition $\widehat{\theta}_0$ to $\widehat{U}_n$ is

$$\frac{1}{\sqrt{q_n}}\sum_{i=n}^{n+q_n-1}\Pi_A(0,i-1)\widehat{\theta}_0.$$

There is a $\beta>0$ such that this last expression is bounded in mean by

$$O(1)\frac{1}{\sqrt{q_n}}\sum_{i=n}^{n+q_n-1}e^{-\beta t_i}=O(1)\sqrt{q_n}e^{-\beta t_n}.$$

The right-hand side goes to zero since both $e^{-\beta t_n}/\epsilon_n$ and $\sqrt{q_n}\epsilon_n$ go to zero. A similar argument can be used for $\widehat{U}_n^0$.

It will be shown next that the nonlinear terms v_n can be ignored. With $\widehat{\theta}_0=0$, we have

$$\widehat{U}_n=\frac{1}{\sqrt{q_n}}\sum_{i=n}^{n+q_n-1}\sum_{l=0}^{i-1}\epsilon_l\Pi_A(l+1,i-1)[Y_l(0)+v_l+(\delta M_l-\delta M_l(0)+\delta\widetilde{M}_l)].$$

We need to show that

$$\frac{1}{\sqrt{q_n}}\sum_{i=n}^{n+q_n-1}\sum_{l=0}^{i-1}\epsilon_l\Pi_A(l+1,i-1)E|v_l|\to 0 \tag{3.12}$$

as $n\to 0$. Note that v_n contains the terms

$$O(|\theta_n|^2),\ \Gamma_{n+1}^d Y_n,\ E_n\Gamma_{n+1}^d\theta_n,\ A\Gamma_n^d\theta_n$$

whose mean values are of the order of ϵ_n by (A3.1) and (3.7). Thus, asymptotically, the left side of (3.12) is of the order of

$$\frac{1}{\sqrt{q_n}}\sum_{i=n}^{n+q_n-1}\int_0^{t_i}e^{A(t_i-s)}\epsilon_{m(s)}ds. \tag{3.13}$$

By (3.8), the integral in (3.13) is of the order of ϵ_i. Hence (3.13) is of the order of

$$\frac{1}{\sqrt{q_n}}\sum_{i=n}^{n+q_n-1}\epsilon_i,$$

which goes to zero as $n\to 0$ by (A3.5) and the fact that $q_n\epsilon_n^2\to 0$. For (3.2b), use (A3.9). Thus the contributions of $\{v_i\}$ are included in Q_n^1 in the statement of the theorem.

The effects of the martingale differences $\{\delta M_n-\delta M_n(0)\}$ and $\{\delta\widetilde{M}_n\}$ are handled in a similar manner. Calculate the mean square value of the left side of (3.12) with $E|v_l|$ replaced by $\delta M_l-\delta M_l(0)$ (resp., by $\delta\widetilde{M}_l$)

and then use $E[\delta M_l - \delta M_l(0)]^2 \to 0$ (resp., $E|\delta \widetilde{M}_l|^2 = O(\mu^2)$, where μ is arbitrarily small) to get that the mean square value goes to zero as $n \to \infty$ (resp., is bounded by $O(\mu^2)$). The details are similar to what will be done in evaluating the S_n and T_n below, with the appropriate replacement for $R(0)$ there. Thus the contributions of the $\{\delta M_i - \delta M_i(0)\}$ and $\{\delta \widetilde{M}_i\}$ to $\widehat{U}_n$ can be supposed to be included in Q_n^1.

The asymptotic covariance. The previous estimates show that the $\widetilde{U}_n$ in the statement of the theorem arises from the linear algorithm

$$\widetilde{\theta}_{n+1} = (I + \epsilon_n A)\widetilde{\theta}_n + \epsilon_n Y_n(0), \tag{3.14}$$

and we need only show that

$$\frac{1}{q_n} \sum_{i,j=n}^{n+q_n-1} E\widetilde{\theta}_i \widetilde{\theta}_j' \tag{3.15}$$

converges to $\bar{V}$ as $n \to \infty$, where $\widetilde{\theta}_n$ is given by the solution to (3.14) with initial condition zero; namely,

$$\widetilde{\theta}_n = \sum_{l=0}^{n-1} \Pi_A(l+1, n-1)\epsilon_l Y_l(0). \tag{3.16}$$

In other words, we need only show that

$$\frac{1}{q_n} \sum_{i,j=n}^{n+q_n-1} \sum_{k=0}^{i-1} \sum_{v=0}^{j-1} \Pi_A(k+1, i-1)\epsilon_k \epsilon_v E Y_k(0) Y_v'(0) \Pi_A'(v+1, j-1) \tag{3.17}$$

converges to $\bar{V}$.

The expression (3.17) will be evaluated directly using (A3.7), and it will be shown that the limit is

$$\bar{V} = \int_0^\infty e^{As} ds \bar{R} + \bar{R} \int_0^\infty e^{A's} ds, \tag{3.18}$$

where

$$\bar{R} = \sum_{l=-\infty}^{\infty} \bar{R}(l), \qquad \bar{R}(l) = \int_0^\infty e^{At} R(l) e^{A't} dt,$$

and $R(l)$ is defined by (3.6b). The matrices $\bar{V}$ and $\bar{R}$ in (3.18) are the same as those defined by (1.3) and (1.5), where Σ in (1.3) and (1.5) takes the explicit value $\sum_{l=-\infty}^{\infty} R(l)$.

The summability of the $\phi(l)$ in (3.6a) implies that (3.17) is bounded uniformly in n and that we can work with each "delay" $l = k - v$ separately in computing the limit in (3.17). The computation is straightforward via

(A3.7), but to see what is being done we start with the simple stationary case, where $EY_l(0)Y_v'(0) = R(l-v)$. The term in (3.17) containing $R(0)$ can be written as $T_n + S_n + S_n'$, where

$$T_n = \frac{1}{q_n} \sum_{i=n}^{n+q_n-1} \sum_{k=0}^{i-1} \Pi_A(k+1, i-1)\epsilon_k^2 R(0)\Pi_A'(k+1, i-1),$$

and

$$S_n = \frac{1}{q_n} \sum_{i=n}^{n+q_n-1} \sum_{j=i+1}^{n+q_n-1} \sum_{k=0}^{i-1} \Pi_A(k+1, i-1)\epsilon_k^2 R(0)\Pi_A'(k+1, i-1)\Pi_A'(i, j-1).$$

By the method of proof of (3.8), it is easily shown that the inner sum of T_n is $O(\epsilon_i)$. Hence $T_n \to 0$ as $n \to \infty$ since $\epsilon_n \to 0$. The term S_n is asymptotically equivalent to (in the sense that the ratio converges to unity as $n \to \infty$)

$$\frac{1}{q_n} \sum_{i=n}^{n+q_n-1} \sum_{j=i+1}^{n+q_n-1} \left[\int_0^{t_i} e^{A(t_i-s)} R(0) e^{A'(t_i-s)} \epsilon_{m(s)} ds \right] e^{A'(t_j-t_i)}.$$

By the argument that led to (3.8), this is asymptotically equivalent to

$$\frac{1}{q_n} \sum_{i=n}^{n+q_n-1} \sum_{j=i+1}^{n+q_n-1} \epsilon_i \left[\int_0^{\infty} e^{As} R(0) e^{A's} ds \right] e^{A'(t_j-t_i)}.$$

Now changing the order of summation

$$\text{from} \sum_{i=n}^{n+q_n-1} \sum_{j=i+1}^{n+q_n-1} \quad \text{to} \quad \sum_{j=n}^{n+q_n-1} \sum_{i=n}^{j-1}$$

and using the definition

$$\bar{R}(0) = \int_0^{\infty} e^{As} R(0) e^{A's} ds$$

yield the asymptotic approximations

$$\bar{R}(0)\frac{1}{q_n} \sum_{j=n}^{n+q_n-1} \sum_{i=n}^{j-1} \epsilon_i e^{A'(t_j-t_i)} \quad \text{and} \quad \bar{R}(0)\frac{1}{q_n} \sum_{j=n}^{n+q_n-1} \int_{t_n}^{t_j} e^{A'(t_j-s)} ds.$$

This last expression converges to $\bar{R}(0) \int_0^{\infty} e^{A's} ds$ as $n \to \infty$. Thus, the limit of $S_n' + S_n$ is

$$\int_0^{\infty} e^{As} ds \bar{R}(0) + \bar{R}(0) \int_0^{\infty} e^{A's} ds.$$

The terms containing $R(l)$, with $l \neq 0$, are handled similarly.

Now, drop the stationarity assumption and use the general condition (A3.7). Following the above procedure for the component $k = v$ of (3.17) yields that S_n is asymptotically equivalent to

$$\frac{1}{q_n} \sum_{i=n}^{n+q_n-1} \sum_{j=i+1}^{n+q_n-1} \int_0^{t_i} e^{A(t_i-s)} EY_{m(s)}(0) Y'_{m(s)}(0) \times e^{A'(t_i-s)} \epsilon_{m(s)} ds \, e^{A'(t_j-t_i)}. \tag{3.19}$$

Due to the continuity of the exponential and the slow (by (A3.4)) rate of change of $\epsilon_{m(t)}$, $EY_{m(s)}(0)Y'_{m(s)}(0)$ can be replaced by $R(0)$ in (3.19) without changing the asymptotic value. The rest of the details for the probability one convergence case are as for the simpler stationary case.

By assuming that (3.9) is bounded with probability one and the other provisos of the theorem, the proof is essentially the same if we replace the probability one convergence assumption by weak convergence for the unconstrained algorithm. The details are left to the reader. □

11.4 The Parameter Identification Problem: An Optimal Algorithm

The advantages of the averaging procedure can be graphically demonstrated by an application to the parameter identification problem of Subsection 3.1.1. The purpose of the development is to illustrate the application of the iterate averaging method, and the conditions are chosen for simplicity of exposition; see also [106, 167, 168] for additional details on this class of problems. Recall that the observation at time n is the real-valued $y_n = \phi_n' \bar{\theta} + \nu_n$. Let $\{\nu_n\}$ be mutually independent with mean zero, constant conditional variance $E_n \nu_n^2 = \sigma^2$, and independent of $\{\phi_n\}$. Suppose that $E\phi_n\phi_n' = Q > 0$ and $n\Phi_n^{-1}$ converges to Q^{-1} in mean, where $\Phi_n = \sum_{i=0}^{n-1} \phi_i \phi_i'$.

The error in the least squares estimator (1.1.11) is

$$(\widehat{\theta}_n - \bar{\theta}) = \Phi_n^{-1} \sum_{i=0}^{n-1} \phi_i \nu_i$$

and

$$nE(\widehat{\theta}_n - \bar{\theta})(\widehat{\theta}_n - \bar{\theta})' = \sigma^2 E n \Phi_n^{-1} \to \sigma^2 Q^{-1}.$$

Now consider the simpler stochastic approximation procedure (1.1.18), where the conditions on $(\nu_n, \phi_n, \epsilon_n)$ are such that Theorem 3.1 holds. The conditions on $\{\phi_n\}$ are basically on the rate of decrease of its correlation function, and are not restrictive. The unconstrained algorithm (1.1.18) can be written as

$$\begin{aligned}\theta_{n+1} &= \theta_n + \epsilon_n \phi_n \left[\nu_n - \phi_n'(\theta_n - \bar{\theta})\right] \\ &= \theta_n + \epsilon_n \left[\phi_n \nu_n - Q(\theta_n - \bar{\theta}) + (Q - \phi_n \phi_n')(\theta_n - \bar{\theta})\right].\end{aligned}$$

We have $A = -Q$ and $\Sigma = \sigma^2 Q$, and $\nu_n \phi_n$ are martingale differences. The term $(Q - \phi_n \phi_n')(\theta_n - \bar{\theta})$ has no effect on the limit normalized mean square error. Thus, $\bar{V} = \sigma^2 Q^{-1}$. Let $q_n = n$. Then, by possibly modifying the algorithm on a set of arbitrarily small probability, Theorem 3.1 asserts that $nE(\Theta_n - \bar{\theta})(\Theta_n - \bar{\theta})'$ converges to $\bar{V}$, where Θ_n is defined by (3.1). Thus, the rate of convergence is the same as that for the more complicated least squares algorithm. The use of iterate averaging for tracking time varying parameters is discussed in [167, 168].

Some data for a constant-step-size algorithm. Data will be presented for the parameter identification procedure of Subsection 3.1.1. The algorithm is

$$\theta_{n+1}^{\epsilon} = \theta_n^{\epsilon} + \epsilon \phi_n [y_n - \phi_n' \theta_n^{\epsilon}],$$

where $y_n = \phi_n \bar{\theta} + \nu_n$. A constant step size is used because it is a common practice in similar applications such as adaptive noise cancellation. Data for the case where the iterate average is fed back into the algorithm appears in [151].

The parameter is $\bar{\theta} = (4.0, -4.2, 3.0, 2.7, -3.0)$. The dimension is $r = 5$ and $\{\nu_n\}$ is a sequence of independent and identically (normally) distributed random variables with mean zero and finite variance, and is independent of $\{\phi_n\}$. The vector ϕ_n is defined by: $\phi_n = (\psi_n, \psi_{n-1}, \psi_{n-2}, \psi_{n-3}, \psi_{n-4})$, where the random variables ψ_n are defined by the equation $\psi_{n+1} = \psi_n/2 + \zeta_n$, and here the ζ_n are mutually independent Gaussian random variables with mean zero and variance 1.0. The standard deviation of the observation noise ν_n is 6.0, a rather large number.

The iterate average is:

$$\Theta_n^{\epsilon} = \frac{1}{\text{window}} \sum_{i=n-\text{window}}^{n} \theta_i^{\epsilon}.$$

The averaging is started after a suitable transient period. Tables 4.1 and 4.2 list the sample mean square errors in the estimates of the five components of $\bar{\theta}$ at 5000 iterations. These numbers are large because the variance of the observation noise is large. The values of the estimators of the parameters are essentially constant after 1000 iterations. The long-run errors decrease as $\epsilon \to 0$, but the data indicates the improvement due to the iterate averaging. In all cases, the ratios should be examined. The results for the iterate average improve further with a larger window. The results for the decreasing-step-size algorithm are better in that all the sample variances are smaller, and they go to zero as the run length increases. Nevertheless, the iterate averaging method still yields substantially better results.

Table 4.1. $\epsilon = 0.05$, Window = 150					
Case	Parameter Component				
	1	2	3	4	5
Variances of the θ_n^ϵ Components	1.133	1.212	1.173	1.251	1.014
Variances of the Θ_n^ϵ Components	0.230	0.297	0.257	0.370	0.207

Table 4.2. $\epsilon = 0.02$, Window = 300					
Case	Parameter Component				
	1	2	3	4	5
Variances of the θ_n^ϵ Components	0.350	0.441	0.385	0.489	0.350
Variances of the Θ_n^ϵ Components	0.086	0.125	0.085	0.194	0.088

12

Distributed/Decentralized and Asynchronous Algorithms

12.0 Outline of Chapter

This chapter is concerned with decentralized and asynchronous forms of the stochastic approximation algorithms, a relatively new area of research. Compared with the rapid progress and extensive literature in stochastic approximation methods, the study of parallel stochastic approximation is still in its infancy. The general ideas of weak convergence theory were applied to a fairly broad class of realistic algorithms in [153, 154]. The general ideas presented there and in [145, 239] form the basis of this chapter. Analogously to the problems in Chapter 8, those methods can handle correlated and state-dependent noise, delays in communication, and asynchronous and distributed networks. Various examples are given in Section 1. In the basic model, there are several processors; each one is responsible for the updating of only a part of the parameter vector. There might be overlaps in that several processors contribute to the updating of the same component of the parameter. Such models were treated in [145, 239, 240]. For a similar model, the problem of finding zeros of a nonlinear function with noisy observations via parallel processing methods and with random truncation bounds was treated in [276]. An attempt to get real-time implementable procedures via pipelining (see Section 1.2) of communication and computation for algorithms with "delayed" observations was in [279]. Pipelining is a potentially useful method for exploiting multiprocessors to accelerate convergence. Suppose that the system observes raw physical data, and then the value of Y_n is computed from that raw data. The actual time required for the computation of Y_n from the raw data might be relatively long in comparison with the potential rate at which the raw data can be made available. Then the computation can be sped up via pipelining. This is discussed further in Section 1.

A survey of some recent developments can be found in [266]. Reviews of decentralized algorithms for traditional numerical analysis problems can be found in [179], and surveys and literature citations for asynchronous (almost entirely) deterministic problems are in the books [20] and [199].

In some applications, the input data are physically distributed; i.e., the underlying computation might use physically separated processors, estimators, trackers, controllers, and so on. For example, in the optimization of queueing networks, each node might be responsible for updating the "local" routing or service speed parameters. In such problems, there are generally communication delays in transferring information from one node to another. This might be because of the physical time required and the fact that different paths are used by each processor, and the path used by each processor is time dependent and random, or it might be due to the time required by other (perhaps higher-priority) demands placed on the network. In other applications, the time required for getting the data needed to update a component of the parameter might be random, and we might not have complete control over which component is to be updated next. An example is the Q-learning problem in Chapter 2.

By means of parallel processing, one may be able to take advantage of state space decomposition. As a result, a large dimensional system may be split into small pieces so that each subsystem can be handled by one of the parallel processors.

The general problem of distributed algorithms can be treated by the methods of the previous chapters, provided they are synchronized or nearly synchronized, even if there are modest delays in the communication among processors. The problem becomes more difficult if the processors are not synchronized. This is particularly important if each processor takes a random length of time to complete its work on a single update but is able to continue processing updates for its set of parameters even if it does not have the latest results from the other processors. Then, in any real-time interval each processor completes a different (random) number of iterations. Because of this, one can no longer use the iterate number $n = 0, 1, \ldots,$ as a clock (time) to study the asymptotic properties. One needs to work in real time, or at least in an appropriately scaled real time.

Suppose there is only one processor (hence no synchronization problems), but that the actual computation of each update takes a random amount of time. This random time is irrelevant to the convergence, which is only concerned with what happens as the number of iterates goes to infinity. Thus, the previous chapters worked with what might be called "iterate time." To introduce the idea of working in real time, this problem is redone in Section 2 in a scaled real time, so that the mean ODE follows the progress not as the iterate number increases, but as real time increases. It will be seen that the analysis is exactly the same except for a simple time transformation, and that the mean ODE is just as before, except for a scale factor that accounts for the average rate at which the iterations progress in time. Analogously to the interpretation of the function $\bar{g}(\theta)$ in the ODE as a mean of the iterate values when the parameter is fixed at θ, the time at the parameter value θ is scaled by the mean rate of updating when the parameter is fixed at that value. If the basic algorithm is of the gradient descent type, then the stable points are the same as for the centralized algorithm.

We concentrate on the weak convergence proofs, although analogs of the probability one results of Chapters 5 and 6 can also be proved. The proofs are actually only mild variations on those of Chapter 8. The notation might get involved at times, but it is constructed for versatility in applications and so that the problem can be set up such that the proofs in Chapter 8 can be used, with only a time rescaling needed at the end to get the desired results. Section 3 deals with the constant step-size algorithm, and the decreasing step-size algorithm is treated in Section 4. Section 5 is concerned with the state-dependent-noise case and is based on the method of Section 8.4. The proofs of the rate of convergence and the stability methods in Sections 6 and 7 follow the lines of development of the analogous sections in Chapters 8 and 10, but work in "real time" rather than iterate time. Iterate averaging such as that used in Chapter 11 has the same advantages here but is omitted for brevity. Finally, we return to the Q-learning problem of Section 2.3, and use the differential inclusions form of the main theorem in Section 3 to readily prove the convergence to the optimal Q-values.

12.1 Examples

12.1.1 Introductory Comments

In the Q-learning example of Section 2.3, the parameter θ is the collection of Q-values, called $\{Q_{i\alpha}; i, \alpha\}$ there. While that algorithm is usually implemented in a centralized way and only one component is updated at each state transition, the algorithm is certainly asynchronous in the sense that the component updated at any time is random and the time between updates of any particular component is random. The methods of this chapter will be appropriate for such problems.

It is easy to construct examples of asynchronous algorithms by simply splitting the work among several processors and allowing them to function asynchronously. For example, in the animal learning example of Section 2.1, there might be several parameters updated at rates that are different and that depend on different aspects of the hunting experience. In the neural net training example of Section 2.2, the actual computation of the derivatives might be quite time-consuming relative to the rate at which new training inputs can be applied and the corresponding outputs measured. This suggests that the algorithm can be improved by "pipelining" in either a synchronized or an asynchronized way (see the example in the next subsection). The relative difficulty of the computation in comparison with the actual speed at which the network can be used suggests a general class of problems where pipelining can be useful. In many applications, one takes a "raw" observation, such as the input and the associated actual and desired outputs in the neural network training procedure. Then a time-consuming computation is done to get the actual Y_n or Y_n^ϵ, which is used in the stochastic approximation. A similar situation holds for the queue optimization problem of Section 2.5. In that case, the raw data are simply the sequences of arrival and service times. The computation of the pathwise derivative might take a (relatively) long time, particularly if the interarrival and service times are very small, the derivatives of the "inverse functions" that are often involved in getting pathwise derivatives are hard to compute, or the processors used must be shared (with perhaps a lower priority) with other tasks.

It is worth emphasizing that asynchronous algorithms commonly arise when the processing is distributed and some of the processors are interrupted at random times to do other work.

The pipelined algorithm described in the next subsection is not asynchronous. It is distributed and uses the notion of "concurrency," which is central to current work in distributed processing. If each processor takes a random length of time for its computation, one might consider the use of an asynchronous form of the pipelined algorithm, and the methods of this chapter can then be applied. In the synchronized case, the proofs of convergence of the pipelined algorithms can be handled by the machinery of Chapters 5–8. Subsection 2.3 concerns the optimization of a queueing network, where the various nodes of the network participate asynchronously in the computation. The example in Section 2.4 arises in telecommunications and is characteristic of a large class of problems. These few examples are illustrative of an ever-expanding set of applications.

12.1.2 Pipelined Computations

This example illustrates the usefulness of multiprocessors in stochastic approximation and is intended to be a simple illustration of the pipelining method. Consider the stochastic approximation algorithm as described in the previous chapters, with constant step size ϵ:

$$\theta_{n+1}^\epsilon = \Pi_H \left(\theta_n^\epsilon + \epsilon Y_n^\epsilon\right) = \theta_n^\epsilon + \epsilon Y_n^\epsilon + \epsilon Z_n^\epsilon. \tag{1.1}$$

The algorithm can often be thought of in terms of the following two-phase computation. In Phase 1, we evaluate the Y_n^ϵ from the raw observational data, and in Phase 2, we update the parameter by adding ϵY_n^ϵ to the current value of θ_n^ϵ (with a projection or a reflection term if needed) and observe new raw data.

Generally, most of the computation time is spent in Phase 1. Frequently, Phase 1 consists of an extensive computation based on physical measurements. Taking this fact into account, consider an alternative algorithm where several processors are lined up as on a production line and update interactively. After each iteration is completed, the new value of the parameter is passed to the next processor in the line.

For a concrete illustration, suppose that the time required for computing Y_n^ϵ is three units of time, after which θ_n^ϵ and ϵY_n^ϵ are added and a new measurement is taken, where the addition and taking the new raw measurement requires one unit of time. Instead of using a single processor as in (1.1), where each iteration requires four units of time, four parallel processors will be used in a pipeline fashion. All processors perform the same kind of computation, but with different starting times. At a given instance n, suppose that Processor 1 has completed its evaluation based on physical data observed three units of time ago when the parameter value was θ_{n-3}^ϵ. In other words, it has computed Y_{n-3}^ϵ. Then, at time $n+1$, Processor 1 computes

$$\theta_{n+1}^\epsilon = \Pi_H(\theta_n^\epsilon + \epsilon Y_{n-3}^\epsilon),$$

where θ_n^ϵ is the current value of the parameter (just provided by Processor 4). Meanwhile Processor 2 is evaluating the increment based on physical measurements taken at time $n-2$, when the parameter was θ_{n-2}^ϵ. Thus, at time $n+2$, Processor 2 computes

$$\theta_{n+2}^\epsilon = \Pi_H(\theta_{n+1}^\epsilon + \epsilon Y_{n-2}^\epsilon)$$

and analogously for Processors 3 and 4.

In general, suppose that the evaluation of each Y_n requires d units of time. Then, using $d+1$ synchronized processors the algorithm is

$$\theta_{n+1}^\epsilon = \Pi_H(\theta_n^\epsilon + \epsilon Y_{n-d}^\epsilon). \tag{1.2}$$

Define the initial condition $\theta_n^\epsilon = \theta_0$ for $n = 0, \ldots, d$. Thus, following the format of the example, Processor 1 takes physical measurements at times $0, d+1, 2(d+1), \ldots$, and computes $\theta_{d+1}^\epsilon, \theta_{2(d+1)}^\epsilon, \ldots$ For $1 \le v \le d+1$, Processor v ($v = 1, \ldots, d+1$) takes physical measurements at times $v-1, v-1+(d+1), v-1+2(d+1), \ldots$, and computes $\theta_{d+v}^\epsilon, \theta_{v-1+2(d+1)}^\epsilon, \ldots$ Once an update θ_n^ϵ is computed, its value is passed to the next processor in the line. The algorithmic form (1.2) is covered by the theorems in Chapters 5–8 without any change, owing to the fact that the update times are synchronized.

The model can easily be extended to allow variable (but bounded or uniformly integrable) delays in communication between processors, so that θ_n^ϵ in (1.2) is replaced by some $\theta_{n-\mu_n}^\epsilon$ where μ_n is the delay. The delay can depend on the destination processor.

12.1.3 A Distributed and Decentralized Network Model

This example is taken from [145], but the notation is slightly different. Consider a system with r processors, the αth one having the responsibility for updating the αth component of θ. We wish to minimize a function $F(\cdot)$ that takes the form $F(\theta) = \sum_{\gamma=1}^{r} F^{\gamma}(\theta)$, for real-valued and continuously differentiable $F^{\gamma}(\cdot)$. Define $F_{\alpha}^{\gamma}(\theta) = \partial F^{\gamma}(\theta)/\partial\theta^{\alpha}$. In our model, for each α and each $\gamma = 1, \ldots, r$, Processor γ produces estimates $Y_{n,\alpha}^{\epsilon,\gamma}$, for $n \geq 0$, which are sent to Processor α for help in estimating $F_{\alpha}^{\gamma}(\cdot)$, at whatever the current parameter value. It also sends the current values of its own component θ^{γ}. The form of the function $F(\cdot)$ is not known.

An important class of examples that fits the above model is optimal adaptive routing or service time control in queueing networks. Let the network have r nodes, with θ being perhaps a routing or service speed parameter, where θ^{α} is the component associated with the αth node. [We note that it is customary in the literature to minimize a mean waiting time, but the mean waiting time is related to the mean queue lengths via Little's formula.] Let $F^{\gamma}(\theta)$ denote the stationary average queue length at node γ under parameter value θ. We wish to minimize the stationary average number of customers in the network $F(\theta) = \sum_{\gamma=1}^{r} F^{\gamma}(\theta)$. The problem arises in the control of telecommunication networks and has been treated in [239, 243]. The controller at node α updates the component θ^{α} of θ based on its own observations and relevant data sent from other nodes. In one useful approach, called the *surrogate estimation method* in the references, each node γ estimates the pathwise derivative of the mean length of its own queue with respect to variations in external inputs to that node. Then one uses the mean flow equations for the system to get acceptable "pathwise" estimates of the derivatives $F_{\alpha}^{\gamma}(\theta)$. These estimates are transmitted to node α for estimating the derivative of $F(\theta)$ with respect to θ^{α} at the current value of θ and then updating the value of θ^{α}. After each transmission, new estimates are taken and the process is repeated. The communication from processor to processor might take random lengths of time, due to the different paths used or the requirements and priorities of other traffic.

The time intervals required for the estimation can depend heavily and randomly on the node; for example, they might depend on the random service and arrival times. The nodes would transmit their estimates in an asynchronous way. Thus the stochastic approximation is both decentralized and asynchronous. More generally, some nodes might require a vector parameter, but the vector case can be reduced to the scalar case by "splitting" the nodes. In a typical application of stochastic approximation, each time a new estimate of $F_{\alpha}^{\gamma}(\theta)$ (at the current value of θ at node γ) is received at node α, that estimate is multiplied by a step-size parameter and subtracted (since we have a gradient descent algorithm) from the current value (at node α) of the state component θ^{α}. This "additive" structure allows us to represent the algorithm in a useful decomposed way, by writing the current value of the component θ^α as the sum of the initial value minus r terms. The γth such term is the product of an appropriate step size times the sum of the past transmissions from node γ to node α, where each transmission is an estimate of $F_\alpha^\gamma(\cdot)$ at the most recent value of the parameter available to node γ when the estimate was made.

As noted, the transmission of information might not be instantaneous. If the delays are not excessively long (say, the sequence of delays is uniformly integrable) and ϵ is small, then they cause no problems in the analysis. It should be mentioned that this insensitivity to delay is an *asymptotic* result and a small step size might be required if it is to hold in any particular application. In any given practical system, the possibility of instabilities due to delays needs to be taken seriously. To simplify the notation in the rest of this example, we will work under the assumption that there are no delays and that the parameters are updated as soon as new information becomes available.

Notation. *The notation is for this example only. The symbols might be given different meanings later in the chapter.* Let the step size be $\epsilon > 0$, a constant. Let $\delta\tau_{n,\alpha}^{\epsilon,\gamma}$ denote the real-time interval between the nth and $(n+1)$st transmissions from node γ to node α. Define

$$\tau_{n,\alpha}^{\epsilon,\gamma} = \epsilon \sum_{i=0}^{n-1} \delta\tau_{i,\alpha}^{\epsilon,\gamma}, \tag{1.3}$$

which is ϵ times the real time required for the first n transmissions from γ to α. Define

$$N_\alpha^{\epsilon,\gamma}(t) = \epsilon \times [\text{number of transmissions from } \gamma \text{ to } \alpha \text{ up to real time } t/\epsilon].$$

Let $\tau_\alpha^{\epsilon,\gamma}(t)$ be the piecewise constant interpolation of the $\tau_{n,\alpha}^{\epsilon,\gamma}$ with interpolation intervals ϵ and initial condition zero. Analogously to the situation to be encountered in Section 2, $N_\alpha^{\epsilon,\gamma}(\cdot)$ and $\tau_\alpha^{\epsilon,\gamma}(\cdot)$ are inverses of each other.

The stochastic approximation algorithm. The notation is a little complex but very natural, as seen later in the chapter. It enables us to use the results of Chapters 5–8 in a much more complex situation via several time change arguments and saves a great deal of work over a direct analysis. As mentioned in the preceding discussion, it is convenient to separate the various updates in the value of θ^α into components that come from the same node. This suggests the following decomposed representation for the stochastic approximation algorithm. Let $\theta_{n,\alpha}^{\epsilon,\gamma}$ represent the contribution to the estimate of the αth component of θ due to the first (n-1) updates from Processor γ. Let $\widehat{\theta}_\alpha^{\epsilon,\gamma}(\cdot)$ denote the interpolation of $\{\theta_{n,\alpha}^{\epsilon,\gamma}\}$ in the real-time (times ϵ) scale. The actual interpolated iterate in the real-time (times ϵ)

scale is defined by

$$\widehat{\theta}^{\epsilon}_{\alpha}(t) = \widehat{\theta}^{\epsilon}_{\alpha}(0) + \sum_{\gamma=1}^{r} \widehat{\theta}^{\epsilon,\gamma}_{\alpha}(t), \quad \widehat{\theta}^{\epsilon}_{\alpha}(0) = \theta^{\epsilon,\alpha}_{0,\alpha}. \tag{1.4}$$

A constraint might be added to (1.4). Define $\widehat{\theta}^{\epsilon}(\cdot) = \{\widehat{\theta}^{\epsilon}_{\alpha}(t), \alpha = 1, \ldots, r\}$.

For each α, γ, let $c^{\gamma}_{\alpha}(\cdot)$ be a continuous and bounded real-valued function. Then the sequence $\theta^{\epsilon,\gamma}_{n,\alpha}$ is defined by

$$\theta^{\epsilon,\gamma}_{n+1,\alpha} = \theta^{\epsilon,\gamma}_{n,\alpha} + \epsilon c^{\gamma}_{\alpha}(\widehat{\theta}^{\epsilon}(\tau^{\epsilon,\gamma,-}_{n+1,\alpha}))Y^{\epsilon,\gamma}_{n,\alpha}, \quad \theta^{\epsilon,\gamma}_{0,\alpha} = 0 \text{ for } \alpha \neq \gamma. \tag{1.5}$$

The role of the functions $c^{\gamma}_{\alpha}(\cdot)$, to be selected by the experimenter, is to adjust the step size to partially compensate for the fact that the frequency of the intervals between updates might depend on θ, α, and γ. If desired, one can always use $c^{\gamma}_{\alpha}(\theta) \equiv 1$. Note that, by the definitions, $\widehat{\theta}^{\epsilon}(\tau^{\epsilon,\gamma,-}_{n+1,\alpha})$ is the value of the state just before the $(n+1)$st updating at Processor α with the contribution from Processor γ. Define

$$\theta^{\epsilon,\gamma}_{\alpha}(t) = \theta^{\epsilon,\gamma}_{n,\alpha} \text{ for } t \in [n\epsilon, (n+1)\epsilon).$$

By the definitions, we have the following important relationships:

$$\begin{aligned} \widehat{\theta}^{\epsilon,\gamma}_{\alpha}(t) &= \theta^{\epsilon,\gamma}_{N^{\epsilon,\gamma}_{\alpha}(t)/\epsilon,\alpha} = \theta^{\epsilon,\gamma}_{\alpha}(N^{\epsilon,\gamma}_{\alpha}(t)), \\ \widehat{\theta}^{\epsilon,\gamma}_{\alpha}(\tau^{\epsilon,\gamma}_{\alpha}(t)) &= \theta^{\epsilon,\gamma}_{\alpha}(t), \end{aligned} \tag{1.6}$$

where $N^{\epsilon,\gamma}_{\alpha}(t)$ is defined after (1.3). Under reasonable conditions on $Y^{\epsilon,\gamma}_{n,\alpha}$, the proofs of convergence are just adaptations of the arguments in the previous chapters modified by the time-change arguments of Section 3 in [145], which are similar to those used in Sections 2 and 3.

12.1.4 Multiaccess Communications

This example is motivated by the work of Hajek concerning the optimization of the ALOHA system [94]. In this system, r independent users share a common communication channel. At each time slot n, each user must decide whether or not to transmit assuming that there is a packet to transmit in that slot. Since there is no direct communication among users, there is a possibility that several users will transmit packets at the same time, in which case all such packets need to be retransmitted. The users can listen to the channel and will know whether or not there were simultaneous transmissions. Then they must decide when to retransmit the packet. One practical and common solution is to transmit or retransmit at random times, where the conditional (given the past) probability of transmission or retransmission at any time is θ^{α} for user α. The determination of the optimal $\theta = (\theta^1, \ldots, \theta^r)$ is to be done adaptively, so that it can change appropriately as the operating conditions change.

Let $\widehat{\theta}_{n,\alpha}$ denote the probability (conditioned on the past) that user α transmits a packet in time slot $n+1$, if a packet is available for transmission. Let I_n^α denote the indicator function of the event that user α transmits in time slot $n+1$, and J_n^α the indicator function of the event that some other user also transmitted in that slot. Let $0 < a_\alpha < b_\alpha < 1, \alpha = 1, \ldots, r$. For appropriate functions $G_\alpha(\cdot)$ the algorithm used in [94] is of the form

$$\widehat{\theta}_{n+1,\alpha} = \Pi_{[a_\alpha, b_\alpha]}\left(\widehat{\theta}_{n,\alpha} + \epsilon G_\alpha(\widehat{\theta}_{n,\alpha}, J_n^\alpha) I_n^\alpha\right). \tag{1.7}$$

Thus, the update times of the transmission probability for each user are random; they are just the random times of transmission for that user. No update clock is synchronized among the users. The only information available concerning the actions of the other users is whether or not there were simultaneous transmissions.

12.2 Real-Time Scale: Introduction

The purpose of this section is to introduce and to motivate the basic issues of scaling for the asynchronous algorithm. To simplify the introduction of the "real-time" notation, assume that there is only one processor. The general case is in the following sections. The essence is to use a random time change argument, which (either implicitly or explicitly) is crucial in analyzing parallel and decentralized algorithms.

The time scale used in the proof of the convergence via the ODE method in Chapters 5–8 was determined by the iteration number and the step size only. For example, for the constant-step-size case, the interpolation $\theta^\epsilon(t)$ was defined to be θ_n^ϵ on the interpolated time interval $[n\epsilon, n\epsilon + \epsilon)$. For the decreasing-step-size case, we used the interpolations $\theta^n(\cdot)$, $n = 0, 1, \ldots,$ defined by $\theta^n(t) = \theta_{n+i}$ on the interpolated time interval $[t_{n+i} - t_n, t_{n+i+1} - t_n)$. These interpolations were natural choices for the problems of concern because our main interest was the characterization of the asymptotic properties of the sequences $\{\theta_n^\epsilon\}$ or $\{\theta_n\}$, and the increment in the iterate at the nth update is proportional to ϵ (ϵ_{n-1}, resp.). Of course, in applications one works in real (clock) time, but this real time is irrelevant for the purpose of proving convergence for the algorithms in Chapters 5–8. As was seen in the examples in Section 1, if the computation is split among several interacting processors that are not fully synchronized, then we might have to work in (perhaps scaled) real time, because there is no "iterate time" common to all the processors.

The situation is particularly acute if some of the interacting processors require (different) random times for the completion of their work on each update but can continue to update their part of the parameter even if there is a modest delay in receiving the latest information from the other processors. Of course, an algorithm that appears at first sight to be asynchronous might be rewritable as a fully synchronized algorithm or perhaps with only "slight" asynchronicities, and the user should be alert to this possibility.

When working in real time, the mean ODE or differential inclusion that characterizes the asymptotic behavior for the asynchronous algorithm will be very similar to what we have derived in Chapters 6–8. Indeed, it is the same except for rescalings that account for the different mean update speeds of the different processors. To prepare ourselves for these rescalings and to establish the type of notations and time scale to be used, we first obtain the mean ODE for a classical fully synchronized system as used in Chapters 5–8, but in a real-time scale.

Consider the algorithm used in Theorem 8.2.1, but without the constraint; namely,

$$\theta_{n+1}^\epsilon = \theta_n^\epsilon + \epsilon Y_n^\epsilon.$$

Assume the conditions of that theorem and that $\{\theta_n^\epsilon; \epsilon, n\}$ is bounded. Then Theorem 8.2.1 holds for the interpolations $\{\theta^\epsilon(\cdot), \theta^\epsilon(\epsilon q_\epsilon + \cdot)\}$, and the mean ODE is $\dot{\theta} = \bar{g}(\theta)$.

Suppose that the computation of Y_n^ϵ requires a random amount of time $\delta\tau_n^\epsilon$, which is the real-time interval between the nth and $(n+1)$st updating. Define the *interpolated real time*

$$\tau_n^\epsilon = \epsilon \sum_{i=0}^{n-1} \delta\tau_i^\epsilon,$$

and define

$$N^\epsilon(t) = \epsilon \times [\text{number of updatings by real time } t/\epsilon].$$

Let $\tau^\epsilon(\cdot)$ denote the interpolation of $\{\tau_n^\epsilon, n < \infty\}$ defined by $\tau^\epsilon(t) = \tau_n^\epsilon$ on $[\epsilon n, \epsilon n + \epsilon)$; see Figures 2.1 and 2.2.

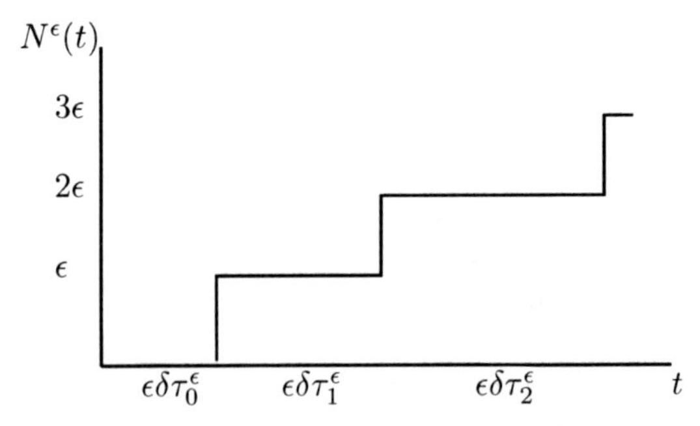

Figure 2.1. The function $N^\epsilon(\cdot)$.

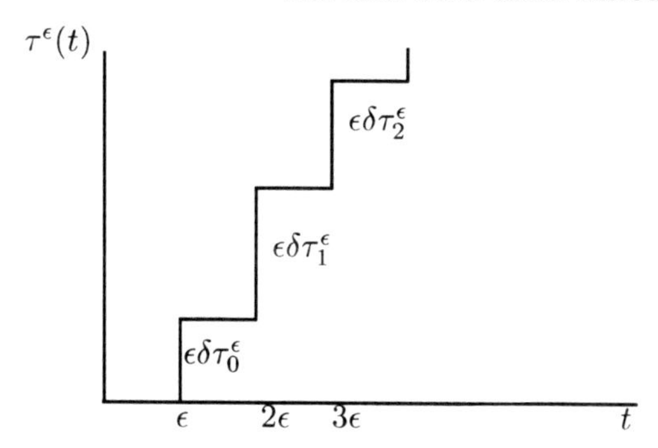

Figure 2.2. The function $\tau^\epsilon(\cdot)$.

Note that $\tau^\epsilon(\cdot)$ is the inverse of $N^\epsilon(\cdot)$ in the sense that

$$N^\epsilon(\tau^\epsilon(t)) = n\epsilon, \text{ for } t \in [n\epsilon, n\epsilon + \epsilon). \tag{2.1}$$

Define $\widehat{\theta}^\epsilon(t) = \theta^\epsilon(N^\epsilon(t))$, which is the interpolation of the iterate sequence in the real-time scale, not in the iterate-time scale. The weak convergence and characterization of the mean ODE for the $\{\widehat{\theta}^\epsilon(\cdot)\}$ sequence is easily done by following the procedure used for Theorem 8.2.1; this will now be illustrated.

Refer to the notation and the structure of the setup for Theorem 8.2.1, and let $\mathcal{F}_n^\epsilon$ measure $\{\delta\tau_i^\epsilon, i < n\}$ as well. The "dot" notation $\dot{N}(t)$ denotes the derivative with respect to the argument. Thus, $\dot{N}(\tau(t))$ denotes the derivative of $N(\cdot)$ evaluated at $\tau(t)$; equivalently,

$$\dot{N}(\tau(t)) = \frac{\partial}{\partial s} N(s)\big|_{s=\tau(t)}.$$

Suppose that

$$\{\delta\tau_n^\epsilon; \epsilon, n\} \text{ is uniformly integrable.} \tag{2.2}$$

Let there be measurable functions $u_n^\epsilon(\cdot)$ and "memory" random variables ξ_n^ϵ such that

$$E_n^\epsilon \delta\tau_n^\epsilon = u_n^\epsilon(\theta_n^\epsilon, \xi_n^\epsilon),$$

where we suppose that

$$\inf_{n,\epsilon,\theta,\xi} u_n^\epsilon(\theta, \xi) > 0, \text{ w.p.1}, \tag{2.3}$$

and $u_n^\epsilon(\cdot, \xi)$ is continuous in θ uniformly in n, ϵ and in ξ on any compact set. Suppose there is a continuous function $\bar{u}(\cdot)$ such that

$$\frac{1}{m} \sum_{i=n}^{n+m-1} E_n^\epsilon u_i^\epsilon(\theta, \xi_i(\theta)) \to \bar{u}(\theta)$$

in mean for each θ as n and m go to infinity. Note that in applications, one often has some influence over the design of the system, and then the assumptions are reasonable requirements on that design.

The set $\{\theta^\epsilon(\cdot), \tau^\epsilon(\cdot), \widehat{\theta}^\epsilon(\cdot), N^\epsilon(\cdot)\}$ is tight. The tightness assertion and the Lipschitz continuity of the weak sense limits of $\{\tau^\epsilon(\cdot)\}$ are consequences of the uniform integrability (2.2). The strict positivity (2.3) implies that all weak sense limits of $\{\tau^\epsilon(\cdot)\}$ are strictly increasing. The tightness assertion and the Lipschitz continuity of the weak sense limits of $\{N^\epsilon(\cdot)\}$ is a consequence of these facts together with the property (2.1). [See the proof of Theorem 3.1 for additional detail.] For notational simplicity, we suppose that the original sequences indexed by ϵ converge weakly; otherwise take an appropriate subsequence. By the assumptions, and the proof of Theorem 8.2.1, there are processes $(\theta(\cdot), \tau(\cdot), \widehat{\theta}(\cdot), N(\cdot))$ such that

$$(\theta^\epsilon(\cdot), \tau^\epsilon(\cdot), \widehat{\theta}^\epsilon(\cdot), N^\epsilon(\cdot)) \Rightarrow (\theta(\cdot), \tau(\cdot), \widehat{\theta}(\cdot), N(\cdot)),$$

where $\widehat{\theta}(t) = \theta(N(t))$ and

$$\tau(t) = \int_0^t \bar{u}(\theta(s))ds. \tag{2.4}$$

Owing to the strict positivity (2.3) and to the fact that $N^\epsilon(\cdot)$ and $\tau^\epsilon(\cdot)$ are inverses of each other, it follows that both $\tau(\cdot)$ and $N(\cdot)$ are strictly increasing and differentiable with probability one and that $N(\tau(t)) = t$. Hence $N(t) = \tau^{-1}(t)$. That is, $N(\cdot)$ is the inverse function to $\tau(\cdot)$. Taking derivatives, we get $\dot{N}(\tau(t))\dot{\tau}(t) = 1$. Then using (2.4) and writing $s = \tau(t)$, the slope of $N(\cdot)$ at s is $\dot{N}(s) = 1/\bar{u}(\theta(\tau^{-1}(s))) = 1/\bar{u}(\widehat{\theta}(s))$. In other words, the mean rate of change of the number of updates is inversely proportional to the mean time between updates. We now have

$$N(t) = \int_0^t \frac{ds}{\bar{u}(\widehat{\theta}(s))}. \tag{2.5}$$

By Theorem 8.2.1, $\theta(\cdot)$ satisfies $\dot{\theta} = \bar{g}(\theta)$. Using the relationships

$$\widehat{\theta}^\epsilon(\cdot) = \theta^\epsilon(N^\epsilon(\cdot)) \Rightarrow \theta(N(\cdot)) \equiv \widehat{\theta}(\cdot) \tag{2.6}$$

and $N(\tau(t)) = t$ yields

$$\dot{\widehat{\theta}}(t) = [\dot{\theta}(N(t))]\dot{N}(t) = \frac{\bar{g}(\theta(N(t)))}{\bar{u}(\widehat{\theta}(t))} = \frac{\bar{g}(\widehat{\theta}(t))}{\bar{u}(\widehat{\theta}(t))}. \tag{2.7}$$

Thus, the derivation of the mean ODE in the real-time scale is just that of Theorem 8.2.1 plus a time-change argument.

Equation (2.7) says that the (asymptotic) progress of the convergence in real time is scaled by the (asymptotic) mean "rate of updating," as measured by the mean rate $1/\bar{u}(\widehat{\theta}(t))$. The purpose of the time-change argument is to avoid dealing with random interpolation intervals and the interaction of Y_n^ϵ and $\delta\tau_n^\epsilon$. The time-change argument exploits the convergence of both the sequence of "time" processes $\{\tau^\epsilon(\cdot), N^\epsilon(\cdot)\}$ and the original sequence of interpolations $\{\theta^\epsilon(\cdot)\}$. The approach can be viewed as a "separation principle," which works with the iterate and real-time scales separately, doing the computations in each that are best suited to it. Similar time scalings will be used in the following sections to enable the general results of Chapter 8 to be carried over to the asynchronous algorithm with minimal effort.

Remark on the analysis for infinite time. The argument leading to (2.7) is for the sequence of processes $\{\theta^\epsilon(\cdot)\}$ that start at time zero and are defined on the interval $[0,\infty)$. This gives us a limit theorem for $\{\widehat{\theta}^\epsilon(\cdot)\}$ for t in any large but bounded interval, or even for an interval $[0,T_\epsilon]$ where $T_\epsilon \to \infty$ slowly enough. It is good enough for most practical algorithms, when ϵ is small and ϵn might be large, but it will not go to infinity in the application.

In Chapters 5–8, to get limits as the interpolated time went to infinity, we worked with a sequence of processes that were actually the "tails" of the original process. For example, in Chapter 8, to get the limits of θ_n^ϵ as $\epsilon \to 0$ but $n\epsilon \to \infty$, we worked with the sequence of interpolations $\theta^\epsilon(\epsilon q_\epsilon + \cdot)$, where $\epsilon q_\epsilon \to \infty$. The natural analog of this procedure in the real time scale uses a sequence of processes starting at a sequence of real times that go to infinity as $\epsilon \to 0$, as follows. Let $T_\epsilon \to \infty$ (real numbers) and work with $\{\widehat{\theta}^\epsilon(T_\epsilon + \cdot)\}$.

The limits of $\widehat{\theta}^\epsilon(T_\epsilon + \cdot)$ as $\epsilon \to 0$ are the limits of the iterate sequence in real time, as real time goes to infinity and $\epsilon \to 0$. Then the analysis is the same as above, except that the value at time zero (the new initial condition) of the real-time interpolated process shifted to the left by T_ϵ is

$$\widehat{\theta}^\epsilon(T_\epsilon) = \theta^\epsilon(N^\epsilon(T_\epsilon)) = \theta^\epsilon_{N^\epsilon(T_\epsilon)/\epsilon}.$$

Recall that $N^\epsilon(t)/\epsilon$ is the index of the last iterate that occurs at or before real-time t and it is a random variable. The uniform integrability and tightness conditions must reflect this change of time origin. In particular, since $N^\epsilon(t)$ is a random variable, we will need to assume that

$$\{\xi^\epsilon_{N^\epsilon(T_\epsilon)/\epsilon+n}, \theta^\epsilon_{N^\epsilon(T_\epsilon)/\epsilon+n}; \epsilon, n \geq 0\} \text{ is tight} \tag{2.8}$$

and

$$\{Y^\epsilon_{N^\epsilon(T^\epsilon)/\epsilon+n}, \delta\tau^\epsilon_{N^\epsilon(T^\epsilon)/\epsilon+n}; \epsilon, n \geq 0\} \text{ is uniformly integrable.} \tag{2.9}$$

For each $\epsilon > 0$, the quantities in (2.9) are used to compute the iterates that occur at or after real-time T_ϵ. In fact, for each $\epsilon > 0$, (2.9) is just the set of observations and time intervals for the process starting at time T_ϵ with initial condition $\widehat{\theta}^\epsilon(T_\epsilon)$. These assumptions would not always hold as required if they held only for $T_\epsilon = 0$. But for the types of renewal processes that commonly model the update times, the verification of (2.8), (2.9), and their analogs in subsequent sections might not be more difficult than the verification for $T_\epsilon = 0$. Indeed, an examination of various cases suggests that (2.8) and (2.9) are not at all restrictive. Analogous conditions will be used in what follows. Unfortunately, the notation needed for the processes that are shifted to the left by an arbitrary real time is rather messy.

Note that in practice, we will start the procedure at some arbitrary time, which suggests that (2.8) and (2.9) are essential for a well-posed problem, whether or not we work in real time, because they are just the required conditions when we start at an arbitrary real time and desire convergence no matter what the starting time is.

12.3 The Basic Algorithms

12.3.1 Constant Step Size: Introduction

In this section, we work with a versatile canonical model, with a constant step size $\epsilon > 0$. The notation is, unfortunately, not simple. This seems to be unavoidable because the actual number of updates at the different processors at any time are only probabilistically related, and we wish to treat a form that covers many interesting cases and exposes the essential issues. The basic format of the convergence proofs is that of weak convergence. Extensions to the decreasing-step-size model are straightforward, being essentially notational, and are dealt with in the next section. In this and the next subsection, we develop the notation. Following the idea of the time-scale changes introduced in Section 2, we first define each of the components of the parameter processes in "iterate time," and then we get the results for the "real-time" evolution from these by the use of appropriate time-scale changes, with each component having its own time-scale process. Thus we also need to deal with the weak convergence of the time-scale processes. The method is first developed for the martingale difference noise case. The extension to the correlated noise case uses the same techniques and appropriate averaging conditions, analogous to what was done in Chapter 8. The basic theorems are for large but bounded time intervals, analogously to the first results in Section 2. The infinite time analysis is done in Subsection 4 and conditions analogous to those of (2.8) and (2.9) will be needed.

For descriptive purposes, it is convenient to use a canonical model in which there are r processors, each responsible for the updating only one component of θ. Let $\delta\tau^\epsilon_{n,\alpha}$ denote the real time between the nth and $(n+1)$st updating of the αth component of θ. It is possible that $\delta\tau^\epsilon_{n,\alpha}$ has the same value for several values of α and all ϵ and n, and that the same processor is responsible for that set of components. However, that case is covered by the current assumption. The network example of Subsection 1.3 is also covered due to the additive form of (1.4), (1.5). Due to the working assumption of a separate processor for the updating of each component of θ, we will simplify further by supposing that the constraint set takes the form

$$H = [a_1, b_1] \times \ldots \times [a_r, b_r], \quad -\infty < a_\alpha < b_\alpha < \infty. \tag{3.1}$$

This assumption on the form of the constraint set is not essential. If some processor is used to update several components, then the general constraints (A4.3.2) or (A4.3.3) can be used on that set. In fact, the algorithm can be written so that the entire vector is constrained, subject to current knowledge at each processor, and the mean ODE will be the correct one for the constrained problem. The proofs of convergence are all very close to those in Chapter 8; only the details concerning the differences will be given.

The algorithm of concern will be written in terms of its components. For $\alpha = 1, \ldots, r$, it takes the form

$$\theta^\epsilon_{n+1,\alpha} = \Pi_{[a_\alpha, b_\alpha]} \left(\theta^\epsilon_{n,\alpha} + \epsilon Y^\epsilon_{n,\alpha}\right) = \theta^\epsilon_{n,\alpha} + \epsilon Y^\epsilon_{n,\alpha} + \epsilon Z^\epsilon_{n,\alpha}. \tag{3.2}$$

Define the scaled *interpolated real time*

$$\tau^\epsilon_{n,\alpha} = \epsilon \sum_{i=0}^{n-1} \delta\tau^\epsilon_{i,\alpha},$$

and, for each $\sigma \geq 0$, define

$$p^\epsilon_\alpha(\sigma) = \min\left\{ n : \sum_{i=0}^{n-1} \delta\tau^\epsilon_{i,\alpha} \geq \sigma \right\},$$

the index of the first update at Processor α at or after (unscaled) real time σ. Define

$$N^\epsilon_\alpha(\sigma) = \epsilon \times [\text{number of updatings of the } \alpha\text{th component by real time } \sigma/\epsilon].$$

Let $\theta^\epsilon_\alpha(\cdot)$ denote the interpolation of $\{\theta^\epsilon_{n,\alpha}, n < \infty\}$ on $[0, \infty)$ defined by $\theta^\epsilon_\alpha(t) = \theta^\epsilon_{n,\alpha}$ on $[n\epsilon, n\epsilon + \epsilon)$, and define $\tau^\epsilon_\alpha(\cdot)$ analogously, but using $\{\tau^\epsilon_{n,\alpha}\}$. The relations among some of the terms are illustrated in Figure 3.1.

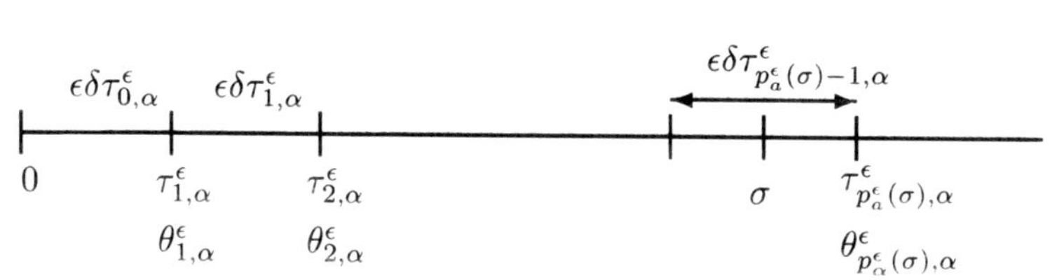

Figure 3.1. The iterates up to real time σ for Processor α.

Analogously to the situation in Section 2, note that $\tau_\alpha^\epsilon(\cdot)$ is the inverse of $N_\alpha^\epsilon(\cdot)$ in the sense that

$$N_\alpha^\epsilon(\tau_\alpha^\epsilon(t)) = n\epsilon \text{ for } t \in [n\epsilon, n\epsilon + \epsilon). \tag{3.3}$$

Define the "real-time" interpolation $\widehat{\theta}_\alpha^\epsilon(t)$ by

$$\widehat{\theta}_\alpha^\epsilon(t) = \theta_{n,\alpha}^\epsilon, \quad t \in [\tau_{n,\alpha}^\epsilon, \tau_{n+1,\alpha}^\epsilon), \tag{3.4}$$

and note the equalities

$$\theta_\alpha^\epsilon(t) = \widehat{\theta}_\alpha^\epsilon(\tau_\alpha^\epsilon(t)), \quad \widehat{\theta}_\alpha^\epsilon(t) = \theta_\alpha^\epsilon(N_\alpha^\epsilon(t)). \tag{3.5}$$

Define the vector $\widehat{\theta}^\epsilon(t) = (\widehat{\theta}_\alpha^\epsilon(t), \alpha \leq r)$.

12.3.2 Martingale Difference Noise

Notation. In Chapter 8, the σ-algebra $\mathcal{F}_n^\epsilon$ measured all the data available until, but not including, the computation of θ_{n+1}^ϵ. With E_n^ϵ denoting the expectation conditioned on $\mathcal{F}_n^\epsilon$, for the martingale difference noise case we had $E_n^\epsilon Y_n^\epsilon = g_n^\epsilon(\theta_n^\epsilon)$ plus possibly an asymptotically negligible error, called β_n^ϵ. By taking this conditional expectation, we were able to write Y_n^ϵ as the sum of the conditional mean and a martingale difference term. This decomposition was fundamental to the analysis, and a key quantity in the analysis was the expectation of Y_n^ϵ conditioned on all the data needed to compute the past iterate values $\{\theta_i^\epsilon, i \leq n\}$. Similar computations, centering the increments about appropriate conditional means, are also fundamental for the analysis of the asynchronous algorithms. Nevertheless, the problem is complicated by the fact that we need to deal with both $\widehat{\theta}_\alpha^\epsilon(\cdot)$ and $\tau_\alpha^\epsilon(\cdot)$ for each α, and the full state can change between the updates at any given α.

The random variable $Y_{n,\alpha}^\epsilon$ is used at real time $\tau_{n+1,\alpha}^\epsilon$ to compute $\theta_{n+1,\alpha}^\epsilon$, as illustrated in Figure 3.2. Hence, when centering $Y_{n,\alpha}^\epsilon$ about a conditional mean (given the "past"), it is appropriate to let the conditioning data for processor α be all the data available on the open real-time interval $[0, \tau_{n+1,\alpha}^\epsilon)$. *Thus, we include $\delta\tau_{n,\alpha}^\epsilon$ in the conditioning data, as well as the value of the full state just before that update. Note that the value of the state just before that update at Processor α is $\widehat{\theta}^\epsilon(\tau_{n+1,\alpha}^{\epsilon,-})$.*

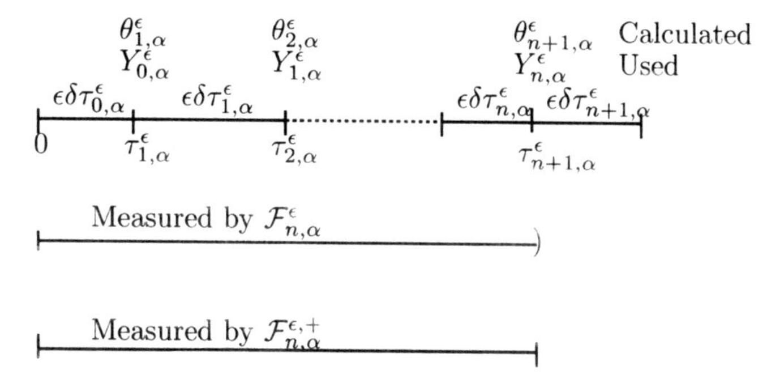

Figure 3.2. The definitions of the σ-algebras.

When centering $\delta\tau^\epsilon_{n,\alpha}$ about its conditional mean (given the "past"), it is appropriate to let the conditioning data for Processor α be all the data available on the closed real-time interval $[0, \tau^\epsilon_{n,\alpha}]$; the data includes $Y^\epsilon_{n-1,\alpha}$. Due to these considerations, different conditioning σ-algebras are used for the centering of the Y and $\delta\tau$ sequences.

For the reasons just explained, for each α, we need the nondecreasing families of σ-algebras $\mathcal{F}^\epsilon_{n,\alpha}$ and $\mathcal{F}^{\epsilon,+}_{n,\alpha}$ defined as follows. The σ-algebra $\mathcal{F}^\epsilon_{n,\alpha}$ measures at least the random variables

$$\theta^\epsilon_0,\ \{Y^\epsilon_{j-1,\gamma}; j,\gamma \text{ such that } \tau^\epsilon_{j,\gamma} < \tau^\epsilon_{n+1,\alpha}\},\ \text{ and}$$
$$\{\tau^\epsilon_{j,\gamma}; j,\gamma \text{ such that } \tau^\epsilon_{j,\gamma} \le \tau^\epsilon_{n+1,\alpha}\}.$$

The σ-algebra $\mathcal{F}^{\epsilon,+}_{n,\alpha}$ measures at least the random variables

$$\theta^\epsilon_0,\ \{Y^\epsilon_{j-1,\gamma}; j,\gamma \text{ such that } \tau^\epsilon_{j,\gamma} \le \tau^\epsilon_{n+1,\alpha}\},\ \text{and}$$
$$\{\tau^\epsilon_{j,\gamma}; j,\gamma \text{ such that } \tau^\epsilon_{j,\gamma} \le \tau^\epsilon_{n+1,\alpha}\}.$$

Thus, $\mathcal{F}^\epsilon_{n,\alpha}$ measures the data available on the real-time interval $[0, \tau^\epsilon_{n+1,\alpha})$, and $\mathcal{F}^{\epsilon,+}_{n,\alpha}$ measures the data available on the real-time interval $[0, \tau^\epsilon_{n+1,\alpha}]$. Let $E^\epsilon_{n,\alpha}$ denote the expectation conditioned on $\mathcal{F}^\epsilon_{n,\alpha}$ and let $E^{\epsilon,+}_{n,\alpha}$ denote the expectation conditioned on $\mathcal{F}^{\epsilon,+}_{n,\alpha}$. The definitions are illustrated in Figure 3.2. The definitions allow considerable versatility, but one might wish to modify them appropriately (depending on the information patterns) in specific applications.

Assumptions. We will need the following analogs of (A8.1.1)–(A8.1.4). The $\Delta^\epsilon_{n,\alpha}$ and $\Delta^{\epsilon,+}_{n,\alpha}$ to be used in conditions (A3.2) and (A3.3) represent communication delays (times ϵ). Conditions (3.8) and (3.10) on these delays hold if the sequence of physical delays $\{\Delta^\epsilon_{n,\alpha}/\epsilon, \Delta^{\epsilon,+}_{n,\alpha}/\epsilon; \epsilon, n, \alpha\}$ is uniformly integrable, a condition that is not restrictive in applications. For notational convenience in the proof of Theorem 3.1, we suppose in (3.6) and (3.9) that for each α all components of the argument $\widehat{\theta}^\epsilon(\cdot)$ are delayed by the same amount, although this is not necessary. Each component can have its own delay as long as it satisfies (3.8) and (3.10). The following conditions are to hold for each α.

(A3.1) $\{Y_{n,\alpha}^{\epsilon}, \delta\tau_{n,\alpha}^{\epsilon}; \epsilon, \alpha, n\}$ is uniformly integrable.

(A3.2) There are real-valued functions $g_{n,\alpha}^{\epsilon}(\cdot)$ that are continuous, uniformly in n and ϵ, random variables $\beta_{n,\alpha}^{\epsilon}$ and non-negative random variables $\Delta_{n,\alpha}^{\epsilon}$ such that

$$E_{n,\alpha}^{\epsilon} Y_{n,\alpha}^{\epsilon} = g_{n,\alpha}^{\epsilon}(\widehat{\theta}^{\epsilon}(\tau_{n+1,\alpha}^{\epsilon,-} - \Delta_{n,\alpha}^{\epsilon})) + \beta_{n,\alpha}^{\epsilon}, \tag{3.6}$$

where

$$\{\beta_{n,\alpha}^{\epsilon}; n, \epsilon, \alpha\} \text{ is uniformly integrable} \tag{3.7}$$

and for each $T > 0$,

$$\sup_{n \le T/\epsilon} \Delta_{n,\alpha}^{\epsilon} \to 0 \tag{3.8}$$

in probability as $\epsilon \to 0$. (The $\Delta_{n,\alpha}^{\epsilon}$ represent the delays in the communication, multiplied by ϵ.)

(A3.3) There are real-valued functions $u_{n,\alpha}^{\epsilon}(\cdot)$ that are strictly positive in the sense of (2.3) and are continuous uniformly in n and ϵ, and non-negative random variables $\Delta_{n,\alpha}^{\epsilon,+}$ such that

$$E_{n,\alpha}^{\epsilon,+}\left[\delta\tau_{n+1,\alpha}^{\epsilon}\right] = u_{n+1,\alpha}^{\epsilon}(\widehat{\theta}^{\epsilon}(\tau_{n+1,\alpha}^{\epsilon} - \Delta_{n+1,\alpha}^{\epsilon,+})) \tag{3.9}$$

and

$$\sup_{n \le T/\epsilon} \Delta_{n,\alpha}^{\epsilon,+} \to 0 \tag{3.10}$$

in probability as $\epsilon \to 0$. (The $\Delta_{n,\alpha}^{\epsilon,+}$ represent the delays in the communication multiplied by ϵ.)

(A3.4) There are continuous and real-valued functions $\bar{g}_{\alpha}(\cdot)$ such that for each $\theta \in H$,

$$\lim_{m,n,\epsilon} \frac{1}{m} \sum_{i=n}^{n+m-1} \left[g_{i,\alpha}^{\epsilon}(\theta) - \bar{g}_{\alpha}(\theta)\right] = 0. \tag{3.11a}$$

(A3.5) There are continuous real-valued functions $\bar{u}_{\alpha}(\cdot)$ (which must be positive by (A3.3)) such that for each $\theta \in H$,

$$\lim_{m,n,\epsilon} \frac{1}{m} \sum_{i=n}^{n+m-1} \left[u_{i,\alpha}^{\epsilon}(\theta) - \bar{u}_{\alpha}(\theta)\right] = 0. \tag{3.11b}$$

(A3.6) $\lim_{m,n,\epsilon} \frac{1}{m} \sum_{i=n}^{n+m-1} E_{n,\alpha}^{\epsilon} \beta_{i,\alpha}^{\epsilon} = 0$ in mean.

The $\lim_{m,n,\epsilon}$ means that the limit is taken as $m \to \infty, n \to \infty$ and $\epsilon \to 0$ simultaneously in any way. We can now write the algorithm for Processor α as

$$\theta^\epsilon_{n+1,\alpha} = \theta^\epsilon_{n,\alpha} + \epsilon g^\epsilon_{n,\alpha}(\widehat{\theta}(\tau^{\epsilon,-}_{n+1,\alpha} - \Delta^\epsilon_{n,\alpha})) + \epsilon\delta M^\epsilon_{n,\alpha} + \epsilon\beta^\epsilon_{n,\alpha} + \epsilon Z^\epsilon_{n,\alpha},$$

where $\delta M^\epsilon_{n,\alpha} = Y^\epsilon_{n,\alpha} - E^\epsilon_{n,\alpha} Y^\epsilon_{n,\alpha}$. There is an analogous decomposition for $\delta\tau^\epsilon_{n,\alpha}$ in terms of the conditional mean and a martingale difference term. Define

$$Z^\epsilon_\alpha(t) = \epsilon \sum_{i=0}^{t/\epsilon - 1} Z^\epsilon_{i,\alpha}.$$

The comments below Theorem 8.2.1 concerning the limit set apply here, as well as in the other theorems of this section.

Theorem 3.1. *Assume* (A3.1)–(A3.6). *Then*

$$\{\theta^\epsilon_\alpha(\cdot), \tau^\epsilon_\alpha(\cdot), \widehat{\theta}^\epsilon_\alpha(\cdot), N^\epsilon_\alpha(\cdot), \alpha \le r\}$$

is tight in $D^{4r}[0,\infty)$. *Let* ϵ *(abusing terminology for simplicity) index a weakly convergent subsequence whose weak sense limit we denote by*

$$(\theta_\alpha(\cdot), \tau_\alpha(\cdot), \widehat{\theta}_\alpha(\cdot), N_\alpha(\cdot), \alpha \le r).$$

Then the limits are Lipschitz continuous with probability one and

$$\theta_\alpha(t) = \widehat{\theta}_\alpha(\tau_\alpha(t)), \quad \widehat{\theta}_\alpha(t) = \theta_\alpha(N_\alpha(t)), \tag{3.12}$$

$$N_\alpha(\tau_\alpha(t)) = t. \tag{3.13}$$

Moreover,

$$\tau_\alpha(t) = \int_0^t \bar{u}_\alpha(\widehat{\theta}(\tau_\alpha(s))ds, \tag{3.14}$$

$$\dot{\theta}_\alpha(t) = \bar{g}_\alpha(\widehat{\theta}(\tau_\alpha(t))) + z_\alpha(t), \tag{3.15}$$

$$\dot{\widehat{\theta}}_\alpha = \frac{\bar{g}_\alpha(\widehat{\theta})}{\bar{u}_\alpha(\widehat{\theta})} + \widehat{z}_\alpha, \; \alpha = 1, \ldots, r, \tag{3.16}$$

where the z_α *and* $\widehat{z}_\alpha$ *serve the purpose of keeping the paths in the interval* $[a_\alpha, b_\alpha]$. *On large intervals* $[0,T]$, *and after a transient period,* $\widehat{\theta}^\epsilon(\cdot)$ *spends nearly all of its time (the fraction going to one as* $\epsilon \to 0$) *in a small neighborhood of* L_H. *Suppose that the solution to* (3.15) *or* (3.16) *is unique for each initial condition. Then* L_H *can be replaced by the set of chain recurrent points in* L_H.

Now, drop the constraint set H *and suppose that* $\{\theta^\epsilon_n; \epsilon, n\}$ *is tight. Then the above conclusions continue to hold with* $z_\alpha(t) = \widehat{z}_\alpha(t) = 0$ *and* L_H *replaced by some bounded invariant set of* (3.16).

Nonstationary or time-varying rates. The representation (3.9) where the function $u_{n,\alpha}^{\epsilon}(\cdot)$ depends on the delayed state is common in applications. If the dependence involves more of the past or is time dependent, then similar results can be obtained. For example, replace (A3.3) and (A3.5) by the assumption that (simply to avoid degeneracy)

$$\inf_{n,\alpha,\omega} E_{n,\alpha}^{\epsilon,+}\left[\delta\tau_{n+1,\alpha}^{\epsilon}\right] > 0.$$

Then $\bar{u}_{\alpha}(\widehat{\theta}(t))$ in (3.16) is replaced by a positive and bounded function $u_{\alpha}(t)$.

Proof. The tightness assertion and the Lipschitz continuity of the paths of the weak sense limit processes of $\{\theta_{\alpha}^{\epsilon}(\cdot), \tau_{\alpha}^{\epsilon}(\cdot), Z_{\alpha}^{\epsilon}(\cdot)\}$ follow from the uniform integrability (A3.1) exactly as was done in the proof of Theorem 8.2.1. The strict positivity of $u_{n,\alpha}^{\epsilon}(\cdot)$ implies that with probability one any weak sense limit of $\{\tau_{\alpha}^{\epsilon}(\cdot)\}$ is strictly monotonically increasing and goes to infinity as $t \to \infty$. In fact, the slope of this weak sense limit process must be at least the lower bound in (2.3). These facts and (3.3) imply the tightness of $\{N_{\alpha}^{\epsilon}(\cdot)\}$ and the Lipschitz continuity of the paths of its weak sense limits. Finally, the tightness of $\{\theta_{\alpha}^{\epsilon}(\cdot), N_{\alpha}^{\epsilon}(\cdot)\}$, together with the Lipschitz property of their weak sense limits and (3.5), imply the tightness of $\{\widehat{\theta}_{\alpha}^{\epsilon}(\cdot)\}$ and the fact that all of its weak sense limits are Lipschitz continuous with probability one.

Equations (3.12) and (3.13) are consequences of the weak convergence and the definitions (3.3) and (3.5). The representation (3.16) follows from (3.12), (3.14), and (3.15) by a change of variable, exactly as (2.7) followed from $\dot{\theta} = \bar{g}(\theta)$, (2.4), and (2.6) in Section 1. In detail,

$$\begin{aligned}\dot{\widehat{\theta}}_{\alpha}(t) &= \dot{\theta}_{\alpha}(N_{\alpha}(t))\dot{N}_{\alpha}(t) \\ &= \frac{\bar{g}_{\alpha}(\widehat{\theta}(\tau_{\alpha}(N_{\alpha}(t)))}{\bar{u}_{\alpha}(\widehat{\theta}(t))} + \text{ reflection term} \\ &= \frac{\bar{g}_{\alpha}(\widehat{\theta}(t))}{\bar{u}_{\alpha}(\widehat{\theta}(t))} + \text{ reflection term.}\end{aligned} \tag{3.17}$$

We need only prove the representations (3.14) and (3.15). We continue to follow the approach of the proof of Theorem 8.2.1. Due to the asymptotic continuity of $\theta_{\alpha}^{\epsilon}(\cdot)$ and $\widehat{\theta}^{\epsilon}(\cdot)$, and (3.8) and (3.10), the delays $\Delta_{n,\alpha}^{\epsilon}$ and $\Delta_{n,\alpha}^{\epsilon,+}$ play no significant roles and only complicate the notation. Without loss of generality, they will be dropped henceforth. Recall the definition $\delta M_{n,\alpha}^{\epsilon} = Y_{n,\alpha}^{\epsilon} - E_{n,\alpha}^{\epsilon} Y_{n,\alpha}^{\epsilon}$ and

$$M_{\alpha}^{\epsilon}(t) = \epsilon \sum_{i=0}^{t/\epsilon - 1} \delta M_{i,\alpha}^{\epsilon}.$$

Define $B_\alpha^\epsilon(t)$ from $\{\beta_{n,\alpha}^\epsilon, n < \infty\}$ analogously, and set

$$\widetilde{G}_\alpha^\epsilon(t) = \epsilon \sum_{i=0}^{t/\epsilon-1} \left[g_{i,\alpha}^\epsilon(\widehat{\theta}^\epsilon(\tau_{i+1,\alpha}^{\epsilon,-})) - \bar{g}_\alpha(\widehat{\theta}^\epsilon(\tau_{i+1,\alpha}^{\epsilon,-}))\right]$$

and

$$\bar{G}_\alpha^\epsilon(t) = \epsilon \sum_{i=0}^{t/\epsilon-1} \bar{g}_\alpha(\widehat{\theta}^\epsilon(\tau_{i+1,\alpha}^{\epsilon,-})).$$

Define $W_\alpha^\epsilon(\cdot)$ by

$$W_\alpha^\epsilon(t) = \widehat{\theta}_\alpha^\epsilon(\tau_\alpha^\epsilon(t)) - \theta_{0,\alpha}^\epsilon - \bar{G}_\alpha^\epsilon(t) - Z_\alpha^\epsilon(t) = \widetilde{G}_\alpha^\epsilon(t) + M_\alpha^\epsilon(t) + B_\alpha^\epsilon(t).$$

Recall the terminology concerning $h(\cdot)$, $s_i \le t$, p, t, and τ used in (8.2.6), where $h(\cdot)$ is an arbitrary bounded, continuous, and real-valued function of its arguments, t and $\tau \ge 0$ are arbitrary real numbers, p is an arbitrary integer, and $s_i \le t, i \le p$, are real numbers. We can write

$$\begin{aligned}
&Eh(\tau_\alpha^\epsilon(s_i), \widehat{\theta}^\epsilon(\tau_\alpha^\epsilon(s_i)), i \le p)\left[W_\alpha^\epsilon(t+\tau) - W_\alpha^\epsilon(t)\right] \\
&\quad - Eh(\tau_\alpha^\epsilon(s_i), \widehat{\theta}^\epsilon(\tau_\alpha^\epsilon(s_i)), i \le p)\left[\widetilde{G}_\alpha^\epsilon(t+\tau) - \widetilde{G}_\alpha^\epsilon(t)\right] \\
&\quad - Eh(\tau_\alpha^\epsilon(s_i), \widehat{\theta}^\epsilon(\tau_\alpha^\epsilon(s_i)), i \le p)\left[M_\alpha^\epsilon(t+\tau) - M_\alpha^\epsilon(t)\right] \\
&\quad - Eh(\tau_\alpha^\epsilon(s_i), \widehat{\theta}^\epsilon(\tau_\alpha^\epsilon(s_i)), i \le p)\left[B_\alpha^\epsilon(t+\tau) - B_\alpha^\epsilon(t)\right] = 0.
\end{aligned} \tag{3.18}$$

Note that the arguments of $h(\cdot)$ involve iterates up to at most the time of the (t/ϵ)th iterate of Processor α, and the terms in the square brackets involve iterates in the time interval from the $(t/\epsilon + 1)$st iterate to the $((t+\tau)/\epsilon)$th iterate of Processor α.

The term involving $M_\alpha^\epsilon(\cdot)$ in (3.18) equals zero by the martingale difference property of the $\delta M_{n,\alpha}^\epsilon$. The term involving $B_\alpha^\epsilon(\cdot)$ goes to zero as $\epsilon \to 0$ by (A3.6). Finally, the term involving $\widetilde{G}_\alpha^\epsilon(\cdot)$ goes to zero as $\epsilon \to 0$ by the uniform continuity of $g_{n,\alpha}^\epsilon(\cdot)$, (A3.4), and the tightness and Lipschitz continuity of the weak sense limits of $\{\widehat{\theta}^\epsilon(\cdot), \tau_\alpha^\epsilon(\cdot), \alpha \le r\}$. Completing the argument exactly as in the proof of Theorem 8.2.1 yields that any weak sense limit $W_\alpha(\cdot), \alpha = 1, \ldots, r$, of $\{W_\alpha^\epsilon(\cdot)\}$ is a martingale (the reader can identify the associated σ-algebras) with Lipschitz continuous paths and the representation

$$W_\alpha(t) = \widehat{\theta}_\alpha(\tau_\alpha(t)) - \theta_\alpha(0) - \bar{G}_\alpha(t) - Z_\alpha(t), \tag{3.19}$$

where

$$\bar{G}_\alpha(t) = \int_0^t \bar{g}_\alpha(\widehat{\theta}(\tau_\alpha(s)))ds,$$

and $Z_\alpha(\cdot)$ is the weak sense limit of $\{Z_\alpha^\epsilon(\cdot)\}$.

Since the value of $W_\alpha(\cdot)$ is zero at time zero, it is identically zero for all time by Theorem 4.1.1. The process $Z_\alpha(t)$ is characterized as in the proof of Theorem 8.2.1. Thus (3.15) holds. Equation (3.14) is proved in a similar manner. □

Differential inclusions and time-varying rates. In Theorems 6.8.1 and 8.2.5, the ODE was replaced by a differential inclusion. Such forms of the mean ODE arise when the local averages of the noise or dynamics vary with time. There is a complete analog for the asynchronous algorithm, and it might be even more important there. Suppose that the mean times between updates does not average out as required by (A3.5) and that the local averages varies over time. This would happen, for example, in the Q-learning problem if the sequence of distributions of the return times to each (state-action) pair has no particular regularity. For reasons of simplicity, the statement is restricted to the variations of the mean times between updates, which seems to be the most important current application. However, the dynamical terms $\bar{g}_\alpha(\cdot)$ can also be elements of a set as in Theorems 6.8.1 and 8.2.5.

Suppose that (3.11b) does not hold in that the i-dependence of $u^\epsilon_{i,\alpha}(\theta)$ cannot be averaged out. Then Theorem 3.2 can be used as follows. Define $\widehat{U}_\alpha(\theta)$ to be the convex hull of the set of limit points of $\sum_{i=n}^{n+m-1} u^\epsilon_{i,\alpha}(\theta)/m$ as $n, m \to \infty$ and $\epsilon \to 0$, and let $U_\alpha(\theta)$ be the uppersemicontinuous regularization (see (1.3.7))

$$U_\alpha(\theta) = \bigcap_{\delta>0} \text{co} \left[\bigcup_{\widehat{\theta} \in N_\delta(\theta)} \widehat{U}_\alpha(\widehat{\theta}) \right].$$

Theorem 3.2. *Assume the conditions of Theorem* 3.1, *but replace* (A3.5) *with the following: There are sets* $U_\alpha(\theta), \alpha = 1, \ldots, r$, *that are upper semi-continuous in the sense of* (4.3.2) *such that for each* $\theta \in H$,

$$\text{distance} \left[\frac{1}{m} \sum_{i=n}^{n+m-1} u^\epsilon_{i,\alpha}(\theta), U_\alpha(\theta) \right] \to 0 \tag{3.20}$$

in mean as n *and* m *go to infinity and* ϵ *goes to zero. Then the conclusions of Theorem* 3.1 *hold with* (3.14) *replaced by*

$$\tau_\alpha(t) = \int_0^t v_\alpha(s) ds, \quad \text{where } v_\alpha(t) \in U_\alpha(\widehat{\theta}(\tau_\alpha(t))), \tag{3.21}$$

and (3.16) *replaced by*

$$\dot{\widehat{\theta}}_\alpha(t) = \frac{\bar{g}_\alpha(\widehat{\theta}(t))}{v_\alpha(t)} + \widehat{z}_\alpha(t), \quad \text{where } v_\alpha(t) \in U_\alpha(\widehat{\theta}(t)). \tag{3.22}$$

Remark on the proof. The proof is essentially the same as that of Theorem 3.1. One need only prove (3.21), because (3.22) follows from (3.13), (3.15), and (3.21) by a time transformation; see also Theorems 6.8.1 and 8.2.5 for related results for the synchronized algorithms.

12.3.3 Correlated Noise

Recall that in Chapter 6 and Subsection 8.2.2, we introduced the "memory" random variables ξ_n^ϵ to account for correlated noise. The situation for the asynchronous algorithm is more complicated due to the asynchronicity and the fact that the appropriate "memories" for the $Y_{n,\alpha}^\epsilon$ and $\delta\tau_{n,\alpha}^\epsilon$ sequences might be quite different. For example, consider the queue optimization problem of Chapters 2 and 9, where the IPA estimators were used to compute pathwise derivatives with respect to the parameter. For the asynchronous form of this problem, the "memory" random variables needed to represent the conditional expectation (given the appropriate past) of the Y-terms might have the basic structure of that used in Chapter 9, but the memory random variables needed to represent the conditional expectation (given the appropriate past) of $\delta\tau_{n,\alpha}^\epsilon$ might not involve any of the terms that were used to compute the pathwise derivatives. They might only involve the minimal part of the past record, which is needed to compute the conditional mean of the time to the next update at α. For this reason, in what follows we introduce different memory random variables for the state and time processes. From the point of view of the proofs, this extra complication is purely notational. Note that the time argument in (A3.7) is the time just before the $(n+1)$st update at Processor α minus the delay, and that in (A3.8) is the time at the $(n+2)$nd update at Processor α minus the delay. There can be a separate delay for each component if we wish.

Keep in mind that the memory random variables need not be known to any of the processors. The use of these memory random variables is a convenient device for obtaining convergence proofs under general conditions.

Assumptions for the correlated noise case. Recall the definitions of $\mathcal{F}_{n,\alpha}^\epsilon$ and $\mathcal{F}_{n,\alpha}^{\epsilon,+}$ given in Subsection 3.2. We adopt the noise model used in Subsection 8.2.2, and enlarge these σ-algebras as follows. For each α let $\xi_{n,\alpha}^\epsilon$ be random variables taking values in some complete separable metric space Ξ, and let $\mathcal{F}_{n,\alpha}^\epsilon$ also measure

$$\{\xi_{j,\gamma}^\epsilon : \gamma, j \text{ such that } \tau_{j,\gamma}^\epsilon < \tau_{n+1,\alpha}^\epsilon\}.$$

Let $\psi_{n,\alpha}^\epsilon$ be random variables taking values in some complete separable metric space Ξ^+ and let $\mathcal{F}_{n,\alpha}^{\epsilon,+}$ measure

$$\{\psi_{j,\gamma}^\epsilon : \gamma, j \text{ such that } \tau_{j,\gamma}^\epsilon \leq \tau_{n+1,\alpha}^\epsilon\}.$$

The $\psi_{n,\alpha}^\epsilon$ are the "memory" random variables for the sequence of update times at α. The following conditions, for $\alpha \leq r$, will be used in addition to (A3.1) and (A3.6). We use A (resp., A^+) to denote an arbitrary compact set in Ξ (resp., in Ξ^+). The number of steps required for the noise process to "settle down" to near its stationary character depends on its initial condition. Owing to the tightness of the noise random variables, we can rerstrict this initial condition to an arbitrary compact set. This is the purpose of the $I_{\{\xi_n^\epsilon \in A\}}$ in (3.13) and (3.14).

(A3.7) There are real-valued measurable functions $g_{n,\alpha}^\epsilon(\cdot)$, random variables $\beta_{n,\alpha}^\epsilon$, and non-negative (delay times ϵ) $\Delta_{n,\alpha}^\epsilon$ such that

$$E_{n,\alpha}^\epsilon Y_{n,\alpha}^\epsilon = g_{n,\alpha}^\epsilon(\widehat{\theta}^\epsilon(\tau_{n+1,\alpha}^{\epsilon,-} - \Delta_{n,\alpha}^\epsilon), \xi_{n,\alpha}^\epsilon) + \beta_{n,\alpha}^\epsilon,$$

and (3.7) and (3.8) hold.

(A3.8) There are strictly positive (in the sense of (2.3)) measurable functions $u_{n,\alpha}^\epsilon(\cdot)$ and non-negative (delay times ϵ) random variables $\Delta_{n,\alpha}^{\epsilon,+}$ such that

$$E_{n,\alpha}^{\epsilon,+} \delta\tau_{n+1,\alpha}^\epsilon = u_{n+1,\alpha}^\epsilon(\widehat{\theta}^\epsilon(\tau_{n+1,\alpha}^\epsilon - \Delta_{n,\alpha}^{\epsilon,+}), \psi_{n+1,\alpha}^\epsilon),$$

and (3.10) holds.

(A3.9) $g_{n,\alpha}^\epsilon(\cdot, \xi)$ is continuous in θ uniformly in n, ϵ, and in $\xi \in A$.

(A3.10) $u_{n,\alpha}^\epsilon(\cdot, \psi)$ is continuous in θ uniformly in n, ϵ, and in $\psi \in A^+$.

(A3.11) The sets $\{\xi_{n,\alpha}^\epsilon, \psi_{n,\alpha}^\epsilon; n, \alpha, \epsilon\}$ are tight.

(A3.12) For each θ,

$$\{g_{n,\alpha}^\epsilon(\theta, \xi_{n,\alpha}^\epsilon), u_{n,\alpha}^\epsilon(\theta, \psi_{n,\alpha}^\epsilon); \epsilon, n\}, \tag{3.23}$$

is uniformly integrable.

(A3.13) There are continuous functions $\bar{g}_\alpha(\cdot)$ such that for each $\theta \in H$,

$$\lim_{m,n,\epsilon} \frac{1}{m} \sum_{i=n}^{n+m-1} E_{n,\alpha}^\epsilon \left[g_{i,\alpha}^\epsilon(\theta, \xi_{i,\alpha}^\epsilon) - \bar{g}_\alpha(\theta) \right] I_{\{\xi_n^\epsilon \in A\}} = 0$$

in probability as n and m go to infinity and $\epsilon \to 0$.

(A3.14) There are continuous, real-valued, and positive functions $\bar{u}_\alpha(\cdot)$ such that for each $\theta \in H$,

$$\lim_{m,n,\epsilon} \frac{1}{m} \sum_{i=n}^{n+m-1} E_{n,\alpha}^{\epsilon,+} \left[u_{i+1,\alpha}^\epsilon(\theta, \psi_{i+1,\alpha}^\epsilon) - \bar{u}_\alpha(\theta) \right] I_{\{\psi_n^\epsilon \in A^+\}} = 0$$

in probability as n and m go to infinity and $\epsilon \to 0$.

An obvious analog of (A8.1.11)–(A8.1.13) can be used in lieu of (A3.9), (A3.10), (A3.13), and (A3.14). We leave it to the reader to write the conditions. We can now state the following theorem.

Theorem 3.3. *Assume* (A3.1) *and* (A3.6)–(A3.14). *Then the conclusions of Theorem* 3.1 *hold. The extension of Theorem* 3.1 *when the constraint set* H *is dropped also holds if* $\{\theta^\epsilon_{n,\alpha}; n, \epsilon\}$ *is tight. The analogs of* (A8.1.11)–(A8.1.13) *can be used in lieu of* (A3.9), (A3.10), (A3.13), and (A3.14).

Comment on the proof. Referring to the proof of Theorem 3.1, redefine $\widetilde{G}^\epsilon_\alpha(\cdot)$ to be

$$\widetilde{G}^\epsilon_\alpha(t) = \epsilon \sum_{i=0}^{t/\epsilon-1} \left[g^\epsilon_{i,\alpha}(\widehat{\theta}^\epsilon(\tau^{\epsilon,-}_{i+1,\alpha} - \Delta^\epsilon_{i,\alpha}), \xi^\epsilon_{i,\alpha}) - \bar{g}_\alpha(\widehat{\theta}^\epsilon(\tau^{\epsilon,-}_{i+1,\alpha} - \Delta^\epsilon_{i,\alpha}))\right],$$

or with the analogous changes required for treating the $\tau^\epsilon_\alpha(\cdot)$. Then the proof is an adaptation of that of Theorem 3.1 in the same way that the proof of Theorem 8.2.2 adapted that of Theorem 8.2.1; the details are omitted.

Differential inclusions and time-varying rates. The next theorem is one useful "differential inclusions" extension. The comments above Theorem 3.2 concerning time-varying rates hold here, with $\widehat{U}_\alpha(\theta)$ defined by the limits in probability of

$$\frac{1}{m} \sum_{i=n}^{n+m-1} E^{\epsilon,+}_{n,\alpha} u^\epsilon_{i+1,\alpha}(\theta, \psi^\epsilon_{i+1,\alpha}).$$

Theorem 3.4. *Assume the conditions of Theorem* 3.3, *but replace* (A3.14) *with the following. There are sets* $U_\alpha(\theta), \alpha = 1, \ldots, r$, *that are upper semi-continuous in the sense of* (4.3.2) *such that for each compact* A^+,

$$\text{distance}\left[\frac{1}{m} \sum_{i=n}^{n+m-1} E^{\epsilon,+}_{n,\alpha} u^\epsilon_{i+1,\alpha}(\theta, \psi^\epsilon_{i+1,\alpha}), U_\alpha(\theta)\right] I_{\{\psi^\epsilon_{n,\alpha} \in A^+\}} \to 0 \quad (3.24)$$

in probability as n and m go to infinity and $\epsilon \to 0$. Then the conclusions of Theorem 3.3 hold with the mean ODEs being (3.21) *and* (3.22).

12.3.4 Analysis for $\epsilon \to 0$ and $T \to \infty$

We now turn to the asymptotic properties of $\widehat{\theta}^\epsilon(T + \cdot)$ as $T \to \infty$ and $\epsilon \to 0$. Recall the discussion in Section 2 concerning the conditions (2.8) and (2.9) that were needed to deal with the convergence of $\widehat{\theta}^\epsilon(t)$ as $t \to \infty$ and $\epsilon \to 0$. Those conditions were needed because the analysis would be done on processes $\{\widehat{\theta}^\epsilon(T_\epsilon + \cdot)\}$ that effectively "start" at large and arbitrary real times. The same issue is more serious in the asynchronous case. We require that the conditions of Theorem 3.3 or 3.4 hold for the processes starting at arbitrary real times. Recall the definition of $p_\alpha^\epsilon(\sigma)$ given below (3.2). The following notation for the process starting at real time σ will be needed.

Definitions. The concepts to be used are as before but under an arbitrary translation of the time origin. This accounts for the various terms that will now be defined. For each $\sigma > 0$, $n \geq 0$, and $\alpha \leq r$, define

$$\theta_{n,\alpha}^{\epsilon,\sigma} = \theta_{p_\alpha^\epsilon(\sigma)+n,\alpha}^\epsilon, \qquad Y_{n,\alpha}^{\epsilon,\sigma} = Y_{p_\alpha^\epsilon(\sigma)+n,\alpha}^\epsilon,$$

and the conditional expectations

$$E_{n,\alpha}^{\epsilon,\sigma} = E_{p_\alpha^\epsilon(\sigma)+n,\alpha}^\epsilon, \qquad E_{n,\alpha}^{\epsilon,\sigma,+} = E_{p_\alpha^\epsilon(\sigma)+n,\alpha}^{\epsilon,+}$$

and define $\delta\tau_{n,\alpha}^{\epsilon,\sigma}$, $\beta_{n,\alpha}^{\epsilon,\sigma}$, $\Delta_{n,\alpha}^{\epsilon,\sigma}$, $\Delta_{n,\alpha}^{\epsilon,\sigma,+}$, $\xi_{n,\alpha}^{\epsilon,\sigma}$, $\psi_{n,\alpha}^{\epsilon,\sigma}$, $g_{n,\alpha}^{\epsilon,\sigma}(\cdot)$, and $u_{n,\alpha}^{\epsilon,\sigma}(\cdot)$, analogously. Define

$$\tau_{n,\alpha}^{\epsilon,\sigma} = \epsilon \sum_{i=0}^{n-1} \delta\tau_{i,\alpha}^{\epsilon,\sigma}.$$

Define the interpolations, for $t \geq 0$:

$$\begin{aligned}
\theta_\alpha^{\epsilon,\sigma}(t) &= \theta_{n,\alpha}^{\epsilon,\sigma}, \quad t \in [n\epsilon, n\epsilon + \epsilon), \\
\widehat{\theta}_\alpha^{\epsilon,\sigma}(t) &= \theta_{n,\alpha}^{\epsilon,\sigma}, \quad t \in [\tau_{n,\alpha}^{\epsilon,\sigma}, \tau_{n+1,\alpha}^{\epsilon,\sigma}), \\
\tau_\alpha^{\epsilon,\sigma}(t) &= \tau_{n,\alpha}^{\epsilon,\sigma}, \quad t \in [n\epsilon, n\epsilon + \epsilon), \\
N_\alpha^{\epsilon,\sigma}(t) &= n\epsilon, \qquad t \in [\tau_{n,\alpha}^{\epsilon,\sigma}, \tau_{n+1,\alpha}^{\epsilon,\sigma}).
\end{aligned}$$

Thus, $\widehat{\theta}_\alpha^\epsilon(\cdot)$ "starts" at the time of the first update at Processor α at or after real time σ. Analogously to (3.3),

$$N_\alpha^{\epsilon,\sigma}(\tau_\alpha^{\epsilon,\sigma}(t)) = n\epsilon, \; t \in [n\epsilon, n\epsilon + \epsilon).$$

Definitions analogous to these are illustrated in Section 4 for the decreasing-step-size case. The conditions are just those for Theorem 3.3, but for arbitrary real starting times σ. The conditions are actually simpler than they might appear at first sight, particularly when the update times have a simple "renewal" character.

The next theorem is simply a restatement of the previous theorems except that time goes to infinity and hence the limit trajectories are supported essentially on the limit set of the ODE. The few extra details of the proof are omitted. The comments below Theorem 8.2.1 concerning the limit set and those above Theorem 3.4 concerning time-varying rates apply here as well.

Theorem 3.5. *Assume the conditions of Theorem 3.3, but with the superscript ϵ replaced by ϵ, σ. Then*

$$\{\theta_\alpha^{\epsilon,\sigma}(\cdot), \tau_\alpha^{\epsilon,\sigma}(\cdot), \widehat{\theta}_\alpha^{\epsilon,\sigma}(\cdot), N_\alpha^{\epsilon,\sigma}(\cdot); \alpha \le r; \epsilon, \sigma\}$$

is tight in $D^{4r}[0,\infty)$. Let (ϵ, σ) (abusing terminology for simplicity) index a weakly convergent (as $\epsilon \to 0$ and $\sigma \to \infty$) subsequence whose weak sense limit we denote by $(\theta_\alpha(\cdot), \tau_\alpha(\cdot), \widehat{\theta}_\alpha(\cdot), N_\alpha(\cdot), \alpha \le r)$. Equations (3.12)–(3.16) hold and for almost all ω, $\widehat{\theta}(\cdot)$ is in the limit set L_H. For any bounded and continuous function $F(\cdot)$ on $D^r(-\infty,\infty)$,

$$EF(\widehat{\theta}^{\epsilon,\sigma}(\epsilon q_\epsilon + \cdot)) \to EF(\widehat{\theta}(\cdot)).$$

There are $T_{\epsilon,\sigma} \to \infty$ such that as $\epsilon \to 0$ and $\sigma \to \infty$,

$$\limsup_{\epsilon} P\left\{\widehat{\theta}^{\epsilon,\sigma}(t) \notin N_\delta(L_H), \text{ some } t \le T_{\epsilon,\sigma}\right\} = 0.$$

Let A denote the set of chain recurrent points in the limit or invariant set and suppose that $\epsilon q_\epsilon \to \infty$ and that the solution to (3.15) or (3.16) is unique for each initial condition. Then A can replace L_H.

Now, drop the constraint set H and assume tightness of $\{\widehat{\theta}_n^{\epsilon,\sigma}; \epsilon, \sigma, n\}$. Then the conclusions continue to hold with $\widehat{z}(t) = 0$ and L_H replaced by a bounded invariant set of the ODE.

For the case of differential inclusions, replace (3.24) by

$$\text{distance}\left[\frac{1}{m}\sum_{i=n}^{n+m-1} E_{n,\alpha}^{\epsilon,\sigma,+} u_{i+1,\alpha}^{\epsilon,\sigma}(\theta, \psi_{i+1,\alpha}^{\epsilon,\sigma}), U_\alpha(\theta)\right] \to 0 \tag{3.25}$$

in probability, where $U_\alpha(\cdot)$ is upper semicontinuous in the sense of (4.3.2). Then the above assertions hold with the ODE given by (3.21) and (3.22).

12.4 Decreasing Step Size

The decreasing-step-size algorithm is treated in virtually the same way as the constant-step-size algorithm. The main differences are either notational or are concerned with the choice of step size. Let $\delta\tau_{n,\alpha}$ denote the real time between the nth and $(n+1)$st updates for Processor α. Define (note that it is not scaled by the step sizes)

$$T_{n,\alpha} = \sum_{i=0}^{n-1} \delta\tau_{i,\alpha}. \tag{4.1}$$

Thus, $T_{n,\alpha}$ is the real time of the nth update for Processor α. For $\sigma \geq 0$, define

$$p_\alpha(\sigma) = \min\{n : T_{n,\alpha} \geq \sigma\}. \tag{4.2}$$

Let $\epsilon_{n,\alpha}$ denote the step size for processor α at update $n+1$. Two classes of step sizes will be used. In the first, there is a function $\epsilon(\cdot)$ such that the step size for an update for Processor α at real time t will be approximately $\epsilon(t)$. It will be seen that such a model is much more versatile than it might appear at first sight. For the second class, we let the step size be $\epsilon_{n,\alpha} = 1/n^a$ for some $a \in (0, 1]$.

The first class of step-size sequences. We will adopt a choice that is versatile and for which the notation is not too complex. There are many variations for which the basic notational structure can be used. Further comments will be made below. The basic step size will be a function of real time. More precisely, we suppose that there is a positive real-valued function $\epsilon(\cdot)$ defined on $[0, \infty)$ such that the step size for the αth processor at the $(n+1)$st update is

$$\epsilon_{n,\alpha} = \frac{1}{\delta\tau_{n,\alpha}} \int_{T_{n,\alpha}}^{T_{n,\alpha}+\delta\tau_{n,\alpha}} \epsilon(s)ds. \tag{4.3}$$

Thus, $\epsilon(\cdot)$ can be used as a measure of real time and the step size is an average of $\epsilon(\cdot)$ over the previous interval. The algorithm can be written as

$$\theta_{n+1,\alpha} = \Pi_{[a_\alpha, b_\alpha]} \left(\theta_{n,\alpha} + \epsilon_{n,\alpha} Y_{n,\alpha}\right), \quad \text{for } \alpha \leq r. \tag{4.4}$$

Assuming the slow rate of variation of $\epsilon(\cdot)$, we have

$$\epsilon_{n,\alpha} \approx \epsilon(T_{n+1,\alpha}). \tag{4.5}$$

The advantage of the form (4.3) is that there is a common measure of real time for all components and the limit ODE does not depend on time.

Definitions. In Chapters 5, 6, and 8, the analysis of the algorithm with decreasing step size used the shifted processes $\theta^n(\cdot)$, which started at the nth iteration. Since the iteration times are not synchronized among the processors, just as for the constant-step-size case, we continue to work with real time. The interpolated process $\theta^n(\cdot)$ will be replaced by the "shifted" process whose time origin is $\sigma > 0$. The following definitions will be needed. Keep in mind that $p_\alpha(\sigma)+n$ is the index of the update, which is the $(n+1)$st update at or after time σ and $T_{p_\alpha(\sigma)+n,\alpha}$ is the real time at which that update occurs. Figures 4.1–4.5 illustrate the following definitions, which are just the basic sequences redefined for the shifted processes. For each

$\sigma \geq 0$ and $n \geq 0$, define

$$\delta\tau^{\sigma}_{n,\alpha} = \delta\tau_{p_\alpha(\sigma)+n,\alpha}, \quad \epsilon^{\sigma}_{n,\alpha} = \epsilon_{p_\alpha(\sigma)+n,\alpha},$$
$$t^{\sigma}_{n,\alpha} = \sum_{i=0}^{n-1} \epsilon^{\sigma}_{i,\alpha}. \tag{4.6}$$

and

$$\tau^{\sigma}_{n,\alpha} = \sum_{i=0}^{n-1} \epsilon^{\sigma}_{i,\alpha}\delta\tau^{\sigma}_{i,\alpha}, \quad Y^{\sigma}_{n,\alpha} = Y_{p_\alpha(\sigma)+n,\alpha}. \tag{4.7}$$

Other functions are defined in terms of these as illustrated by the graphs.

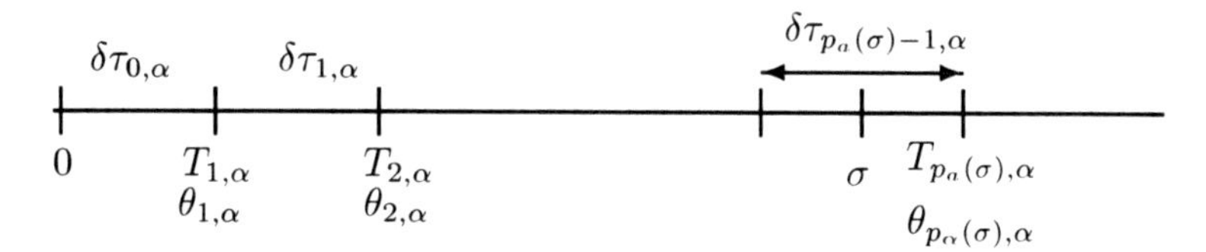

Figure 4.1. The iterates up to time σ for Processor α.

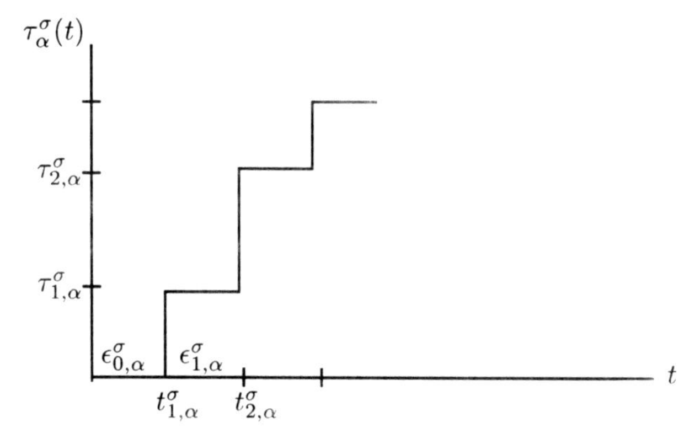

Figure 4.2. The function $\tau^{\sigma}_{\alpha}(t)$.

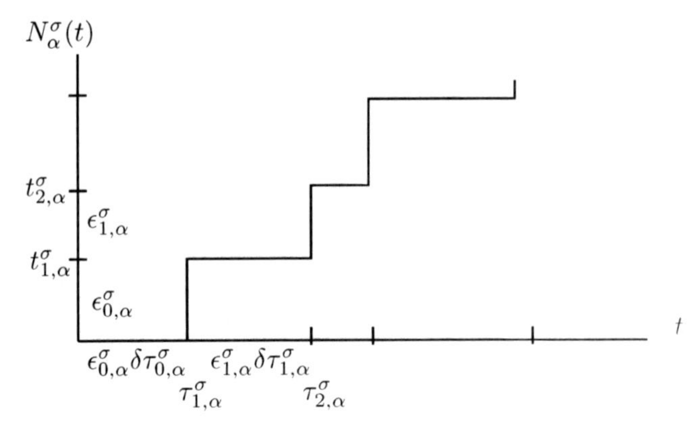

Figure 4.3. The function $N_\alpha^\sigma(t)$.

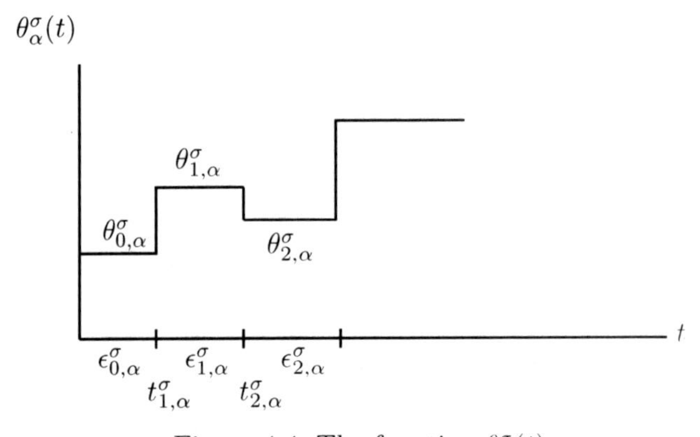

Figure 4.4. The function $\theta_\alpha^\sigma(t)$.

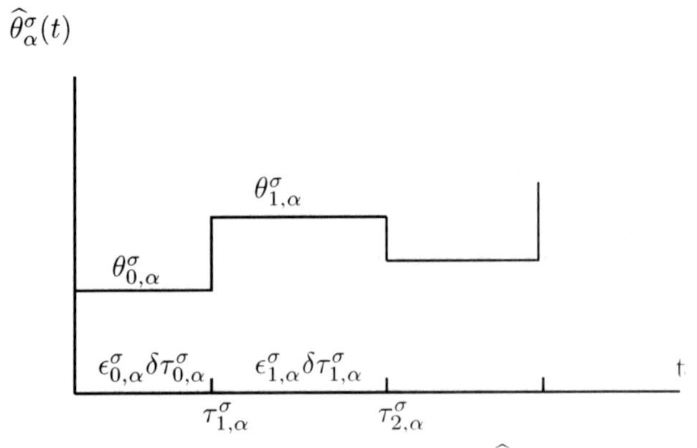

Figure 4.5. The function $\widehat{\theta}_\alpha^\sigma(t)$.

Thus, $\delta\tau^\sigma_{0,\alpha}, \delta\tau^\sigma_{1,\alpha}, \ldots,$ are the inter-update times for Processor α starting at the first update at or after σ, and $\epsilon^\sigma_{0,\alpha}, \epsilon^\sigma_{1,\alpha}, \ldots,$ are the associated step sizes. Also, $\tau^\sigma_{n,\alpha}$ is the integral of $\epsilon(\cdot)$ from the time of the first update at Processor α at or after time σ until the time of the nth update there. In general, for each $\sigma \geq 0$ and a sequence $\{X_{n,\alpha}\}$, we define $X^\sigma_{n,\alpha} = X_{p_\alpha(\sigma)+n,\alpha}$.

The sequence of updates for Processor α starting at the first update at or after time σ can be written as follows. Define the sequence $\{\theta^\sigma_{n,\alpha}\}$ by $\theta^\sigma_{0,\alpha} = \theta_{p_\alpha(\sigma),\alpha}$, its value at the first update at or after time σ. For $n \geq 0$, we have

$$\theta^\sigma_{n+1,\alpha} = \Pi_{[a_\alpha,b_\alpha]}\left(\theta^\sigma_{n,\alpha} + \epsilon^\sigma_{n,\alpha} Y^\sigma_{n,\alpha}\right). \tag{4.8}$$

Define the interpolations

$$\theta^\sigma_\alpha(t) = \theta^\sigma_{n,\alpha}, \quad t \in [t^\sigma_{n,\alpha}, t^\sigma_{n+1,\alpha}), \tag{4.9}$$

$$\widehat{\theta}^\sigma_\alpha(t) = \theta^\sigma_{n,\alpha}, \quad t \in [\tau^\sigma_{n,\alpha}, \tau^\sigma_{n+1,\alpha}), \tag{4.10}$$

$$N^\sigma_\alpha(t) = t^\sigma_{n,\alpha}, \quad t \in [\tau^\sigma_{n,\alpha}, \tau^\sigma_{n+1,\alpha}), \tag{4.11}$$

$$\tau^\sigma_\alpha(t) = \tau^\sigma_{n,\alpha}, \quad t \in [t^\sigma_{n,\alpha}, t^\sigma_{n+1,\alpha}). \tag{4.12}$$

Define the vector $\widehat{\theta}^\sigma(\cdot) = \{\widehat{\theta}^\sigma_\alpha(\cdot), \alpha \leq r\}$.

Analogously to the case in Section 3,

$$N^\sigma_\alpha(\tau^\sigma_\alpha(t)) = t^\sigma_{n,\alpha}, \quad t \in [t^\sigma_{n,\alpha}, t^\sigma_{n+1,\alpha}), \tag{4.13}$$

$$\theta^\sigma_\alpha(t) = \widehat{\theta}^\sigma_\alpha(\tau^\sigma_\alpha(t)),\ \widehat{\theta}^\sigma_\alpha(t) = \theta^\sigma_\alpha(N^\sigma_\alpha(t)). \tag{4.14}$$

The processes $\theta^\sigma_\alpha(\cdot)$ and $\widehat{\theta}^\sigma_\alpha(\cdot)$ are interpolations of the sequences $\{\theta^{\epsilon,\sigma}_{n,\alpha}, n \geq 0\}$, starting at the first update at or after σ. The interpolation $\theta^\epsilon_\alpha(\cdot)$ is in the "iterate" scale (scaled by $\epsilon(\cdot)$), and the interpolation $\widehat{\theta}^\epsilon_\alpha(\cdot)$ is in a scaled (by the $\epsilon(\cdot)$ process) real time. Since the real time at which the first update at or after σ occurs might differ among the components, the processes $\widehat{\theta}^\sigma_\alpha(\cdot)$, $\alpha = 1, \ldots, r$, are not quite properly aligned with each other in time, but the misalignment goes to zero as $\sigma \to \infty$.

Definitions. Analogously to what was done in Subsection 3.2, for each α we will need the nondecreasing families of σ-algebras $\mathcal{F}_{n,\alpha}$ and $\mathcal{F}^+_{n,\alpha}$ defined as follows. The σ-algebra $\mathcal{F}_{n,\alpha}$ measures at least the random variables

$$\begin{aligned}&\theta_0,\ \{Y_{j-1,\gamma}; j,\gamma \text{ such that } T_{j,\gamma} < T_{n+1,\alpha}\}, \text{ and}\\ &\{T_{j,\gamma}; j,\gamma \text{ such that } T_{j,\gamma} \leq T_{n+1,\alpha}\}.\end{aligned}$$

The σ-algebra $\mathcal{F}^+_{n,\alpha}$ measures at least the random variables

$$\begin{aligned}&\theta_0,\ \{Y_{j-1,\gamma}; j,\gamma \text{ such that } T_{j,\gamma} \leq T_{n+1,\alpha}\}, \text{ and}\\ &\{T_{j,\gamma}; j,\gamma \text{ such that } T_{j,\gamma} \leq T_{n+1,\alpha}\}.\end{aligned}$$

Let $E_{n,\alpha}$ and $E^+_{n,\alpha}$ denote the expectations conditioned on $\mathcal{F}_{n,\alpha}$ and $\mathcal{F}^+_{n,\alpha}$, respectively.

Assumptions. Except for those concerning the step size, the assumptions are obvious modifications of those used in the previous theorems, and the reader should have little difficulty in making the notational changes. The assumptions on the step size are:

(A4.1) $\int_0^\infty \epsilon(s)ds = \infty$, where $0 < \epsilon(s) \to 0$ as $s \to \infty$.

(A4.2) There are real $T(s) \to \infty$ as $s \to \infty$ such that

$$\lim_{s\to\infty} \sup_{0 \le t \le T(s)} \left| \frac{\epsilon(s)}{\epsilon(s+t)} - 1 \right| = 0.$$

Theorem 4.1. Assume (A4.1)–(A4.2) *and the obvious modifications of the conditions of Theorem* 3.5. *Then the conclusions of that theorem hold. In particular,* (3.16) *holds, where* $\widehat{\theta}(\cdot)$ *is a weak sense limit of the sequence* $\widehat{\theta}^\sigma(\cdot)$, *where* $\sigma \to \infty$.

Comments on the step-size sequence (4.3). By suitable transformations, the step-size form (4.3) will model or approximate other reasonable sequences. For example, suppose that a separate sequence $\{\epsilon_{n,\alpha}\}$ is given *a priori* for Processor $\alpha, \alpha \le r$. Hence, the step size is indexed by the processor and the update number at that processor. Then the ratio of the step sizes used at the first update at or after real time s for Processors α and γ is

$$\frac{\epsilon_{p_\alpha(s),\alpha}}{\epsilon_{p_\gamma(s),\gamma}} = \frac{\epsilon_{s[p_\alpha(s)/s],\alpha}}{\epsilon_{s[p_\gamma(s)/s],\gamma}}. \tag{4.15}$$

Let there be positive numbers v_α such that $\lim_s p_\alpha(s)/s = v_\alpha$ in the sense of convergence in probability; i.e., the number of updates at Processor α by real time t, divided by t, converges in probability to a constant as $t \to \infty$. Then, for large s, (4.15) is approximately $\epsilon_{v_\alpha s,\alpha}/\epsilon_{v_\gamma s,\gamma}$. If we use the form $\epsilon_{n,\alpha} = 1/n^a, a \in (0,1]$, then this last ratio will not depend on s. Suppose that the ratios in (4.15) do not (asymptotically) depend on s for any pair (α,γ), and let the limits be denoted by $K_{\gamma,\alpha}$. Then for large real time, the step sizes for different components essentially differ by a constant factor. By incorporating these constant factors into the Y^α_n terms, then for all practical purposes there is an $\epsilon(\cdot)$ whose use in (4.3) or (4.5) will yield the same asymptotic results as the use of the separate $\epsilon_{n,\alpha}$ sequences. In particular, define $\epsilon(s) = \epsilon_{v_\gamma s,\gamma}$ for some fixed value of γ. Then the ODE is

$$\dot{\widehat{\theta}}_\alpha = \frac{K_{\gamma,\alpha}\bar{g}_\alpha(\widehat{\theta})}{\bar{u}_\alpha(\widehat{\theta})} + \widehat{z}_\alpha, \ \alpha = 1,\ldots,r. \tag{4.16}$$

Comment on different time scales for different components and time-varying limits. Suppose that for some (but not all) α as $s \to \infty$, with positive probability either $p_\alpha(s)/s \to 0$ or $[p_\alpha(T_s+s)-p_\alpha(T_s)]/s \to 0$ for some sequence $T_s \to \infty$. Then the sequence $\{\delta\tau_{n,\alpha}\}$ will not be uniformly integrable. The problem takes on a singular character in that some components are on a different time scale than others. The algorithm can still be analyzed and convergence proved under conditions similar to what we have used. The techniques are similar, but they require taking into account the fact that some processors work much faster than others. Such systems are discussed in Section 8.6, and the methods used there can be readily extended to the asynchronous problem. One first shows that if the slower components were actually held fixed, then the faster components would converge to some unique limit point, depending on the fixed values of the other components. Then one shows that one can work with the modified system involving only the slower components, where the faster components are replaced by the limit values. See [132] for the analysis of many types of singularly perturbed stochastic systems and [111] for the analysis of deterministic singularly perturbed systems.

Suppose that the ratios of the number of updates per unit time at the various processors are bounded above and below by positive real numbers, but vary with time. Suppose also that the ratios (4.15) are bounded in s, α, γ and the $\epsilon_{n,\alpha}$ vary slowly. Then we can get a nonhomogeneous result. The next paragraph contains some additional detail.

An alternative step-size sequence. Use the assumptions of Theorem 4.1, but let the step-size sequence be $\epsilon_{n,\alpha} = 1/n^a$, where $a \in (0,1]$. In addition, suppose that $\sup_{n,\alpha}[\delta\tau_{n,\alpha}]^2 < \infty$. Let $u_{n+1,\alpha}(\cdot)$ denote the conditional expectation of $\delta\tau_{n+1,\alpha}$, that is analogous to that defined in (A3.8) and suppose that there are positive real numbers $k_{0,\alpha}, k_{1,\alpha}$ such that the conditional expectation lies in $[k_{0,\alpha}, k_{1,\alpha}]$.

In general, the limit ODE will be time dependent. The main issue concerns the function that is used to record the evolution of real time. For simplicity, define $\epsilon(s) = 1/s^a$, and redefine the interpolations in Figures 4.2–4.5 with this $\epsilon(\cdot)$ used to get the $\epsilon^\sigma_{n,\alpha}$ via (4.3). For an update at time σ, this leads to a step size that is approximately $1/\sigma^a$ instead of the correct value $1/p^a_\alpha(\sigma)$ and this needs to be accounted for in obtaining the limit ODE. In particular, the dynamics for $\widehat{\theta}^\alpha(\cdot)$ must be multiplied by the ratio $[1/p^a_\alpha(\sigma)]/[1/\sigma^a]$ for large σ. Under the additional hypotheses supposed here, asymptotically $\sigma/p_\alpha(\sigma)$ lies in $[k_{0,\alpha}, k_{1,\alpha}]$ (the values of the function $\bar{u}_\alpha(\cdot)$ also lie in this interval). Hence, the correct limit ODE has the form

$$\dot{\widehat{\theta}}_\alpha = k_\alpha(t)\frac{\bar{g}_\alpha(\widehat{\theta})}{\bar{u}_\alpha(\widehat{\theta})} + \widehat{z}_\alpha, \ \alpha = 1, \ldots, r, \tag{4.17}$$

where $k_\alpha(t) \in [k^a_{0,\alpha}, k^a_{1,\alpha}]$.

12.5 State-Dependent Noise

The state dependent noise case is treated by the methods of Section 8.4, modified by the techniques of this chapter to deal with the asynchronicity. The notation and assumptions will be stated. The proof is a straightforward combination of the methods of Section 8.4 and Section 3, and the details are omitted. The setup will be based on the approach for Theorem 8.4.1, but the approach of Theorem 8.4.3–8.4.6 can also be used.

As noted in Section 3, there are two "memory random variables," called $\xi^\epsilon_{n,\alpha}$ and $\psi^\epsilon_{n,\alpha}$. The first (resp., second) was used to represent the expectation of the observation (resp., time between the observations) at Processor α, given the "past." The same arrangement is useful for the state dependent noise case. However, to adapt the setup in Section 8.4 to this situation there must be two sets of "fixed-θ processes," namely, $\xi_{n,\alpha}(\theta)$ and $\psi_{n,\alpha}(\theta)$, $\alpha \le r$. This helps explain the following conditions.

We will work with the constant-step-size algorithm, use the terminology of Subsection 3.3, and use (A3.1), (A3.6)–(A3.8) and (A3.11), as well as the following assumptions, where $\alpha = 1, \ldots, r$. The delays in (5.1) and (A5.4) can depend on the component of $\widehat{\theta}$.

(A5.1) There are transition functions $P^\epsilon_{n,\alpha}(\cdot, \cdot|\theta)$ such that $P^\epsilon_{n,\alpha}(\cdot, A|\cdot)$ is measurable for each Borel set A in the range space Ξ of $\xi^\epsilon_{n,\alpha}$ and

$$P\left\{\xi^\epsilon_{n+1,\alpha} \in \cdot|\mathcal{F}^\epsilon_{n,\alpha}\right\} = P^\epsilon_{n,\alpha}(\xi^\epsilon_{n,\alpha}, \cdot|\widehat{\theta}^\epsilon(\tau^{\epsilon,-}_{n+1,\alpha} - \Delta^\epsilon_{n,\alpha})). \tag{5.1}$$

(A5.2) For each fixed θ, there is a transition function $P_\alpha(\xi, \cdot|\theta)$ such that

$$P^\epsilon_{n,\alpha}(\xi, \cdot|\theta) \Rightarrow P_\alpha(\xi, \cdot|\theta), \quad \text{as } n \to \infty \text{ and } \epsilon \to 0, \tag{5.2}$$

where the limit is uniform on each compact (θ, ξ) set; that is, for each bounded and continuous real-valued function $F(\cdot)$ on Ξ,

$$\int F(\widetilde{\xi}) P^\epsilon_{n,\alpha}(\xi, d\widetilde{\xi}|\theta) \to \int F(\widetilde{\xi}) P_\alpha(\xi, d\widetilde{\xi}|\theta) \tag{5.3}$$

uniformly on each compact (θ, ξ)-set, as $n \to \infty$ and $\epsilon \to 0$.

(A5.3) $P_\alpha(\xi, \cdot|\theta)$ is weakly continuous in (θ, ξ).

(A5.4) The analogs of (A5.1)–(A5.3) hold for the $\psi^\epsilon_{n,\alpha}$, where the delays (multiplied by ϵ) are $\Delta^{\epsilon,+}_{n,\alpha}$. Let $P^+(\cdot|\theta)$ denote the analog of $P(\cdot|\theta)$, with a similar modification for $P^\epsilon_{n,\alpha}$, etc.

The fixed-θ Markov chains. For each α and fixed θ, $P_\alpha(\cdot, \cdot|\theta)$ is the transition function of a Markov chain with state space Ξ, analogous to the situation in Section 8.4. This chain is referred to as the *fixed-θ chain* and the random variables of the chain are denoted by $\xi_{n,\alpha}(\theta)$. Unless otherwise specified, when using the expression $E^\epsilon_{n,\alpha} F(\xi_{n+j,\alpha}(\theta))$ for $j \ge 0$, the conditional expectation is for the fixed-θ chain

$$\{\xi_{n+j,\alpha}(\theta), j \geq 0;\ \xi_{n,\alpha}(\theta) = \xi^\epsilon_{n,\alpha}\}, \tag{5.4}$$

which starts at time n with initial condition $\xi_{n,\alpha}(\theta) = \xi^\epsilon_{n,\alpha}$ and θ is given. These fixed-θ chains play the same role here that they do in the synchronized case of Section 8.4. The fixed-θ chains $\psi_{n,\alpha}(\theta)$ are defined analogously, using $P^+(\cdot|\theta)$ and with state space Ξ^+.

(A5.5) $g^\epsilon_{n,\alpha}(\cdot)$ and $u^\epsilon_{n,\alpha}(\cdot)$ are continuous on each compact (θ, ξ)-set, uniformly in ϵ and n.

(A5.6) For any compact sets $A_\alpha \subset \Xi$ and $A^+_\alpha \subset \Xi^+$, and for each $\theta \in H$,

$$\{g^\epsilon_{n+j,\alpha}(\theta, \xi_{n+j,\alpha}(\theta)); j \geq 0, \text{ all } \xi_{n,\alpha}(\theta) \in A_\alpha;\ n \geq 0\} \text{ and} \tag{5.5}$$

$$\{u^\epsilon_{n+j,\alpha}(\theta, \psi_{n+j,\alpha}(\theta)); j \geq 0, \text{ all } \psi_{n,\alpha}(\theta) \in A^+_\alpha:\ n \geq 0\} \tag{5.6}$$

are uniformly integrable.

(A5.7) There are continuous functions $\bar{g}_\alpha(\cdot)$ and $\bar{u}_\alpha(\cdot)$ such that for any compact sets $A_\alpha \subset \Xi$ and $A^+_\alpha \subset \Xi^+$,

$$\frac{1}{m} \sum_{i=n}^{n+m-1} E^\epsilon_{n,\alpha} \left[g^\epsilon_{i,\alpha}(\theta, \xi_{i,\alpha}(\theta)) - \bar{g}_\alpha(\theta)\right] I_{\{\xi^\epsilon_{n,\alpha} \in A_\alpha\}} \to 0, \tag{5.7}$$

$$\frac{1}{m} \sum_{i=n}^{n+m-1} E^{\epsilon,+}_{n,\alpha} \left[u^\epsilon_{i,\alpha}(\theta, \psi_{i,\alpha}(\theta)) - \bar{u}_\alpha(\theta)\right] I_{\{\psi^\epsilon_{n,\alpha} \in A^+_\alpha\}} \to 0 \tag{5.8}$$

in the mean for each θ, as n and m go to infinity and $\epsilon \to 0$.

(A5.8) For each compact set $A_\alpha \subset \Xi$ and $A^+_\alpha \subset \Xi^+$ and each $\mu > 0$, there are compact sets $A_{\alpha,\mu} \subset \Xi$ and $A^+_{\alpha,\mu} \subset \Xi^+$ such that

$$P\left\{\xi_{n+j,\alpha}(\theta) \in A_{\alpha,\mu} \middle| \xi_{n,\alpha}(\theta)\right\} \geq 1 - \mu,$$

for all $\theta \in H, j > 0, n > 0$, and $\xi_{n,\alpha}(\theta) \in A_\alpha$, and

$$P\left\{\psi_{n+j,\alpha}(\theta) \in A^+_{\alpha,\mu} \middle| \psi_{n,\alpha}(\theta)\right\} \geq 1 - \mu,$$

for all $\theta \in H, j > 0, n > 0$, and $\psi_{n,\alpha}(\theta) \in A^+_\alpha$.

The proof of the following theorem uses the methods of Section 8.4 adjusted by the scaling techniques of this chapter; the details are left to the reader. The comments below Theorem 8.2.1 concerning the limit set and those above Theorem 3.5 concerning time-varying rates hold here as well.

Theorem 5.1. *Assume* (A3.1), (A3.6)–(A3.8), (A3.11), *and* (A5.1)–(A5.8). *Then the conclusions of Theorem* 3.3 *hold. The conclusions of Theorem* 3.5 *hold if the assumptions* (A3.1), (A3.6)–(A3.8), *and* (A3.11), *are* *replaced by those on the σ-shifted sequences analogously to what was done in Theorem 3.5. The replacement for (3.24) is: For each compact set $A_\alpha^+ \subset \Xi^+$,*

$$\text{distance}\left[\frac{1}{m}\sum_{i=n}^{n+m-1} E_{n,\alpha}^{\epsilon,+} u_{i,\alpha}^{\epsilon}(\theta, \psi_{i,\alpha}(\theta)), U_\alpha(\theta)\right] I_{\{\psi_{n,\alpha}^{\epsilon} \in A_\alpha^+\}} \to 0 \tag{5.9}$$

in the mean as n and m go to infinity.

12.6 Rate of Convergence: Limit Rate Equations

The rate of convergence analysis of Chapter 10 can be applied to the distributed and asynchronous algorithms. The basic methods of proof combine the techniques of Chapter 10 with those used previously in this chapter. The previous results of this chapter were proved by working basically in "iterate time" for each component and then using a time change argument. To obtain the rate of convergence results, it is simpler if we work directly in (scaled) real time. Because of this, it is convenient to restrict the problem slightly. The restriction is of no practical consequence.

It will be supposed that there is some small number so that the updates at any processor can occur only at times that are integral multiples of this number. There is no loss of generality in doing this since the number can be as small as desired, and the procedure is insensitive to small variations. Having made this assumption, without loss of generality we can suppose that the basic interval is one unit of time. The basic structure of Subsection 3.3 will be retained. Thus we work with a constant step size and weak convergence. There are also analogs of what was done in Chapter 10 for the decreasing-step-size case under either weak or probability one convergence. In Section 10.1, we were concerned with the asymptotic behavior of $(\theta_n^\epsilon - \bar{\theta})/\sqrt{\epsilon}$ as $\epsilon \to 0$ where $\bar{\theta}$ is the limit point, and $n\epsilon \to \infty$ fast enough, and we will treat the analogous situation here.

With the above assumption that the updates can occur only at integral units of time, let $\widehat{\theta}_n^\epsilon = \{\widehat{\theta}_{n,\alpha}^\epsilon, \alpha = 1, \ldots, r\}$ denote the value of the iterate at real time n (not to be confused with the iteration number). We retain the other notations of the previous sections. Recall, in particular, that $\tau_{n,\alpha}^\epsilon$ is ϵ times the real time of the nth update at Processor α. Let $I_{n,\alpha}^\epsilon$ denote the indicator of the event that the αth component is updated at time $n+1$; that is, when $\widehat{\theta}_{n+1,\alpha}^\epsilon$ differs from $\widehat{\theta}_{n,\alpha}^\epsilon$. Define the random variables $\widehat{\xi}_{n,\alpha}^\epsilon$, $\widehat{Y}_{n,\alpha}^\epsilon$ and the functions $\widehat{g}_{n,\alpha}^\epsilon(\cdot)$ by

$$\left.\begin{aligned} &\widehat{\xi}_{n,\alpha}^\epsilon = \xi_{i,\alpha}^\epsilon, \quad \widehat{Y}_{n,\alpha}^\epsilon = Y_{i,\alpha}^\epsilon, \quad \widehat{g}_{n,\alpha}^\epsilon(\cdot) = g_{i,\alpha}^\epsilon(\cdot) \\ &\delta\widehat{M}_{n,\alpha}^\epsilon = \delta M_{i,\alpha}^\epsilon \end{aligned}\right\} \text{ for } n\epsilon \in (\tau_{i,\alpha}^\epsilon, \tau_{i+1,\alpha}^\epsilon]. \tag{6.0}$$

Thus, these newly defined discrete parameter random variables are constants in the unscaled random interval $(\tau^\epsilon_{i,\alpha}/\epsilon, \tau^\epsilon_{i+1,\alpha}/\epsilon]$.

Following the usage in Subsection 10.1.2, it is convenient to define the observation $Y^\epsilon_{n,\alpha}(\bar\theta)$ that would be obtained if the parameter were fixed at $\bar\theta$. Let $\widehat{\mathcal{F}}^\epsilon_n$ denote the σ-algebra that measures all the data that is used to calculate the iterates up to and including $\widehat\theta^\epsilon_n$, as well as $I^\epsilon_{n,\alpha}, \alpha \le r$ and $\{\widehat Y^\epsilon_{i,\alpha}(\bar\theta); \alpha \le r, i < n\}$. Let $\widehat E^\epsilon_n$ denote the expectation conditioned on $\widehat{\mathcal{F}}^\epsilon_n$. Then the algorithm can be written as

$$\widehat\theta^\epsilon_{n+1,\alpha} = \Pi_{[a_i,b_i]}\left(\widehat\theta^\epsilon_{n,\alpha} + \epsilon I^\epsilon_{n,\alpha}\widehat Y^\epsilon_{n,\alpha}\right), \quad \alpha = 1, \ldots, r,$$

which we rewrite as

$$\widehat\theta^\epsilon_{n+1,\alpha} = \widehat\theta^\epsilon_{n,\alpha} + \epsilon I^\epsilon_{n,\alpha}\left(\widehat Y^\epsilon_{n,\alpha} + \widehat Z^\epsilon_{n,\alpha}\right), \quad \alpha = 1, \ldots, r. \tag{6.1}$$

Assumptions. Following the "division of labor" approach that was used in Chapter 10, which was used there to reduce the proofs and ideas to their essential components, we will start by assuming (A6.2), the tightness of the normalized iterates, and then we will give sufficient conditions for this tightness in the next section.

The notation concerning the shift by σ in Subsection 3.4 will be used. The physical delays in communication $(\Delta^\epsilon_{n,\alpha}/\epsilon, \Delta^{\epsilon,+}_{n,\alpha}/\epsilon)$ are dropped for notational simplicity only. The theorem remains true if they are retained and are uniformly integrable. Compare the following assumptions with those used in Subsection 10.1.2.

(A6.1) Let $\bar\theta$ be an isolated stable point of the ODE (3.16) in the interior of H. Let N_ϵ be a sequence of integers such that $\epsilon N_\epsilon \to \infty$. Then $\{\widehat\theta^\epsilon(\epsilon N_\epsilon + \cdot)\}$ converges weakly to the process with constant value $\bar\theta$.

(A6.2) There are integers $p_\epsilon \to \infty$, which can be taken to be $\ge N_\epsilon$, such that

$$\{(\widehat\theta^\epsilon_{p_\epsilon+n} - \bar\theta)/\sqrt{\epsilon};\ \epsilon > 0, n \ge 0\} \quad \text{is tight.}$$

(A6.3) Define $\delta M^\epsilon_{n,\alpha}(\bar\theta) = Y^\epsilon_{n,\alpha}(\bar\theta) - E^\epsilon_{n,\alpha}Y^\epsilon_{n,\alpha}(\bar\theta)$. Then

$$E\left|\delta M^{\epsilon,\sigma}_{n,\alpha} - \delta M^{\epsilon,\sigma}_{n,\alpha}(\bar\theta)\right|^2 \to 0$$

as $n, \sigma \to \infty$ and $\epsilon \to 0$.

(A6.4) For each $\alpha \le r$, the sequence of real-valued processes $W^{\epsilon,\sigma}_\alpha(\cdot)$ defined on $(-\infty, \infty)$ by

$$W^{\epsilon,\sigma}_\alpha(t) = \sqrt{\epsilon}\sum_{i=0}^{t/\epsilon-1} Y^{\epsilon,\sigma}_{i,\alpha}(\bar\theta), \text{ with } Y^{\epsilon,\sigma}_{i,\alpha}(\bar\theta) = g^{\epsilon,\sigma}_{i,\alpha}(\bar\theta, \xi^{\epsilon,\sigma}_{i,\alpha}) + \delta M^{\epsilon,\sigma}_{i,\alpha}(\bar\theta)$$

(for $t \geq 0$, with the analogous definition for $t < 0$) converges weakly (as $\sigma \to \infty$ and $\epsilon \to 0$) to a real-valued Wiener process $W_\alpha(\cdot)$ with variance parameter denoted by Σ_α. Also, for each positive t and sequence of random times N_ϵ that go to infinity in probability,

$$\sum_{i=N_\epsilon}^{N_\epsilon+t/\epsilon-1} \sqrt{\epsilon}\, E^\epsilon_{N_\epsilon,\alpha} Y^\epsilon_{i,\alpha}(\bar{\theta}) \to 0 \tag{6.2}$$

in mean as $\epsilon \to 0$.

(A6.5) $g^\epsilon_{n,\alpha}(\cdot,\xi)$ is continuously differentiable for each n, ϵ, α, and ξ and can be expanded as

$$g^\epsilon_{n,\alpha}(\theta,\xi) = g^\epsilon_{n,\alpha}(\bar{\theta},\xi) + (g^\epsilon_{n,\alpha,\theta}(\bar{\theta},\xi))'(\theta-\bar{\theta}) + (y^\epsilon_{n,\alpha}(\theta,\xi))'(\theta-\bar{\theta}),$$

where

$$y^\epsilon_{n,\alpha}(\theta,\xi) = \int_0^1 \left[g^\epsilon_{n,\alpha,\theta}(\bar{\theta}+s(\theta-\bar{\theta}),\xi) - g^\epsilon_{n,\alpha,\theta}(\bar{\theta},\xi)\right] ds,$$

and if $\delta^\epsilon_n \to 0$ as $\epsilon \to 0$ and $n \to \infty$, then

$$E\left|y^{\epsilon,\sigma}_{n,\alpha}(\theta^{\epsilon,\sigma}_{n,\alpha},\xi^{\epsilon,\sigma}_{n,\alpha})\right| I_{\{|\theta^{\epsilon,\sigma}_{n,\alpha}-\bar{\theta}|\leq\delta^\epsilon_n\}} \to 0$$

as $\epsilon \to 0$ and $n,\sigma \to \infty$.

(A6.6) The set

$$\{g^{\epsilon,\sigma}_{n,\alpha,\theta}(\bar{\theta},\xi^{\epsilon,\sigma}_{n,\alpha});\ \alpha,\epsilon,\sigma\} \quad \text{is uniformly integrable.}$$

(A6.7) There is a Hurwitz matrix A with rows A_α (considered as row vectors), $\alpha = 1,\ldots,r$, such that

$$\frac{1}{m}\sum_{i=n}^{n+m-1} E^{\epsilon,\sigma}_{n,\alpha}\left[(g^{\epsilon,\sigma}_{i,\alpha,\theta}(\bar{\theta},\xi^{\epsilon,\sigma}_{i,\alpha}))' - A_\alpha\right] \to 0$$

in probability as $\epsilon \to 0$ and n, m, and $\sigma \to \infty$, where the gradient $g^{\epsilon,\sigma}_{i,\alpha,\theta}(\bar{\theta},\xi^{\epsilon,\sigma}_{i,\alpha})$ is a column vector.

Following the approach taken in Subsection 10.1.2, let q_ϵ be integers such that $\epsilon(q_\epsilon - p_\epsilon) \to \infty$ as $\epsilon \to 0$. Define $\widehat{U}^\epsilon_n = (\widehat{\theta}^\epsilon_{q_\epsilon+n} - \bar{\theta})/\sqrt{\epsilon}$, and define the process $\widehat{U}^\epsilon(\cdot)$ on $[0,\infty)$ by $\widehat{U}^\epsilon(t) = \widehat{U}^\epsilon_n$ on $[n\epsilon, n\epsilon+\epsilon)$. Define the process $\widehat{W}^\epsilon(\cdot)$ by

$$\widehat{W}^\epsilon_\alpha(t) = \sqrt{\epsilon}\sum_{i=q_\epsilon}^{q_\epsilon+t/\epsilon-1} \widehat{Y}^\epsilon_{i,\alpha}(\bar{\theta}).$$

Theorem 6.1. *With the shift superscript* (ϵ, σ) *replacing* ϵ, *assume conditions* (A3.1), (A3.7), (A3.8), (*with the delays and the* $\beta_{n,\alpha}^\epsilon$ *dropped*), *and* (A3.10), (A3.11), *and* (A3.14) (*for* $\theta = \bar{\theta}$ *only*). *Assume also* (A6.1)–(A6.7). *Then the sequence* $\{\widehat{U}^\epsilon(\cdot), \widehat{W}^\epsilon(\cdot)\}$ *converges weakly in* $D^r[0,\infty) \times D^r(-\infty,\infty)$ *to* $(\widehat{U}(\cdot), \widehat{W}(\cdot))$, *where* $\widehat{W}(\cdot) = \{\widehat{W}_\alpha(\cdot), \alpha \le r\}$ *is a Wiener process with* $E[\widehat{W}_\alpha(1)]^2 = \Sigma_\alpha / \bar{u}_\alpha(\bar{\theta})$, *and* $\widehat{U}(\cdot) = \{\widehat{U}_\alpha(\cdot), \alpha \le r\}$ *is the stationary solution to*

$$d\widehat{U}_\alpha = \frac{A_\alpha \widehat{U}}{\bar{u}_\alpha(\bar{\theta})} dt + d\widehat{W}_\alpha, \ \ \alpha = 1, \ldots, r. \tag{6.3}$$

Proof: Part 1. Formulation of the problem. By the assumed weak convergence (A6.1), for any $T > 0$ and $\mu > 0$,

$$\limsup_\epsilon P \left\{ \sup_{-T/\epsilon \le i \le T/\epsilon} \left| \widehat{\theta}_{q_\epsilon + i}^\epsilon - \bar{\theta} \right| \ge \mu/2 \right\} = 0. \tag{6.4}$$

Hence, by modifying $\{I_{q_\epsilon+n,\alpha}^\epsilon \widehat{Y}_{q_\epsilon+n,\alpha}^\epsilon; \alpha, |n| \le T/\epsilon\}$ on a set of arbitrarily small measure for each ϵ, without altering the assumptions, it can be supposed (justified by Theorem 7.3.6) that

$$\sup_{-T/\epsilon \le i \le T/\epsilon} \left| \widehat{\theta}_{q_\epsilon+i}^\epsilon - \bar{\theta} \right| < \mu \tag{6.5}$$

for small ϵ and for any $T > 0$ and $\mu > 0$ of concern. Thus, without loss of generality, the Z-term in (6.1) can be dropped, and we can suppose that (6.5) holds for any given T and μ.

Let us use $\widehat{y}$ for the integral in (A6.5) if $\widehat{g}$ replaces g in the integrand. Now we can rewrite (6.1) in the form

$$\begin{aligned} \widehat{\theta}_{n+1,\alpha}^\epsilon = \widehat{\theta}_{n,\alpha}^\epsilon &+ \epsilon I_{n,\alpha}^\epsilon \left[\widehat{g}_{n,\alpha}^\epsilon(\bar{\theta}, \widehat{\xi}_{n,\alpha}^\epsilon) + \delta \widehat{M}_{n,\alpha}^\epsilon(\bar{\theta}) + (\widehat{g}_{n,\alpha,\theta}^\epsilon(\bar{\theta}, \widehat{\xi}_{n,\alpha}^\epsilon))'(\widehat{\theta}_n^\epsilon - \bar{\theta}) \right] \\ &+ \epsilon I_{n,\alpha}^\epsilon \left[(\widehat{y}_{n,\alpha}^\epsilon(\widehat{\theta}_n^\epsilon, \widehat{\xi}_{n,\alpha}^\epsilon))'(\widehat{\theta}_n^\epsilon - \bar{\theta}) + \left(\delta \widehat{M}_{n,\alpha}^\epsilon(\bar{\theta}) - \delta \widehat{M}_{n,\alpha}^\epsilon \right) \right], \end{aligned}$$

where $\widehat{g}_{n,\alpha}^\epsilon(\cdot)$, $\delta \widehat{M}_{n,\alpha}^\epsilon$, and $\widehat{\xi}_{n,\alpha}^\epsilon$ are defined in (6.0). Recalling that $\widehat{U}_n^\epsilon = (\widehat{\theta}_{q_\epsilon+n}^\epsilon - \bar{\theta})/\sqrt{\epsilon}$, we can write for each α,

$$\begin{aligned} \widehat{U}_{n+1,\alpha}^\epsilon &= \widehat{U}_{n,\alpha}^\epsilon + \epsilon I_{q_\epsilon+n,\alpha}^\epsilon A_\alpha \widehat{U}_n^\epsilon + \sqrt{\epsilon} I_{q_\epsilon+n,\alpha}^\epsilon \widehat{Y}_{q_\epsilon+n,\alpha}^\epsilon(\bar{\theta}) \\ &\quad + \epsilon I_{q_\epsilon+n,\alpha}^\epsilon \left[(\widehat{g}_{q_\epsilon+n,\alpha,\theta}^\epsilon(\bar{\theta}, \widehat{\xi}_{q_\epsilon+n,\alpha}^\epsilon))' - A_\alpha \right] \widehat{U}_n^\epsilon \\ &\quad + \epsilon I_{q_\epsilon+n,\alpha}^\epsilon (\widehat{y}_{q_\epsilon+n,\alpha}^\epsilon(\widehat{\theta}_{q_\epsilon+n}^\epsilon, \widehat{\xi}_{q_\epsilon+n,\alpha}^\epsilon))' \widehat{U}_n^\epsilon \\ &\quad + \sqrt{\epsilon} I_{q_\epsilon+n,\alpha}^\epsilon \left[\delta \widehat{M}_{q_\epsilon+n,\alpha}^\epsilon(\bar{\theta}) - \delta \widehat{M}_{q_\epsilon+n,\alpha}^\epsilon \right]. \end{aligned}$$

Recall the truncation function $q_M(\cdot)$ and truncated process $U^{\epsilon,M}(\cdot)$ used in Theorem 10.1.1. Introducing the same idea here leads to the following equation for the truncated process:

$$\begin{aligned}\widehat{U}^{\epsilon,M}_{n+1,\alpha} &= \widehat{U}^{\epsilon,M}_{n,\alpha} + \epsilon I^{\epsilon}_{q_\epsilon+n,\alpha} A_\alpha \widehat{U}^{\epsilon,M}_n q_M(\widehat{U}^{\epsilon,M}_n) + \sqrt{\epsilon} I^{\epsilon}_{q_\epsilon+n,\alpha} \widehat{Y}^{\epsilon}_{q_\epsilon+n,\alpha}(\bar{\theta}) \\ &+\epsilon I^{\epsilon}_{q_\epsilon+n,\alpha}\left[(\widehat{g}^{\epsilon}_{q_\epsilon+n,\alpha,\theta}(\bar{\theta},\widehat{\xi}^{\epsilon}_{q_\epsilon+n,\alpha}))' - A_\alpha\right]\widehat{U}^{\epsilon,M}_n q_M(\widehat{U}^{\epsilon,M}_n) \\ &+\epsilon I^{\epsilon}_{q_\epsilon+n,\alpha}(\widehat{y}^{\epsilon}_{q_\epsilon+n,\alpha}(\widehat{\theta}^{\epsilon}_{q_\epsilon+n},\widehat{\xi}^{\epsilon}_{q_\epsilon+n,\alpha}))'\widehat{U}^{\epsilon,M}_n q_M(\widehat{U}^{\epsilon,M}_n) \\ &+\sqrt{\epsilon} I^{\epsilon}_{q_\epsilon+n,\alpha}\left[\delta\widehat{M}^{\epsilon}_{q_\epsilon+n,\alpha}(\bar{\theta}) - \delta\widehat{M}^{\epsilon}_{q_\epsilon+n,\alpha}\right].\end{aligned} \tag{6.6}$$

Let $\widehat{U}^{\epsilon,M}_{\alpha}(\cdot)$ denote the interpolated process with values $\widehat{U}^{\epsilon,M}_{\alpha}(t) = \widehat{U}^{\epsilon,M}_{n,\alpha}$ on the interval $[\epsilon n, \epsilon n+\epsilon)$.

Part 2. Tightness of $\{\widehat{U}^{\epsilon,M}_{\alpha}(\cdot)\}$, and the weak sense limit process. The set of initial conditions $\{\widehat{U}^{\epsilon}_0\}$ is tight by assumption (A6.2). The sequence of processes defined by

$$\epsilon \sum_{i=0}^{t/\epsilon-1} I^{\epsilon}_{q_\epsilon+i,\alpha} A_\alpha \widehat{U}^{\epsilon,M}_i q_M(\widehat{U}^{\epsilon,M}_i)$$

is tight by the uniform boundedness of $\widehat{U}^{\epsilon,M}_n q_M(\widehat{U}^{\epsilon,M}_n)$.

Define the processes $\widehat{W}^{\epsilon}_{\alpha}(\cdot)$ by

$$\widehat{W}^{\epsilon}_{\alpha}(t) = \sqrt{\epsilon} \sum_{i=q_\epsilon}^{q_\epsilon+t/\epsilon-1} I^{\epsilon}_{i,\alpha} \widehat{Y}^{\epsilon}_{i,\alpha}(\bar{\theta}) \tag{6.7}$$

and (see Subsection 3.4 for the σ-notation)

$$W^{\epsilon,\sigma}_{\alpha}(t) = \sqrt{\epsilon} \sum_{i=0}^{t/\epsilon-1} Y^{\epsilon,\sigma}_{i,\alpha}(\bar{\theta}). \tag{6.8}$$

Then, with $\sigma = q_\epsilon$ we have the following relationship:

$$\widehat{W}^{\epsilon}_{\alpha}(t) = W^{\epsilon,q_\epsilon}_{\alpha}(N^{\epsilon,q_\epsilon}_{\alpha}(t)). \tag{6.9}$$

Now $\{\tau^{\epsilon,q_\epsilon}_{\alpha}(\cdot)\}$ is tight by (A3.1). Since $\widehat{\theta}^{\epsilon,q_\epsilon}(\cdot)$ converges weakly to the process with constant value $\bar{\theta}$ by (A6.1), $\tau^{\epsilon,q_\epsilon}_{\alpha}(\cdot)$ converges weakly to the process with values $\bar{u}_\alpha(\bar{\theta})t$. Similarly, $N^{\epsilon,q_\epsilon}_{\alpha}(\cdot)$ converges weakly to the process with values $t/\bar{u}_\alpha(\bar{\theta})$. These convergences, the representation (6.9), and the weak convergence assumed in (A6.4), imply the tightness of $\{\widehat{W}^{\epsilon}_{\alpha}(\cdot)\}$ and that the weak sense limit $\widehat{W}_{\alpha}(\cdot)$ has the representation $\widehat{W}_{\alpha}(t) = W_\alpha(t/\bar{u}_\alpha(\bar{\theta}))$, which is a real-valued Wiener process with variance parameter $\Sigma_\alpha/\bar{u}_\alpha(\bar{\theta})$.

Define the process

$$\widehat{y}^\epsilon_\alpha(t) = \epsilon \sum_{i=q_\epsilon}^{q_\epsilon+t/\epsilon-1} I^\epsilon_{i,\alpha} \widehat{y}^\epsilon_{i,\alpha}(\widehat{\theta}^\epsilon_{i,\alpha}, \widehat{\xi}^\epsilon_i). \tag{6.10}$$

Since, for purposes of the proof we can suppose that $\widehat{\theta}^\epsilon_{q_\epsilon+n} \to \bar{\theta}$ as $\epsilon \to 0$ uniformly in n, the last part of condition (A6.5) implies that the sequence defined by (6.10) converges weakly to the "zero" process.

The tightness of the sequence defined by

$$\epsilon \sum_{i=0}^{t/\epsilon-1} I^\epsilon_{q_\epsilon+i,\alpha} \left[\left(\widehat{g}^\epsilon_{q_\epsilon+i,\alpha,\theta} \left(\bar{\theta}, \widehat{\xi}^\epsilon_{q_\epsilon+i,\alpha} \right) \right)' - A_\alpha \right] \widehat{U}^{\epsilon,M}_i q_M(\widehat{U}^{\epsilon,M}_i) \tag{6.11}$$

is a consequence of the uniform integrability in (A6.6) and the boundedness of the $\widehat{U}^{\epsilon,M}_n q_M(\widehat{U}^{\epsilon,M}_n)$. By (A6.3), the interpolations of the last term on the right of (6.6) converges weakly to the "zero process." Putting these pieces together, we have the tightness of $\{\widehat{U}^\epsilon_\alpha(\cdot)\}$ for each α. Then the fact that the weak sense limit of the processes defined in (6.11) is the "zero" process follows from this tightness and (A6.7).

Let $(\widehat{U}^M(\cdot), \widehat{W}(\cdot))$, with $\widehat{U}^M(\cdot) = \{\widehat{U}^M_\alpha(\cdot), \alpha \le r\}$ and $\widehat{W}(\cdot) = \{\widehat{W}_\alpha(\cdot), \alpha \le r\}$, denote the weak sense limit of a weakly convergent subsequence of $\{\widehat{U}^{\epsilon,M}(\cdot), \widehat{W}^\epsilon(\cdot)\}$. Then the process defined by

$$\sum_{i=0}^{t/\epsilon-1} \epsilon I^\epsilon_{q_\epsilon+i,\alpha} A_\alpha \widehat{U}^{\epsilon,M}_i q_M(\widehat{U}^{\epsilon,M}_i)$$

converges weakly (along the chosen subsequence) to

$$\int_0^t \frac{A_\alpha \widehat{U}^M_\alpha(s) q_M(\widehat{U}^M(s))}{\bar{u}_\alpha(\bar{\theta})} ds.$$

While we have shown that $\widehat{W}_\alpha(\cdot)$ is a real-valued Wiener process for each α, we have not shown that $\{\widehat{W}_\alpha(\cdot), \alpha \le r\}$ is a Wiener process. The joint distributions are Gaussian since the marginal distributions are Gaussian, but we need to show that the increments are mutually independent. This independence (equivalently, orthogonality, due to the Gaussian property and the zero mean value) follows from the assumption on the conditional expectation in (6.2) and the weak convergence (which implies that $\widehat{W}(\cdot)$ is a martingale); the details are left to the reader.

The rest of the details are a combination of those used in the proofs of Theorems 10.1.1, 10.1.2, 3.1, and 3.3. As in the proof of Theorem 10.1.1, one shows that the truncation is not needed. The stationarity is proved exactly the same as in Theorem 10.1.1. □

12.7 Stability and Tightness of the Normalized Iterates

12.7.1 Unconstrained Algorithms

The result in this section is an extension of that in Section 10.5.1 for the unconstrained algorithm. It illustrates one typical Liapunov function-based approach for getting stability for the asynchronous problem. For simplicity, the delays will be dropped. Delays can be handled, but one needs to obtain bounds on the changes in the Liapunov functions during the "delay" intervals. A stability analysis for a synchronous problem, where delays need to be accounted for, is in [40]. Without loss of generality, we retain the structure and notation of the last section, where the updates occur only at the integer times $1, 2, \ldots$

Write the algorithm (6.1) in unconstrained form as

$$\widehat{\theta}^\epsilon_{n+1,\alpha} = \widehat{\theta}^\epsilon_{n,\alpha} + \epsilon I^\epsilon_{n,\alpha} \widehat{Y}^\epsilon_{n,\alpha}. \tag{7.1}$$

The mean ODE for the unconstrained algorithm is

$$\dot{\widehat{\theta}}_\alpha = \frac{\bar{g}_\alpha(\widehat{\theta})}{\bar{u}_\alpha(\widehat{\theta})}, \quad \alpha = 1, \ldots, r. \tag{7.2}$$

Define the vector $\bar{\gamma}(\theta) = \{\bar{g}_\alpha(\theta)/\bar{u}_\alpha(\theta), \alpha = 1, \ldots, r\}$.

Assumptions. The assumptions are to hold for each α and are analogs of those used in Subsection 10.5. The K_i are arbitrary positive real numbers whose values might be different in different usages.

(A7.1) $\bar{\theta}$ is an asymptotically stable point of the ODE $\dot{\theta} = \bar{\gamma}(\theta)$. There is a non-negative continuously differentiable function $V(\cdot)$ that is a Liapunov function for the ODE; its second-order partial derivatives are bounded, and $V(\theta) \to \infty$ as $|\theta| \to \infty$.

(A7.2) $|V_\theta(\theta)|^2 \le K_1(V(\theta) + 1)$.

(A7.3) There is a $\lambda > 0$ such that

$$V_\theta'(\theta)\bar{\gamma}(\theta) \le -\lambda V(\theta).$$

(A7.4) There is a $K_1 > 0$ such that for each $K > 0$,

$$\sup_n E|I^\epsilon_{n,\alpha}\widehat{Y}^\epsilon_{n,\alpha}|^2 I_{\{|\widehat{\theta}^\epsilon_n - \bar{\theta}| \le K\}} \le K_1 E(V(\widehat{\theta}^\epsilon_n) + 1) I_{\{|\widehat{\theta}^\epsilon_n - \bar{\theta}| \le K\}}.$$

(A7.5) Condition (A3.7) holds with the delay and $\beta^\epsilon_{n,\alpha}$ dropped: equivalently,

$$\widehat{E}^\epsilon_n \widehat{Y}^\epsilon_{n,\alpha} I^\epsilon_{n,\alpha} = \widehat{g}^\epsilon_{n,\alpha}(\widehat{\theta}^\epsilon_n, \widehat{\xi}^\epsilon_{n,\alpha}) I^\epsilon_{n,\alpha}.$$

where $\widehat{E}_n^\epsilon$ is defined above (6.1).

(A7.6) For each n, let $I_{i,\alpha}^\epsilon(\theta, n)$ denote the indicator of the event that there is an update at time $i \geq n$ given that the parameter is held fixed at the value θ starting at iterate n. [Note that $I_{n,\alpha}^\epsilon(\widehat{\theta}_n^\epsilon, n) = \widehat{E}_{n,\alpha}^\epsilon I_{n,\alpha}^\epsilon$.] Let the random function

$$\Gamma_{n,\alpha}^{\epsilon,d}(\theta) = \epsilon \sum_{i=n}^{\infty} (1-\epsilon)^{i-n} \widehat{E}_n^\epsilon \left[\widehat{g}_{i,\alpha}^\epsilon(\theta, \widehat{\xi}_{i,\alpha}^\epsilon) - \bar{g}_\alpha(\theta) \right] I_{i,\alpha}^\epsilon(\theta, n) \tag{7.3}$$

be well defined for each θ, α, n and small $\epsilon > 0$ in that the sum of the absolute values of the summands is integrable. Define $\Gamma_n^{\epsilon,d}(\theta) = \{\Gamma_{n,\alpha}^{\epsilon,d}(\theta), \alpha \leq r\}$. Then

$$E|\Gamma_{n,\alpha}^{\epsilon,d}(\widehat{\theta}_n^\epsilon)|^2 = O(\epsilon^2)\left(EV(\widehat{\theta}_n^\epsilon) + 1\right). \tag{7.4}$$

(A7.7) $E\left|\Gamma_{n+1,\alpha}^{\epsilon,d}(\widehat{\theta}_{n+1}^\epsilon) - \Gamma_{n+1,\alpha}^{\epsilon,d}(\widehat{\theta}_n^\epsilon)\right|^2 = O(\epsilon^4)\left(EV(\widehat{\theta}_n^\epsilon) + 1\right).$

(A7.8) Let the random function

$$\Lambda_{n,\alpha}^{\epsilon,d}(\theta) = \epsilon \sum_{i=n}^{\infty} (1-\epsilon)^{i-n} \widehat{E}_n^\epsilon \left[I_{i,\alpha}^\epsilon(\theta, n) - \frac{1}{\bar{u}_\alpha(\theta)} \right] \bar{g}_\alpha(\theta) \tag{7.5}$$

be well defined for each θ, α, n, and small $\epsilon > 0$ in that the sum of the absolute values of the summands is integrable. Also,

$$E|\Lambda_{n,\alpha}^{\epsilon,d}(\widehat{\theta}_n^\epsilon)|^2 = O(\epsilon^2)\left(EV(\widehat{\theta}_n^\epsilon) + 1\right). \tag{7.6}$$

Define $\Lambda_n^{\epsilon,d}(\theta) = \{\Lambda_{n,\alpha}^{\epsilon,d}(\theta), \alpha \leq r\}$.

(A7.9) $E\left|\Lambda_{n+1,\alpha}^{\epsilon,d}(\widehat{\theta}_{n+1}^\epsilon) - \Lambda_{n+1,\alpha}^{\epsilon,d}(\widehat{\theta}_n^\epsilon)\right|^2 = O(\epsilon^4)\left(EV(\widehat{\theta}_n^\epsilon) + 1\right).$

Comment on $\Gamma_{n,\alpha}^{\epsilon,d}(\theta)$ and $\Lambda_{n,\alpha}^{\epsilon,d}(\theta)$. The proof will use a perturbed Liapunov function method. There are two random effects to average via the perturbations. The first are those of $g_{n,\alpha}^\epsilon(\theta, \xi_{n,\alpha}^\epsilon)$; they are handled with a perturbation using $\Gamma_{n,\alpha}^{\epsilon,d}(\theta)$, which will help us to replace this "noise" term with $\bar{g}_\alpha(\theta)$ plus a "small" error. The indicator function $I_{i,\alpha}^\epsilon(\theta, n)$ in the perturbation function is used simply as a way to keep track of the actual (random) number of updates on any real-time interval. Otherwise the perturbation is similar to what has been used before (say in Theorem 10.5.1). The second random effects to be averaged are those due to the random time between updates at any of the processors. This is handled by a perturbation based on $\Lambda_{n,\alpha}^{\epsilon,d}(\theta)$, which works "on top of" the first perturbation, in that it uses the mean $\bar{g}_\alpha(\cdot)$ and not $g_{n,\alpha}^\epsilon(\cdot)$. This perturbation is "centered" at $1/\bar{u}_\alpha(\theta)$, which is the mean rate of increase of $\sum I_{i,\alpha}^\epsilon(\theta)$.

Theorem 7.1. *Consider algorithm* (7.1) *and assume conditions* (A7.1)–(A7.9). *Let there be a symmetric and positive definite matrix P such that*

$$V(\theta) = (\theta - \bar{\theta})' P (\theta - \bar{\theta}) + o(|(\theta - \bar{\theta})|^2) \tag{7.7}$$

for small $|\theta - \bar{\theta}|$. Then there are $n_\epsilon < \infty$ such that $\{(\widehat{\theta}_n^\epsilon - \bar{\theta})/\sqrt{\epsilon}; n \geq n_\epsilon, \epsilon\}$ is tight. In fact, n_ϵ satisfies $e^{-\lambda_1 \epsilon n_\epsilon}|\theta_0 - \bar{\theta}| = O(\sqrt{\epsilon})$, where $\lambda_1 < \lambda$, but is arbitrarily close to it. If there is a symmetric and positive definite matrix P_1 such that

$$V(\theta) \geq (\theta - \bar{\theta})' P_1 (\theta - \bar{\theta}), \tag{7.8}$$

then

$$\limsup_{\epsilon \to 0} \sup_{n \geq n_\epsilon} E \frac{|\theta_n^\epsilon - \bar{\theta}|^2}{\epsilon} < \infty. \tag{7.9}$$

Proof. The proof is very similar to that of Theorem 10.5.1. Recall that $I_{n,\alpha}^\epsilon(\widehat{\theta}_n^\epsilon, \widehat{\xi}_{n,\alpha}^\epsilon) = \widehat{E}_n^\epsilon I_{n,\alpha}^\epsilon$. Start, as usual, by expanding $V(\cdot)$ as

$$\begin{aligned} \widehat{E}_n^\epsilon V(\theta_{n+1}^\epsilon) - V(\theta_n^\epsilon) &= \epsilon \sum_\alpha V_{\theta^\alpha}(\widehat{\theta}_n^\epsilon) \widehat{E}_n^\epsilon \widehat{Y}_{n,\alpha}^\epsilon I_{n,\alpha}^\epsilon \\ &\quad + O(\epsilon^2) \sum_\alpha \widehat{E}_n^\epsilon |\widehat{Y}_{n,\alpha}^\epsilon I_{n,\alpha}^\epsilon|^2. \end{aligned} \tag{7.10}$$

Two perturbations will be used to handle the noise and random update times. The first is

$$\delta V_{n,1}^\epsilon(\widehat{\theta}_n^\epsilon) = V_\theta'(\widehat{\theta}_n^\epsilon) \Gamma_n^{\epsilon,d}(\widehat{\theta}_n^\epsilon).$$

We can write

$$\begin{aligned} &\widehat{E}_n^\epsilon \delta V_{n+1,1}^\epsilon(\widehat{\theta}_{n+1}^\epsilon) - \delta V_{n,1}^\epsilon(\widehat{\theta}_n^\epsilon) \\ &\quad = -\epsilon \sum_\alpha V_{\theta^\alpha}(\widehat{\theta}_n^\epsilon) \left[\widehat{g}_{n,\alpha}^\epsilon(\widehat{\theta}_n^\epsilon, \widehat{\xi}_{n,\alpha}^\epsilon) - \bar{g}_\alpha(\widehat{\theta}_n^\epsilon) \right] I_{n,\alpha}^\epsilon \\ &\quad + \widehat{E}_n^\epsilon \left[V_\theta'(\widehat{\theta}_{n+1}^\epsilon) \Gamma_{n+1}^{\epsilon,d}(\widehat{\theta}_{n+1}^\epsilon) - V_\theta'(\widehat{\theta}_n^\epsilon) \Gamma_{n+1}^{\epsilon,d}(\widehat{\theta}_n^\epsilon) \right] \\ &\quad + \epsilon V_\theta'(\widehat{\theta}_n^\epsilon) \widehat{E}_n^\epsilon \Gamma_{n+1}^{\epsilon,d}(\widehat{\theta}_n^\epsilon). \end{aligned} \tag{7.11}$$

Define the first perturbed Liapunov function $V_{n,1}^\epsilon(\widehat{\theta}_n^\epsilon) = V(\widehat{\theta}_n^\epsilon) + \delta V_{n,1}^\epsilon(\widehat{\theta}_n^\epsilon)$. As in the proof of Theorem 10.5.1, adding (7.10) and (7.11) yields

$$\begin{aligned} \widehat{E}_n^\epsilon V_{n+1,1}^\epsilon(\widehat{\theta}_{n+1}^\epsilon) - V_{n,1}^\epsilon(\widehat{\theta}_n^\epsilon) &= \epsilon \sum_\alpha V_{\theta^\alpha}(\widehat{\theta}_n^\epsilon) \bar{g}_\alpha(\widehat{\theta}_n) I_{n,\alpha}^\epsilon \\ &\quad + \text{error terms}, \end{aligned} \tag{7.12}$$

where the error terms are the last term on the right side of (7.10) and the last two lines of (7.11).

Next, define the second perturbation

$$\delta V_{n,2}^\epsilon(\widehat{\theta}_n^\epsilon) = V_\theta'(\widehat{\theta}_n^\epsilon) \Lambda_n^\epsilon(\widehat{\theta}_n^\epsilon),$$

and define the final perturbed Liapunov function

$$V_n^\epsilon(\widehat{\theta}_n^\epsilon) = V(\widehat{\theta}_n^\epsilon) + \delta V_{n,1}^\epsilon(\widehat{\theta}_n^\epsilon) + \delta V_{n,2}^\epsilon(\widehat{\theta}_n^\epsilon) = V_{n,1}^\epsilon(\widehat{\theta}^\epsilon) + \delta V_{n,2}^\epsilon(\widehat{\theta}_n^\epsilon).$$

We can write

$$\begin{aligned} &\widehat{E}_n^\epsilon \delta V_{n+1,2}^\epsilon(\widehat{\theta}_{n+1}^\epsilon) - \delta V_{n,2}^\epsilon(\widehat{\theta}_n^\epsilon) \\ &\quad = -\epsilon \sum_\alpha V_{\theta^\alpha}(\widehat{\theta}_n^\epsilon)\left[\bar{g}_\alpha(\widehat{\theta}_n) I_{n,\alpha}^\epsilon - \bar{\gamma}_\alpha(\widehat{\theta}_n^\epsilon)\right] \\ &\qquad + \widehat{E}_n^\epsilon\left[V_\theta'(\widehat{\theta}_{n+1}^\epsilon)\Lambda_{n+1}^{\epsilon,d}(\widehat{\theta}_{n+1}^\epsilon) - V_\theta'(\widehat{\theta}_n^\epsilon)\Lambda_{n+1}^{\epsilon,d}(\widehat{\theta}_n^\epsilon)\right] \\ &\qquad + \epsilon V_\theta'(\widehat{\theta}_n^\epsilon)\widehat{E}_n^\epsilon \Lambda_{n+1}^{\epsilon,d}(\widehat{\theta}_n^\epsilon), \end{aligned} \tag{7.13}$$

where $\bar{\gamma}_\alpha(\cdot)$ is defined below (7.2). Finally, putting the above expansions together and using the bounds in (A7.1)–(A7.9) yield for small ϵ (analogously to what was done in the proof of Theorem 10.5.1),

$$\widehat{E}_n^\epsilon V_{n+1}^\epsilon(\widehat{\theta}_{n+1}^\epsilon) - V_n^\epsilon(\widehat{\theta}_n^\epsilon) \le -\epsilon\lambda_1 V(\widehat{\theta}_n^\epsilon) + O(\epsilon^2), \tag{7.14}$$

where $\lambda_1 < \lambda$ but is arbitrarily close to it for small ϵ. Follow the proof of Theorem 10.5.1 from this point on to get the final result. □

The constrained algorithm and other extensions. There are more or less obvious versions of the results in Subsection 10.5.2, and of the results based on probability one convergence in Section 10.4. Similarly, the soft constraint idea of Chapter 5 can be used.

12.8 Convergence for Q-Learning: Discounted Cost

The differential inclusions form of Theorem 3.5 will be applied to prove convergence of the Q-learning algorithm of Section 2.4 for the discounted case with $\beta < 1$. The unconstrained algorithm for the decreasing-step-size case is (2.4.3) or (2.4.4); the reader is referred to that section for the notation. We will work with a practical constrained form and let the step size be ϵ, a constant. The proof for the decreasing-step-size case is virtually the same. The values of $Q_{n,id}$ will be truncated at $\pm B$ for large B. Thus $H = [-B, B]^r$, where r is the number of possible (state, action) pairs. Define $C = \max\{\bar{c}_{id}\}$. Then the Q-values under any policy are bounded by $C/(1-\beta)$. [This does not imply that the Q-values given by an *unconstrained* stochastic approximation will be no greater than $C/(1-\beta)$.] Let $B > C/(1-\beta)$.

Let ψ_n^ϵ denote the state of the controlled chain at time n. The state will generally depend on the step size ϵ, since it depends on the control policy that, in turn, depends on the current estimate of the optimal Q-values.

Recall that the Q-value for the (state, action) pair (i, d) is updated at time $n+1$ (after the next state value is observed) if that pair (i, d) occurs at time n. Thus, we can suppose that the updates are at times $n = 1, 2, \ldots$ Let $\widehat{Q}^\epsilon_{n,id}$ denote the Q-values at real time n and let $I^\epsilon_{n,id}$ be the indicator of the event that (i, d) occurs at real time n.

The algorithm can be written as

$$\widehat{Q}^\epsilon_{n+1,id} = \Pi_{[-B,B]}\left[\widehat{Q}^\epsilon_{n,id} + \epsilon\left(c_{n,id} + \beta \min_{v\in U(\psi^\epsilon_{n+1})} \widehat{Q}^\epsilon_{n,\psi^\epsilon_{n+1}v} - \widehat{Q}^\epsilon_{n,id}\right) I^\epsilon_{n,id}\right] \tag{8.1}$$

or, equivalently, as

$$\widehat{Q}^\epsilon_{n+1,id} = \Pi_{[-B,B]}\left[\widehat{Q}^\epsilon_{n,id} + \epsilon\left(T_{id}(\widehat{Q}^\epsilon_n) - \widehat{Q}^\epsilon_{n,id} + \delta M^\epsilon_n\right) I^\epsilon_{n,id}\right], \tag{8.2}$$

where the operator T_{id} is defined below (2.4.3) and the noise δM^ϵ_n is defined above (2.4.4).

In the absence of the constraint, the general stability results of Section 7 can be used to prove the tightness of $\{\widehat{Q}^\epsilon_{n,id}; n, i, d, \epsilon\}$, which can then be used to get the convergence.

Suppose that there are $n_0 < \infty$ and $\delta_0 > 0$ such that for each state pair i, j,

$$\inf P\left\{\psi^\epsilon_{n+k} = j, \text{ for some } k \le n_0 \middle| \psi^\epsilon_n = i\right\} \ge \delta_0, \tag{8.3}$$

where the inf is over the time n and all ways in which the experimenter might select the controls. Let $\delta\tau^\epsilon_{n,id}$ denote the time interval between the nth and $(n+1)$st occurrences of the (state, action) pair (i, d), and let $E^{\epsilon,+}_{n,id}$ denote the expectation, conditioned on all the data up to and including the $(n+1)$st update for the pair (i, d). Define $u^\epsilon_{n,id}$ by

$$E^{\epsilon,+}_{n,id}\delta\tau^\epsilon_{n+1,id} = u^\epsilon_{n+1,id},$$

and suppose that the (possibly random) $\{u^\epsilon_{n,id}; n, \epsilon\}$ are uniformly bounded by a real number $\bar{u}_{id}$ (which must be ≥ 1) and $\{\delta\tau_{n,id}; n, i, d\}$ is uniformly integrable.

Before proceeding, let us note the following. By (8.3), the conditional mean times between returns to any state i is bounded and the intervals are uniformly integrable. But this will not always be the case for return to any pair (i, d), since, depending on the random evolution, it is possible that some action value d might never be chosen again if the Q-value for that pair becomes too large. To avoid this, and to assure uniform integrability of $\{\delta\tau_{n,id}; n, i, d\}$, we can modify the algorithm slightly as follows. Let $\delta > 0$ be small. At each time select a control at random with probability δ. The analysis is similar with such a modification, and the limit point will differ from $\bar{Q}$ by $O(\delta)$. To assure that the limit point is $\bar{Q}$, one can let the intervals between the randomizations go to infinity, but this has little practical benefit.

Continuing with the analysis, let $\widehat{Q}^\epsilon_{id}(\cdot)$ denote the piecewise constant interpolation of $\{\widehat{Q}^\epsilon_{n,id}; n < \infty\}$ in scaled real time; that is, with interpolation intervals of width ϵ. Define $\widehat{Q}^\epsilon(\cdot) = \{\widehat{Q}^\epsilon_{id}(\cdot);\ i,d\}$. The problem is relatively easy to deal with since the noise terms δM^ϵ_n are martingale differences. Under the given conditions, for any sequence of real numbers T_ϵ, $\{Q^\epsilon_{id}(T_\epsilon + \cdot)\}$ is tight, and (if $T_\epsilon \to \infty$) it will be seen to converge weakly (as $\epsilon \to 0$) to the process with constant value $\bar{Q}$, the optimal value defined by (2.4.2).

The mean ODE will be shown to be

$$\dot{Q}_{id} = D_{id}(t)\,(T_{id}(Q) - Q_{id}) + z_{id}, \quad \text{all } i, d, \tag{8.4}$$

where the $z_{id}(\cdot)$ serve the purpose of keeping the values in the interval $[-B, B]$ and will be shown to be zero. The values of $D_{id}(\cdot)$ lie in the intervals $[1/\bar{u}_{id}, 1]$. We next show that all solutions of (8.4) tend to the unique limit point $\bar{Q}$. Suppose that $Q_{id}(t) = B$ for some t and i, d pair. Note that, by the lower bound on B,

$$\sup_{Q:Q_{id}=B} [T_{id}(Q) - Q_{id}] \le C + \beta \sum_j p_{ij}(d)B - B = C + (\beta - 1)B < 0.$$

This implies that $\dot{Q}_{id}(t) < 0$ if $Q_{id}(t) = B$. Analogously, $\dot{Q}_{id}(t) > 0$ if $Q_{id}(t) = -B$. Thus, the boundary of the constraint set H is not accessible by a trajectory of (8.4) from any interior point. Now, dropping the z_{id} terms, by the contraction property of $T(\cdot)$, $\bar{Q}$ is the unique limit point of (8.4).

We need only prove that (8.4) is the mean ODE by verifying the conditions of the differential inclusions part of Theorem 3.5. The expectation of the observation in (8.2) used at time $(n+1)$, given the past and that the pair (i, d) occurred at time n, is

$$\bar{g}_{id}(\widehat{Q}^\epsilon_n) \equiv T_{id}(\widehat{Q}^\epsilon_n) - \widehat{Q}^\epsilon_{id}.$$

Thus, the $g^\epsilon_{n,id}(\cdot)$ of Theorem 3.5 are all $\bar{g}_{id}(\cdot)$. As a result (A3.7) holds, with the $\beta^\epsilon_{n,\alpha}$-terms and the delays being zero. [Delays can be allowed, provided that they are uniformly integrable, but in current Q-learning applications, delays are of little significance.]

Condition (A3.9) is obvious (since the noise terms are martingale differences, the memory random variable $\xi^\epsilon_{n,\alpha}$ is not needed). Also the conditions for the differential inclusion result in Theorem 3.5 hold, where $U_{id}(\theta) = U_{id} = [1, \bar{u}_{id}]$. All the other conditions in the theorem are either obvious or not applicable. Thus, Theorem 3.5 holds.

References

[1] J. Abounadi, D. Bertsekas, and V.S. Borkar. Learning algorithms for Markov decision processes with average cost. *SIAM J. Control Optim.*, 40:681–698, 2002.

[2] M. Andrews, K. Kumaran, K. Ramanan, A. Stolyar, R. Vijayakumar, and P. Whiting. CDMA data QoS scheduling on the forward link with variable channel conditions. Bell Labs. preprint, 2000.

[3] L. Arnold. *Stochastic Differential Equations.* Wiley, New York, 1974.

[4] G. Bao, C.C. Cassandras, and M.A. Zazanis. First and second derivative estimators for cyclic closed queueing networks. *IEEE Trans. Automatic Control*, AC-41:1106–1124, 1996.

[5] A.G. Barto, S.J. Bradtke, and S.P. Singh. Learning to act using real-time dynamic programming. *Artificial Intelligence*, 72:81–138, 1995.

[6] J.A. Bather. Stochastic approximation: A generalization of the Robbins-Monro procedure. In P. Mandl and M. Hušková, editors, *Proc. Fourth Prague Symp. Asymptotic Statistics*, pages 13–27. Charles University, Prague, 1989.

[7] M. Benaïm. The off-line learning approximation in continuous time neural networks: An adiabatic theorem. *Neural Networks*, 6:655–665, 1993.

[8] M. Benaïm. Global asymptotic properties of learning processes. In *Proc. World Congress on Neural Networks, 1995.*

[9] M. Benaïm. A dynamical systems approach to stochastic approximations. *SIAM J. Control Optim.*, 34:437–472, 1996.

[10] M. Benaïm. Dynamics of stochastic approximation algorithms. In *Lecture Notes in Mathematics: 1769: Séminaire de Probabilities*, pages 1–69. Springer-Verlag, New York and Berlin, 1999.

[11] M. Benaïm and M.W. Hirsch. Mixed equilibria and dynamical systems arising from fictitious play in perturbed games. *Games Economic Behavior*, 29:36–72, 1999.

[12] M. Benaïm and M.W. Hirsch. Stochastic approximation algorithms with constant step size whose average is cooperative. *Ann. Appl. Probab.*, 9:216–241, 1999.

[13] M. Benaïm and L. Tomasini. Competitive and self organizing algorithms based on the minimization of an information criterion. In O. Simula T. Kohonen, K. Mäkisara and J. Kangas, editors, *Artificial Neural Networks*, pages 391–396. North-Holland, Amsterdam, 1991.

[14] P. Bender, P. Black, M. Grob, R. Padovani, N. Sindhushyana, and S. Viterbi. CDMA/HDR: a bandwidth efficient high speed wireless data service for nomadic users. *IEEE Comm. Magazine*, 38:70–77, 2000.

[15] A. Benveniste, M. Goursat, and G. Ruget. Robust identification of a nonminimum phase system: Blind adjustment of a linear equalizer in data communications. *IEEE Trans. Automatic Control*, AC-25:385–399, 1980.

[16] A. Benveniste, M. Metivier, and P. Priouret. *Adaptive Algorithms and Stochastic Approximation.* Springer-Verlag, Berlin and New York, 1990.

[17] E. Berger. Asymptotic behavior of a class of stochastic approximation procedures. *Probab. Theory Related Fields*, 71:517–552, 1986.

[18] L. Berkovitz. *Optimal Control Theory*, volume 12 of *Appl. Math. Sciences.* Springer-Verlag, Berlin and New York, 1974.

[19] J. Bertran. Optimisation stochastique dans un espace de Hilbert. *C. R. Acad. Sci. Paris Ser A*, 276:613–616, 1973.

[20] D. Bertsekas and J.N. Tsitsiklis. *Parallel and Distributed Computation.* Prentice-Hall, Englewood Cliffs, 1989.

[21] D.P. Bertsekas. *Constrained Optimization and Lagrange Multiplier Methods.* Academic Press, New York, 1982.

[22] D.P. Bertsekas. *Dynamic Programming: Deterministic and Stochastic Models*. Prentice-Hall, Englewood Cliffs, NJ, 1987.

[23] S. Bhatnagar and V.S. Borkar. Multiscale stochastic approximation for parametric optimization of hidden Markov models. *Probab. Eng. Inform. Sci.*, 11:509–522, 1997.

[24] S. Bhatnagar and V.S. Borkar. A two-time-scale stochastic approximation scheme for simulation based parametric optimization. *Probab. Eng. Inform. Sci.*, 12:519–531, 1998.

[25] P. Billingsley. *Convergence of Probability Measures*. Wiley, New York, 1968.

[26] G. Blankenship and G.C. Papanicolaou. Stability and control of systems with wide band noise disturbances. *SIAM J. Appl. Math.*, 34:437–476, 1978.

[27] J.R. Blum. Multidimensional stochastic approximation. *Ann. Math. Statist.*, 9:737–744, 1954.

[28] V.S. Borkar. Stochastic approximation with two time scales. *Systems Control Lett.*, 29:291–294, 1997.

[29] O. Brandière. The dynamical system method and traps. *Adv. in Appl. Probab.*, 30:137–151, 1998.

[30] O. Brandière. Some pathological traps for stochastic approximation. *SIAM J. Control Optim.*, 36:1293–1314, 1998.

[31] O. Brandière and M. Duflo. Les algorithmes stochastique contournent-ils les pieges? *Ann. Inst. Henri Poincaré*, 32:395–427, 1996.

[32] M. Bray. Convergence to rational expectations equilibrium. In R. Frydman and E.S. Phelps, editors, *Individual Forecasting and Aggregate Outcomes*, pages 123–137. Cambridge University Press, Cambridge, UK, 1983.

[33] M.M. Bray and N.E. Savin. Rational expectations equilibria, learning and model specification. *Econometrica*, 54:1129–1160, 1986.

[34] L. Breiman. *Probability Theory*, volume 7 of *Classics in Applied Mathematics*, A reprint of the 1968 Addison-Wesley edition. SIAM, Philadelphia, 1992.

[35] P. Brémaud and F.J. Váquez-Abad. On the pathwise computation of derivatives with respect to the rate of a point process: The phantom RPA method. *Queueing Systems Theory Appl.*, 10:249–270, 1992.

[36] J.-M. Brossier. *Egalization Adaptive et Estimation de Phase: Application aux Communications Sous-Marines.* PhD thesis, Institut National Polytechnique de Grenoble, 1992.

[37] R. Buche. *Stochastic approximation: Rate of convergence for constrained algorithms; asynchronous algorithms and analysis of a competitive resource sharing system.* PhD thesis, Brown University, Appl. Math. Dept., 2000.

[38] R. Buche and H.J. Kushner. Stochastic approximation and user adaptation in a competitive resource sharing system. *IEEE Trans. Automatic Control*, 45:844–853, 2000.

[39] R. Buche and H.J. Kushner. Rate of convergence for constrained stochastic approximation algorithms. *SIAM. J. Control Optim.*, 40:1011–1041, 2001.

[40] R. Buche and H.J. Kushner. Stability and control of mobile communications with time varying channels. Brown Univ., Div. of Applied Math. report. A short version is in Proc. IEEE Internat. Conf. on Communications-2002, 687–693, Vol 2, IEEE Press, New York., 2001.

[41] R. Buche and H.J. Kushner. Adaptive optimization of least squares tracking algorithms: with applications to adaptive antennas arrays for randomly time-varying mobile communications systems. Brown Univ. LCDS Report, 2003.

[42] J.A. Bucklew. *Large Deviations Techniques in Decision, Simulation and Estimation.* Wiley, New York, 1990.

[43] R.S. Bucy. Stability and positive supermartingales. *J. Differential Equations*, 1:151–155, 1965.

[44] F. Campillo and E. Pardoux. Numerical methods in ergodic optimal stochastic control. In I. Karatzas and D. Ocone, editors, *Applied Stochastic Analysis*, pages 59–73, Berlin and New York, 1991. Springer-Verlag.

[45] M.D. Canon, C.D. Cullum, and E. Polak. *The Theory of Optimal Control and Mathematical Programming.* McGraw-Hill, New York, 1970.

[46] M. Caramanis and G. Liberopoulos. Perturbation analysis for the design of flexible manufacturing system flow controllers. *Oper. Res.*, 40:1107–1125, 1992.

[47] H.-F. Chen. Asymptotically efficient stochastic approximation. *Stochastics Stochastics Rep.*, 45:1–16, 1993.

[48] H.-F. Chen. *Stochastic Approximation and Its Applications.* Kluwer Academic, Dordrecht, Netherlands, 2003.

[49] H.-F. Chen and L. Guo. *Identification and Stochastic Adaptive Control.* Birkhäuser, Boston, 1991.

[50] H.-F. Chen and Y.M. Zhu. Stochastic approximation procedures with randomly varying truncations. *Sci. Sinica Ser. A*, 29:914–926, 1986.

[51] B. Cheng and D.M. Titterington. Neural networks: A review from a statistical perspective. *Statistical Sci.*, 9:2–54, 1994.

[52] E.K.P. Chong and P.J. Ramadge. Optimization of queues using an infinitesimal perturbation analysis-based stochastic algorithm with general update times. *SIAM J. Control Optim.*, 31:698–732, 1993.

[53] K.-L. Chung. *A Course in Probability Theory,* (second edition). Academic Press, New York, 1974.

[54] R.T. Compton. *Adaptive Antennas: Concepts and Performance.* Prentice Hall, Englewood Cliffs, 1988.

[55] B. Delyon and A. Juditsky. Stochastic optimization with averaging of trajectories. *Stochastics Stochastic Rep.*, 39:107–118, 1992.

[56] B. Delyon and A. Juditsky. Asymptotical study of parameter tracking algorithms. *SIAM J. Control Optim.*, 33:323–345, 1995.

[57] M. Duflo. *Random Iterative Algorithms.* Springer-Verlag, Berlin and New York, 1997.

[58] N. Dunford and J.T. Schwartz. *Linear Operators, Part 1: General Theory.* Wiley-Interscience, New York, 1966.

[59] V. Dupač. On the Kiefer–Wolfowitz approximation method. *Casopis Pest. Mat.*, 82:47–75, 1957.

[60] P. Dupuis and R. Ellis. *A Weak Convergence Approach to the Theory of Large Deviations.* Wiley, New York, 1997.

[61] P. Dupuis and H.J. Kushner. Stochastic approximation via large deviations: Asymptotic properties. *SIAM J. Control Optim.*, 23:675–696, 1985.

[62] P. Dupuis and H.J. Kushner. Asymptotic behavior of constrained stochastic approximations via the theory of large deviations. *Probab. Theory Related Fields*, 75:223–244, 1987.

[63] P. Dupuis and H.J. Kushner. Stochastic approximation and large deviations: Upper bounds and w.p.1 convergence. *SIAM J. Control Optim.*, 27:1108–1135, 1989.

[64] P. Dupuis and R.J. Williams. Lyapunov functions for semimartingale reflecting Brownian motions. *Ann. Probab.*, 22:680–702, 1994.

[65] A. Dvoretzky. On stochastic approximation. In *Proc. of the Third Berkeley Symposium on Mathematical Statistics and Probability*, pages 39–55, 1956.

[66] A. Dvoretzky. Stochastic approximation revised. *Adv. in Appl. Math.*, 7:220–227, 1986.

[67] Yu. Ermoliev. Stochastic quasigradient methods and their applications to system optimization. *Stochastics*, 9:1–36, 1983.

[68] S.N. Ethier and T.G. Kurtz. *Markov Processes: Characterization and Convergence.* Wiley, New York, 1986.

[69] G.W. Evans and S. Honkapohja. Local convergence of recursive learning to steady states and cycles in stochastic nonlinear models. *Econometrica*, 63:195–206, 1995.

[70] V. Fabian. Stochastic approximation methods. *Czechoslovak Math. J.*, 10:123–159, 1960.

[71] V. Fabian. On asymptotic normality in stochastic approximation. *Ann. Math. Statist.*, 39:1327–1332, 1968.

[72] V. Fabian. Stochastic approximation. In J.S. Rustagi, editor, *Optimizing Methods in Statistics.* Academic Press, New York, 1971.

[73] J.-C. Fort and G. Pagès. On the a.s. convergence of the Kohonen algorithm with a general neighborhood function. *Ann. Appl. Probab.*, 5:1177–1216, 1995.

[74] G.J. Foschini. Layered space-time architecture for wireless communication in a fading environment when using multi-element antennas. *Bell Labs Technical J.*, autumn 1966.

[75] M.I. Freidlin. The averaging principle and theorems on large deviations. *Russian Math. Surveys*, 33:117–176, 1978.

[76] M.I. Freidlin and A.D. Wentzell. *Random Perturbations of Dynamical Systems.* Springer-Verlag, New York, 1984.

[77] A. Friedman. *Stochastic Differential Equations and Applications.* Academic Press, New York, 1975.

[78] M.C. Fu. Convergence of the stochastic approximation algorithm for the GI/G/1 queue using infinitesimal perturbation analysis. *J. Optim. Theory Appl.*, 65:149–160, 1990.

[79] D. Fudenberg and D.M. Kreps. Learning mixed equilibria. *Games Economic Behavior*, 5:320–367, 1993.

[80] C.R. Fuller and A.H. von Flotow. Active control of sound and vibration. *IEEE Control Systems Magazine*, 6:9–19, 1995.

[81] S.B. Gelfand and S.K. Mitter. Recursive stochastic algorithms for global optimization in $\mathbb{R}^d$. *SIAM J. Control Optim.*, 29:999–1018, 1991.

[82] I.I. Gihman and A.V. Skorohod. *Introduction to the Theory of Random Processes.* Saunders, Philadelphia, 1965.

[83] E.G. Gladyshev. On stochastic approximation. *Theory Probab. Appl.*, 10:275–278, 1965.

[84] P. Glasserman. *Gradient Estimation via Perturbation Analysis.* Kluwer Academic, 1991.

[85] P. Glasserman. Filtered monte carlo. *Math. Oper. Res.*, 18:610–634, 1993.

[86] P. Glasserman and S. Tayur. Sensitivity analysis for base-stock levels in multiechelon production-inventory systems. *Management Sci.*, 41:263–281, 1995.

[87] L. Goldstein. Minimizing noisy functionals in Hilbert space: An extension of the Kiefer–Wolfowitz procedure. *J. Theoretical Probab.*, 1:189–294, 1988.

[88] U. Grenander and G. Szego. *Toeplitz Forms and Their Applications.* University of California Press, Berkeley, 1958.

[89] J. Guckenheimer and P. Holmes. *Nonlinear Oscillations, Dynamical Systems, and Bifurcations of Vector Fields.* Springer-Verlag, Berlin and New York, 1983.

[90] S. Gunnarsson and L. Ljung. Frequency domain tracking characteristics of adaptive algorithms. *IEEE Trans. Acoust. Speech Signal Process.*, ASSP-37:1072–1089, 1989.

[91] L. Guo. Stability of recursive tracking algorithms. *SIAM J. Control Optim.*, 32:1195–1225, 1994.

[92] L. Guo and L. Ljung. Performance analysis of general tracking algorithms. *IEEE Trans. Automatic Control*, AC-40:1388–1402, 1995.

[93] L. Guo, L. Ljung, and P. Priouret. Tracking performance analyses of the forgetting factor RLS algorithm. In *Proc. of the* 31*st Conference on Decision and Control,* Tucson, Arizona, pages 688–693, New York, 1992. IEEE.

[94] B. Hajek. Stochastic approximation methods for decentralized control of multiaccess communications. *IEEE Trans. Inform. Theory*, IT-31:176–184, 1985.

[95] W.K. Härdle and R. Nixdorf. Nonparametric sequential estimation of zeros and extrema of regression functions. *IEEE Trans. Inform. Theory*, IT-33:367–372, 1987.

[96] S. Haykin. *Adaptive Filter Theory.* Prentice-Hall, Englewood Cliffs, NJ, 1990.

[97] S. Haykin. *Neural Networks: A Comprehensive Foundation.* Macmillan, New York, 1994.

[98] M.W. Hirsch. Systems of differential equations that are competitive and cooperative: Convergence almost everywhere. *SIAM J. Math. Anal.*, 16:423–439, 1985.

[99] Y.-C. Ho. Performance evaluation and perturbation analysis of discrete event dynamic systems. *IEEE Trans. Automatic Control*, AC-32:563–572, 1987.

[100] Y.-C. Ho and X.-R Cao. Performance sensitivity to routing changes in queueing networks and flexible manufacturing systems using perturbation analysis. *IEEE J. Robotics Automation*, RA-1:165–172, 1985.

[101] Y.-C. Ho and X.-R. Cao. *Perturbation Analysis of Discrete Event Dynamical Systems.* Kluwer, Boston, 1991.

[102] P.J. Huber. *Robust Statistics.* Wiley, New York, 1981.

[103] I. Iscoe, P. Ney, and E. Nummelin. Large deviations of uniformly recurrent Markov additive processes. *Adv. in Appl. Math.*, 6:373–412, 1985.

[104] J. Jacod and A.N. Shiryaev. *Limit Theorems for Stochastic Processes.* Springer-Verlag, Berlin and New York, 1987.

[105] J.A. Joslin and A.J. Heunis. Law of the iterated logarithm for constant-gain linear stochastic gradient algorithm. *SIAM J. Control Optim.*, 39:533–570, 2000.

[106] A. Juditsky. A stochastic estimation algorithm with observation averaging. *IEEE Trans. Automatic Control*, 38:794–798, 1993.

[107] F.P. Kelly, A.K. Maulloo, and D.K.H. Tan. Rate control in communication networks: Shadow prices, proportional fairness and stability. *J. Operational Research Soc.*, 1998.

[108] R.Z. Khasminskii. Application of random noise to optimization and recognition problems. *Problems Inform. Transmission*, 1:89–93, 1965.

[109] R.Z. Khasminskii. *Stochastic Stability of Differential Equations.* Sijthoff, Noordhoff, Alphen aan den Rijn, Amsterdam, 1982.

[110] J. Kiefer and J. Wolfowitz. Stochastic estimation of the maximum of a regression function. *Ann. Math. Statist.*, 23:462–466, 1952.

[111] P. Kokotovic, H. Khalil, and J. O'Reilly. *Singular Perturbation Methods in Control: Analysis and Design.* Academic Press, New York, 1986.

[112] V.R. Konda and V. S. Borkar. Actor-critic like learning algorithms for Markov processes. *SIAM J. Control Optim.*, 38:94–123, 1999.

[113] V.R. Konda and J.N. Tsitsikls. Actor-critic algorithms. Submitted to SIAM J. Control Optim., 2001.

[114] A.P. Korostelev. *Stochastic Recurrent Processes.* Nauka, Moscow, 1984.

[115] V. Krishnamurthy, G. Yin, and S. Singh. Adaptive step-size algorithms for blind interference suppression in DS/CDMA systems. *IEEE Trans. Signal Process.*, 49:190–201, 2001.

[116] C.M. Kuan and H. White. Adaptive learning with nonlinear dynamics driven by dependent processes. *Econometrica*, 62:1087–1114, 1994.

[117] T.G. Kurtz. Semigroups of conditional shifts and approximation of Markov processes. *Ann. Probab.*, 4:618–642, 1975.

[118] T.G. Kurtz. *Approximation of Population Processes*, volume 36 of *CBMS-NSF Regional Conf. Series in Appl. Math.* SIAM, Philadelphia, 1981.

[119] J. Kurzweil. On the inversion of Liapunov's second theorem on stability of motion. *Amer. Math. Soc. Translations, Ser 2*, 24:19–77, 1963.

[120] H.J. Kushner. A simple iterative procedure for the identification of the unknown parameters of a linear time varying discrete system. *J. Basic Eng.*, 85:227–235, 1963.

[121] H.J. Kushner. On the stability of stochastic dynamical systems. *Proc. Nat. Acad. Sci.*, 53:5–12, 1965.

[122] H.J. Kushner. *Stochastic Stability and Control.* Academic Press, New York, 1967.

[123] H.J. Kushner. General convergence results for stochastic approximations via weak convergence theory. *J. Math. Anal. Appl.*, 61:490–503, 1977.

[124] H.J. Kushner. *Probability Methods for Approximations in Stochastic Control and for Elliptic Equations.* Academic Press, New York, 1977.

[125] H.J. Kushner. Jump-diffusion approximations for ordinary differential equations with wideband random right hand sides. *SIAM J. Control Optim.*, 17:729–744, 1979.

[126] H.J. Kushner. An averaging method for stochastic approximations with discontinuous dynamics, constraints and state dependent noise. In H.H. Rizvi, J. Rustagi, and D. Siegmund, editors, *Recent Advances in Statistics*, pages 211–235. Academic Press, New York, 1983.

[127] H.J. Kushner. *Approximation and Weak Convergence Methods for Random Processes with Applications to Stochastic Systems Theory.* MIT Press, Cambridge, Mass., 1984.

[128] H.J. Kushner. Asymptotic behavior of stochastic approximation and large deviations. *IEEE Trans. Automatic Control*, AC-29:984–990, 1984.

[129] H.J. Kushner. Direct averaging and perturbed test function methods for weak convergence. In V.I. Arkin, A. Shiryaev, and R. Wets, editors, *Lecture Notes in Control and Information Sciences, Vol 81, Stochastic Optimization*, pages 412–426. Springer-Verlag, Berlin and New York, 1985.

[130] H.J. Kushner. Asymptotic global behavior for stochastic approximation and diffusions with slowly decreasing noise effects: Global minimization via monte carlo. *SIAM J. Appl. Math.*, 47:169–185, 1987.

[131] H.J. Kushner. Numerical methods for stochastic control problems in continuous time. *SIAM J. Control Optim.*, 28:999–1048, 1990.

[132] H.J. Kushner. *Weak Convergence Methods and Singularly Perturbed Stochastic Control and Filtering Problems*, volume 3 of *Systems and Control.* Birkhäuser, Boston, 1990.

[133] H.J. Kushner. A note on closed loop adaptive noise cancellation. Brown University, Lefschetz Center for Dynamical Systems Report, 1997.

[134] H.J. Kushner and R. Buche. Stochastic approximation: Rate of convergence for constrained problems and applications to Lagrangian algorithms. In *Proc. IEEE Conf. on Decision and Control.* IEEE Press, 1999.

[135] H.J. Kushner and D.S. Clark. *Stochastic Approximation for Constrained and Unconstrained Systems.* Springer-Verlag, Berlin and New York, 1978.

[136] H.J. Kushner and H. Huang. Averaging methods for the asymptotic analysis of learning and adaptive systems, with small adjustment rate. *SIAM J. Control Optim.*, 19:635–650, 1981.

[137] H.J. Kushner and M.L. Kelmanson. Stochastic approximation algorithms of the multiplier type for the sequential monte carlo optimization of systems. *SIAM J. Control Optim.*, 14:827–841, 1976.

[138] H.J. Kushner and R. Kumar. Convergence and rate of convergence of recursive identification and adaptive control methods which use truncated estimators. *IEEE Trans. Automatic Control*, AC-27:775–782, 1982.

[139] H.J. Kushner and S. Lakshimivarahan. Numerical studies of stochastic approximation procedures for constrained problems. *IEEE Trans. Automatic Control*, 22, 1977.

[140] H.J. Kushner and E. Sanvicente. Penalty function methods for constrained stochastic approximation. *J. Math. Anal. Appl.*, 46:499–512, 1974.

[141] H.J. Kushner and E. Sanvicente. Stochastic approximation for constrained systems with observation noise on the system and constraint. *Automatica*, 11:375–380, 1975.

[142] H.J. Kushner and A. Shwartz. An invariant measure approach to the convergence of stochastic approximations with state dependent noise. *SIAM J. Control Optim.*, 22:13–27, 1984.

[143] H.J. Kushner and A. Shwartz. Weak convergence and asymptotic properties of adaptive filters with constant gains. *IEEE Trans. Inform. Theory*, IT-30:177–182, 1984.

[144] H.J. Kushner and A. Shwartz. Stochastic approximation and optimization of linear continuous parameter systems. *SIAM J. Control Optim.*, 23:774–793, 1985.

[145] H.J. Kushner and F.J. Vázquez-Abad. Stochastic approximation algorithms for systems over an infinite horizon. *SIAM J. Control Optim.*, 34:712–756, 1996.

[146] H.J. Kushner and P.A. Whiting. Asymptotic properties of proportional-fair sharing algorithms. In *Proc. 2002 Allerton Conf.*, Champaigne-Urbana, IL, 2002. Univ. of Illinois Press.

[147] H.J. Kushner and P.A. Whiting. Convergence of proportional-fair sharing algorithms under general conditions. To appear in *IEEE Trans. Wireless Comm.*, 2003.

[148] H.J. Kushner and J. Yang. A monte carlo method for sensitivity analysis and parametric optimization of nonlinear stochastic systems: The ergodic case. *SIAM J. Control Optim.*, 30:440–464, 1992.

[149] H.J. Kushner and J. Yang. Stochastic approximation with averaging of the iterates: Optimal asymptotic rates of convergence for general processes. *SIAM J. Control Optim.*, 31:1045–1062, 1993.

[150] H.J. Kushner and J. Yang. Analysis of adaptive step size SA algorithms for parameter tracking. *IEEE Trans. Automatic Control*, AC-40:1403–1410, 1995.

[151] H.J. Kushner and J. Yang. Stochastic approximation with averaging and feedback: faster convergence. In G.C. Goodwin K. Åström and P.R. Kumar, editors, *IMA Volumes in Mathematics and Applications, Volume 74, Adaptive Control, Filtering and Signal Processing*, pages 205–228. Springer-Verlag, Volume 74, the IMA Series, Berlin and New York, 1995.

[152] H.J. Kushner and J. Yang. Stochastic approximation with averaging and feedback: Rapidly convergent "on line" algorithms. *IEEE Trans. Automatic Control*, AC-40:24–34, 1995.

[153] H.J. Kushner and G. Yin. Asymptotic properties of distributed and communicating stochastic approximation algorithms. *SIAM J. Control Optim.*, 25:1266–1290, 1987.

[154] H.J. Kushner and G. Yin. Stochastic approximation algorithms for parallel and distributed processing. *Stochastics*, 22:219–250, 1987.

[155] S. Lakshmivarahan. *Learning Algorithms: Theory and Applications*. Springer-Verlag, Berlin and New York, 1981.

[156] J.P. LaSalle and S. Lefschetz. *Stability by Liapunov's Direct Method with Applications*. Academic Press, New York, 1961.

[157] A. Le Breton. About the averaging approach in Gaussian schemes for stochastic approximation. *Math. Methods Statist.*, 2:295–315, 1993.

[158] A. Le Breton. Averaging with feedback in Gaussian schemes in stochastic approximation. *Math. Methods Statist.*, 6:313–331, 1997.

[159] A. Le Breton and A.A. Novikov. Some results about averaging in stochastic approximation. *Metrica*, 42:153–171, 1995.

[160] P. L'Ecuyer, N. Giroux, and P.W. Glynn. Stochastic optimization by simulation: Numerical experiments with the $M/M/1$ queue in steady state. *Management Sci.*, 1994.

[161] P. L'Ecuyer and P.W. Glynn. Stochastic optimization by simulation: Convergence proofs for the GI/G/1 queue. *Management Sci.*, 40:1562–1578, 1994.

[162] P. L'Ecuyer and G. Yin. Budget-dependent convergence rate of stochastic approximation. *SIAM J. Optim.*, 8:217–247, 1998.

[163] R. Liptser and A.N. Shiryaev. *Statistics of Random Processes.* Springer-Verlag, Berlin and New York, 1977.

[164] L. Ljung. Analysis of recursive stochastic algorithms. *IEEE Trans. Automatic Control*, AC-22:551–575, 1977.

[165] L. Ljung. On positive real transfer functions and the convergence of some recursive schemes. *IEEE Trans. Automatic Control*, AC-22:539–551, 1977.

[166] L. Ljung. *System Identification Theory for the User.* Prentice-Hall, Englewood Cliffs, NJ, 1986.

[167] L. Ljung. Recursive least squares and accelerated convergence in stochastic approximation schemes. *Int. J. Adaptive Control Signal Process.*, 15:169–178, 2001.

[168] L. Ljung and A.V. Nazin. Asymptotically optimal smoothing of averaged LMS estimates for regression parameter tracking. In *Proc. 15th IFAC World Congress, Barcelona*, Amsterdam, 2002. Klewer.

[169] L. Ljung and T. Söderström. *Theory and Practice of Recursive Identification.* MIT Press, Cambridge, Mass, 1983.

[170] J. T-H. Lo. Synthetic approach to nonlinear filtering. *IEEE Trans. Neural Networks*, 5:803–811, 1994.

[171] Y. Maeda, H. Hirano, and Y. Kanata. A learning rule of neural networks via simultaneous perturbation and its hardware implementation. *Neural Networks*, 8:251–259, 1995.

[172] P. Marbach, O. Mihatsch, and J.N. Tsitsiklis. Call admission control and routing in integrated services networks using neuro-dynamic programming. *IEEE Trans. Selected Areas in Comm.*, 18:197–208, 2000.

[173] P. Marback and J.N. Tsitsiklis. Gradient-based optimization of Markov reward processes: Practical variants. LIDS Tech. Report, MIT, 2001.

[174] P. Marback and J.N. Tsitsiklis. Simulation-based optimization of Markov reward processes. *IEEE Trans. Automatic Control*, 46:191–209, 2001.

[175] A. Marcet and T.J. Sargent. Convergence of least squares learning in environments with hidden variables and private information. *J. Political Economy*, 97:1306–1322, 1989.

[176] R. Marimon and E. McGrattan. On adaptive learning in strategic games. In A. Kirman and M. Salmon, editors, *Learning and Rationality in Economics*, pages 63–101. Basil Blackwell, Oxford, 1995.

[177] F. Menczer, W.E. Hart, and M.L. Littman. Appendix to: The effect of memory length on individual fitness in a lizard. In R. K. Belew and M. Mitchell, editors, *Adaptive Individuals in Evolving Populations: SFI Studies in the Sciences of Complexity, Vol. XXII.* Addison-Wesley, 1995.

[178] S.P. Meyn and R.I. Tweedie. *Markov Chains and Stochastic Stability.* Springer-Verlag, Berlin and New York, 1994.

[179] W.L. Miranker. A survey of parallelism in numerical analysis. *SIAM Rev.*, 13:524–547, 1971.

[180] A.V. Nazin. Generalized averaging along stochastic approximation trajectories. In *Proceedings of the Second European Control Conference, ECC'93, Groningen, The Netherlands, Vol. 4*, pages 2112–2116, 1993.

[181] A.V. Nazin, B.T. Polyak, and A.B. Tsybakov. Passive stochastic approximation. *Automat. Remote Control*, 50:1563–1569, 1989.

[182] M.B. Nevelson and R.Z. Khasminskii. *Stochastic Approximation and Recursive Estimation.* Amer. Math. Soc, Providence, RI, 1976. Translation of Math. Monographs, Vol. 47.

[183] J. Neveu. *Mathematical Foundations of the Calculus of Probability.* Holden-Day, San Francisco, 1965.

[184] H. Niederreiter. *Random Number Generation and Monte-Carlo Methods.* SIAM, Philadelphia, 1992.

[185] R. Nixdorf. An invariant principle for a finite dimensional stochastic approximation method in a Hilbert space. *J. Multivariate Anal.*, 15:252–260, 1984.

[186] J. Pan and C.G. Cassandras. Flow control of bursty traffic using infinitesimal perturbation analysis. *J. Discrete Event Dynamical Systems*, 4:325–358, 1993.

[187] G.C. Papanicolaou, D. Stroock, and S.R.S. Varadhan. Martingale approach to some limit theorems. In *Proc. of the 1976 Duke University Conference on Turbulence*, 1976.

[188] M. Pelletier. On the almost sure asymptotic behavior of stochastic algorithms. *Stochastic Process Appl.*, 78:217–244, 1998.

[189] R. Pemantle. Nonconvergence to unstable points in urn models and stochastic approximation. *Ann. Probab.*, 18:698–712, 1990.

[190] B. Perthame. Perturbed dynamical systems with an attracting singularity and weak viscosity limits in Hamilton–Jacobi equations. Technical Report 18, Ecole Normale Supérieure, 1987.

[191] H. Pezeshki-Esfanahani and A.J. Heunis. Strong diffusion approximations for recursive stochastic algorithms. *IEEE Trans. Inform. Theory*, 43:312–323, 1997.

[192] G.Ch. Pflug. *Optimization of Stochastic Models.* Kluwer, Boston, MA, 1996.

[193] D.T. Pham and X. Liu. *Neural Networks for Identification, Prediction and Control.* Springer-Verlag, Berlin and New York, 1995.

[194] B.T. Polyak. New stochastic approximation type procedures. *Automat. Remote Control*, 7:98–107, 1990.

[195] B.T. Polyak and A.B. Juditsky. Acceleration of stochastic approximation by averaging. *SIAM J. Control Optim.*, 30:838–855, 1992.

[196] B.T. Polyak and Ya.Z. Tsypkin. Optimal pseudogradient adaptation procedures. *Automat. Remote Control*, 41:1101–1110, 1981.

[197] B.T. Polyak and Ya.Z. Tsypkin. Robust pseudogradient adaptation algorithms. *Automat. Remote Control*, 41:1404–1409, 1981.

[198] M.L. Puterman. *Markov Decision Processes.* Wiley, New York, 1994.

[199] M.L. Quinn. *Designing Efficient Algorithms for Parallel Computers.* McGraw-Hill, New York, 1987.

[200] M.I. Reiman and A. Weiss. Sensitivity analysis for simulation via likelihood ratios. *Oper. Res.*, 37:930–844, 1989.

[201] P. Révész. Robbins-Monro procedure in a Hilbert space and its application in the theory of learning processes I,II. *Studia Sci. Math. Hungar.*, 8:391–398, 469–472, 1973.

[202] P. Révész. How to apply the method of stochastic approximation in the non-parametric estimation of regression function. *Matem. Operations Stat. Ser. Statist.*, 8:119–126, 1977.

[203] D. Revuz. *Markov Chains.* North Holland, Amsterdam, 1984.

[204] J.A. Rice. *Mathematical Statistics and Data Analysis.* Duxbury Press, Belmont, CA, 1995.

[205] B.D. Ripley. *Pattern Recognition and Neural Networks.* Cambridge University Press, Cambridge, UK, 1996.

[206] J. Rissanen. Minimum description length principles. In S. Kotz and N. L. Johnson, editors, *Encyclopedia of Statistical Sciences,* Vol. 5. Wiley, New York, 1985.

[207] H. Robbins and S. Monro. A stochastic approximation method. *Ann. Math. Statist.*, 22:400–407, 1951.

[208] B. Van Roy. Neuro-dynamic programming: Overview and recent results. In E.A . Feinberg and A. Shwartz, editors, *Handbook of Markov Decision Processes: Methods and Applications*, pages 431–460. Kluwer, Boston, 2002.

[209] H.L. Royden. *Real Analysis, second edition.* Macmillan, New York, 1968.

[210] R.Y. Rubinstein. Sensitivity analysis and performance extrapolation for computer simulation models. *Oper. Res.*, 37:72–81, 1989.

[211] D. Ruppert. Stochastic approximation. In B.K. Ghosh and P.K. Sen, editors, *Handbook in Sequential Analysis*, pages 503–529. Marcel Dekker, New York, 1991.

[212] P. Sadegh. Constrained optimization via stochastic approximation with a simultaneous perturbation gradient approximation. *Automatica*, 33:889–892, 1997.

[213] P. Sadegh and J.C. Spall. Optimal random perturbations for stochastic approximation using a simultaneous perturbation gradient approximation. *IEEE Trans. Automat. Control*, 43:1480–1484, 1998.

[214] G.I. Salov. Stochastic approximation theorem in a Hilbert space and its application. *Theory Probab. Appl.*, 24:413–419, 1979.

[215] L. Schmetterer. Stochastic approximation. In *Proc. of the Fourth Berkeley Symposium on Mathematical Statistics and Probability*, pages 587–609, Berkeley, 1960. Univ. of California.

[216] R. Schwabe. On Bather's stochastic approximation algorithm. *Kybernetika*, 30:301–306, 1994.

[217] R. Schwabe and H. Walk. On a stochastic approximation procedure based on averaging. *Metrica*, 44:165–180, 1996.

[218] S. Shafir and J. Roughgarden. The effect of memory length on individual fitness in a lizard. In R. K. Belew and M. Mitchell, editors, *Adaptive Individuals in Evolving Populations: SFI Studies in the Sciences of Complexity, Vol. XXII*. Addison-Wesley, 1995.

[219] A. Shwartz and N. Berman. Abstract stochastic approximations and applications. *Stochastic Process. Appl.*, 28:133–149, 1989.

[220] A. Shwartz and A. Weiss. *Large Deviations for Performance Analysis*. Chapman & Hall, London, 1995.

[221] J. Si and Y.-T. Wang. On-line learning control by association and reinforcement. *IEEE Trans. Neural Networks*, 12:264–276, 2001.

[222] A.V. Skorohod. Limit theorems for stochastic processes. *Theory Probab. Appl.*, pages 262–290, 1956.

[223] H.L. Smith. *Monotone Dynamical Systems: An Introduction to Competitive and Cooperative Systems, AMS Math. Surveys and Monographs, Vol. 41*. Amer. Math. Soc., Providence RI, 1995.

[224] V. Solo. The limit behavior of LMS. *IEEE Trans Acoust. Speech Signal Process.*, ASSP-37:1909–1922, 1989.

[225] V. Solo and X. Kong. *Adaptive Signal Processing Algorithms*. Prentice-Hall, Englewood Cliffs, NJ, 1995.

[226] J.C. Spall. Multivariate stochastic approximation using a simultaneous perturbation gradient approximation. *IEEE Trans. Automatic Control*, AC-37:331–341, 1992.

[227] J.C. Spall. A one measurement form of simultaneous perturbation stochastic approximation. *Automatica*, 33:109–112, 1997.

[228] J.C. Spall. Adaptive stochastic approximation by the simultaneous perturbation method. *IEEE Trans. Automat. Control*, 45:1839–1853, 2000.

[229] J.C. Spall and J.A. Cristion. Nonlinear adaptive control using neural networks: estimation with a smoothed form of simultaneous perturbation gradient approximation. *Statist. Sinica*, 4:1–27, 1994.

[230] D.W. Stroock. *Probability Theory, An Analytic View: Revised Edition*. Cambridge University Press, Cambridge, 1994.

[231] R. Suri and M. Zazanis. Perturbation analysis gives strongly consistent sensitivity estimates for the M/M/1 queue. *Management Sci.*, 34:39–64, 1988.

[232] R.S. Sutton and A.G. Barto. *Reinforcement Learning*. MIT Press, Cambridge, MA, 1998.

[233] V. Tarokh, N. Seshadri, and A.R. Calderbank. Space-time codes for high data rate wireless communication: Performance criterion and code construction. *IEEE Trans. Inform. Theory*, 44:744–765, 1998.

[234] J.N. Tsitsiklis. Asynchronous stochastic approximation and Q-learning. *Machine Learning*, 16:185–202, 1994.

[235] J.N. Tsitsiklis and V. Van Roy. An analysis of temporal difference learning with function approximation. *IEEE Trans. Automatic Control*, 42:674–690, 1997.

[236] G.V. Tsoulos, Editor. *Adaptive Antenna Arrays for Wireless Communications*. IEEE Press, New York, 2001.

[237] Ya.Z. Tsypkin. *Adaptation and Learning in Automatic Systems*. Academic Press, New York, 1971.

[238] S.R.S. Varadhan. *Large Deviations and Applications*. CBMS-NSF Regional Conference Series in Mathematics. SIAM, Philadelphia, 1984.

[239] F. J. Vázquez-Abad. *Stochastic recursive algorithms for optimal routing in queueing networks*. PhD thesis, Brown University, 1989.

[240] F.J. Vázquez-Abad, C.C. Cassandras, and V. Julka. Centralized and decentralized asynchronous optimization of stochastic discrete event systems. *IEEE Trans. Automatic Control*, 43:631–655, 1998.

[241] F.J. Vázquez-Abad and K. Davis. Strong points of weak convergence: A study using RPA gradient estimation for automatic learning. Technical report, Report 1025, Dept. IRO, Univ. of Montreal, 1996.

[242] F.J. Vázquez-Abad and H.J. Kushner. Estimation of the derivative of a stationary measure with respect to a control parameter. *J. Appl. Probab.*, 29:343–352, 1992.

[243] F.J. Vázquez-Abad and H.J. Kushner. The surrogate estimation approach for sensitivity analysis in queueing networks. In G.W. Evans, M. Mollaghasemi, E.C. Russel, and W.E. Biles, editors, *Proceedings of the Winter Simulation Conference. 1993*, pages 347–355, 1993.

[244] F.J. Vázquez-Abad and L. Mason. Adaptive control of deds under non-uniqueness of the optimal control. *J. Discrete Event Dynamical Syst.*, 6:323–359, 1996.

[245] F.J. Vázquez-Abad and L. Mason. Decentralized isarithmic flow control for high speed data networks. *Oper. Res.*, 47:928–942, 1999.

[246] H. Walk. An invariant principle for the Robbins Monro process in a Hilbert space. *Z. Wahrsch. verw. Gebiete*, 62:135–150, 1977.

[247] H. Walk. Martingales and the Robbins-Monro procedure in $D[0, 1]$. *J. Multivariate Anal.*, 8:430–452, 1978.

[248] H. Walk and L. Zsidó. Convergence of Robbins–Monro method for linear problem in a Banach space. *J. Math. Anal. Appl.*, 139:152–177, 1989.

[249] I.J. Wang, E.K.P. Chong, and S.R. Kulkarni. Equivalent necessary and sufficient conditions on noise sequences for stochastic approximation algorithms. *Adv. in Appl. Probab.*, 28:784–801, 1996.

[250] M.T. Wasan. *Stochastic Approximation.* Cambridge University Press, Cambridge, UK, 1969.

[251] C.I.C.H. Watkins. *Learning from delayed rewards.* PhD thesis, University of Cambridge, Cambridge, UK, 1989.

[252] C.I.C.H. Watkins and P. Dayan. Q-learning. *Machine Learning*, 8:279–292, 1992.

[253] H. White. *Artificial Neural Networks.* Blackwell, Oxford, UK, 1992.

[254] B. Widrow, P.E. Mantey, L.J. Griffiths, and B.B. Goode. Adaptive antenna systems. *Proc. IEEE*, 55:2143–2159, December 1967.

[255] B. Widrow and S.D. Stearns. *Adaptive Signal Processing.* Prentice-Hall, Englewood Cliffs, NJ, 1985.

[256] F.W. Wilson. Smoothing derivatives of functions and applications. *Trans. Amer. Math. Soc.*, 139:413–428, 1969.

[257] J.H. Winters. Signal acquisition and tracking with adaptive arrays in digital mobile radio system IS-54 with flat fading. *IEEE Trans. Vehicular Technology*, 42:377–393, 1993.

[258] S. Yakowitz. A globally convergent stochastic approximation. *SIAM J. Control Optim.*, 31:30–40, 1993.

[259] S. Yakowitz, P. L'Ecuyer, and F. Vaźquez-Abad. Global stochastic optimization with low-discrepancy point sets. *Oper. Res.*, 48:939–950, 2000.

[260] H. Yan, G. Yin, and S.X.C. Lou. Using stochastic approximation to determine threshold values for control of unreliable manufacturing systems. *J. Optim. Theory Appl.*, 83:511–539, 1994.

[261] H. Yan, X.Y. Zhou, and G. Yin. Approximating an optimal production policy in a continuous flow line: Recurrence and asymptotic properties. *Oper. Res.*, 47:535–549, 1999.

[262] J. Yang and H.J. Kushner. A monte carlo method for the sensitivity analysis and parametric optimization of nonlinear stochastic systems. *SIAM J. Control Optim.*, 29:1216–1249, 1991.

[263] G. Yin. Asymptotic properties of an adaptive beam former algorithm. *IEEE Trans. Inform. Theory*, IT-35:859–867, 1989.

[264] G. Yin. A stopping rule for least squares identification. *IEEE Trans. Automatic Control*, AC-34:659–662, 1989.

[265] G. Yin. On extensions of Polyak's averaging approach to stochastic approximation. *Stochastics Stochastics Rep.*, 36:245–264, 1991.

[266] G. Yin. Recent progress in parallel stochastic approximation. In L Gerencér and P.E. Caines, editors, *Topics in Stochastic Systems: Modelling, Estimation and Adaptive Control*, pages 159–184. Springer-Verlag, Berlin and New York, 1991.

[267] G. Yin. Stochastic approximation via averaging: Polyak's approach revisited. In G. Pflug and U. Dieter, editors, *Lecture Notes in Economics and Math. Systems 374*, pages 119–134. Springer-Verlag, Berlin and New York, 1992.

[268] G. Yin. Adaptive filtering with averaging. In G.C. Goodwin, K. Åström, and P.R. Kumar, editors, *Adaptive Control, Filtering and Signal Processing*, pages 375–396. Springer-Verlag, Berlin and New York, 1995. Volume 74, the IMA Series.

[269] G. Yin. Rates of convergence for a class of global stochastic optimization algorithms. *SIAM J. Optim.*, 10:99–120, 1999.

[270] G. Yin and P. Kelly. Convergence rates of digital diffusion network algorithms for global optimization with applications to image estimation. *J. Global Optim.*, 23:329–358, 2002.

[271] G. Yin, R.H. Liu, and Q. Zhang. Recursive algorithms for stock liquidation: A stochastic optimization approach. *SIAM J. Optim.*, 13:240–263, 2002.

[272] G. Yin, H. Yan, and S.X.C. Lou. On a class of stochastic approximation algorithms with applications to manufacturing systems. In H. P. Wynn W. G. Müller and A. A. Zhigljavsky, editors, *Model Oriented Data Analysis*, pages 213–226. Physica-Verlag, Heidelberg, 1993.

[273] G. Yin and K. Yin. Asymptotically optimal rate of convergence of smoothed stochastic recursive algorithms. *Stochastics Stochastics Rep.*, 47:21–46, 1994.

[274] G. Yin and K. Yin. A class of recursive algorithms using nonparametric methods with constant step size and window width: a numerical study. In *Advances in Model-Oriented Data Analysis*, pages 261–271. Physica-Verlag, 1995.

[275] G. Yin and K. Yin. Passive stochastic approximation with constant step size and window width. *IEEE Trans. Automatic Control*, AC-41:90–106, 1996.

[276] G. Yin and Y.M. Zhu. On w.p.1 convergence of a parallel stochastic approximation algorithm. *Probab. Eng. Inform. Sci.*, 3:55–75, 1989.

[277] G. Yin and Y.M. Zhu. On H-valued Robbins-Monro processes. *J. Multivariate Anal.*, 34:116–140, 1990.

[278] W.I. Zangwill. *Nonlinear Programming: A Unified Approach.* Prentice-Hall, Englewood Cliffs, 1969.

[279] Y.M. Zhu and G. Yin. Stochastic approximation in real time: A pipeline approach. *J. Computational Math.*, 12:21–30, 1994.

Symbol Index

Subject Index

Applications of Mathematics

(continued from page ii)

35 Kushner/Yin, **Stochastic Approximation and Recursive Algorithms and Applications,** Second Ed. (2003)
36 Musiela/Rutkowski, **Martingale Methods in Financial Modeling: Theory and Application** (1997)
37 Yin/Zhang, **Continuous-Time Markov Chains and Applications** (1998)
38 Dembo/Zeitouni, **Large Deviations Techniques and Applications**, Second Ed. (1998)
39 Karatzas/Shreve, **Methods of Mathematical Finance** (1998)
40 Fayolle/Iasnogorodski/Malyshev, **Random Walks in the Quarter Plane** (1999)
41 Aven/Jensen, **Stochastic Models in Reliability** (1999)
42 Hernández-Lerma/Lasserre, **Further Topics on Discrete-Time Markov Control Processes** (1999)
43 Yong/Zhou, **Stochastic Controls: Hamiltonian Systems and HJB Equations** (1999)
44 Serfozo, **Introduction to Stochastic Networks** (1999)
45 Steele, **Stochastic Calculus and Financial Applications** (2000)
46 Chen/Yao, **Fundamentals of Queueing Networks: Performance, Asymptotics, and Optimization** (2001)
47 Kushner, **Heavy Traffic Analysis of Controlled Queueing and Communication Networks** (2001)
48 Fernholz, **Stochastic Portfolio Theory** (2002)
49 Kabanov/Pergamenshchikov, **Two-Scale Stochastic Systems** (2003)
50 Han, **Information-Spectrum Methods in Information Theory** (2003)
51 Asmussen, **Applied Probability and Queues** (2003)

CPSIA information can be obtained at www.ICGtesting.com
Printed in the USA
LVOW13*1719181213

365921LV00013B/670/P

9 780387 008943